Modulators of Oxidative Stress

Modulators of Oxidative Stress

Chemical and Pharmacological Aspects

Editors

Luciano Saso
Hande Gurer-Orhan
Višnja Stepanić

MDPI • Basel • Beijing • Wuhan • Barcelona • Belgrade • Manchester • Tokyo • Cluj • Tianjin

Editors
Luciano Saso
Sapienza University of Rome
Italy

Hande Gurer-Orhan
Ege University
Turkey

Višnja Stepanić
Ruđer Bošković Institute
Croatia

Editorial Office
MDPI
St. Alban-Anlage 66
4052 Basel, Switzerland

This is a reprint of articles from the Special Issue published online in the open access journal *Antioxidants* (ISSN 2076-3921) (available at: https://www.mdpi.com/journal/antioxidants/special_issues/modulators_oxidative_stress).

For citation purposes, cite each article independently as indicated on the article page online and as indicated below:

LastName, A.A.; LastName, B.B.; LastName, C.C. Article Title. *Journal Name* **Year**, *Article Number*, Page Range.

ISBN 978-3-03943-228-8 (Hbk)
ISBN 978-3-03943-229-5 (PDF)

Contents

About the Editors

Luciano Saso is a member of the Faculty of Pharmacy and Medicine, Sapienza University of Rome, Italy (http://en.uniroma1.it/). He is the author of more than 250 original scientific articles published in peer-reviewed international journals (H-index Google Scholar = 47, H-index SCOPUS = 38), mainly in the field of oxidative stress and antioxidants. He has coordinated several international research projects and has been a referee for many national and international funding agencies and international scientific journals in the last 25 years. He has been a guest editor of several Special Issues in the fields of chemistry, biology, and pharmacology of modulators of oxidative stress.

Hande Gurer-Orhan obtained her BSc degree in Pharmacy in 1992, and her MSc degree (1994) and PhD degree (1999) in Toxicology from the University of Hacettepe, Turkey. She received a TUBITAK/NATO A2 grant and performed the laboratory work of her PhD thesis at the Department of Chemistry, University of Missouri-Rolla, USA. She worked as a post-doctoral fellow at the Medical School of Washington University, St Louis, MO in 2001 and at the Department of Molecular Toxicology, Vrije Universiteit during the period 2001–2003. She (https://avesis.ege.edu.tr/hande.gurer.orhan) is currently working as a full professor and the head of the Toxicology Department at the Faculty of Pharmacy, Ege University, İzmir, Turkey. She is a registered toxicologist in the European Union (ERT), a member of the Executive Committee of the Turkish Society of Toxicology, and a member of IUTOX and EUROTOX as well as the member of the Education Sub-Committee of EUROTOX. Her research interests are screening novel natural and synthetic endocrine modulating compounds, the assessment of their potential safety and efficiency for the treatment of hormone-related cancers and investigating the efficiency and toxicity of novel synthetic oxidative stress modulators, mainly melatonin analogues.

Višnja Stepanić graduated and completed her MSc and PhD at the Faculty of Science of the University of Zagreb (Croatia). Afterwards, she joined the PLIVA Research Institute in 2002 and after its GlaxoSmithKline's (GSK) acquisition in 2006, she was a member of GSK Research Centre Zagreb. Since 2009, she has been a researcher at the Ruder Bošković Institute. She has participated in international scientific networks like COST actions. Her current main research interest includes (bio)chemical modulation of redox status with a focus on small molecular weight redox active natural and (semi)synthetic compounds with an optimized ADMET profile.

Editorial

Modulators of Oxidative Stress: Chemical and Pharmacological Aspects

Luciano Saso [1], Hande Gurer-Orhan [2] and Višnja Stepanić [3,*]

[1] Department of Physiology and Pharmacology "Vittorio Erspamer", Sapienza University, 00185 Rome, Italy; luciano.saso@uniroma1.it

[2] Toxicology Department, Faculty of Pharmacy, Ege University, Izmir 35040, Turkey; hande.gurer.orhan@ege.edu.tr

[3] Ruđer Bošković Institute, Bijenička cesta 54, 10000 Zagreb, Croatia

* Correspondence: stepanic@irb.hr; Tel.: +385-1-457-1356

Received: 13 July 2020; Accepted: 22 July 2020; Published: 24 July 2020

Oxidative stress is represented as an imbalance between reactive oxygen species (ROS) production and the response of antioxidant proteins. Oxidative stress is implicated in the pathogenesis of a variety of conditions, including cardiovascular diseases, neurodegenerative diseases, and cancer. Unfortunately, antioxidant therapies have not been proven to be effective in most clinical studies [1]. One of the reasons for this disappointing failure was testing antioxidant potential of compounds only through their capacity to reduce the number of free radicals directly by their free radical scavenging and metal chelating activities. It took some time to realize that some small molecules can achieve a more significant antioxidant effect by targeted modulation of activities of proteins included in the antioxidant system [2]. Relatively recently modulation of oxidative stress through regulation of gene transcription has been discovered to be a promising strategy for the development of new drugs for cardiovascular diseases and cancer types that are resistant to other treatments [3]. This Special Issue of the "Antioxidants" collects seven original research manuscripts and six reviews addressing different pharmacological aspects of the modulation of oxidative stress by natural and synthetic small molecular weight compounds in regard to various therapeutic approaches in cardiovascular and neurodegenerative diseases, cancer and diabetes. Understanding the relationship between oxidative stress and underlying anti- and pro-oxidant mechanisms is crucial to understand mechanisms of diseases and also for the discovery and development of innovative, targeted therapeutic strategies.

The two research articles present the results of cellular studies on contribution of antioxidants in alleviating ischemia–reperfusion (IR) injuries in myocardial cells and cardiac fibroblasts (CF). IR injury is a common clinical problem that occurs in the heart (in myocardial infarction) as well as other organ systems (e.g., in the brain in stroke and in limb ischemia) and may also be a result of reperfusion strategy or occur during surgeries [4]. IR occurs when blood supply to an organ is diminished and then re-established. Although blood flow is necessary for survival, reperfusion induces necrotic and apoptotic cell death. One of the key mediators of reperfusion injury is an oxidative stress. A huge generation of ROS at the beginning of reperfusion initiates lipid peroxidation and protein/nucleic acid oxidation, consequently activating pro-apoptotic pathways associated with p38-MAPK and JNK proteins [5]. In their manuscript, Parra-Flores et al. [6] showed for the first time synergistic cytoprotective effect of ascorbic acid, deferoxamine, and *N*-acetylcysteine in simulated IR in CF. When added at low concentrations (10 μM) each at the onset of simulated reperfusion, the three antioxidants synergistically reduced intracellular ROS production and increased cell viability, induced phosphorylation of pro-survival kinases ERK1/2 (extracellular-signal-regulated kinases 1/2) and PKB (protein kinase B)/Akt, and decreased phosphorylation of the pro-apoptotic kinases p38-MAPK (proteins p38-mitogen-activated protein kinase) and JNK (c-Jun-N-terminal kinase). The mixture also recovered the CF function associated with wound repair by restoring serum-induced migration,

TGF-1-mediated differentiation of CF into CMF and angiotensin II-induced pro-collagen I synthesis. The three compounds have different mechanism of antioxidant activity and act at concentrations higher than 10 μM. However, in the mix, the oxidized form of ascorbate—dehydroascorbate, can be transformed back in the reduced form by intracellular glutathione (GSH) [7] and the decreased intracellular concentration of GSH may be replenished by *N*-acetylcysteine which is a GSH precursor. Deferoxamine can chelate an excess of catalytic free iron which increases during ischemia and cardiac reperfusion from cell lysis, and in this way the production of OH radicals through the Fenton reaction is lowered, thus preventing the pro-oxidant interaction of iron with ascorbate [8]. In another study, Peserico et al. [9] pointed to the repurposing of ezetimibe from a lipid-lowering compound to the antioxidant with diminishing effects of the oxidative stress induced by the simulated IR in the human monocyte-like cells THP-1 and cardiomyocytes [10]. Overnight pre-incubation of the cells with ezetimibe (50 μM) reduced ROS formation, NF-kB activation and apoptosis, induced by simulated IR. Ezetimibe induced up-regulation of NRF2 (Nuclear factor erythroid 2-related factor 2)/ARE (antioxidant response element) signalling and the pro-survival UPR (unfolded protein response) pathway by up-regulation of the pro-survival activating transcription factor 6 (ATF6) and down-regulation of the pro-apoptotic CCAAT-enhancer-binding protein homologous protein (CHOP). In such a way, misfolded protein mass may be reduced and cells convert to a pro-survival state. The NRF2 activation was found to be dependent on AMPK phosphorylation.

Oxidative stress and depletion of intracellular antioxidants, together with a systemic inflammatory response and ultimately organ failure are characteristics of the sepsis [11]. Since organ dysfunction associated with sepsis is due, at least in part, to mitochondrial dysfunction and mitochondrial dysfunction is considered as main cause of oxidative stress in sepsis, antioxidants targeted to mitochondria have been considered as an efficient antioxidant protection in sepsis [12]. Minter et al. [13] compared the positively charged antioxidant MitoVitE, which accumulates within mitochondria, with the membrane-bound α-tocopherol and water-soluble cytosolic Trolox with regard to their effects on oxidative stress, mitochondrial function and expression of key genes and proteins involved in the toll-like receptor (TLR)-2 and -4 inflammatory signalling pathways in the umbilical vein endothelial cell line HUVEC-C under conditions mimicking acute bacterial sepsis. Each of these three forms of vitamin E (5 μM each) was found to have similar reducing effects against induced oxidative stress and NFκB activation caused by exposure to lipopolysaccharide/peptidoglycan G. However, in difference to α-tocopherol and Trolox, MitoVitE had also widespread downregulatory effects on gene expression involved in TLR pathways, resulting in anti-inflammatory profile. Many of these downregulated genes have been shown to be upregulated in mononuclear cells from patients with sepsis.

Mitochondrial dysfunction and consequential oxidative stress are also connected with organ injuries caused by exposure to ionizing radiation in cancer therapy [14]. A serious limitation to radiation therapy for some cancers is the possibility of irradiating the liver and induced liver injury. When cells are exposed to radiation, the redox system begins producing free radicals a few hours after exposure, but with the potential to last for years and cause pathological long-term persistent liver responses to ionizing radiation, such as radiation-induced liver disease (RILD) [15]. It has been known that major mitochondrial deacetylase sirtuin 3 (SIRT3) contributes to the development of ionizing radiation-induced acute liver injury [16]. The results of the in vivo study conducted by LoBianco et al. [17] show that loss of or decrease in SIRT3 levels could be an important factor contributing to a damage-permissive phenotype in murine liver long after exposure to ionising radiation. Their study on the two groups of Sirt3−/− and Sirt3+/+ male mice six months after their liver was exposed to 24 Gy radiation additionally suggested that superoxide radical anion-driven acute liver injury following radiation exposure appears to shift towards a peroxide-mediated long-term injury and is no longer limited to the mitochondria. On the other hand, there is a possibility to use radiation to stimulate cellular mechanisms and facilitate the removal of oxidative stress through the modulation of various signalling pathways in the presence of nonionizing light at a specific wavelength. In their review article, Rajendran et al. [18] described and discussed the effects of photobiomodulation on the progress of cancer, diabetes and wound healing through the crosstalk between ROS and the first discovered redox-regulated transcription factor, NF-κB (nuclear factor kappa-light-chain-enhancer of activated B cells).

A master regulator of cellular redox homeostasis is the transcription factor NRF2 [1,3]. NRF2 is a regulator of the transactivation of over 500 cytoprotective genes, many of which are included in the antioxidant defence against intracellular ROS increase. The major signalling pathway that regulates oxidative stress is NRF2–KEAP1 (Kelch-like ECH-associated protein 1). Panieri et al. [19] reviewed the activation mechanisms and the therapeutic modulation of NRF2 within the specific biological context. In the early stages of carcinogenesis, pharmacological induction of the NRF2 pathway might be chemopreventive. The NRF2 inducers protect normal cells from carcinogen effects by reducing the risk of cancer initiation and development in normal cells by ROS scavenging, decreasing DNA damage and preventing genomic instability. However, in advanced stages of cancer, the adverse effects of the NRF2 induction might occur due to the development of therapy resistance [20]. Over-activation of the NRF2–KEAP1 pathway in many tumours has been found to promote cancer growth and survival, metastasis formation and therapy resistance. Sustained NRF2 activation in cancer cells lead to metabolic changes that frequently develop "NRF2 addiction" [21]. Panieri et al. pointed to disruption of NRF2 signalling in cancer cells by using NRF2 inhibitors as a promising therapeutic approach against NRF2-addicted cancers.

NRF2 activators have also been proposed as antioxidants in neurodegenerative and neuropsychiatric disorders (Alzheimer's disease, Parkinson's disease, Huntington's disease, autism, schizophrenia, and depression) to counteract the increase in oxidative stress and inflammation. Alvarez-Arellano et al. [22] made an overview of the role of inflammation and oxidative stress in attention-deficit/hyperactivity disorder (ADHD). The amelioration of inflammation and oxidative stress by natural antioxidants sulforaphane, *N*-acetylcysteine and omega-3 fatty acids (docosahexaenoic acid (DHA) and eicosapentaenoic acid (EPA)) in potential adjuvant therapy of ADHD is described, and the observed effects of these antioxidants on the activation of the NRF2 pathway in neurodegenerative and neuropsychiatric disorders are summarized.

García-Guede et al. [23] reviewed the effects of oxidative stress on the epigenetic machinery of DNA methylation, histone modifications and non-coding RNA, particularly microRNAs (miRNAs) and long non-coding RNAs (lncRNAs). The influence of these epigenetic modifications is extensively described in the case of the specific mechanisms for developing cisplatin resistance.

In the four studies reported in this Special Issue, the effects of food components on human health through modulation of oxidative status of the body cells are investigated. One of the irreplaceable food component necessary for maintaining health and preventing diseases for humans is ascorbic acid/vitamin C [24]. By employing untargeted mass spectrometry and metabolomics analysis of the C2C12 myoblast cells pre-treated with ascorbic acid (100 μM), Magaña et al. [25] suggested that co-supplementation of ascorbate with folic acid may activate the folate-mediated one-carbon cycle, thereby promoting the conversion of homocysteine into methionine while reducing cellular oxidative stress. The pre-treatment of the C2C12 myoblast causes a change in the concentration of tetrahydrofolate (THF) metabolites. Ascorbate appears to spare NAD(P)H and thus increases availability of reducing equivalents for the reductive steps from 10-formyl-THF to 5-methyl-THF via 5,10-methylene-THF. In agreement, an increase in the levels of methionine, whose formation from homocysteine is 5-methyl-THF dependent, and an increase in thymidine, whose formation from deoxyuridine monophosphate (dUMP) is dependent on 5,10-methylene-THF, were observed.

Castejón et al. [26] made a comprehensive review of secoiridoid compounds of the olive tree (*Olea europaea* L. (Oleaceae)) in regard to the chemical, biosynthetic and pharmacological (including pharmacokinetic) aspects. They have tabulated and discussed the effects of the main olive phenolics, namely, oleuropein and its aglycon, oleocanthal and oleacein, which were observed recently in in vitro and in vivo studies in relation to the control and progression of different types of cancer, cardiovascular diseases and neurodegeneration processes in autoimmune diseases, as well as their potential roles in anti-aging.

Overconsumption of some type of food has been associated with the induction of oxidative stress and consequentially the development of pathological transformations. In the review conducted by Cherkas et al. [27], current epidemiological evidence and redox biology advancements are discussed with the purpose of defining the linkage between the observed disruptive metabolic effects, detrimental health consequences of chronic nutritional carbohydrate overload and the metabolic role of glucose in maintaining redox homeostasis. Glucose is the most important energy source and glycolysis is a main pathway for obtaining energy when glucose is abundant and redox conditions are favourable. The authors hypothesized that under extreme oxidative and toxic challenges, glucose provides stress resistance by the reduction of nicotinamide adenine dinucleotide phosphate $NADP^+$ to NADPH through its metabolism in the pentose phosphate pathway (PPP). The reduced form NADPH is needed for the reduction of oxidized glutathione and protein thiols, the synthesis of lipids and DNA as well as for xenobiotic detoxification, regulatory redox signalling and counteracting infections. In regard to this aspect, they discussed the role of glycogen stores in resistance to oxidative challenges.

The important approach to study ageing and age-linked pathologies is to investigate sex dimorphism in the defence against metabolic stressors. In most mammals, females show lower incidence of some age-related pathologies linked with oxidative stress conditions, and this sex difference disappears after menopause, which led to the conclusion that this protection is attributed to the effect of sex hormones [28]. By conducting in vivo experiments on Sirt3 wild-type (WT) and knockout (KO) mice of both sexes fed with a standard or high-fat diet (HFD) for ten weeks, Pinterić et al. [29] found significant sex differences in male and female mice responses at the metabolic level, oxidant/antioxidant status, and liver mitochondrial parameters. Their study points towards a different role of SIRT3 in males and females under the conditions of nutritive stress. For example, the male-specific effects of an HFD include reduced SIRT3 expression in WT and alleviated lipid accumulation and reduced glucose uptake in KO mice.

The research articles and reviews collected in this Special Issue clearly contribute to further appreciation and enlightening of the underlying complexity of the investigation of chemical and pharmacological aspects of modulators of oxidative stress and the importance of studying the modulation of NRF2–KEAP1 and other relevant signalling pathways in order to develop effective drugs for diseases in which oxidative stress plays a significant etiopathogenetic role.

Funding: This research received no external funding.

Conflicts of Interest: The authors declare no conflict of interest.

References

1. Saso, L.; Firuzi, O. Pharmacological Applications of Antioxidants: Lights and Shadows. *Curr. Drug Targets* **2014**, *15*, 1177–1199. [CrossRef] [PubMed]
2. Stepanić, V.; Čipak Gašparović, A.; Gall Trošelj, K.; Amić, D.; Žarković, N. Selected Attributes of Polyphenols in Targeting Oxidative Stress in Cancer. *Curr. Top. Med. Chem.* **2015**, *15*, 496–509. [CrossRef] [PubMed]
3. Telkoparan-Akillilar, P.; Suzen, S.; Saso, L. Pharmacological Applications of Nrf2 Inhibitors as Potential Antineoplastic Drugs. *Int. J. Mol. Sci.* **2019**, *20*, 2025. [CrossRef] [PubMed]
4. Eltzschig, H.; Eckle, T. Ischemia and Reperfusion- From Mechanism to Translation. *Nat. Med.* **2011**, *17*, 1391–1401. [CrossRef]
5. Guo, W.; Liu, X.; Li, J.; Shen, Y.; Zhou, Z.; Wang, M.; Xie, Y.; Feng, X.; Wang, L.; Wu, X. Prdx1 Alleviates Cardiomyocyte Apoptosis Through ROS-activated MAPK Pathway During Myocardial Ischemia/Reperfusion Injury. *Int. J. Biol. Macromol.* **2018**, *112*, 608–615. [CrossRef]
6. Parra-Flores, P.; Riquelme, J.A.; Valenzuela-Bustamante, P.; Leiva-Navarrete, S.; Vivar, R.; Cayupi-Vivanco, J.; Castro, E.; Espinoza-Pérez, C.; Ruz-Cortés, F.; Pedrozo, Z.; et al. The Association of Ascorbic Acid, Deferoxamine and N-Acetylcysteine Improves Cardiac Fibroblast Viability and Cellular Function Associated with Tissue Repair Damaged by Simulated Ischemia/Reperfusion. *Antioxidants* **2019**, *8*, 614. [CrossRef]

7.   May, J.M.; Qu, Z.C.; Neel, D.R.; Li, X. Recycling of Vitamin C from its Oxidized Forms by Human Endothelial Cells. *Biochim. Biophys. Acta (BBA)-Mol. Cell Res.* **2003**, *1640*, 153–161. [CrossRef]

8.   Levine, M.; Padayatty, S.J.; Espey, M.G. Vitamin C: A Concentration-Function Approach Yields Pharmacology and Therapeutic Discoveries. *Adv. Nutr.* **2011**, *2*, 78–88. [CrossRef]

9.   Peserico, D.; Stranieri, C.; Garbin, U.; Mozzini C, C.; Danese, E.; Cominacini, L.; Fratta Pasini, A.M. Ezetimibe Prevents Ischemia/Reperfusion-Induced Oxidative Stress and Up-Regulates Nrf2/ARE and UPR Signaling Pathways. *Antioxidants* **2020**, *9*, 349. [CrossRef]

10.   Jia, L.; Betters, J.L.; Yu, L. Niemann-pick C1-like1 (NPC1L1) Protein in Intestinal and Hepatic Cholesterol Transport. *Annu. Rev. Physiol.* **2011**, *73*, 239–259. [CrossRef]

11.   Singer, M.; Deutschman, C.S.; Seymour, C.W.; Shankar-Hari, M.; Annane, D.; Bauer, M.; Bellomo, R.; Bernard, G.R.; Chiche, J.D.; Coopersmith, C.M.; et al. The Third International Consensus Definitions for Sepsis and Septic Shock (Sepsis-3). *JAMA* **2016**, *315*, 801–810. [CrossRef] [PubMed]

12.   Galley, H.F. Bench to Bedside Review: Targeting Antioxidants to Mitochondria in Sepsis. *Crit. Care* **2010**, *14*, 230. [CrossRef] [PubMed]

13.   Minter, B.E.; Lowes, D.A.; Webster, N.R.; Galley, H.F. Differential Effects of MitoVitE, α-Tocopherol and Trolox on Oxidative Stress, Mitochondrial Function and Inflammatory Signalling Pathways in Endothelial Cells Cultured under Conditions Mimicking Sepsis. *Antioxidants* **2020**, *9*, 195. [CrossRef] [PubMed]

14.   Siegel, R.L.; Miller, K.D.; Jemal, A. Cancer Statistics. *CA Cancer J. Clin.* **2020**, *70*, 7–30. [CrossRef]

15.   Guha, C.; Kavanagh, B.D. Hepatic Radiation Toxicity: Avoidance and Amelioration. *Semin. Radiat. Oncol.* **2011**, *21*, 256–263. [CrossRef]

16.   Tao, R.; Coleman, M.C.; Pennington, J.D.; Ozden, O.; Park, S.H.; Jiang, H.; Kim, H.S.; Flynn, C.R.; Hill, S.; McDonald, W.H.; et al. Sirt3-Mediated Deacetylation of Evolutionarily Conserved Lysine 122 Regulates MnSOD Activity in Response to Stress. *Mol. Cell* **2010**, *40*, 893–904. [CrossRef]

17.   LoBianco, F.V.; Krager, K.J.; Carter, G.S.; Alam, S.; Yuan, Y.; Lavoie, E.G.; Dranoff, J.A.; Aykin-Burns, N. The Role of Sirtuin 3 in Radiation-Induced Long-Term Persistent Liver Injury. *Antioxidants* **2020**, *9*, 409. [CrossRef]

18.   Kumar Rajendran, N.; George, B.P.; Chandran, R.; Tynga, I.M.; Houreld, N.; Abrahamse, H. The Influence of Light on Reactive Oxygen Species and NF-κB in Disease Progression. *Antioxidants* **2019**, *8*, 640. [CrossRef]

19.   Panieri, E.; Buha, A.; Telkoparan-Akillilar, P.; Cevik, D.; Kouretas, D.; Veskoukis, A.; Skaperda, Z.; Tsatsakis, A.; Wallace, D.; Suzen, S.; et al. Potential Applications of NRF2 Modulators in Cancer Therapy. *Antioxidants* **2020**, *9*, 193. [CrossRef]

20.   Milković, L.; Siems, W.; Siems, R.; Zarković, N. Oxidative Stress and Antioxidants in Carcinogenesis and Integrative Therapy of Cancer. *Curr. Pharm. Des.* **2014**, *20*, 6529–6542. [CrossRef]

21.   Kitamura, H.; Motohashi, H. NRF2 Addiction in Cancer Cells. *Cancer Sci.* **2018**, *109*, 900–911. [CrossRef] [PubMed]

22.   Alvarez-Arellano, L.; González-García, N.; Salazar-García, M.; Corona, J.C. Antioxidants as a Potential Target against Inflammation and Oxidative Stress in Attention-Deficit/Hyperactivity Disorder. *Antioxidants* **2020**, *9*, 176. [CrossRef] [PubMed]

23.   García-Guede, Á.; Vera, O.; Ibáñez-de-Caceres, I. When Oxidative Stress Meets Epigenetics: Implications in Cancer Development. *Antioxidants* **2020**, *9*, 468. [CrossRef] [PubMed]

24.   Higdon, J.V.; Frei, B. *Vitamin C: An Introduction, in the Antioxidant Vitamins C and E*; Packer, L., Traber, M.G., Kraemer, K., Frei, B., Eds.; AOAC Press: Champaign, IL, USA, 2002; pp. 11–16.

25.   Alcazar Magana, A.; Reed, R.L.; Koluda, R.; Miranda, C.L.; Maier, C.S.; Stevens, J.F. Vitamin C Activates the Folate-Mediated One-Carbon Cycle in C2C12 Myoblasts. *Antioxidants* **2020**, *9*, 217. [CrossRef] [PubMed]

26.   Castejón, M.L.; Montoya, T.; Alarcón-de-la-Lastra, C.; Sánchez-Hidalgo, M. Potential Protective Role Exerted by Secoiridoids from Olea europaea L. in Cancer, Cardiovascular, Neurodegenerative, Aging-Related, and Immunoinflammatory Diseases. *Antioxidants* **2020**, *9*, 149. [CrossRef]

27.   Cherkas, A.; Holota, S.; Mdzinarashvili, T.; Gabbianelli, R.; Zarkovic, N. Glucose as a Major Antioxidant: When, What for and Why It Fails? *Antioxidants* **2020**, *9*, 140. [CrossRef]

28. Lejri, I.; Grimm, A.; Eckert, A. Mitochondria, Estrogen and Female Brain Aging. *Front. Aging Neurosci.* **2018**, *10*, 124. [CrossRef]

29. Pinterić, M.; Podgorski, I.I.; Hadžija, M.P.; Tartaro Bujak, I.; Dekanić, A.; Bagarić, R.; Farkaš, V.; Sobočanec, S.; Balog, T. Role of Sirt3 in Differential Sex-Related Responses to a High-Fat Diet in Mice. *Antioxidants* **2020**, *9*, 174. [CrossRef]

Article

The Association of Ascorbic Acid, Deferoxamine and N-Acetylcysteine Improves Cardiac Fibroblast Viability and Cellular Function Associated with Tissue Repair Damaged by Simulated Ischemia/Reperfusion

Pablo Parra-Flores [1,2], Jaime A Riquelme [2], Paula Valenzuela-Bustamante [1,2], Sebastian Leiva-Navarrete [2], Raúl Vivar [3], Jossete Cayupi-Vivanco [1,2], Esteban Castro [1,2], Claudio Espinoza-Pérez [1,2], Felipe Ruz-Cortés [1,2], Zully Pedrozo [2,3], Sergio Lavandero [2,3,4], Ramon Rodrigo [3] and Guillermo Diaz-Araya [1,2,*]

[1] Laboratorio de Farmacología Molecular, Departamento de Química Farmacológica y Toxicológica, Facultad de Ciencias Químicas y Farmacéuticas, Universidad de Chile, Santos Dumont 964, Independencia, Santiago 8380492, Chile; qfpablo@gmail.com (P.P.-F.); p.valenzuelabustamante@gmail.com (P.V.-B.); jossete.c.v@gmail.com (J.C.-V.); ecastro.chl@gmail.com (E.C.); claudio.espinoza06@gmail.com (C.E.-P.); felipe.ruz.cortes@gmail.com (F.R.-C.)

[2] Advanced Center for Chronic Diseases (ACCDiS), Facultad de Ciencias Químicas y Farmacéuticas and Facultad de Medicina, Universidad de Chile, Santos Dumont 964, Independencia, Santiago 8380492, Chile; riquelme@ciq.uchile.cl (J.A.R.); srleiva.navarrete@gmail.com (S.L.-N.); zpedrozo@med.uchile.cl (Z.P.); slavander@uchile.cl (S.L.)

[3] Instituto de Ciencias Biomédicas, Facultad de Medicina, Universidad de Chile, Independencia 1027, Independencia, Santiago 8380453, Chile; raulvivar@med.uchile.cl (R.V.); rrodrigo@med.uchile.cl (R.R.)

[4] Cardiology Division, Department of Internal Medicine, University of Texas Southwestern Medical Center, Dallas, TX 75390-8573, USA

[*] Correspondence: gadiaz@ciq.uchile.cl

Received: 12 October 2019; Accepted: 26 November 2019; Published: 3 December 2019

Abstract: Acute myocardial infarction is one of the leading causes of death worldwide and thus, an extensively studied disease. Nonetheless, the effects of ischemia/reperfusion injury elicited by oxidative stress on cardiac fibroblast function associated with tissue repair are not completely understood. Ascorbic acid, deferoxamine, and N-acetylcysteine (A/D/N) are antioxidants with known cardioprotective effects, but the potential beneficial effects of combining these antioxidants in the tissue repair properties of cardiac fibroblasts remain unknown. Thus, the aim of this study was to evaluate whether the pharmacological association of these antioxidants, at low concentrations, could confer protection to cardiac fibroblasts against simulated ischemia/reperfusion injury. To test this, neonatal rat cardiac fibroblasts were subjected to simulated ischemia/reperfusion in the presence or absence of A/D/N treatment added at the beginning of simulated reperfusion. Cell viability was assessed using trypan blue staining, and intracellular reactive oxygen species (ROS) production was assessed using a 2′,7′-dichlorofluorescin diacetate probe. Cell death was measured by flow cytometry using propidium iodide. Cell signaling mechanisms, differentiation into myofibroblasts and pro-collagen I production were determined by Western blot, whereas migration was evaluated using the wound healing assay. Our results show that A/D/N association using a low concentration of each antioxidant increased cardiac fibroblast viability, but that their separate administration did not provide protection. In addition, A/D/N association attenuated oxidative stress triggered by simulated ischemia/reperfusion, induced phosphorylation of pro-survival extracellular-signal-regulated kinases 1/2 (ERK1/2) and PKB (protein kinase B)/Akt, and decreased phosphorylation of the pro-apoptotic proteins p38- mitogen-activated protein kinase (p38-MAPK) and c-Jun-N-terminal kinase (JNK). Moreover, treatment with A/D/N also reduced reperfusion-induced apoptosis, evidenced by a decrease in the sub-G1 population, lower fragmentation of pro-caspases 9 and 3, as well as increased B-cell lymphoma-extra large protein (Bcl-xL)/Bcl-2-associated X protein (Bax) ratio. Furthermore, simulated ischemia/reperfusion abolished serum-induced migration, TGF-β1 (transforming growth factor beta 1)-mediated cardiac fibroblast-to-cardiac myofibroblast differentiation, and angiotensin II-induced pro-collagen I synthesis, but these effects were prevented by treatment with A/D/N. In conclusion, this is the first study where a pharmacological combination of A/D/N, at low concentrations, protected cardiac fibroblast viability and function after simulated ischemia/reperfusion, and thereby represents a novel therapeutic approach for cardioprotection.

Keywords: ascorbic acid; deferoxamine; N-acetylcysteine; ischemia/reperfusion; cardiac fibroblasts; reactive oxygen species

---

1. Introduction

Ischemic heart disease is still one of the leading causes of mortality in the world [1]. Prolonged myocardial ischemia due to partial or complete coronary artery occlusion generates an imbalance between the supply and demand of O_2, which subsequently leads to cardiac cell death and thus, to development of acute myocardial infarction (MI) [2,3]. Restoration of blood flow is of utmost importance for myocardial salvage. Nevertheless, this procedure itself induces necrotic and apoptotic cell death. This paradoxical effect is known as lethal reperfusion injury and can contribute up to 50% of final infarct size [4,5]. One of the key mediators of reperfusion injury is oxidative stress, characterized by a massive release of reactive oxygen species (ROS) at the beginning of reperfusion, which triggers cellular lipid peroxidation, as well as protein/nucleic acid oxidation, and consequently, activates pro-apoptotic pathways associated with p38-MAPK and JNK proteins [6,7]. In addition, ROS of enzymatic origin can be produced by xanthine oxidase, reduced nicotinamide adenine dinucleotide phosphate hydrogen (NADPH) oxidase, and uncoupled endothelial nitric oxide synthase (eNOS), whereas non-enzymatic generation of ROS may arise from uncoupled mitochondrial electron transport chain and Fenton and Haber-Weiss reactions mediated by free iron from cell lysis [6,8].

After myocardial damage caused by ischemia/reperfusion (I/R), the tissue repair processes are mainly coordinated by cardiac fibroblasts (CF) [9], which are the second most numerous non-myocyte population in the mouse heart [10]. Under physiological conditions, these cells can synthesize and degrade collagen to provide a constant replacement of the myocardial extracellular matrix (ECM). Moreover, CF secrete a variety of paracrine factors that may regulate cellular function in cardiomyocytes, the endothelium, vascular smooth muscle cells, and immune system cells [11]. In addition, upon pathophysiological stimuli, CF migrate, proliferate, secrete pro-inflammatory mediators, metalloproteases, and differentiate into cardiac myofibroblasts (CMF) to promote scar tissue formation [9].

Although previous studies have evaluated the effect of simulated I/R (sI/R) on CF death [12–15], the impact of oxidative stress triggered by reperfusion injury on the functional capacity associated with fibroblast-mediated tissue repair, remains to be fully elucidated. In this context, the use of antioxidants to confer cardioprotection in patients with MI subjected to percutaneous coronary angioplasty has been previously reviewed [6,8,16]. Ascorbic acid (A), deferoxamine (D), and N-acetylcysteine (N) are antioxidants that can separately act as a free radical scavenger, a free iron chelator and a reduced glutathione (GSH) precursor, respectively. In ex vivo and in vivo models of MI these three antioxidants can protect against myocardial reperfusion injury and prevent the death of cardiomyocytes exposed to sI/R [17–22]. However, the effects of these antioxidants, either separate or combined, on the viability of CF after sI/R and the potential mechanisms associated with this protection have not been thoroughly explored.

The aim of this study was to determine whether the pharmacological association of A/D/N, at low concentrations, protects CF from death and recovers cellular function associated with tissue repair processes after damage by sI/R; along with elucidating the signaling pathways that mediate this effect.

2. Material and Methods

2.1. Reagents

Dulbecco's Modified Eagle's medium F12 (DMEM-F12; No 12500062), alamarBlue™ Cell Viability Reagent (Resazurin; No DAL1100), and Collagenase, Type II, powder (No 17101015) were obtained from Thermo Fisher Scientific (Waltham, MA, USA). Fetal Bovine Serum (FBS; No 04-001-1A), Trypsin ethylenediaminetetraacetic acid (EDTA; 0.5%), EDTA 0.2% (10X Solution (No 03-051-5B), penicillin-streptomycin-amphotericin B Solution (No 03-033-1B), and Trypan Blue (0.5%) Solution (No 03-102-1B) were obtained from Biological Industries (Cromwell, CT, USA). Recombinant human TGF-beta 1 protein (TGF-β1; No 240-B) was obtained from R & D Systems, Inc. (Minneapolis, MN, USA). Pascorbin (vitamin C infusion bottle) was obtained from Pascoe Naturmedizin (Giessen, Germany). M199 medium (No M2520), pancreatin from porcine pancreas (No P3292), RNAse A (No 10109142001), propidium iodide (No P4170), Bradford reagent (No B6916), RIPA lysis and extraction buffer (No 89900), Halt™ Protease Inhibitor Cocktail (100X, No 78438), Halt™ Phosphatase Inhibitor Cocktail (No 78427), enhanced chemiluminescence (ECL) western blotting detection reagents (No GERPN2209), deferoxamine mesylate salt (No D9533), N-Acetyl-L-cysteine (No A7250), 2′,7′-Dichlorofluorescin diacetate (DCFH-DA; No D6883), 5-Bromo-2′-deoxyuridine (BrdU; No B5002), angiotensin II human (No A9525), and crystal violet (No C0775) were obtained from Sigma-Aldrich Corp. (St. Louis, MI, USA). All inorganic salt products and methanol (No 106035) were obtained from Merck (Darmstadt, Germany). Prestained Protein Ladder-Broad molecular weight (10–245 kDa; No ab116028) was obtained from Abcam (Cambridge, MA, USA). Nitrogen gas (N_2) cylinders were obtained from Linde Group (Santiago, Chile). Primary antibodies for phospho-ERK 1/2 (p-ERK1/2, No 9101), ERK1/2 (No 9102), phospho-Akt (p-Akt, No 9271), Akt (No. 9272), phospho-p38-MAPK (p-p38-MAPK, No 9211), p38 (No 9212), phospho-JNK (p-JNK, No 9251), JNK (No 9252), Pro-caspase 3 (No 9662), Pro-caspase 9 (No 9508), Bax (No 2772), Bcl-xl (No 2764), alpha-smooth muscle actin (α-SMA; No 19245), alpha-1 type 1 collagen (COL1A1; No 84336), glyceraldehyde 3-phosphate dehydrogenase (GAPDH; No 5174), and secondary antibodies for anti-rabbit IgG (No 7074) and anti-mouse IgG (No 7076) conjugated with horseradish peroxidase (HRP) were obtained from Cell Signaling Technology (Danvers, MA, USA). All plastic material was obtained from Corning Incorporated (Corning, NY, USA).

2.2. Animals

Sprague-Dawley neonatal rats (1 to 3-days-old) were obtained from Animal Breeding Facilities of the Facultad de Ciencias Químicas y Farmacéuticas (Universidad de Chile, Santiago, Chile). All studies followed the Guide for the Care and Use of Laboratory Animals (8th edition, 2011) [23], and International Guiding Principles for Biomedical Research Involving Animals, as issued by the Council for the International Organizations of Medical Sciences [24]. Experimental protocols were approved by the Bioethics Committee for Animal Research from the Facultad de Ciencias Químicas y Farmacéuticas, Universidad de Chile (CBE2017-08).

2.3. Isolation and Culture of Cardiac Fibroblasts

Neonatal rat CF were isolated as previously described [12]. In a sterile zone, rats were swiftly decapitated and their hearts rapidly excised. After removing the atrium, ventricles were minced and digested with collagenase (0.05% *w/v*) and pancreatin (0.05% *w/v*). Cells were pre-cultured on plastic dishes of 100 mm diameter, with DMEM-F12/M199 (4:1) medium containing FBS (10%) and penicillin-streptomycin-amphotericin B and kept in an incubator with O_2 (95%) and CO_2 (5%) at 37 °C. CF differentially adhered on plastic dishes, which separated them from cardiomyocytes. After

3 h, CF were washed three times with sterile phosphate buffer saline (PBS) and culture medium was replenished. Next, CF were incubated during 3–5 days, until confluence was reached. Afterwards, medium was replaced by DMEM-F12 containing FBS (3%) and penicillin-streptomycin-amphotericin B. Cells underwent up to a maximum of two passages and detachment was performed using trypsin EDTA (0.5%), EDTA 0.2% (1X), followed by protease inhibition with DMEM-F12 containing FBS (10%). CF were then collected and seeded on plastic plates in DMEM-F12 medium without FBS for 24 h before experiments.

2.4. Isolation and Culture of Cardiomyocytes

Isolation of neonatal rat ventricular cardiomyocytes was performed using one- to three-day-old Sprague-Dawley rats, as described previously [25]. After enzymatic digestion of the myocardium with collagenase 0.02% and pancreatin 0.06%, cells were pre-plated to discard non-myocyte cells and the myocyte-enriched fraction was plated on gelatin-precoated 35 mm dishes and grown in DMEM/M199 (4:1) medium with 10% (*w/v*) fetal bovine serum (FBS) and 100 mM bromodeoxyuridine for 24 h before the experiments.

2.5. Simulated Ischemia/Reperfusion (sI/R) and Antioxidant Treatment

We used a modified protocol from Vivar et al. [13]. Following 24 h serum deprivation, cells were washed two times with PBS before sI/R. Cell medium was replaced with ischemic buffer, which contained: NaCl (115 mM), KCl (12 mM), $MgCl_2$ (1.2 mM), $CaCl_2$ (2 mM), (4-(2-hydroxyethyl)-1-piperazineethanesulfonic acid (HEPES) (25 mM), and lactic acid (20 mM), pH 6.2. For simulated ischemia, hypoxia was achieved by placing the cells in a customized chamber (STEMCELL Technologies Inc., Vancouver, Canada) with < 1% O_2 and 99% N_2 environment at 37 °C for 6 h. Simulated reperfusion was performed replacing the ischemic medium with DMEM-F12 medium and placing cells with O_2 (95%) and CO_2 (5%) at 37 °C for 16 h. At the onset of simulated reperfusion CF, were treated separately with A, D, and N using 10,000, 1000, 100, 10, and 1 μM each for preliminary cell viability studies. Antioxidants were then, administered in different combinations at 10 or 1 μM each. In later experiments, the A/D/N association was added using 10 μM of each antioxidant and simulated reperfusion time was variable, depending on the experiment. Control cells were incubated under normoxic conditions in DMEM-F12 medium and kept in an incubator with O_2 (95%) and CO_2 (5%) at 37 °C for the exact duration of simulated ischemia and sI/R experiments.

2.6. Cell Viability with Trypan Blue

CF at a 156 cells/mm^2 density on 35 mm plastic dishes were used to assess viability using trypan blue (0.5%) staining. After 16 h simulated reperfusion, cells were washed two times with PBS and treated with trypsin EDTA (0.5%), EDTA 0.2% (1×) to detach cells, followed by administration of FBS (10%) to induce inactivation. After collecting the cells, aliquots of 20 μL of sample, plus 20 μL of trypan blue (0.5%) reagent were homogenized, and then 8 μL were transferred to a Neubauer chamber (Paul Marienfeld GmbH & Co. KG, Lauda-Königshofen, Germany) to count viable cells (unstained) using optic microscopy. Cell viability was quantified as a percentage (%) of number of cells after 6 h normoxia (100%). A minimum of 1000 cells/mL was counted in each sample.

2.7. Cell Viability with the Resazurin Reduction Assay

Cells were seeded in clear bottom 24-well plates with a density of 263 cells/mm^2. After 16 h simulated reperfusion, culture medium was replaced by DMEM-F12, followed by incubation with resazurin (10%) for 4 h. Viable cells with active metabolism can reduce the non-fluorescent resazurin to fluorescent resorufin. Fluorescence was measured at 585 nm (λ excitation) and 570 nm (λ emission) in a BioTek™ Synergy™ 2 Multi-Mode Microplate Reader (BioTek Instruments, Inc., Winooski, VT, USA).

2.8. Determination of Intracellular ROS

Cells were seeded in clear bottom 24-well plates with a density of 421 cells/mm^2. DCFH-DA is a non-fluorescent cell membrane permeable probe which is de-esterified intracellularly and then oxidized to the fluorescent 2′,7′-dichlorofluorescein. At the end of 6 h simulated ischemia, cells were incubated with DCFH-DA (20 μM) dissolved in fresh ischemic medium for 45 min at 37 °C. Rapidly, cells were washed two times with PBS, and then DMEM-F12, with or without antioxidants, was added. Fluorescence intensity was measured during the first 30 min of simulated reperfusion at 485 nm (λ excitation) and 535 nm (λ emission) in a BioTek™ Synergy™ 2 Multi-Mode Microplate Reader (BioTek Instruments, Inc., Winooski, VT, USA).

2.9. Necrosis Assessment by Flow Cytometer

CF were seeded with a 106 cells/mm^2 density on 60 mm plastic dishes. After 16 h of simulated reperfusion, dead cells were collected from medium, centrifuged at 252× g for 5 min, and kept at 4 °C. Live cells were detached from plates using trypsin EDTA (0.5%), EDTA 0.2% (1X), and mixed with the pellet of dead cells. Subsequently, propidium iodide (1 mg/mL) was added and necrotic cell death was evaluated by flow cytometry in a BD FACSCantoA (Becton Dickinson & Company, Franklin Lakes, NJ, USA). A total of 5000 cells/sample were analyzed.

2.10. Sub-G1 Population Determination by Flow Cytometry

Cells were seeded at 106 cells/mm^2 density on 60 mm plastic dishes. After 16 h of simulated reperfusion, dead and live cells were collected according to the same protocol used in Section 2.8. Next, to permeabilize cell membranes, cold methanol was added to the live and dead cell mixture for 24 h, at −20 °C. RNAse (0.1 mg/mL) was then added to the samples for 1 h at room temperature. Finally, propidium iodide (1 mg/mL) was added to cells and apoptosis was determined by flow cytometry using a BD FACSCanto (Becton Dickinson & Company, Franklin Lakes, NJ, USA). Propidium iodide marks condensed chromatin and/or fragmented DNA in apoptotic bodies giving a low intensity signal (sub-G1 population), under the prominent G1 signal of living cells with integral DNA. A total of 5000 cells/sample were analyzed.

2.11. Western Blot Analysis

For protein content analysis, CF were seeded at a 106 cells/mm^2 density on 60 mm plastic dishes. At the end of simulated reperfusion, cells were washed three times with cold PBS, followed by addition of RIPA lysis buffer with protease and phosphatase inhibitors. Samples were centrifuged at 252× g for 10 min at 4 °C, and supernatants were collected. Total protein concentration of samples was determined using the Bradford reagent, and absorbance was measured to 595 nm in an Epoch UV-Vis Spectrophotometer (BioTek Instruments, Inc., Winooski, VT, USA). We used 25 μg of total protein sample, which was separated by sodium dodecyl sulfate polyacrylamide gel electrophoresis (SDS-PAGE) using 10–20% acrylamide/bis-acrylamide gels run for 1.5–2 h. at constant 70–100 V. Proteins were then electro-transferred to a nitrocellulose membrane for 1–1.5 h with a 0.35 A constant current. Membrane was blocked with non-fat milk (5% *w/v*) for 1 h. Primary antibodies against p-ERK1/2, ERK1/2, p-Akt, Akt, p-p38, p38, p-JNK, JNK, pro-caspase 3, pro-caspase 9, Bax, Bcl-xl, α-SMA, COL1A1 (dilution 1:1000), or GAPDH (dilution 1:2000) were incubated overnight at 4 °C. Secondary antibodies for anti-rabbit IgG or anti-mouse IgG conjugated with HRP (dilution 1:5000) were incubated for 1.5 h at room temperature. Membrane was exposed to the ECL reagent and revealed in the C-DiGit Chemiluminescent Western Blot Scanner (LI-COR Biosciences, Lincoln, NE, USA). Images and blots were analyzed and quantified using the Image Studio™ software (LI-COR Biosciences, Lincoln, NE, USA).

2.12. Evaluation of Cell Migration by Wound Healing Assay

CF were seeded 156 cells/mm^2 density on 35 mm plastic dishes in DMEM-F12 containing FBS (10%) and penicillin-streptomycin-amphotericin B and incubated during 24 h to allow proliferation until confluence was reached. Subsequently, cells were washed two times with PBS and incubated with DMEM-F12 for 24 h. Next, simulated ischemia was performed for 6 h, followed by a scratch made using a 200 µL tip in fibroblast monolayers at the beginning of simulated reperfusion. BrdU (100 µM) was used as a proliferation inhibitor, allowing CF to migrate without proliferating in the presence of FBS (10%). After 24 h, cells were stained with crystal violet (0.3% *w/v*) for 20 min at room temperature. After washing and drying plates, images were obtained from four fields per plate using an optic microscope. Scratched areas per image were analyzed with ImageJ software (LI-COR Biosciences, Lincoln, NE, USA). and the mean of four fields of scratched areas per plate was used for data analysis. All values of mean scratched area were normalized with respect to the value of mean scratched area of control groups with cells incubated in DMEM-F12.

2.13. Statistical Analysis

All data are presented as mean ± standard error of the mean (S.E.M.), of at least three independent experiments, and were analyzed using GraphPad Prism (GraphPad, San Diego, CA, USA) version 5.01 software. The differences between two experimental groups were evaluated by paired Student's t-test. The differences between three or more experimental groups were evaluated by a one-way analysis of variance (ANOVA) followed by a Tukey post-test. The differences between two experimental groups at each time were evaluated by two-way ANOVA, followed by a Bonferroni post-test. Statistical significance was accepted at $p < 0.05$.

3. Results

3.1. Individual Effects of Ascorbic Acid, Deferoxamine, and N-Acetylcysteine on Viability of Cardiac Fibroblasts after Simulated Ischemia/Reperfusion

In order to study the cytoprotective effects of antioxidants, we first validated our model by subjecting neonatal rat CF to 6 h of simulated ischemia, followed by 16 h of simulated reperfusion and we measured cell viability using trypan blue. The results show that after sI/R, cell viability was significantly reduced compared to normoxic controls ($p < 0.001$; Figure 1A), which was corroborated using the resazurin reduction assay ($p < 0.01$; Figure 1B). We then tested administration ascorbic acid (A), deferoxamine (D), and N-acetylcysteine (N) at 10,000; 1000; 100; 10; and 1 µM at the beginning of reperfusion and evaluated cell viability using trypan blue. None of the three antioxidants had any effect on CF viability at 10 and 1 µM after sI/R. (Figure 1C–E). However, treatment with D increased cell viability at 10,000; 1000; and 100 µM, while addition of N increased cell viability at 1000 and 100 µM, but not at 10,000 µM (Figure 1D,E, respectively). Administration of A increased cell viability only at 100 µM (non-significant), whereas 10,000 µM further reduced CF viability after sI/R, compared to untreated cells ($p < 0.001$; Figure 1C).

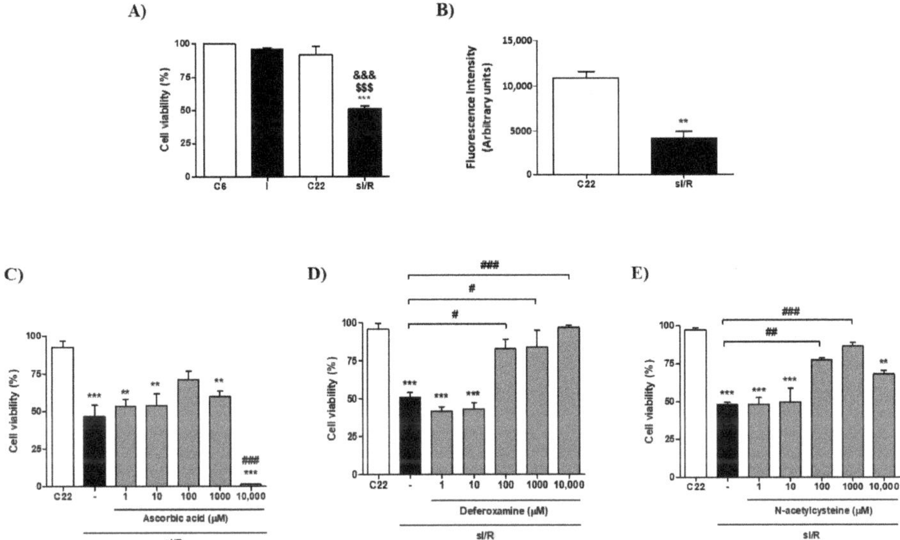

Figure 1. Effects of ascorbic acid, deferoxamine and N-acetylcysteine, at different concentrations, on viability of cardiac fibroblasts exposed to simulated ischemia/reperfusion. Cardiac fibroblasts were exposed to 6 h simulated ischemia, followed by 16 h simulated reperfusion (sI/R). (**A**) Cell viability was quantified as a percentage (%) of number of cells after 6 h normoxia (C6, 100%) by cell count after trypan blue staining ($n = 3$). (**B**) Cell viability was quantified as fluorescence intensity using the resazurin reduction assay ($n = 3$). At the beginning of simulated reperfusion, ascorbic acid (**C**), deferoxamine (**D**) and N-acetylcysteine (**E**) were added using 10,000; 1000; 100; 10; and 1 µM. Cell viability was quantified as the percentage (%) of number of cells after 6 h normoxia (100%) by cell count after trypan blue staining ($n = 3$). The results are expressed as mean ± S.E.M. &&& $p < 0.001$ vs. C6; \$\$\$ $p < 0.001$ vs. I (cells after 6 h simulated ischemia); *** $p < 0.001$ and ** $p < 0.01$ vs. C22 (control cells after 22 h normoxia); ### $p < 0.001$, ## $p < 0.01$, and # $p < 0.05$ vs. sI/R.

3.2. Ascorbic Acid, Deferoxamine, and N-Acetylcysteine Association Increased Cell Viability and Reduced Intracellular ROS Production in Cardiac Fibroblasts Subjected to Simulated Ischemia/Reperfusion

Next, we sought to evaluate whether the pharmacological association of A, D, and N could provide synergistic protection by using a low concentration of each antioxidant that had no effect on cell viability when administered separately. To achieve this, CF were exposed to sI/R and we simultaneously administered combinations of the three antioxidants at the onset of reperfusion at 1 and 10 µM, and then measured cell viability using trypan blue. Our results show that treatments with the associations of A/D, A/N, and A/D/N, but not D/N, increased cell viability after sI/R at 10 µM each, when compared to untreated conditions (Figure 2A). However, only the A/D association protected at 1 µM (Supplementary Figure S1). Based on these findings, we decided to study the cytoprotective effect of the A/D/N association at 10 µM given its potential to provide more robust protection by reducing oxidative stress by three different mechanisms, unlike double associations of these antioxidants. In addition, we observed that the A/D/N association at 10 µM each significantly reduced intracellular ROS production after sI/R, compared to untreated cells, as measured with a DCFH-DA probe (Figure 2B). Therefore, we used the joint administration of 10 µM of A/D/N for the rest of the experiments.

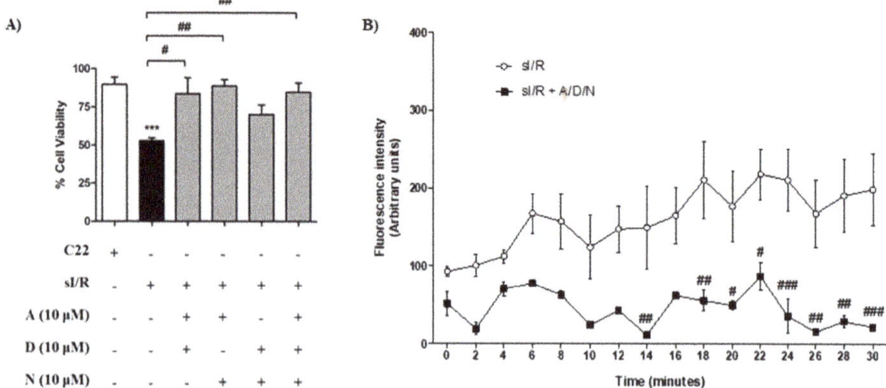

Figure 2. The association of ascorbic acid, deferoxamine, and N-acetylcysteine, at 10 μM each, increased cell viability and decreased intracellular reactive oxygen species (ROS) generation in cardiac fibroblasts exposed to simulated ischemia/reperfusion. (**A**) Cardiac fibroblasts were exposed to 6 h simulated ischemia followed by 16 h simulated reperfusion (sI/R). Associations between ascorbic acid (A), deferoxamine (D), and N-acetylcysteine (N), using 10 μM of each antioxidant, were added at the beginning of simulated reperfusion. Cell viability was quantified as a percentage (%) of number of cells after 6 h normoxia (100%) by cell count after trypan blue staining ($n = 5$). (**B**) Cardiac fibroblasts were exposed to 6 h simulated ischemia followed by 30 min simulated reperfusion (sI/R). At the end of ischemia, cells were incubated with 2′,7′-dichlorofluorescin diacetate and then treated with A/D/N using 10 μM of each antioxidant at the onset of simulated reperfusion. Intracellular ROS generation was measured as the fluorescence intensity in a time course during the first 30 min of simulated reperfusion ($n = 3$). The results are expressed as mean ± S.E.M. *** $p < 0.001$ vs. C22 (control cells after 22 h normoxia); ### $p < 0.001$, ## $p < 0.01$ and # $p < 0.05$ vs. sI/R. Symbol "+" represents presence of condition and symbol "-" represents absence of condition.

3.3. Association of Ascorbic Acid, Deferoxamine, and N-Acetylcysteine Reduced Apoptosis of CF Exposed to Simulated Ischemia/Reperfusion

To confirm our findings of cell death types, we measured necrosis and apoptosis after sI/R using flow cytometry analysis with propidium iodide staining. The results indicate that antioxidant association had no effect on necrosis induced by sI/R (Figure 3A). However, treatment with A/D/N induced a decrease in the sub-G1 population of CF exposed to sI/R, compared to untreated conditions ($p < 0.05$; Figure 3B). To further corroborate that the A/D/N association can inhibit apoptosis induced by sI/R in CF, we determined the protein levels of pro-caspase 9 and pro-caspase 3, as well as the Bcl-xl/Bax ratio after treatment with A/D/N. Our data shows that cells exposed to sI/R presented lower levels of pro-caspases 9 and 3, as compared with normoxic cells (both $p < 0.001$). This result suggests an induction of apoptosis (Figure 3C,D), but administration of the A/D/N association prevented this effect and increased the Bcl-xl/Bax ratio ($p < 0.05$), compared to untreated CF after sI/R (Figure 3C–E). In addition, to test whether these findings were reproduced in a different cell type, we subjected primary neonatal rat cardiomyocytes to sI/R and then treated them with A/D/N. Our results show that A/D/N increased viability of cardiomyocytes (Supplementary Figure S2A) and increased the levels of pro-caspases 9 and 3 (Supplementary Figure S2B,C).

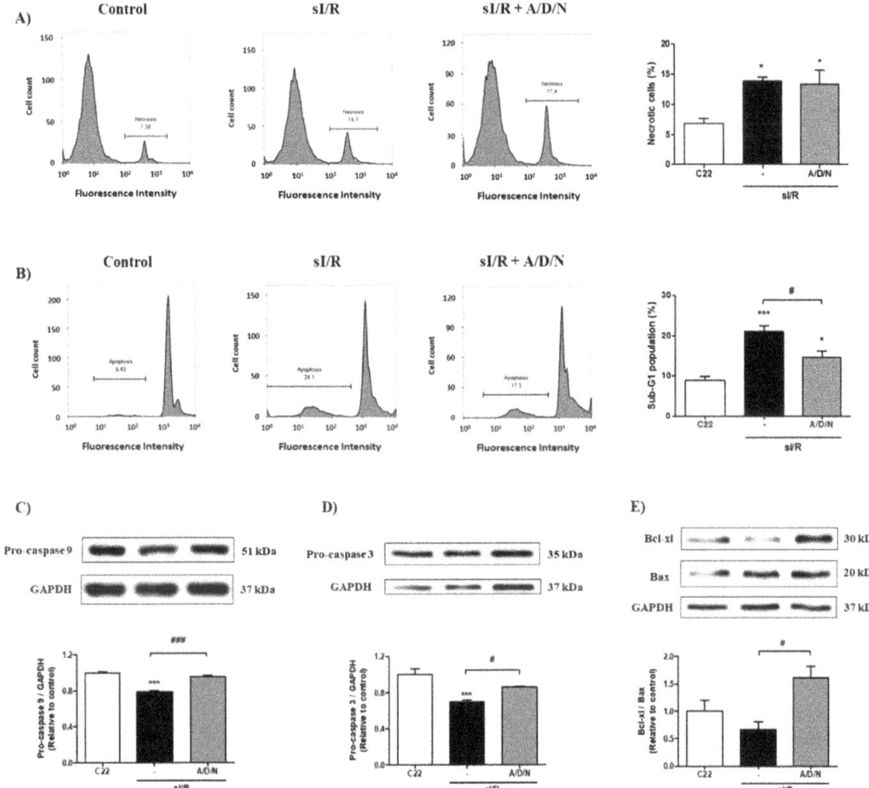

Figure 3. The association of ascorbic acid, deferoxamine and N-acetylcysteine reduced apoptosis induced by simulated ischemia/reperfusion in CF. Cells were exposed to 6 h simulated ischemia, followed by 16 h simulated reperfusion (sI/R). Cells were treated with the association of ascorbic acid, deferoxamine, and n-acetylcysteine (A/D/N) using 10 μM of each antioxidant at the onset of simulated reperfusion. (**A**) The percentage (%) of necrotic cells was quantified by flow cytometry using propidium iodide (right panel), with representative histograms of each experimental group (left panel; $n = 4$). (**B**) The percentage (%) of the sub-G1 population was quantified by flow cytometry using propidium iodide (right panel), with representative histograms of each experimental group (left panel; $n = 5$). (**C–E**) show representative Western blots (upper panel) and densitometric analysis (lower panel) of pro-caspase 9 ($n = 4$), pro-caspase-3 ($n = 5$), and Bcl-xl/Bax ratio ($n = 3$), respectively. Glyceraldehyde 3-phosphate dehydrogenase (GAPDH) was used as a loading control. The results are expressed as mean ± S.E.M. *** $p < 0.001$ and * $p < 0.05$ vs. C22 (control cells after 22 h normoxia); ### $p < 0.001$ and # $p < 0.05$ vs. sI/R.

3.4. Association of Ascorbic Acid, Deferoxamine and N-Acetylcysteine Activated the Pro-Survival Kinases ERK1/2 and Akt and Reduced the Phosphorylation of the Pro-Apoptotic Proteins p38.MAPK and JNK Induced by Simulated Ischemia/Reperfusion in CFs

In order to pursue the mechanism by which the association of antioxidants conferred its protective effect, we sought to evaluate the signaling pathways associated with cell survival. To test this, we determined by Western blot the early activation of ERK1/2 and Akt in response to administration of A/D/N in CF exposed to sI/R. The results show that sI/R significantly increased phosphorylation of ERK1/2 and Akt after 10 min of simulated reperfusion, compared to normoxic conditions, but this effect was further potentiated by the treatment with the antioxidant association (Figure 4A,B). In addition, sI/R also elicited early phosphorylation of the pro-apoptotic proteins p38 and JNK in comparison with normoxic conditions, but this effect was diminished by administration of A/D/N ($p < 0.05$; Figure 4C,D).

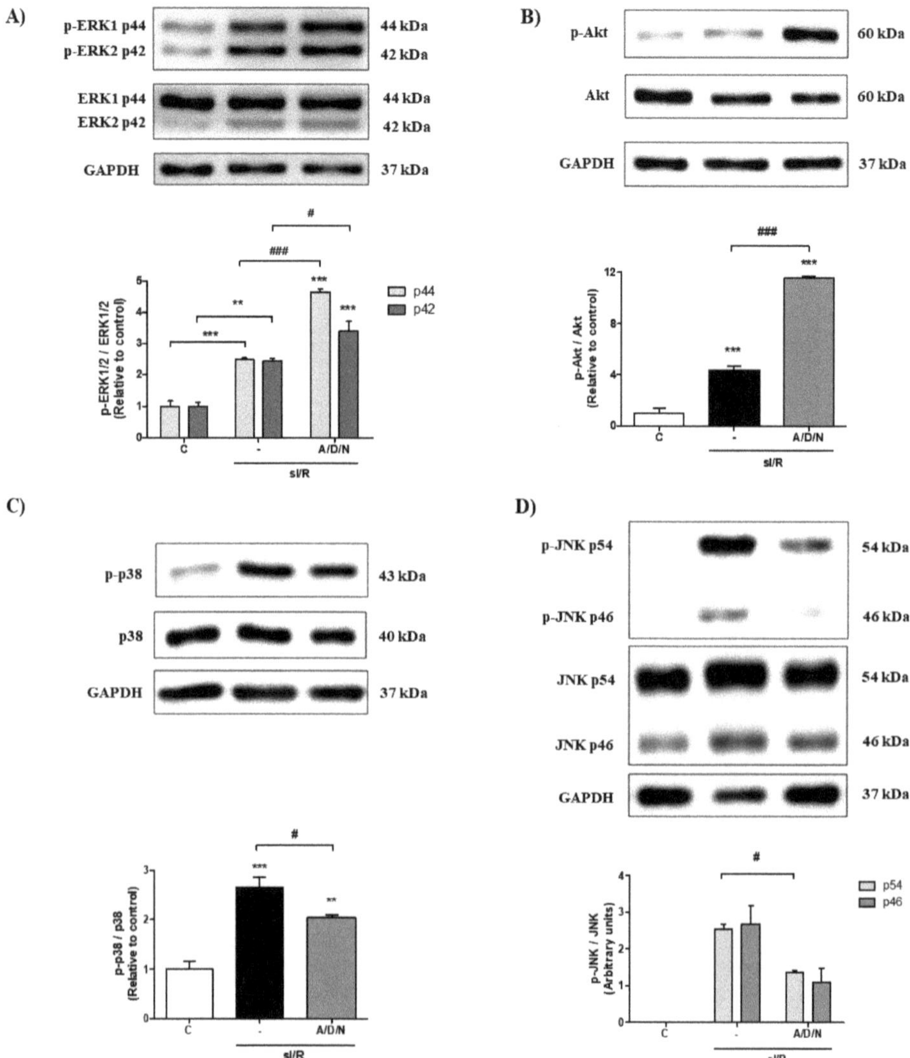

Figure 4. The association of ascorbic acid, deferoxamine, and N-acetylcysteine increased the activation of the pro-survival kinases ERK1/2 and Akt, and reduced activation of the pro-apoptotic proteins p38 and JNK in cardiac fibroblasts exposed to simulated ischemia/reperfusion. Cardiac fibroblasts were exposed to 6 h simulated ischemia, followed by 10 min of simulated reperfusion (sI/R). Cells were treated with the association of ascorbic acid, deferoxamine and N-acetylcysteine (A/D/N) using 10 μM of each antioxidant at the onset of simulated reperfusion. (**A–D**) show representative Western blots (upper panel) and densitometric analysis (lower panel) of p-ERK1/2 and ERK1/2 ($n = 3$), p-Akt and Akt ($n = 3$), p-p38 and p38 ($n = 4$), and p-JNK and JNK ($n = 3$), respectively. GAPDH was used as a loading control. The results are expressed as mean ± S.E.M. *** $p < 0.001$ and ** $p < 0.01$ vs. C (control cells after 70 min normoxia); ### $p < 0.001$ and # $p < 0.05$ vs. sI/R.

3.5. Association of Ascorbic Acid, Deferoxamine, and N-Acetylcysteine Prevented the Loss of Function Associated with Tissue Repair Induced by Simulated Ischemia/Reperfusion in CF

Finally, we studied whether the A/D/N association can protect CF function associated with detriments in migration, differentiation and collagen secretion, after sI/R. To assess cell migration, we performed the wound healing assay using FBS (10%) to induce migration of CF to the scratched area during simulated reperfusion, and we applied BrdU to inhibit the proliferation induced by FBS. Figure 5A shows that, in normoxic conditions, CF migrated in the presence of FBS (10%), with or without BrdU, reducing the scratched area over 50%, compared with control cells ($p < 0.001$). However, cells did not migrate in the presence of the A/D/N association. Then, we tested whether CF exposed to sI/R can migrate in the presence of FBS (10%) + BrdU. Figure 5A shows that, after 6 h simulated ischemia followed by 24 h simulated reperfusion, CF did not migrate in the presence of FBS (10%) + BrdU. In addition, the A/D/N association by itself did not modify the migration of CF exposed to sI/R. However, when the cells were exposed to sI/R and treated with A/D/N in the presence of FBS (10%) + BrdU, migration was significantly increased compared to normoxic control cells without FBS (10%; $p < 0.01$), and compared to cells under sI/R and in presence of FBS (10%) + BrdU ($p < 0.01$; Figure 5A).

In addition, we induced CF-to-CMF differentiation by incubating CF in the presence or absence of TGF-β1 (10 ng/mL) during simulated reperfusion, and measured α-SMA (alpha smooth muscle actin)—a differentiation protein level marker—by Western blot after 48 h. Figure 5B shows that α-SMA protein content is increased after 6 h simulated ischemia, followed by 48 h simulated reperfusion with or without TGF-β1 administration, compared to control conditions ($p < 0.001$), but these effects were inhibited after sI/R. Addition of A/D/N alone did not increase α-SMA protein levels in CF after sI/R, but co-administration of antioxidants with TGF-β1 restored the cytokine's differentiating effect, compared to control conditions ($p < 0.001$; Figure 5B).

Furthermore, pro-collagen I synthesis was assessed by stimulation of CF with angiotensin II (100 nM) during simulated reperfusion for 48 h. Western blot analysis revealed increased angiotensin II induced-pro-collagen I protein levels with respect to control conditions ($p < 0.05$), which were significantly inhibited after 6 h simulated ischemia followed by 48 h simulated reperfusion ($p < 0.001$; Figure 5C). A/D/N alone also had no effect in pro-collagen I protein levels in CF after sI/R, but joint administration with angiotensin II restored the production of pro-collagen I induced by this peptide, in comparison to normoxic conditions ($p < 0.05$), and to cells under sI/R treated with angiotensin II ($p < 0.01$; Figure 5C).

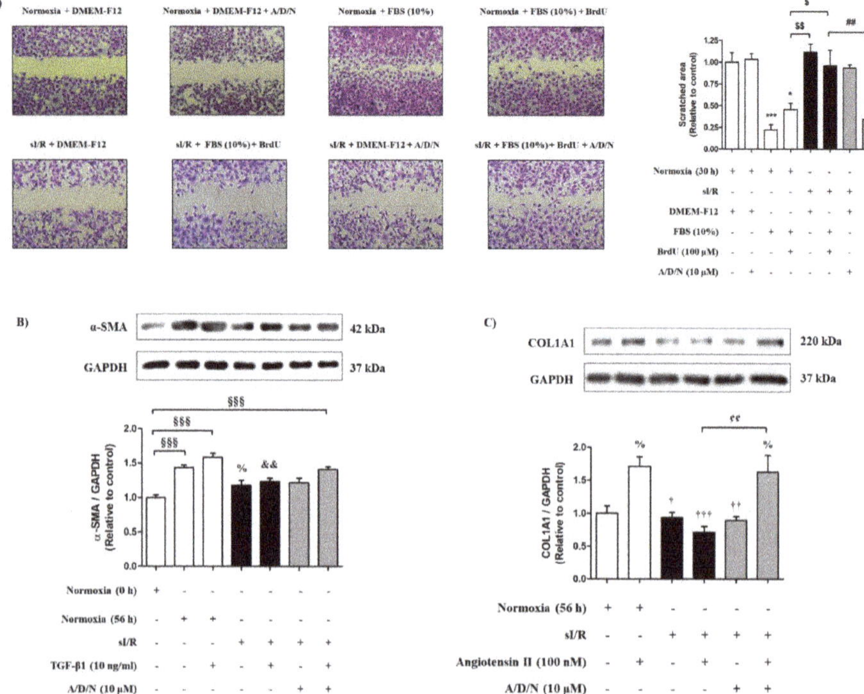

Figure 5. The association of ascorbic acid, deferoxamine, and N-acetylcysteine prevents the loss of serum-induced migration, TGF-β1-mediated differentiation and angiotensin II-induced pro-collagen I synthesis induced by simulated ischemia/reperfusion in CF. (**A**) Cells were exposed to 6 h simulated ischemia, followed by 24 h simulated reperfusion (sI/R). Cells were treated with the association of ascorbic acid, deferoxamine, and n-acetylcysteine (A/D/N) using 10 μM of each antioxidant at the onset of simulated reperfusion. A scratch was made on a cell monolayer prior to simulated reperfusion with DMEM-F12, FBS (10%), or FBS (10%) + BrdU (100 μM). Representative images of each experimental group (left panel) and quantification of area (right panel) are shown (*n* = 3). (**B**) Cardiac fibroblasts were exposed to 6 h simulated ischemia, followed by 48 h simulated reperfusion (sI/R). At the onset of simulated reperfusion, cells were treated with the association of A/D/N using 10 μM of each antioxidant and stimulated with or without TGF-β1 (10 ng/mL). Representative Western blots (upper panel) and densitometric analysis (lower panel) of α-SMA and GAPDH as loading control are shown (*n* = 5). (**C**) Cardiac fibroblasts were exposed to 6 h simulated ischemia, followed by 48 h simulated reperfusion (sI/R). At the beginning of simulated reperfusion, cells were treated with the association of A/D/N using 10 μM of each antioxidant and stimulated with or without angiotensin II (100 nM). Ascorbic acid (100 nM) was added to all experimental groups as a co-factor in pro-collagen type I synthesis. Representative Western blots (upper panel) and densitometric analysis (lower panel) of COL1A1 and GAPDH as loading control are shown. The results are expressed as mean ± S.E.M. *** $p < 0.001$, ** $p < 0.01$ and * $p < 0.05$ vs. Normoxia (30 h) + DMEM-F12; $$ $p < 0.01$ and $ $p < 0.05$ vs. Normoxia (30 h) + FBS (10%) + BrdU; ## $p < 0.01$ vs. sI/R + FBS (10%) + BrdU; §§§ $p < 0.001$ vs. Normoxia (0 h); % $p < 0.05$ vs. Normoxia (56 h); && $p < 0.01$ vs. Normoxia (56 h) + TGF-β1; ††† $p < 0.001$, †† $p < 0.01$ and † $p < 0.05$ vs. Normoxia (56 h) + angiotensin II; ¢¢ $p < 0.01$ vs. sI/R + angiotensin II. COL1A1 = alpha-1 type 1 collagen; TGF-β1 = transforming growth factor beta 1; α-SMA = alpha smooth muscle actin; FBS = fetal bovine serum; BrdU = 5-bromo-2′-deoxyuridine. Symbol "+" represents presence of condition and symbol "-" represents absence of condition.

4. Discussion

The main findings of the present study showed that the pharmacological association of A, D, and N: (a) increased cell viability in CF exposed to sI/R using a lower concentration than those of each antioxidant which, separately, did not show cell viability protection, (b) reduced intracellular ROS production and decreased apoptotic cell death induced by sI/R, (c) activated the pro-survival kinases ERK1/2 and Akt, but inhibited the pro-apoptotic p38 and JNK kinases, and (d) recovered CF function associated with wound repair induced by sI/R by restoring serum-induced migration, TGF-β1-mediated differentiation of CF into CMF and angiotensin II-induced pro-collagen I synthesis.

The deleterious effects of reperfusion injury and oxidative stress on the functional capacity of CF could negatively affect the optimal development of myocardial repairing after tissue damage [26]; therefore, cytoprotection of fibroblasts is of the utmost importance. Currently, there are multiple strategies to protect the myocardium from I/R injury, but the translation of these therapies from bench to bedside has proved challenging. Combinations of therapies with synergistic effects, as well as protection of all cardiac cell types and not just cardiomyocytes, are believed to be essential to achieve cardioprotection in the clinical arena [27,28].

Our preliminary studies revealed that A, D, and N separately increase the viability of CF exposed to sI/R in a concentration-dependent manner, at high concentrations (≥100 μM). However, 10 mM of A was cytotoxic due to its pro-oxidant activity at higher concentrations, which has been demonstrated in murine tumors [29]. Moreover, these antioxidants, separately, did not protect against sI/R injury at 1 and 10 μM, but when they were associated (dual or triple combinations) at 10 μM each, showed a synergistic cytoprotective effect. Previous studies have thoroughly established that A, D, and N -alone or in dual combination- yield cardioprotective effects [17–22,30–33]. Nonetheless, this is the first study that shows protection elicited by A/D/N administration in CF after sI/R. In addition, in vivo assays of MI will also be necessary to clarify whether the A/D/N association could prevent cardiac cell death, in order to reduce infarct size and finally, improve wound healing. Conversely, a study reported by Nikas et al., [34] showed that combined intravenous administration of A, D, and N, 15 min before and 5 min after reperfusion, with the same dose, was unable to reduce infarct size, the deleterious effects on ventricular function parameters, or oxidative stress markers in an in vivo model of myocardial reperfusion injury in pigs. Although the authors use a reliable I/R model tested in large animals, they only tested a single dose of each antioxidant (3–3.5 g A, 1.8–2.1 g D, and 3–3.5 g N) and did not measure the plasma levels reached by A, N, and D to compare with the results of the present study.

We speculated that the synergistic cytoprotective effects of the associations between A/D, A/N, and A/D/N, but not D/N (using each antioxidant at 10 μM) on the viability of CF exposed to sI/R, could be intrinsically related to a favorable pharmacological interaction between A, N and D, in the context of myocardial reperfusion injury elicited by oxidative stress [6,8]. Intracellular GSH -an endogenous antioxidant molecule- can reduce oxidized dehydroascorbate to ascorbate (an ionized form of A), which increases the bioavailability of ascorbate to interact against ROS [35]. Moreover, the decrease in intracellular GSH levels may be replenished by N, and D can chelate an excess of catalytic free iron (which increases during ischemia and cardiac reperfusion from cell lysis), thereby decreasing the production of hydroxyl radicals through the Fenton reaction, and preventing the pro-oxidant interaction of iron with A [36]. Thus, the A/D/N association has the advantage of improving the antioxidant response against the increase of intracellular ROS during simulated reperfusion, by providing three different mechanisms of antioxidant action. Finally, the use of lower concentrations of each antioxidant, compared to those used in independent administration to achieve a pharmacological effect in CF, ensured us that possible toxic effects could be low or absent.

Moreover, the A/D/N combination also activated ERK1/2 and Akt, which are key components of the reperfusion injury salvage kinases (RISK) pathway and therefore, essential to protect cardiac cells exposed to sI/R from reperfusion-induced cell death [37,38]. Currently, there are a few studies linking the activation of the RISK pathway to A and N [18,31,39], and therefore, our findings further support a nexus between this well-known survival signaling pathway and the cytoprotective effect of antioxidant association. However, additional studies are necessary to establish cause-effect mechanisms in order to evaluate whether inhibition of the RISK pathway results in the loss of protection conferred by the A/D/N triad. Additionally, treatment with A/D/N triggered anti-apoptotic effects in CF exposed to sI/R by reducing the phosphorylation of p38 and JNK, which are important players in apoptosis in the context of cardiac I/R [7,40]. This signaling pathway may be triggered by the apoptosis signal-regulating kinase 1 (ASK1), which can be activated by ROS, and consequently, induce apoptosis [41]. Further research is also required to thoroughly assess the molecular mechanisms by which the A/D/N antioxidant association activates the RISK pathway and also inactivates p38 and JNK.

The A/D/N association reduced the activation of caspases 9, 3 and increased the Bcl-xl/Bax ratio. Moreover, using the same sI/R protocol, protection was also observed in cardiomyocytes treated with A/D/N. These effects were further confirmed by the assessment of pro-apoptotic proteins, suggesting that our pharmacological approach may protect the whole myocardium and not only fibroblasts. These results are consistent with the previously reported anti-apoptotic effects conferred by A, D, and N [18,19,31,42], and contribute to a better characterization of the role of oxidative stress as a pharmacological target in apoptosis induced by I/R. Furthermore, we did not observe a protective effect of A/D/N against necrosis, which could initially suggest that there are factors other than oxidative stress (e.g., decreased energy metabolism) that might contribute most importantly to this type of cell death during simulated reperfusion. However, the RISK pathway protects against cell death by apoptosis and necrosis, and in our in vitro model this survival pathway is activated by the A/D/N association in cardiac fibroblasts; therefore, future studies will be required to understand this differential protection against these two types of cell death. Moreover, given the normal limitations of the cell death assays we used, future research should thoroughly assess the exact type of cell death prevented by the A/D/N association.

CF migration is a key step in the wound repair process, allowing CF located in close areas to the infarction zone to migrate and repopulate the necrotic area. Chemokines (Fractalkine/CX3CL1), growth factors (TGF-β and fibroblast growth factor), and cytokines (interleukin-1β, tumor necrosis factor α and cardiotrophin-1) secreted from other cells types further secrete ECM, cytokines and chemokines (as MCP-1) to induce immune cell migration and ensure fast tissue repair [9,43]. Due to its high content of embryonic growth-promoting factors, FBS is used to induce in vitro migration in CF and other cell types [44,45]. In our study, cellular injury caused by sI/R triggered the impairment of the FBS-induced migratory capacity of CF. Similar results were found in adult rat CF, where stimulation with hydrogen peroxide 10 and 100 μM showed that migration induced by a fibronectin gradient was delayed, compared to untreated control cells [46]; these results highlight the deleterious effect of oxidative stress on migration of these cells and suggest the protective molecular mechanism of the A/D/N association.

One of the main characteristics of CF is their ability to differentiate into CMF, which are characterized by a pro-fibrotic phenotype [47]. These cellular changes are induced by several stimuli, such as TGF-β1, interleukin-10, thrombospondin-1, angiotensin II, stimulation of injury-site cardiomyocytes, and vascular cells, among others [26]. CMF are the main secretory source of ECM proteins, as well as matrix metalloproteases, in cardiac fibrotic remodeling [26]. In our study, we induced spontaneous CF-to-CMF differentiation, which is due probably to the autocrine effects of TGF-β1 secreted by CF in culture, as well as treatment with TGF-β1. Both methods induced an increase in α-SMA levels in CF after 48 h of simulated reperfusion. Although TGF-β1 stimulation was not significantly greater than spontaneous differentiation, we and others have shown that TGF-β1 increases α-SMA levels in CF in a time-dependent manner [45,48]. Interestingly, this inhibitory effect of sI/R on the increase of α-SMA levels induced by TGF-β1 or spontaneous differentiation in CF has not been previously described. A similar effect has been observed in H9c2 cardiomyocytes and in corneal keratocytes, where hypoxia prevented the transformation to myofibroblasts induced by, an effect which was associated with changes in TGF-β1 signaling pathways [49,50]. Further studies are necessary to elucidate the effects of the A/D/N association on signaling pathways implicated in the effects of differentiation of CF to MCF induced by TGF-β1 after sI/R.

CF and CMF secrete and degrade various types of collagen to maintain ECM homeostasis. Among these, type I collagen is widely expressed in cardiac tissue of mammals and forms thick and stiff fibers [26]. Angiotensin II is a peptide known to elicit cardiac fibrosis [51,52] by stimulating CF collagen production and secretion [53,54]. In our CF in vitro model, sI/R prevented the increase of pro-collagen I synthesis triggered by angiotensin II and did not induce pro-collagen I production by itself. These results were corroborated by Siwik et al., [55] who demonstrated that oxidative stress decreases fibrillar collagen synthesis in CF. Interestingly, previous reports have shown that 72 h of hypoxia induce an increase in pro-collagen type I α mRNA and protein levels in human CF [56], while 6 h of hypoxia also increased collagen I levels in adult rat CF [57]. Additionally, another study found that 1 h of hypoxia followed by 12 h reoxygenation increases the secretion of soluble collagen from neonatal rat CF [58]. These differences can be attributed to various factors, such as duration of hypoxia/reoxygenation, age, and species from which these cells originate.

Our observations that the A/D/N association prevents cell death, reduces oxidative stress and recovers cellular functions associated with tissue repair induced by sI/R in CF certainly supports the previously described cardioprotective effects of A, D, and N; either separately or combined, on ventricular function in animal models of myocardial I/R [17–19,30,32,33].

5. Conclusions

Overall, our findings indicate, for the first time, that the association of A, D, and N protects CF from cell death and recovers pro-wound healing function damaged by sI/R. We used a low concentration of each antioxidant, which did not increase cell viability when administered separately. Moreover, this effect may be mediated by activation of the RISK pathway and inhibition of the pro-apoptotic proteins p38 and JNK, suggesting that pharmacological association of these antioxidants may be a novel therapeutic strategy to protect the myocardium from reperfusion injury elicited by oxidative stress.

Supplementary Materials: The following are available online at http://www.mdpi.com/2076-3921/8/12/614/s1. Figure S1: Effects of the association between ascorbic acid, deferoxamine and N-acetylcysteine, at 1 μM each, on viability of cardiac fibroblasts exposed to simulated ischemia/reperfusion. Cardiac fibroblasts were exposed to 6 h simulated ischemia, followed by 16 h simulated reperfusion (sI/R). Associations between ascorbic acid (A), deferoxamine (D) and N-acetylcysteine (N), using 1 μM of each antioxidant, were added at the beginning of simulated reperfusion. Cell viability was quantified as a percentage (%) of the number of cells after 6 h normoxia (100%) by cell count after trypan blue staining ($n = 3$). The results are expressed as mean ± S.E.M. *** $p < 0.001$ and * $p < 0.05$ vs. C22 (control cells after 22 h normoxia); ## $p < 0.01$ vs. sI/R. Figure S2: Effects of associations between ascorbic acid, deferoxamine and N-acetylcysteine, at 10 μM each, on viability and pro-caspases 9 and 3 protein levels of cardiomyocytes exposed to simulated ischemia/reperfusion. Cardiomyocytes were exposed to 6 h simulated ischemia, followed by 16 h simulated reperfusion (sI/R). Associations between ascorbic acid (A), deferoxamine (D) and N-acetylcysteine (N), using 10 μM of each antioxidant, were added at the beginning of simulated reperfusion. (**A**) Cell viability was quantified as a percentage (%) of the number of cells after 6 h normoxia (100%) by cell count after trypan blue staining ($n = 3$). (**B**) and (**C**) show representative Western blots (upper panel) and densitometric analysis (lower panel) of pro-caspase 9 ($n = 4$) and pro-caspase-3 ($n = 4$), respectively. GAPDH was used as a loading control. The results are expressed as mean ± S.E.M. *** $p < 0.001$ and ** $p < 0.01$ vs. C22 (control cells after 22 h normoxia). ### $p < 0.001$, ## $p < 0.01$ and # $p < 0.05$ vs. sI/R.

Author Contributions: For research articles with several authors, a short paragraph specifying their individual contributions must be provided. The following statements should be used "conceptualization, P.P.-F., G.D.-A. and R.R. methodology, P.P.-F., P.V.-B., S.L.-N. and R.V.; validation, J.C.-V., E.C., C.E.-P. and F.R.-C.; investigation, P.P.-F., C.E.-P. and F.R.-C.; resources, G.D.-A., Z.P. and S.L.; writing—original draft preparation, P.P.-F. and J.A.R.; writing—review and editing, G.D.-A.; project administration, P.P.-F. and G.D.-A.; funding acquisition, G.D.-A., Z.P. and S.L.

Funding: This work was supported by Fondo Nacional de Desarrollo Científico y Tecnológico, FONDECYT [grant number 1170425] to G.D-A.; FONDECYT [grant number 11181000] to J.A.R.; FONDECYT [grant number 1180613] to Z.P.; CONICYT [grant number 21151215] to P.P-F.; FONDEF [grant number ID15I10285] to R.R. and FONDAP ACCDiS [grant number 15130011] to S.L.

Acknowledgments: The authors thank José Riquelme, Gindra Latorre and Fidel Albornoz for their important technical assistance during this investigation. We also wish to thank Marcelo Kogan (Universidad de Chile,

Chile) for the provision of measurement equipment and Ana María Avalos for proofreading and editing the final manuscript.

Conflicts of Interest: The authors declare that they have no conflict of interests.

Chemical compounds studied in this article: Angiotensin II (PubChem CID: 172198); Ascorbic acid (PubChem CID: 54670067); 5-Bromo-2′-deoxyuridine (PubChem CID: 6035); Deferoxamine mesylate (PubChem CID: 62881); 2,7-Dichlorodihydrofluorescein diacetate (2′,7′-dichlorofluorescin diacetate, PubChem CID: 77718); N-acetylcysteine (PubChem CID: 12035); Propidium iodide (PubChem CID: 104981); Resazurin (PubChem CID: 11077).

Abbreviations

α-SMA = alpha smooth muscle actin; A/D/N = ascorbic acid, deferoxamine and N-acetylcysteine; A = ascorbic acid; BrdU = 5-bromo-2′-deoxyuridine; CF = cardiac fibroblast; CMF = cardiac myofibroblast; COL1A1 = alpha-1 type 1 collagen; DCFH-DA = 2′,7′-dichlorofluorescin diacetate; D = deferoxamine; ECM = extracellular matrix; ERK1/2 = extracellular-signal-regulated kinase1/2; FBS = fetal bovine serum; I/R = ischemia/reperfusion; JNK = c-Jun-N-terminal kinase; MI = myocardial infarction; N = N-acetylcysteine; RISK = reperfusion injury salvage kinases; ROS = reactive oxygen species; sI/R = simulated ischemia/reperfusion; TGF-β1 = transforming growth factor beta 1.

References

1. GBD 2015 Mortality and Causes of Death Collaborators. Global, regional, and national life expectancy, all-cause mortality, and cause-specific mortality for 249 causes of death, 1980–2015: A systematic analysis for the Global Burden of Disease Study 2015. *Lancet* **2016**, *388*, 1459–1544. [CrossRef]
2. Thygesen, K.; Alpert, J.S.; White, H.D.; Jaffe, A.S.; Katus, H.A.; Apple, F.S.; Lindahl, B.; Morrow, D.A.; Chaitman, B.R.; Clemmensen, P.M.; et al. Third universal definition of myocardial infarction. *J. Am. Coll. Cardiol.* **2012**, *60*, 1581–1598. [CrossRef] [PubMed]
3. White, H.D.; Chew, D.P. Acute myocardial infarction. *Lancet* **2008**, *372*, 570–584. [CrossRef]
4. Hausenloy, D.J.; Yellon, D.M. Myocardial ischemia-reperfusion injury: A neglected therapeutic target. *J. Clin. Investig.* **2013**, *123*, 92–100. [CrossRef]
5. Vilahur, G.; Juan-Babot, O.; Peña, E.; Oñate, B.; Casaní, L.; Badimon, L. Molecular and cellular mechanisms involved in cardiac remodeling after acute myocardial infarction. *J. Mol. Cell. Cardiol.* **2011**, *50*, 522–533. [CrossRef]
6. Braunersreuther, V.; Jaquet, V. Reactive oxygen species in myocardial reperfusion injury: From physiopathology to therapeutic approaches. *Curr. Pharm. Biotechnol.* **2012**, *13*, 97–114. [CrossRef]
7. Guo, W.; Liu, X.; Li, J.; Shen, Y.; Zhou, Z.; Wang, M.; Xie, Y.; Feng, X.; Wang, L.; Wu, X. Prdx1 alleviates cardiomyocyte apoptosis through ROS-activated MAPK pathway during myocardial ischemia/reperfusion injury. *Int. J. Biol. Macromol.* **2018**, *112*, 608–615. [CrossRef]
8. González-Montero, J.; Brito, R.; Gajardo, A.I.; Rodrigo, R. Myocardial reperfusion injury and oxidative stress: Therapeutic opportunities. *World J. Cardiol.* **2018**, *10*, 74–86. [CrossRef]
9. Chistiakov, D.A.; Orekhov, A.N.; Bobryshev, Y.V. The role of cardiac fibroblasts in post-myocardial heart tissue repair. *Exp. Mol. Pathol.* **2016**, *101*, 231–240. [CrossRef]
10. Pinto, A.R.; Ilinykh, A.; Ivey, M.J.; Kuwabara, J.T.; D'antoni, M.L.; Debuque, R.; Chandran, A.; Wang, L.; Arora, K.; Rosenthal, N.A.; et al. Revisiting cardiac cellular composition. *Circ. Res.* **2016**, *118*, 400–409. [CrossRef]
11. Talman, V.; Ruskoaho, H. Cardiac fibrosis in myocardial infarction—From repair and remodeling to regeneration. *Cell Tissue Res.* **2016**, *365*, 563–581. [CrossRef] [PubMed]
12. Vivar, R.; Humeres, C.; Varela, M.; Ayala, P.; Guzmán, N.; Olmedo, I.; Catalán, M.; Boza, P.; Muñoz, C.; Araya, G.D. Cardiac fibroblast death by ischemia/reperfusion is partially inhibited by IGF-1 through both PI3K/Akt and MEK–ERK pathways. *Exp. Mol. Pathol.* **2012**, *93*, 1–7. [CrossRef] [PubMed]
13. Vivar, R.; Humeres, C.; Ayala, P.; Olmedo, I.; Catalán, M.; García, L.; Lavandero, S.; Díaz-Araya, G. TGF-β1 prevents simulated ischemia/reperfusion-induced cardiac fibroblast apoptosis by activation of both canonical and non-canonical signaling pathways. *Biochim. Biophys. Acta Mol. Basis Dis.* **2013**, *1832*, 754–762. [CrossRef] [PubMed]
14. Zhou, Y.; Richards, A.M.; Wang, P. Characterization and standardization of cultured cardiac fibroblasts for ex vivo models of heart fibrosis and heart ischemia. *Tissue Eng. Part C Methods* **2017**, *23*, 422–433. [CrossRef]

15. Lefort, C.; Benoist, L.; Chadet, S.; Piollet, M.; Heraud, A.; Babuty, D.; Baron, C.; Ivanes, F.; Angoulvant, D. Stimulation of P2Y11 receptor modulates cardiac fibroblasts secretome toward immunomodulatory and protective roles after Hypoxia/Reoxygenation injury. *J. Mol. Cell. Cardiol.* **2018**, *121*, 212–222. [CrossRef]

16. Ekelof, S.; Jensen, S.E.; Rosenberg, J.; Gogenur, I. Reduced oxidative stress in STEMI patients treated by primary percutaneous coronary intervention and with antioxidant therapy: A systematic review. *Cardiovasc. Drugs Ther.* **2014**, *28*, 173–181. [CrossRef]

17. Williams, R.E.; Zweier, J.L.; Flaherty, J.T. Treatment with deferoxamine during ischemia improves functional and metabolic recovery and reduces reperfusion-induced oxygen radical generation in rabbit hearts. *Circulation* **1991**, *83*, 1006–1014. [CrossRef]

18. Hao, J.; Li, W.W.; Du, H.; Zhao, Z.F.; Liu, F.; Lu, J.C.; Yang, X.C.; Cui, W. Role of vitamin C in cardioprotection of ischemia/reperfusion injury by activation of mitochondrial KATP channel. *Chem. Pharm. Bull.* **2016**, *64*, 548–557. [CrossRef]

19. Peng, Y.W.; Buller, C.L.; Charpie, J.R. Impact of N-acetylcysteine on neonatal cardiomyocyte ischemia-reperfusion injury. *Pediatr. Res.* **2011**, *70*, 61–66. [CrossRef]

20. Abe, M.; Takiguchi, Y.; Ichimaru, S.; Tsuchiya, K.; Wada, K. Comparison of the protective effect of N-acetylcysteine by different treatments on rat myocardial ischemia-reperfusion injury. *J. Pharmacol. Sci.* **2008**, *106*, 571–577. [CrossRef]

21. Nishinaka, Y.; Sugiyama, S.; Yokota, M.; Saito, H.; Ozawa, T. The effects of a high dose of ascorbate on ischemia-reperfusion-induced mitochondrial dysfunction in canine hearts. *Heart Vessel.* **1992**, *7*, 18–23. [CrossRef] [PubMed]

22. Reddy, B.R.; Kloner, R.A.; Przyklenk, K. Early treatment with deferoxamine limits myocardial ischemic/reperfusion injury. *Free Radic. Biol. Med.* **1989**, *7*, 45–52. [CrossRef]

23. National Research Council (US) Committee for the Update of the Guide for the Care and Use of Laboratory Animals. *Guide for the Care and Use of Laboratory Animals*, 8th ed.; National Academies Press: Washington, DC, USA, 2011. Available online: https://www.ncbi.nlm.nih.gov/books/NBK54050/ (accessed on 01 March 2016). [CrossRef]

24. Bankowski, Z.; Howard-Jones, N. (Eds.) *International Guiding Principles for Biomedical Research Involving Animals*; CIOMS: Geneva, Switzerland, 1985.

25. Mendoza-Torres, E.; Riquelme, J.A.; Vielma, A.; Sagredo, A.R.; Gabrielli, L.; Bravo-Sagua, R.; Jalil, J.E.; Rothermel, B.A.; Sanchez, G.; Ocaranza, M.P.; et al. Protection of the myocardium against ischemia/reperfusion injury by angiotensin-(1–9) through an AT_2R and Akt-dependent mechanism. *Pharmacol. Res.* **2018**, *135*, 112–121. [CrossRef] [PubMed]

26. Frangogiannis, N.G. Cardiac fibrosis: Cell biological mechanisms, molecular pathways and therapeutic opportunities. *Mol. Asp. Med.* **2019**, *65*, 70–99. [CrossRef] [PubMed]

27. Davidson, S.M.; Ferdinandy, P.; Andreadou, I.; Bøtker, H.E.; Heusch, G.; Ibáñez, B.; Ovize, M.; Schulz, R.; Yellon, D.M.; Hausenloy, D.J.; et al. Multitarget strategies to reduce myocardial ischemia/reperfusion injury: JACC review topic of the week. *J. Am. Coll. Cardiol.* **2019**, *73*, 89–99. [CrossRef] [PubMed]

28. Chen, M.; Zhang, M.; Zhang, X.; Li, J.; Wang, Y.; Fan, Y.; Shi, R. Limb ischemic preconditioning protects endothelium from oxidative stress by enhancing nrf2 translocation and upregulating expression of antioxidases. *PLoS ONE* **2015**, *10*, e0128455. [CrossRef]

29. Wang, G.; Yin, T.; Wang, Y. In vitro and in vivo assessment of high-dose vitamin C against murine tumors. *Exp. Ther. Med.* **2016**, *12*, 3058–3062. [CrossRef]

30. Gao, F.; Yao, C.L.; Gao, E.; Mo, Q.Z.; Yan, W.L.; McLaughlin, R.; Lopez, B.L.; Christopher, T.A.; Ma, X.L. Enhancement of glutathione cardioprotection by ascorbic acid in myocardial reperfusion injury. *J. Pharmacol. Exp. Ther.* **2002**, *301*, 543–550. [CrossRef]

31. Guaiquil, V.H.; Golde, D.W.; Beckles, D.L.; Mascareno, E.J.; Siddiqui, M. Vitamin C inhibits hypoxia-induced damage and apoptotic signaling pathways in cardiomyocytes and ischemic hearts. *Free Radic. Biol. Med.* **2004**, *37*, 1419–1429. [CrossRef]

32. Karahaliou, A.; Katsouras, C.; Koulouras, V.; Nikas, D.; Niokou, D.; Papadopoulos, G.; Nakos, G.; Sideris, D.; Michalis, L. Ventricular arrhythmias and antioxidative medication: Experimental study. *Hell. J. Cardiol.* **2008**, *49*, 320–328.

33. Phaelante, A.; Rohde, L.E.; Lopes, A.; Olsen, V.; Tobar, S.A.L.; Cohen, C.; Martinelli, N.; Biolo, A.; Dal-Pizzol, F.; Clausell, N.; et al. *N*-acetylcysteine plus deferoxamine improves cardiac function in wistar rats after non-reperfused acute myocardial infarction. *J. Cardiovasc. Transl. Res.* **2015**, *8*, 328–337. [CrossRef] [PubMed]

34. Nikas, D.N.; Chatziathanasiou, G.; Kotsia, A.; Papamichael, N.; Thomas, C.; Papafaklis, M.; Naka, K.K.; Kazakos, N.; Milionis, H.J.; Vakalis, K.; et al. Effect of intravenous administration of antioxidants alone and in combination on myocardial reperfusion injury in an experimental pig model. *Curr. Ther. Res. Clin. Exp.* **2008**, *69*, 423–439. [CrossRef] [PubMed]

35. May, J.M.; Qu, Z.C.; Neel, D.R.; Li, X. Recycling of vitamin C from its oxidized forms by human endothelial cells. *Biochim. Biophys. Acta* **2003**, *1640*, 153–161. [CrossRef]

36. Levine, M.; Padayatty, S.J.; Espey, M.G. Vitamin C: A concentration-function approach yields pharmacology and therapeutic discoveries. *Adv. Nutr.* **2011**, *2*, 78–88. [CrossRef] [PubMed]

37. Rossello, X.; Yellon, D.M. The RISK pathway and beyond. *Basic Res. Cardiol.* **2018**, *113*, 2. [CrossRef]

38. Rossello, X.; Riquelme, J.A.; Davidson, S.M.; Yellon, D.M. Role of PI3K in myocardial ischaemic preconditioning: Mapping pro-survival cascades at the trigger phase and at reperfusion. *J. Cell. Mol. Med.* **2018**, *22*, 926–935. [CrossRef]

39. Wang, T.; Mao, X.; Li, H.; Qiao, S.; Xu, A.; Wang, J.; Lei, S.; Liu, Z.; Ng, K.F.; Wong, G.T.; et al. *N*-Acetylcysteine and allopurinol up-regulated the Jak/STAT3 and PI3K/Akt pathways via adiponectin and attenuated myocardial postischemic injury in diabetes. *Free Radic. Biol. Med.* **2013**, *63*, 291–303. [CrossRef]

40. Wang, D.; Chen, T.; Liu, F. Betulinic acid alleviates myocardial hypoxia/reoxygenation injury via inducing Nrf2/HO-1 and inhibiting p38 and JNK pathways. *Eur. J. Pharmacol.* **2018**, *838*, 53–59. [CrossRef]

41. Tobiume, K.; Matsuzawa, A.; Takahashi, T.; Nishitoh, H.; Morita, K.I.; Takeda, K.; Minowa, O.; Miyazono, K.; Noda, T.; Ichijo, H. ASK1 is required for sustained activations of JNK/p38 MAP kinases and apoptosis. *EMBO Rep.* **2001**, *2*, 222–228. [CrossRef]

42. Dobšák, P.; Siegelova, J.; Wolf, J.; Rochette, L.; Eicher, J.; Vasku, J.; Kuchtickova, S.; Horky, M. Prevention of apoptosis by deferoxamine during 4 h of cold cardioplegia and reperfusion: In vitro study of isolated working rat heart model. *Pathophysiology* **2002**, *9*, 27–32. [CrossRef]

43. Humeres, C.; Vivar, R.; Boza, P.; Muñoz, C.; Bolivar, S.; Anfossi, R.; Osorio, J.M.; Olivares-Silva, F.; García, L.; Díaz-Araya, G. Cardiac fibroblast cytokine profiles induced by proinflammatory or profibrotic stimuli promote monocyte recruitment and modulate macrophage M1/M2 balance in vitro. *J. Mol. Cell. Cardiol.* **2016**, *101*, 69–80. [CrossRef] [PubMed]

44. Wu, M.F.; Stachon, T.; Seitz, B.; Langenbucher, A.; Szentmáry, N. Effect of human autologous serum and fetal bovine serum on human corneal epithelial cell viability, migration and proliferation in vitro. *Int. J. Ophthalmol.* **2017**, *10*, 908–913. [CrossRef] [PubMed]

45. Olmedo, I.; Muñoz, C.; Guzmán, N.; Catalán, M.; Vivar, R.; Ayala, P.; Humeres, C.; Aránguiz, P.; García, L.; Velarde, V.; et al. EPAC expression and function in cardiac fibroblasts and myofibroblasts. *Toxicol. Appl. Pharmacol.* **2013**, *272*, 414–422. [CrossRef] [PubMed]

46. Colston, J.T.; Sam, D.; Freeman, G.L. Impact of brief oxidant stress on primary adult cardiac fibroblasts. *Biochem. Biophys. Res. Commun.* **2004**, *316*, 256–262. [CrossRef]

47. Shinde, A.V.; Humeres, C.; Frangogiannis, N.G. The role of α-smooth muscle actin in fibroblast-mediated matrix contraction and remodeling. *Biochim. Biophys. Acta Mol. Basis Dis.* **2017**, *1863*, 298–309. [CrossRef]

48. Driesen, R.B.; Nagaraju, C.K.; Abi-Char, J.; Coenen, T.; Lijnen, P.J.; Fagard, R.H.; Sipido, K.R.; Petrov, V.V. Reversible and irreversible differentiation of cardiac fibroblasts. *Cardiovasc. Res.* **2013**, *101*, 411–422. [CrossRef]

49. Xing, D.; Bonanno, J.A. Hypoxia reduces TGFβ1-induced corneal keratocyte myofibroblast transformation. *Mol. Vis.* **2009**, *15*, 1827–1834.

50. Yan, Z.; Shen, D.; Liao, J.; Zhang, Y.; Chen, Y.; Shi, G.; Gao, F. Hypoxia suppresses TGF-B1-induced cardiac myocyte myofibroblast transformation by inhibiting Smad2/3 and rhoa signaling pathways. *Cell. Physiol. Biochem.* **2018**, *45*, 250–257. [CrossRef]

51. Shu, Q.; Lai, S.; Wang, X.M.; Zhang, Y.L.; Yang, X.L.; Bi, H.L.; Li, H.H. Administration of ubiquitin-activating enzyme UBA1 inhibitor PYR-41 attenuates angiotensin II-induced cardiac remodeling in mice. *Biochem. Biophys. Res. Commun.* **2018**, *505*, 317–324. [CrossRef]

52. Wang, H.X.; Li, W.J.; Hou, C.L.; Lai, S.; Zhang, Y.L.; Tian, C.; Yang, H.; Du, J.; Li, H.H. CD1d-dependent natural killer T cells attenuate angiotensin II-induced cardiac remodelling via IL-10 signalling in mice. *Cardiovasc. Res.* **2018**, *115*, 83–93. [CrossRef]
53. Tian, H.P.; Sun, Y.H.; He, L.; Yi, Y.F.; Gao, X.; Xu, D.L. Single-stranded DNA-binding protein 1 abrogates cardiac fibroblast proliferation and collagen expression induced by angiotensin II. *Int. Heart J.* **2018**, *59*, 1398–1408. [CrossRef] [PubMed]
54. Yu, C.; Wang, W.; Jin, X. Hirudin protects Ang II-induced myocardial fibroblasts fibrosis by inhibiting the extracellular signal-regulated Kinase1/2 (ERK1/2) pathway. *Med. Sci. Monit.* **2018**, *24*, 6264–6272. [CrossRef] [PubMed]
55. Siwik, D.A.; Pagano, P.J.; Colucci, W.S. Oxidative stress regulates collagen synthesis and matrix metalloproteinase activity in cardiac fibroblasts. *Am. J. Physiol. Cell Physiol.* **2001**, *280*, C53–C60. [CrossRef] [PubMed]
56. Agocha, A.; Lee, H.W.; Eghbali-Webb, M. Hypoxia regulates basal and induced DNA synthesis and collagen type I production in human cardiac fibroblasts: Effects of transforming growth factor-β1, thyroid hormone, angiotensin II and basic fibroblast growth factor. *J. Mol. Cell. Cardiol.* **1997**, *29*, 2233–2244. [CrossRef] [PubMed]
57. Shyu, K.G.; Wang, B.W.; Chen, W.J.; Kuan, P.; Lin, C.M. Angiotensin II mediates urotensin II expression by hypoxia in cultured cardiac fibroblast. *Eur. J. Clin. Investig.* **2012**, *42*, 17–26. [CrossRef]
58. Grobe, J.L.; der Sarkissian, S.; Stewart, J.M.; Meszaros, J.G.; Raizada, M.K.; Katovich, M.J. ACE2 overexpression inhibits hypoxia-induced collagen production by cardiac fibroblasts. *Clin. Sci.* **2007**, *113*, 357–364. [CrossRef]

 *antioxidants*

Article

Ezetimibe Prevents Ischemia/Reperfusion-Induced Oxidative Stress and Up-Regulates Nrf2/ARE and UPR Signaling Pathways

Denise Peserico [1], Chiara Stranieri [1], Ulisse Garbin [1], Chiara Mozzini C [1], Elisa Danese [2], Luciano Cominacini [1] and Anna M. Fratta Pasini [1,*]

[1] Department of Medicine, Section of General Medicine and Atherothrombotic and Degenerative Diseases, University of Verona, Policlinico G.B. Rossi, Piazzale L.A. Scuro 10, 37134 Verona, Italy; denise.peserico@univr.it (D.P.); chiara.stranieri@univr.it (C.S.); ulisse.garbin@univr.it (U.G.); chiara.mozzini@univr.it (C.M.C.); luciano.cominacini@univr.it (L.C.)

[2] Department of Neurosciences, Biomedicine and Movement Sciences, University of Verona, 37134 Verona, Italy; elisa.danese@univr.it

* Correspondence: annamaria.frattapasini@univr.it; Tel.: +39-045-812-4749

Received: 23 March 2020; Accepted: 17 April 2020; Published: 23 April 2020

Abstract: Background: While reperfusion is crucial for survival after an episode of ischemia, it also causes oxidative stress. Nuclear factor-E2-related factor 2 (Nrf2) and unfolded protein response (UPR) are protective against oxidative stress and endoplasmic reticulum (ER) stress. Ezetimibe, a cholesterol absorption inhibitor, has been shown to activate the AMP-activated protein kinase (AMPK)/Nrf2 pathway. In this study we evaluated whether Ezetimibe affects oxidative stress and Nrf2 and UPR gene expression in cellular models of ischemia-reperfusion (IR). Methods: Cultured cells were subjected to simulated IR with or without Ezetimibe. Results: IR significantly increased reactive oxygen species (ROS) production and the percentage of apoptotic cells without the up-regulation of Nrf2, of the related antioxidant response element (ARE) gene expression or of the pro-survival UPR activating transcription factor 6 (ATF6) gene, whereas it significantly increased the pro-apoptotic CCAAT-enhancer-binding protein homologous protein (CHOP). Ezetimibe significantly decreased the cellular ROS formation and apoptosis induced by IR. These effects were paralleled by the up-regulation of Nrf2/ARE and ATF6 gene expression and by a down-regulation of CHOP. We also found that Nrf2 activation was dependent on AMPK, since Compound C, a pan inhibitor of p-AMPK, blunted the activation of Nrf2. Conclusions: Ezetimibe counteracts IR-induced oxidative stress and induces Nrf2 and UPR pathway activation.

Keywords: oxidative stress; Nrf2; ER stress; Ezetimibe; ischemia-reperfusion

---

1. Introduction

Ischemia-reperfusion (IR) injury takes place when blood provision to an organ is diminished or interrupted and then reestablished, and underlies many pathological situations such as heart attacks, strokes and peripheral artery disease (PAD). While reperfusion is crucial for survival, it also causes oxidative stress and cell damage through the production of reactive oxygen species (ROS) [1–3].

Nuclear factor-E2-related factor 2 (Nrf2) is a master transcription factor that target genes coding for antioxidant proteins and detoxification enzymes [4–6]. Nrf2 controls the basal and induced expression of an array of antioxidant response element (ARE)-dependent genes, such as heme-oxygenase (HO)-1 and glutamate-cysteine ligase catalytic (GCLC) subunit, to regulate the physiological and pathophysiological outcomes of oxidant exposure [4–6]. Under basal conditions, Nrf2-dependent transcription is repressed by its negative regulator Kelch-like enoyl-CoA hydratase- associated protein

1 (Keap-1); when cells are exposed to oxidative stress or electrophiles, Nrf2 accumulates in the nucleus and drives the expression of its target genes [4–6].

The canonical nuclear factor (NF)-kB complex is composed of a dimer of p65 and p50 that is involved in various processes such as inflammation, apoptosis and the immune response [7]. In unstimulated cells, NF-kB is retained in the cytoplasm and bound to IkB-α; during oxidative stress, IkB kinase is activated and causes the phosphorylation of IkB-α, resulting in the nuclear translocation of p65 with the subsequent transcription of pro-inflammatory mediators [7]. Interestingly, activation of the Nrf2 and NF-kB pathways has been recently reported in different models of IR [8–10].

Stresses like ischemia and oxidative stress that perturb the folding of nascent endoplasmic reticulum (ER) proteins activate ER stress, which, in turn, triggers the unfolded protein response (UPR) that deals with unfolded and misfolded proteins [11–13]. In this context, it has been shown that the activating transcription factor 6 (ATF6) branch of the UPR may induce the expression of proteins that can reduce IR injury in the hearts of transgenic mice [12] and in cardiac myocytes [13]. Moreover, we have recently demonstrated in cultured cells that IR promoted a considerable increment in ROS, which was not followed by Nrf2 and UPR signaling pathway activation [14].

There is evidence showing that oxidative stress, ER stress and inflammation are inseparably linked, as each causes and amplifies the others: in particular, it has been shown that direct or indirect activation and inhibition occur between members of the Nrf2 and NF-kB pathways [15] and that the UPR-induced Nrf2 activation participates in maintaining redox homeostasis and cell survival [16]. Furthermore, we have previously shown, in the circulating cells of smokers [17] and of patients with type 2 diabetes [18] and coronary artery disease [19], that the degree of oxidative stress greatly influences the Nrf2, NF-κB and UPR signaling pathways.

Ezetimibe regulates exogenous dietary cholesterol absorption in the small intestine, blocking the transmembrane protein Niemann Pick C1-like 1 (NPC1L1) [20,21]. Beyond its powerful cholesterol lowering effects, the latest evidence shows that Ezetimibe has additional pleiotropic effects. In fact, Ezetimibe has been demonstrated to reduce oxidative stress through the activation of the AMP-activated protein kinase (AMPK)/Nrf2 pathway in animal models of nonalcoholic steatohepatitis [22] and ischemic stroke [23,24]. Furthermore, there are data showing that Ezetimibe can reduce the expression of ER stress markers in liver extracts of Zucker obese fatty rats [25] and that Ezetimibe may increase the glutathione content in rat livers subjected to IR [26].

Therefore, in cellular models of IR (using monocyte-like THP-1 cells and human cardiomyocytes), we aimed to investigate: (1) whether Ezetimibe reduces oxidative stress, NF-kB activation and apoptosis; (2) the effect of Ezetimibe on the adaptive response to oxidant injuries through Nrf2/ARE and UPR pathway up-regulation; and (3) the involvement of AMPK phosphorylation in Nrf2 activation.

2. Materials and Methods

2.1. Cell Cultures

Monocyte-like THP-1 cells were cultured in RPMI 1640 (Invitrogen Carlsbad, CA, USA), as previously described [19]. Human cardiomyocyte ventricular primary cells (Celprogen, Torrance, CA, USA) were cultured following the recommended manufacturer's protocols in extracellular matrix pre-coated flasks and/or well plates with ready-to-use complete growth medium (Celprogen, Torrance, CA, USA). The choice of such human cell lines was guided by previous evidence showing their high reliability for evaluating oxidative stress, antioxidant signaling pathways and IR injuries [19,27,28]. Human hepatocarcinoma HepG2 cells (Sigma-Aldrich, St. Louis, MI, USA) were cultured in EMEM medium (Invitrogen, Carlsbad, CA, USA) and used as control cells to test the presence of NPC1L1 receptor. All cell cultures were maintained in a humidified incubator with 95% air and 5% CO_2 at 37 °C. The endotoxin contamination of cells was routinely excluded with the chromogenic Limulus amoebocyte lysate assay.

2.2. Dose-Response Effect of Ezetimibe on Cell Viability and Toxicity

We performed preliminary dose-dependent tests with Ezetimibe (Cayman Chemical Company, Ann Arbor, MI, USA) in order to identify the appropriate experimental conditions. THP-1 cells and human cardiomyocytes were incubated overnight with increasing amounts of Ezetimibe (from 5 to 50 μM). Cellular viability was evaluated with the Annexin V staining assay (BD Bioscience, Franklin Lakes, NJ, USA) as previously reported [19] and analyzed by flow cytometry on FACSCantoII (Becton Dickinson). The number of each type of cell was expressed as a percentage of the number of total stained cells.

2.3. Oxidative Stress Assessment

To evaluate intracellular ROS formation, THP-1 cells and human cardiomyocytes were pretreated overnight with increasing amounts of Ezetimibe (from 5 to 50 μM), then incubated for 45 min at 37 °C with 100 μM of tert-butyl hydroperoxide solution (TBHP), followed by incubation with the fluorogenic CellROX™ Deep Red Reagent (Life Technologies, Eugene, CA, USA) (1 μM) for 45 min [29]. The cells were immediately analyzed by flow cytometry on FACSCantoII (Becton Dickinson). After IR experiments, CellROX™ Deep Red reagent (1 μM) was added to the cells, which were then incubated for 45 min at 37 °C and immediately analyzed by flow cytometry.

Then, 8-iso Prostaglandin F2α (8-iso) was measured in the medium of cell cultures using the 8-isoprostane ELISA kit (Cayman Chemical Company, Ann Arbor, MI, USA) following the manufacturer's instructions.

2.4. Quantification of Ezetimibe and of Its Derivative Ezetimibe-Glucuronide in Cells

The quantification of ezetimibe (EZE) and its derivative glucuronate (ezetimibe glucuronide, EZE-G) was performed with a liquid chromatography–tandem mass spectrometry-based method (LC-MS/MS). The sample preparation was carried by protein precipitation extraction. Briefly, 100 μL of lysate cell sample were spiked with 10 μL of IS-mixture (containing EZE-D4 and EZE-G-D4 at a final concentration of 400 nmol/L), mixed with 150 μL of acetonitrile and then vortexed for 1 min. After 15 min of centrifugation at $13,000\times g$, 300 μL of the supernatant were transferred to autosampler vials for LC-MS/MS analysis. The injection volume was 10 μL. A mixed stock solution of each standard was prepared in methanol and then diluted with methanol/water (50:50, *v/v*) to prepare an 8-point calibration curve (range: 0.25–200 nM). Standard solutions of EZE and EZE-G were purchased from Cayman Chemical Company (Ann Arbor, MI, USA), while their labeled internal standards were purchased from Santa Cruz Biotecnology (Santa Cruz, CA, USA). Chromatographic separation was performed on a Nexera X2 series UHPLC (Shimadzu, Kyoto, Japan), followed by detection on a 4500 MD triple quadrupole Mass Spectrometer (AB Sciex, Darmstadt, Germany). The electrospray ionization was completed in negative mode. Data were recorded in multiple reaction monitoring mode (MRM). System operation, data acquisition and quantification were performed using the Analyst 1.6.2. software and Multiquant 3.0.2. (AB Sciex, Darmstadt, Germany). The limit of quantitation (LOQ), determined as where the signal-to-noise ratio is ten, was below 0.20 nmol/L for both compounds. The intra-assay imprecision was between 3% and 3.5%, whereas the inter-assay imprecision was between 4.6% and 6.7%. The cellular concentrations of Ezetimibe were expressed as pmol/μg protein.

2.5. Induction of IR in THP-1 Cells and Cardiomyocytes

For IR experiments, the EVOS FL Auto Imaging System (Thermo Fisher, Waltham, MA, USA), equipped with the EVOS Onstage Incubator, was used, according to the manufacturer's instructions and as previously reported [14]. The EVOS FL system provides an environmental chamber, allowing for the precise control of temperature, humidity and gases (N_2, CO_2 and O_2), and hypoxia can be monitored by long-term fluorescence live-cell imaging, using Invitrogen™ Image-iT™ Hypoxia Reagent (Thermo Fisher, Waltham, MA, USA). THP-1 cells were cultured in RPMI 1640 with L-glutamine at 37 °C in an incubator set at normoxic conditions (20% O_2). Then, THP-1 cells were placed on the EVOS FL Auto Imaging System and incubated for 60 min to allow the system to reach the required temperature (37 °C), humidity (>80%) and CO_2 level (5%) under normoxic conditions (20% O_2). Under normoxic conditions, there was no signal from the Image-iT™ Hypoxia Reagent (Thermo Fisher, Waltham, MA, USA) but in response to the decrease in oxygen levels, the signal from the Image-iT™ Hypoxia Reagent increased, with nearly all the cells being hypoxic after 60 min at 5% O_2 levels. The increase in signal from the Image-iT™ Hypoxia Reagent was reversible, and when oxygen levels returned to normal, the signal decreased back to baseline. After reaching hypoxic conditions, THP-1 cells and cardiomyocytes were subjected to a single period (2 h at 0% O_2 and 24 h at 1% O_2, respectively) of ischemia followed by reperfusion (1 h and 24 h at 20% O_2, respectively), as previously described [14,30], in the presence or absence of Ezetimibe. Oxidative stress markers, apoptosis, Nrf2/ARE and UPR gene expression, as well AMPKα, phosphorylated (p)-AMPKα, p65 and p-p65, were evaluated. Moreover, in some experiments, THP-1 cells were pre-treated with Compound C (CC, 10 μM, Abcam, Cambridge, UK), a pan inhibitor of AMPK phosphorylation, for 30 min, then incubated overnight with or without Ezetimibe and subjected to IR. Control cells were maintained at 37 °C in an incubator set at normoxic conditions for the same period of time. The endotoxin contamination of cells was routinely excluded with the chromogenic Limulus amoebocyte lysate assay.

2.6. RNA Isolation and Quantitative Real-Time PCR

Total RNA was isolated with the RNEasy Mini Kit (Qiagen, Hilden, Germany). The concentration and quality of RNA were evaluated using the RNA 6000 Nano LabChip Kit (Agilent 2100 Bioanalyzer, Agilent Technologies Inc., Santa Clara, CA, USA). Reverse transcription of total RNA was carried out using the IScript cDNA Synthesis Kit (Bio-Rad, Hercules, CA, USA) according to the manufacturer's recommendations. The relative mRNA expression levels of ACTB (β-actin), NFE2L2 (Nrf2), HMOX1 (HO-1), GCLC, ATF6 and DDIT3 (CHOP) were measured in triplicate using the PrimePCR Probe Assay (ACTB 5′ HEX, NFE2L2 5′ 6-FAM, HMOX1 5′ TEX615, GCLC 5′ Cy5, ATF6 5′ 6-FAM, DDIT3 5′ Cy5 and MAPLC3B 5′ TEX615) and SsoAdvanced Universal Supermix, with the CFX96 RealTime System C1000 Touch Thermal Cycler instrument (Bio-Rad). Data were analyzed using the CFX Maestro software (Bio-Rad). Normalized gene expression levels are given as the ratio between the mean value for the target gene and that for β-actin in each sample.

2.7. Western Blot Analysis

Western blots were performed as previously described [31] with some modifications. Cytoplasmic and nuclear extracts were prepared using the Nuclear Extraction Kit (Cayman Chemical Company, Ann Arbor, MI, USA), and the protein concentration was determined using the Pierce™ BCA Protein Assay Kit (Thermo Fisher, Waltham, MA, USA). Western blots were performed using the XCell SureLock™ Mini-Cell and XCell II™ Blot Module (Thermo Fisher) devices, following the recommended manufacturer's protocols. iBright Prestained Protein Ladder and MagicMark™ XP Western Protein Standard (Thermo Fisher) markers were included in at least two lanes on all gels. Samples (8 μg) were loaded into NuPAGE™ Bis-Tris protein gels (4–12% and 10% gel) (Thermo Fisher) and resolved by SDS-PAGE using NuPAGE™ MES SDS Running Buffer (Thermo Fisher), then transferred onto a polyvinylidene difluoride (PVDF) Transfer Membrane (0.45 μm, Thermo Fisher) using NuPAGE™ Transfer Buffer (Thermo Fisher). Blots were blocked in BSA prepared in Tris-buffered saline with 1% Tween 20 (TBS-T) for 1 h at room temperature. The following primary antibodies were used overnight at 4 °C: anti-AMPKα (#5831) (1:1000), anti-AMPKα (phospho Thr172) (#2535) (1:1000), anti-NF-κB p65 (#8242) (1:1000), anti-NF-κB p65 (phospho Ser536) (#3033) (1:1000) and anti-β-tubulin (#2128) (1:1000) from Cell Signaling Technology (Danvers, MA, USA); anti-Nrf2 (ab62352) (1:500), anti-HO-1 (ab52947) (1:2000), anti-GCLC (ab190685) (1:1000) from Abcam; anti-β-actin (sc-47778) (1:200) and anti-NPC1L1 (sc-166802)(1:100) from Santa Cruz Biotecnology; anti-ATF6 (NBP1-40256) (1:500) from Novus Biologicals (Centennial, CO, USA); anti-CHOP (MA1-250) (1:500) from Thermo Fisher. After 1 h incubation at room temperature with goat anti-rabbit IgG H&L (HRP) (ab6721) (1:3000) (Abcam) or anti-mouse IgG HRP-linked (#7076) (1:2000) (Cell Signaling Technology) antibodies, probed blots were treated with SuperSignal™ WestPico PLUS Chemiluminescent Substrate (Thermo Fisher) and visualized using ImageQuant™ LAS 4000 (GE Healthcare). Bands were quantified with ImageJ [32].

2.8. Nuclear and Cytoplasmic Assays of Nrf2

Nuclear and cytoplasmic Nrf2 were quantified using sandwich enzyme-linked immunosorbent assay (ELISA) kits (LifeSpan BioScience, Seattle, WA, USA), following the manufacturer's instructions.

2.9. Statistical Analysis

Data are expressed as mean ± SD values. Differences between groups were analyzed by a two-tailed Student's *t*-test or by ANOVA followed by post hoc analysis. A probability value (p) of 0.05 was considered to be statistically significant. All data were analyzed with SPSS Statistic Version 20 (IBM Corp., New York, NY, USA).

3. Results

3.1. Effects of Ezetimibe on NPC1L1 Protein Expression, Intracellular ROS Formation and Cell Viability

Preliminarily, we assessed the presence of NPC1L1 receptor on THP-1, HepG2 cells and cardiomyocytes and excluded any effect of Ezetimibe and IR on NPCL1 expression (Figure 1a,b). Ezetimibe dose-dependently reduced ROS formation ($p < 0.01$), without affecting cell viability, in THP-1 cells and cardiomyocytes exposed to TBHP (Figure 1c–e). Based on these results, all subsequent experiments were performed by incubating cells overnight with Ezetimibe 50 μmol, determining a cellular concentration of 0.34 ± 0.02 nmol/μg protein as assessed by LC-MS/MS. Although it is always difficult to compare the drug concentrations used in in vitro studies with those found in vivo [33], the cellular concentrations found in our study are of the same order of magnitude as those found in patients treated with Ezetimibe.

3.2. Effect of Ezetimibe on the Oxidative Stress, Apoptosis and NF-kB Activation Induced by IR

Our results show that IR induced a significant rise in intracellular ROS formation ($p < 0.01$), (Figure 2a) and 8-iso in the culture medium ($p < 0.01$) of THP-1 cells (Figure 2b). Interestingly, both ROS and 8-iso were significantly reduced ($p < 0.01$) in the cells pre-incubated with Ezetimibe (Figure 2a,b).

Our results also show that the percentage increase of apoptotic THP-1 cells ($p < 0.01$) induced by IR was almost abolished when the cells were pre-incubated with Ezetimibe (Figure 2c).

We next evaluated the expression of p65 and p-p65 in the cytoplasmic and nuclear extracts of THP-1 cells in the presence or absence of Ezetimibe. As shown in Figure 2d,f, IR induced a significant nuclear translocation of p-p65 ($p < 0.01$) that was reduced ($p < 0.01$) in THP-1 cells pre-incubated with Ezetimibe. To the contrary, no p65 nuclear translocation was observed (Figure 2e,f). Taken together these results indicate the ability of Ezetimibe to counteract oxidative stress, apoptosis and the activation of the NF-kB pathway induced by IR.

3.3. Ezetimibe Up-Regulates Nrf2/ARE and Pro-Survival UPR Gene Expression in THP-1 Cells Subjected to IR

In our cellular models of IR, we also assessed whether Ezetimibe affects Nrf2/ARE and UPR gene expression. After IR, there were no significant changes in Nrf2 (mRNA and protein) expression (Figure 3a,c,d) and in that of the ARE-correlated genes HO-1 and GCLC (Figure 3b,e,f). To the contrary, when THP-1 cells were preincubated with Ezetimibe, we found a significant up-regulation of the Nrf2 mRNA and nuclear protein ($p < 0.01$) under basal conditions that was even more evident in THP-1 cells subjected to IR (Figure 3a,c,d). Accordingly, the mRNA and protein expression of HO-1 and GCLC was significantly up-regulated by Ezetimibe (Figure 3b,e,f).

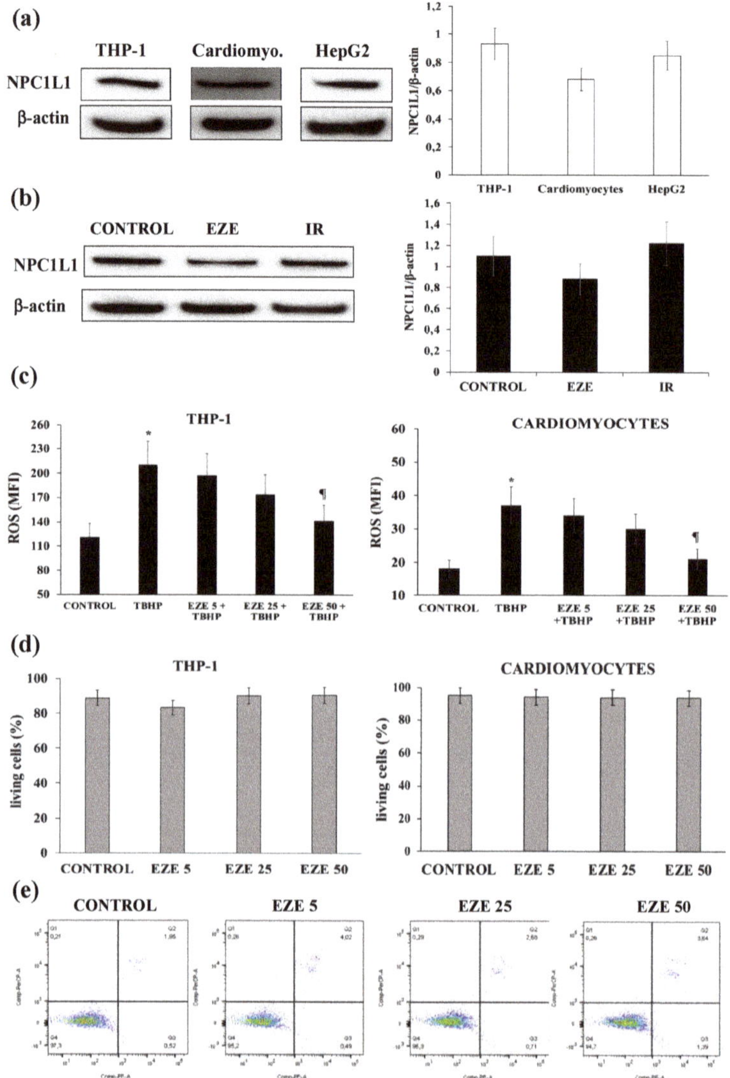

Figure 1. The effects of Ezetimibe (EZE) on Niemann Pick C1-like 1 (NPC1L1) protein expression, intracellular reactive oxygen species (ROS) formation and cell viability. (**a**) Representative Western blot analyses for NPC1L1 protein expression in THP-1 cells, cardiomyocytes (Cardiomyo.) and HepG2 cells and the average quantification of NPC1L1 obtained by the densitometric analysis of three independent experiments. (**b**) Representative Western blot analyses for NPC1L1 protein expression in THP-1 cells under basal conditions, pre-treated with EZE or subjected to ischemia-reperfusion (IR) and the average quantification of NPC1L1 obtained by the densitometric analysis of three independent experiments. (**c**) The dose-response effect of EZE on tert-butyl hydroperoxide (TBHP)-induced ROS formation in THP-1 cells and cardiomyocytes. (**d**) The dose-response effect of EZE on cell viability in THP-1 cells and cardiomyocytes. (**e**) Representative Fluorescence-activated cell sorter (FACS) analysis on cell viability. Data represent the mean ± SD of measurements performed in triplicate in three different experiments; * $p < 0.01$ vs. control; ¶ $p < 0.01$ vs. TBHP.

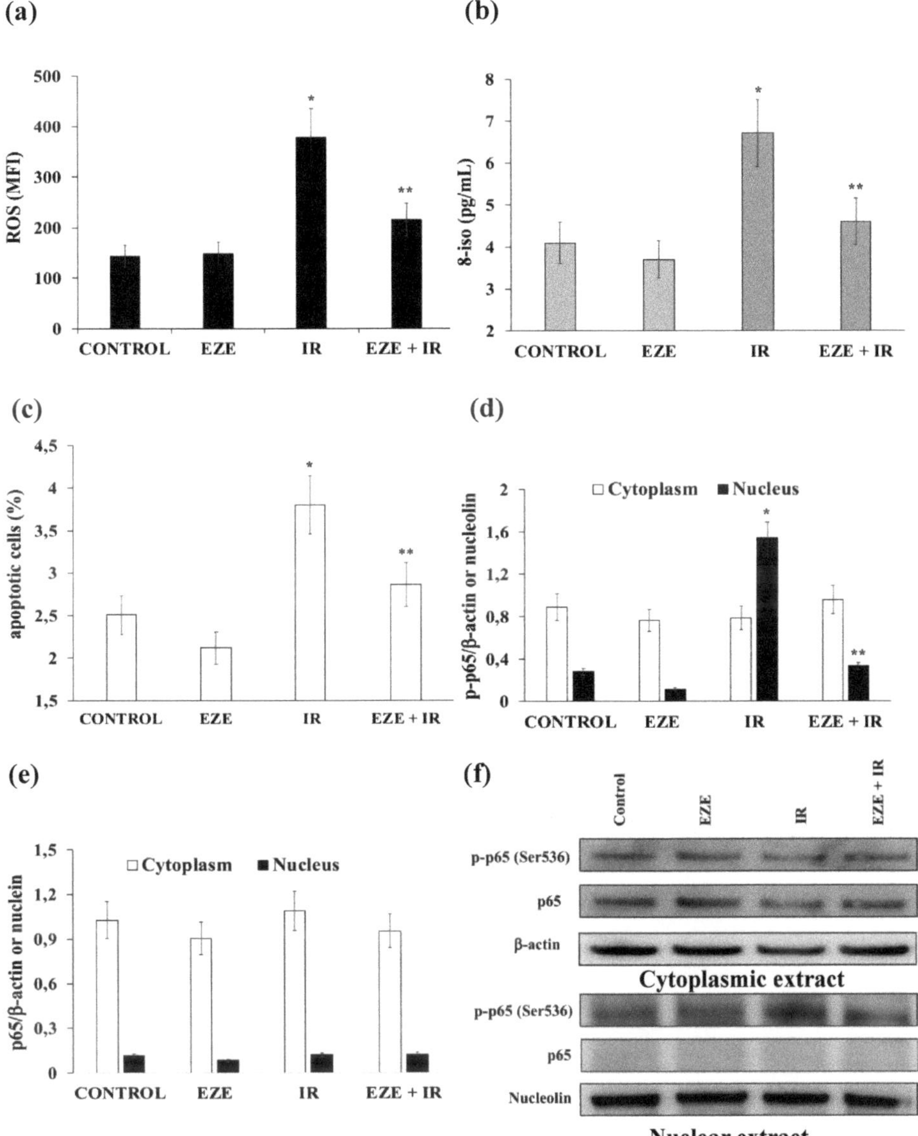

Figure 2. The effect of Ezetimibe on markers of oxidative stress, apoptosis and p65 and p-p65 protein expression in THP-1 cells subjected to ischemia-reperfusion (IR). (**a**) IR-induced ROS formation in THP-1 cells. (**b**) 8-iso concentrations in the culture medium of THP-1 cells. (**c**) The percentages apoptotic THP-1 cells upon exposure to IR. (**d**) The average quantification of nuclear and cytoplasmic p-p65 obtained by the densitometric analysis of three independent experiments. (**e**) The average quantification of nuclear and cytoplasmic p65 obtained by the densitometric analysis of three independent experiments. (**f**) Representative cytoplasmic and nuclear Western blot analyses for the indicated proteins. Data represent the mean ± SD of measurements performed in triplicate in three different experiments; * $p < 0.01$ vs. control; ** $p < 0.01$ vs. IR.

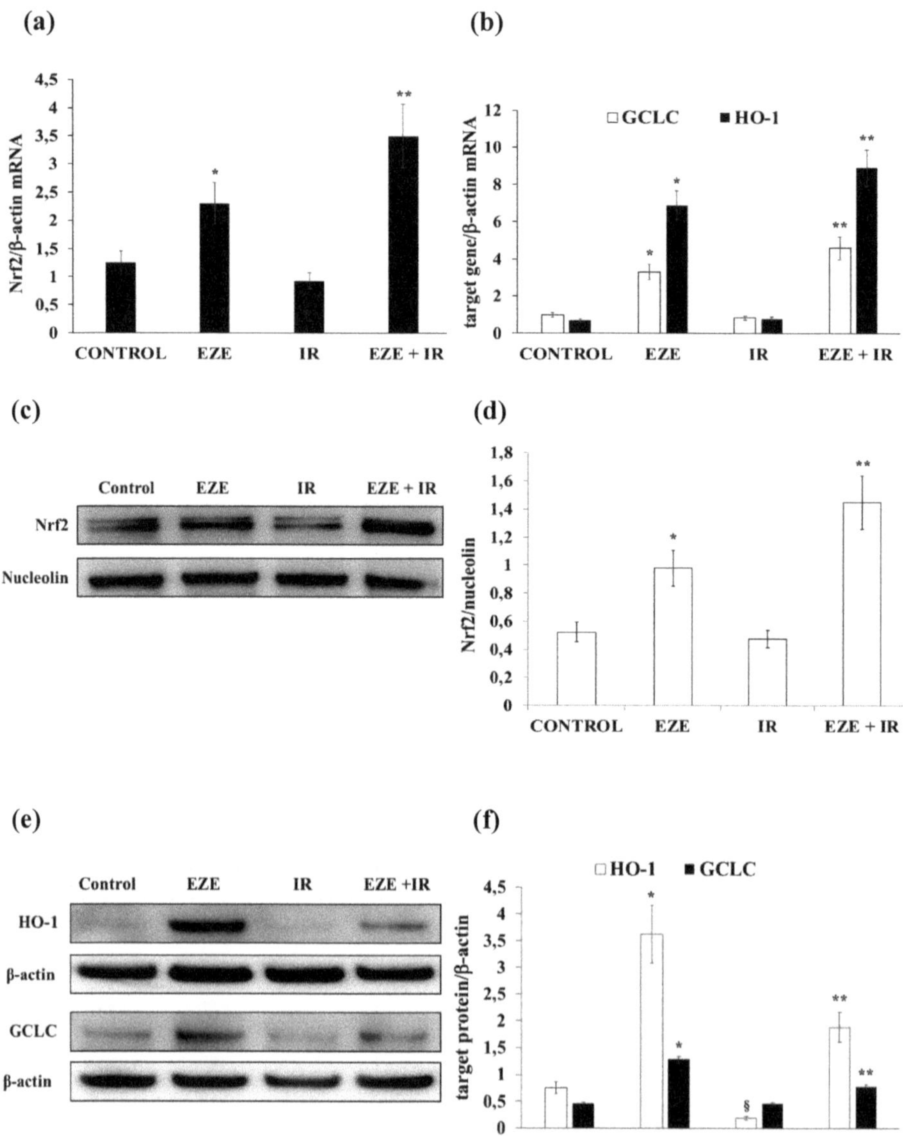

Figure 3. Ezetimibe induces nuclear factor-E2-related factor 2 (Nrf2)/antioxidant response element (ARE) gene expression in THP-1 cells subjected to ischemia-reperfusion (IR). (**a**) The mRNA expression of Nrf2. (**b**) The mRNA expression of HMOX1 (HO-1) and GCLC. (**c**) Representative Western blot analysis for Nrf2 nuclear protein. (**d**) The Nrf2 average quantification obtained by the densitometric analysis of three independent experiments. (**e**) Representative Western blot analyses for the indicated proteins. (**f**) The average quantification of the HO-1 and GCLC proteins obtained by the densitometric analysis of three independent experiments. mRNA was analyzed by quantitative real-time PCR; normalized gene expression levels are given as the ratio between the mean value for the target gene and that for β-actin in each sample. Data are expressed as mean ± SD; * $p < 0.01$ vs. control; ** $p < 0.01$ vs. IR.

Concerning UPR gene expression, IR did not modify ATF6 mRNA and protein expression (Figure 4a–c), whereas it significantly up-regulated ($p < 0.01$) the mRNA and nuclear protein of the pro-apoptotic CHOP (Figure 4a,b,d). IR did not modify the expression of the IRE1-XBP1 pathway (data not shown). Intriguingly, Ezetimibe was able to significantly increase ATF6 gene expression ($p < 0.01$) and to significantly down-regulate ($p < 0.01$) CHOP gene expression (Figure 4a–d) in THP-1 cells subjected to IR.

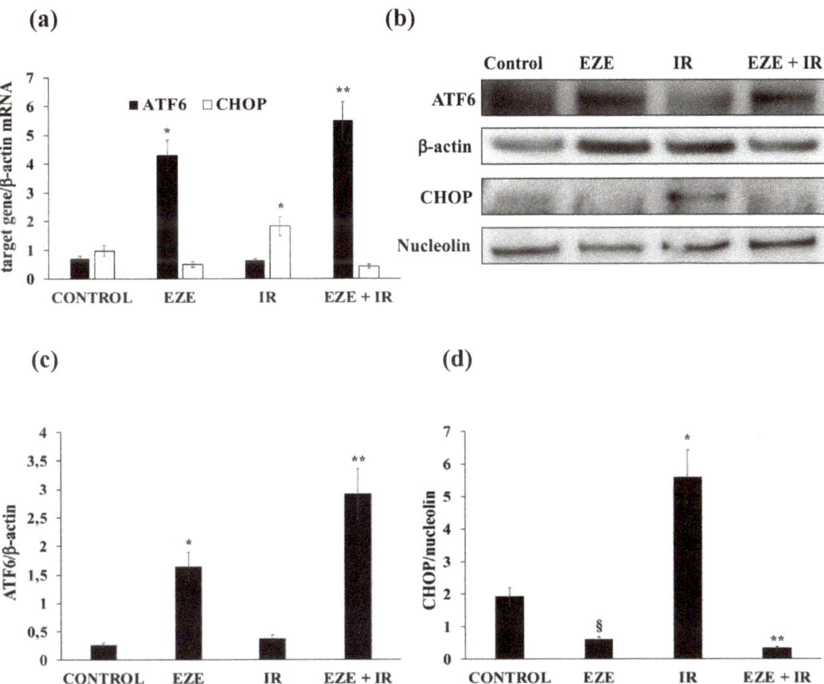

Figure 4. The effect of Ezetimibe on unfolded protein response (UPR) gene expression in THP-1 cells exposed to ischemia-reperfusion (IR). (**a**) The mRNA expression of activating transcription factor 6 (ATF6) and CCAAT-enhancer-binding protein homologous protein (CHOP). (**b**) Representative Western blot analyses for the indicated proteins. (**c,d**) The average quantification of ATF6 and CHOP obtained by the densitometric analysis of three independent experiments. mRNA was analyzed by quantitative real-time PCR; normalized gene expression levels are given as the ratio between the mean value for the target gene and that for β-actin in each sample. Data are expressed as mean ± SD. * $p < 0.01$ vs. control (up-regulation); ** $p < 0.01$ vs. IR; § $p < 0.01$ vs. control (down-regulation).

3.4. Effect of Ezetimibe on AMPK Activation in Our Model of IR

To assess whether AMPK activation has a role in the Nrf2 activation induced by Ezetimibe, we first evaluated the effect of Ezetimibe on phosphorylated and non-phosphorylated AMPK expression. We found that Ezetimibe was able to induce a significant increase in the p-AMPK/AMPK ratio, both in control and in THP-1 cells subjected to IR (Figure 5a,c). Then, we performed further experiments using Compound C, a pan inhibitor of AMPK phosphorylation. Our results show that Compound C significantly decreased the Ezetimibe- and Ezetimibe plus IR-mediated increases in the p-AMPK/AMPK ratio (Figure 5a,c). Moreover, the pre-incubation of THP-1 cells with Compound C almost abolished the nuclear Nrf2 translocation induced by Ezetimibe ($p < 0.001$), both in THP-1 cells subjected to IR and in control cells (Figure 5b,c).

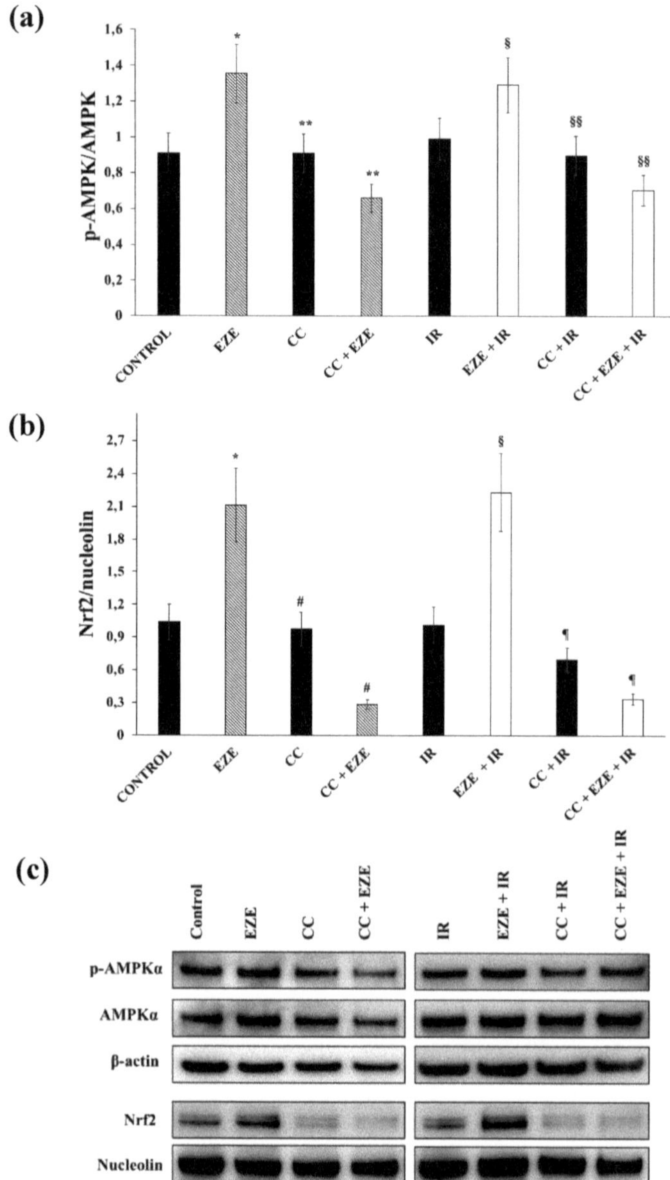

Figure 5. The effect of the AMP-activated protein kinase (AMPK) inhibitor Compound C (CC) on Ezetimibe-induced Nrf2 activation in THP-1 cells exposed to ischemia-reperfusion (IR). (**a**) The average quantification of the p-AMPK/AMPK ratio obtained by the densitometric analysis of three independent experiments. (**b**) The average quantification of nuclear Nrf2 obtained by the densitometric analysis of three independent experiments. (**c**) Representative cytoplasmic and nuclear Western blot analyses for the indicated proteins. Data represent the mean ± SD of measurements performed in triplicate in three different experiments; * $p < 0.01$ vs. control; ** $p < 0.001$ vs. EZE; [§] $p < 0.01$ vs. IR; [§§] $p < 0.01$ vs. EZE + IR; [#] $p < 0.001$ vs. EZE; [¶] $p < 0.001$ vs. EZE + IR.

Furthermore, the pre-incubation with Compound C induced a significant increment in ROS generation (Figure 6a) and an increase in the percentage of apoptotic THP-1 cells (Figure 6b,c).

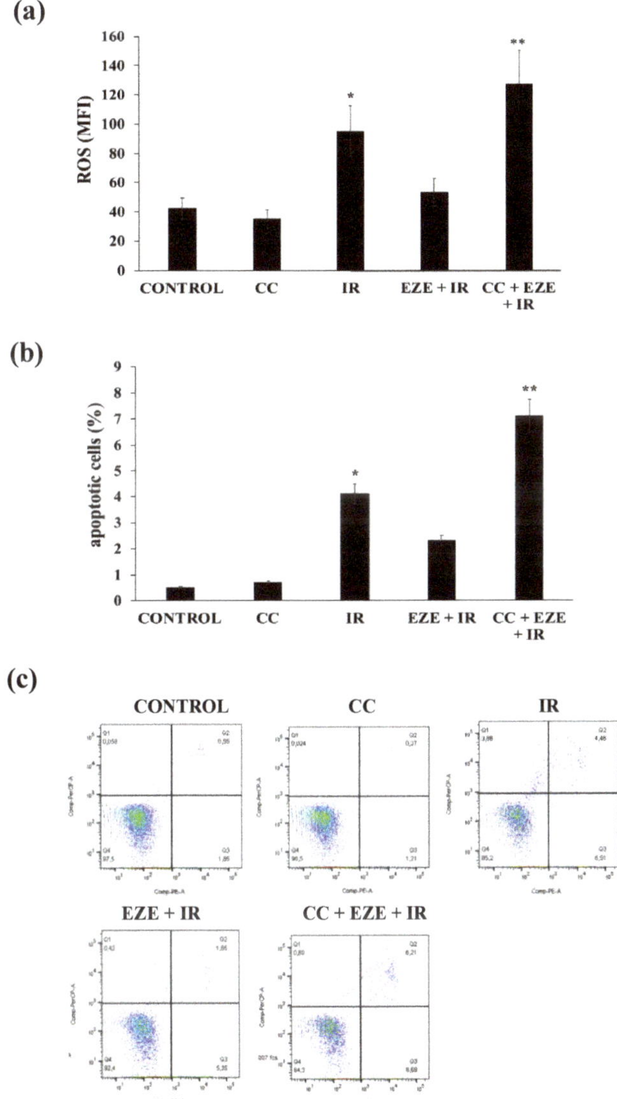

Figure 6. The effect of Compound C (CC) on ROS formation and apoptosis in THP-1 cells pretreated with Ezetimibe (EZE) and subjected to IR. (**a**) ROS formation. (**b**) The percentages of apoptotic cells. (**c**) Representative FACS analyses of the percentages of apoptotic cells. Data are represented as the mean ± SD of measurements performed in triplicate in three different experiments; * $p < 0.01$ vs control; ** $p < 0.01$ vs. IR and EZE + IR.

3.5. Effect of Ezetimibe on IR-Induced Oxidative Stress, Apoptosis and Nrf2/ARE Activation in Human Cardiomyocytes

Finally, in order to additionally confirm our data in primary cells directly subjected to IR, we performed further experiments using cardiomyocytes. IR induced a significant rise in intracellular ROS formation ($p < 0.01$), which was reduced ($p < 0.01$) in cardiomyocytes pre-incubated with Ezetimibe (Figure 7a). Moreover, we found a significant increment in the percentage of apoptotic cells ($p < 0.01$) induced by IR that was completely abolished when human cardiomyocytes were pre-incubated with Ezetimibe (Figure 7b,c).

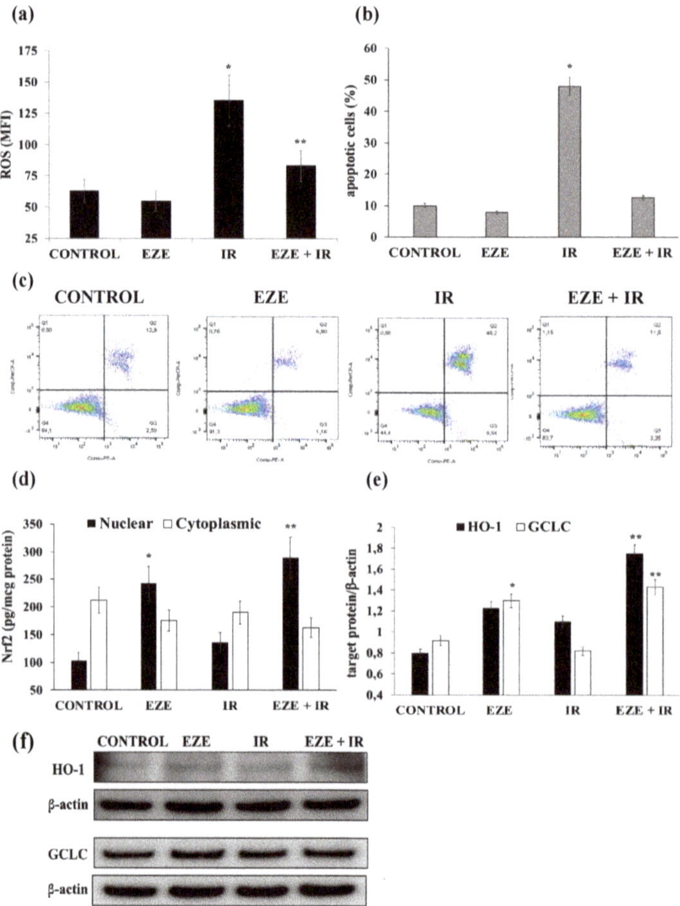

Figure 7. The effect of Ezetimibe on markers of oxidative stress, apoptosis and Nrf2/ARE gene expression in cardiomyocytes subjected to ischemia-reperfusion (IR). (**a**) IR-induced ROS formation. (**b**) The percentages of apoptotic cardiomyocytes upon exposure to IR. (**c**) Representative FACS analyses of cell viability. (**d**) Cytoplasmic and nuclear assays of Nrf2. (**e**) The average quantification of the HO-1 and GCLC proteins obtained by the densitometric analysis of three independent experiments. (**f**) Representative Western blot analyses for the indicated proteins. mRNA was analyzed by quantitative real-time PCR; normalized gene expression levels are given as the ratio between the mean value for the target gene and that for β-actin in each sample. Data are expressed as mean ± SD; * $p < 0.01$ vs. control; ** $p < 0.01$ vs. IR.

Regarding Nrf2 signaling pathway activation, IR did not modify cytoplasmic or nuclear protein concentrations and was not associated with variations of BACH1 gene expression (data not shown), whereas when cardiomyocytes were pre-incubated with Ezetimibe, Nrf2 nuclear protein was significantly increased ($p < 0.01$), both in control cells and under IR conditions (Figure 7d). Accordingly, the protein expression of the Nrf2-correlated genes HO-1 and GCLC was significantly increased ($p < 0.01$) by Ezetimibe (Figure 7e,f).

Finally, we tested whether Ezetimibe was also able to phosphorylate AMPK in cardiomyocytes. Similarly to THP-1, we found that Ezetimibe significantly increased the p-AMPK/AMPK ratio, both in controls and in cardiomyocytes subjected to IR (Figure 8a,b).

(a)

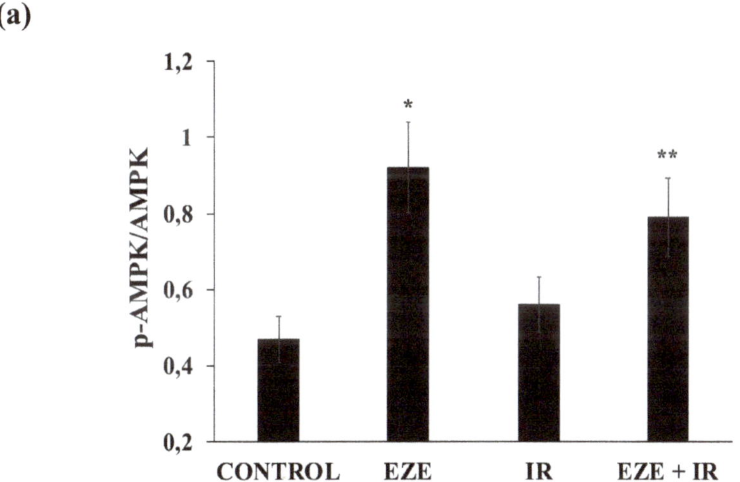

(b)

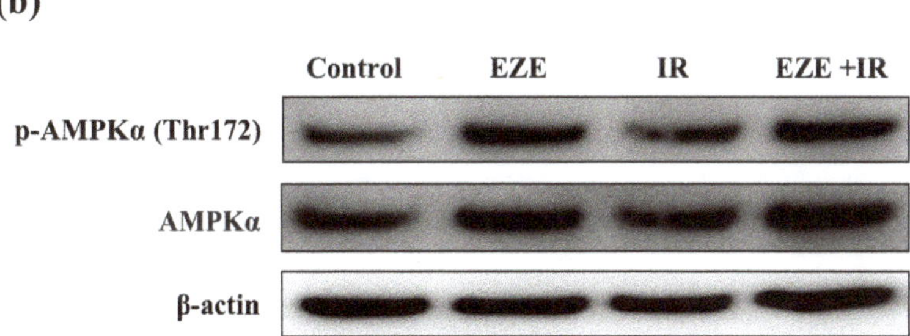

Figure 8. The Ezetimibe (EZE)-induced phosphorylation of AMPK in cardiomyocytes subjected to ischemia-reperfusion (IR). (**a**) The average quantification of the p-AMPK/AMPK ratio obtained by the densitometric analysis of three independent experiments. (**b**) Representative Western blot analyses for the indicated proteins. Data represent the mean ± SD of measurements performed in triplicate in three different experiments; * $p < 0.01$ vs. control; ** $p < 0.01$ vs. IR.

4. Discussion

IR injury is a common and important clinical problem that occurs in many different organ systems, including the brain (in stroke) and heart (in myocardial infarction), and in limb ischemia, and it occurs with the respective reperfusion strategies (thrombolytic therapy, angioplasty and operative revascularization) but also in routine surgical procedures (organ transplantation, free tissue-transfer, cardiopulmonary bypass, vascular surgery and the treatment of major trauma/shock) [3]. Over the past three decades, many ischemic and pharmacological cardioprotective tools, derived from experimental studies, have been examined in the clinical setting of acute pathological situations characterized by IR [34]. The results of these studies have been disappointing, and no effective cardioprotective therapy is currently used in clinical practice [34].

In this study, we found that after IR, there was a significant increase in intracellular ROS formation and in 8-iso in the culture media, together with an increased percentage of apoptotic cells. As recently reviewed by Cadenas S., the mitochondrial electron transport chain, as well as some NOX isoforms, contribute to ROS generation during IR injury [35]. This augmented oxidative stress induced by IR was not associated with an up-regulation of Nrf2/ARE gene expression. This apparently paradoxical result may be related to the fact that oxidative stress activates I-κB kinase, which phosphorylates NF-kB, leading to its translocation into the nucleus and the activation of pro-inflammatory cytokines [7]. Furthermore, when NF-kB binds to cAMP-response-element binding protein (CBP) in a competitive manner, it inhibits the binding of CBP to Nrf2, which leads to the inhibition of Nrf2 transactivation [7]. In addition, NF-kB increases the recruitment of histone deacetylase to the ARE region, and hence, Nrf2 transcriptional activation is prevented [7]. The fact that in this study, IR caused a significant increment in nuclear p-p65—an indicator of NF-kB activation—supports this conclusion.

The lack of Nrf2/ARE gene up-regulation in our model of IR is in line with the data recently published by our group showing that in peripheral blood mononuclear cells derived from patients with peripheral artery disease subjected to one cycle of IR, there was no up-regulation of Nrf2/ARE gene expression [14]. At variance with our results, previous studies have shown a protective role of the Nrf2/ARE pathway in animal models of myocardial ischemia and reperfusion injury [36,37]. It is likely that these differences are related to the different experimental conditions.

Concerning UPR, the results of this study also show that IR did not modify the expression of the pro-survival genes ATF6 and IRE1, but it increased the expression of CHOP, i.e., the pro-apoptotic gene. As a matter of fact, CHOP has been shown to play an important role in the induction of apoptosis [11] and therefore in causing the cellular injuries caused by IR.

Ezetimibe was able to counteract both the effect on oxidative stress and apoptosis caused by IR. In particular, Ezetimibe was associated with a substantial decline in intracellular ROS and 8-iso in the culture media and also led to a substantial reduction in apoptotic cells.

In our study, the positive effect of Ezetimibe on oxidative stress and apoptosis was paralleled by an up-regulation of Nrf2, HO-1 and GCLC expression. Although Ezetimibe did not affect nuclear p-p65 translocation in control cells, it strongly reduced nuclear p-p65 and the p-p65/p65 ratio in THP-1 cells subjected to IR, confirming the inverse relationship between Nrf2 and NF-kB activation [7]. Previous reports have shown that Nrf2 activation attenuates the injuries caused by IR [8,9]; since the generation of ROS has been implicated in the cell injury caused by IR [1–3], reducing oxidative stress may be a potential therapeutic approach to the prevention of IR injuries. Here, for the first time, we show that Ezetimibe protects cells subjected to IR by up-regulating Nrf2 and the related ARE gene expression. The mechanism involved in Ezetimibe-induced Nrf2 gene activation in our cellular model of IR is unclear. However, Ezetimibe has recently been found to protect the mouse liver from oxidative injury caused by diet-induced nonalcoholic fatty liver disease [22]. In particular, it has been shown that Ezetimibe induced Nrf2 activation in a p62-dependent manner and that this induction can be attributed to Nrf2 activation [22], since p62 has been shown to be an Nrf2 target gene [38]. As previously reported, the increment in p62 phosphorylation induced by Ezetimibe was demonstrated to be dependent on p-AMPK activation [22]. The N-terminus PB1 domain of p62 was shown to specifically bind AMPKα1 and AMPKα2 [22]. In agreement with these data, our study shows that Ezetimibe, but not IR, increased the phosphorylation of AMPK and that Compound C, a specific inhibitor of AMPK phosphorylation, inhibited this phosphorylation. Moreover, our evidence that Compound C was also able to inhibit Nrf2 nuclear translocation suggests that Ezetimibe activates

AMPK, leading to Nrf2 activation. This indirect effect of Compound C on Nrf2 nuclear translocation was associated with an increment in ROS generation and a decline in the percentage of apoptotic cells. A likely mechanism for Ezetimibe-induced AMPK phosphorylation was elucidated in a recent report suggesting that Ezetimibe increases the oxygen consumption rate (OCR) and reduces the amount of adenosine triphosphate (ATP) in subjects with hypercholesterolemia [39]. AMPK activity is governed by adenosine diphosphate (ADP) and adenosine monophosphate (AMP) levels by phosphorylation when the ratio of intracellular AMP or ADP to ATP increases [40]. Therefore, an Ezetimibe-induced rise in OCR may cause an elevated ADP/ATP ratio followed by the activation of AMPK.

Furthermore, the results we obtained using human cardiomyocytes (primary cell line directly subjected to IR) are in line with those found in THP-1 cells. Concerning oxidative stress, we showed that Ezetimibe dose-dependently decreased TBHP-induced intracellular ROS formation. Moreover, IR induced a significant rise in ROS accompanied by a significant increment in apoptosis that were almost abolished when cardiomyocytes were pre-incubated with Ezetimibe. Interestingly, Nrf2/ARE gene expression was also significantly up-regulated by Ezetimibe both in the control cells and in the cardiomyocytes exposed to IR.

Finally, the results of this study also show that Ezetimibe affected UPR gene expression. In particular, Ezetimibe was shown to increase the pro-survival ATF6 gene's expression both after IR and in control cells; to the contrary, the increased up-regulation of the pro-apoptotic CHOP induced by IR was almost abolished by Ezetimibe. The mechanism underlying this effect of Ezetimibe is presently unclear. However, recent studies showed that oxidative stress can elicit a reduction–oxidation imbalance and worsen ER stress through reducing the efficiency of protein folding pathways and increasing the production of misfolded proteins [41]. The abnormal amount of misfolded proteins may drive the UPR towards cellular apoptosis through the activity of CHOP. Ezetimibe, by reducing oxidative stress, may reduce misfolded protein mass and therefore convert the cells to a pro-survival state.

5. Conclusions

In these cellular models of IR, we describe new pleiotropic effects of Ezetimibe and the molecular mechanism that cause a series of adaptations that may contribute to the modification of the cells towards a "resistant phenotype". In particular, we show that Ezetimibe prevents IR-induced oxidative stress and apoptosis and up-regulates the Nrf2 and UPR signaling pathways. Nevertheless, the well-known difficulties in comparing the drug concentrations used in in vitro studies with those found in vivo must be considered a limitation of the study. Hence, although these preliminary results need to be confirmed in patients, we suggest that Ezetimibe, beyond its cholesterol absorption inhibitor effect, may be beneficial in IR injury, a common important clinical problem that affects many different organ systems.

Author Contributions: Conceptualization: D.P. and A.M.F.P., methodology: D.P., C.S., E.D. and U.G.; data curation: C.M.C.; Writing—Original Draft Preparation: L.C., and A.M.F.P.; Writing—Review & Editing: L.C., D.P. and A.M.F.P.; supervision: L.C. and A.M.F.P.; funding acquisition: A.M.F.P. All Authors have approved the submitted version and agree to be personally accountable for the author's own contributions and for ensuring that questions related to the accuracy or integrity of any part of the work, even ones in which the author was not personally involved, are appropriately investigated, resolved, and documented in the literature. All authors have read and agreed to the published version of the manuscript.

Funding: This work was support in part by grants from University of Verona, School of Medicine.

Conflicts of Interest: The authors declare no conflict of interest.

References

1. Murphy, E.; Steenbergen, C. Mechanisms underlying acute protection from cardiac ischemia-reperfusion injury. *Physiol. Rev.* **2007**, *88*, 581–609. [CrossRef] [PubMed]
2. Yellon, D.M.; Hausenloy, D.J. Myocardial reperfusion injury. *N. Engl. J. Med.* **2007**, *357*, 1121–1135. [CrossRef] [PubMed]

3. Eltzschig, H.; Eckle, T. Ischemia and reperfusion-from mechanism to translation. *Nat. Med.* **2011**, *17*, 1391–1401. [CrossRef] [PubMed]

4. Kensler, T.W.; Wakabayashi, N.; Biswal, S. Cell survival responses to environmental stresses via the Keap1-Nrf2-ARE pathway. *Annu. Rev. Pharmacol. Toxicol.* **2007**, *47*, 89–116. [CrossRef]

5. Komatsu, M.; Kurokawa, H.; Waguri, S.; Taguchi, K.; Kobayashi, A.; Ichimura, Y.; Sou, Y.S.; Ueno, I.; Sakamoto, A.; Tong, K.I.; et al. The selective autophagy substrate p62 activates the stress responsive transcription factor Nrf2 through inactivation of Keap1. *Nat. Cell. Biol.* **2010**, *12*, 213–223. [CrossRef]

6. Taguchi, K.; Motohashi, H.; Yamamoto, M. Molecular mechanisms of the Keap1- Nrf2 pathway in stress response and cancer evolution. *Genes Cells* **2011**, *16*, 123–140. [CrossRef]

7. Schoonbroodt, S.; Piette, J. Oxidative stress interference with the nuclear factor-kappa B activation pathways. *Biochem. Pharmacol.* **2000**, *60*, 1075–1083. [CrossRef]

8. Calvert, J.W.; Elston, M.; Jha, S.; Gundewar, S.; Aragon, J.P.; Grinsfelder, D.B.; Ramachandran, A.; Elrod, J.W.; Lefer, D.J. Genetic and pharmacologic hydrogen sulfide therapy attenuates ischemia-induced heart failure in mice. *Circulation* **2010**, *122*, 11–19. [CrossRef]

9. Ashrafian, H.; Czibik, G.; Bellahcene, M.; Aksentijević, D.; Smith, A.C.; Mitchell, S.J.; Dodd, M.S.; Kirwan, J.; Byrne, J.J.; Ludwig, C.; et al. Fumarate is cardioprotective via activation of the Nrf2 antioxidant pathway. *Cell. Metab.* **2012**, *15*, 361–371. [CrossRef]

10. Diamanti, M.A.; Gupta, J.; Bennecke, M.; De Oliveira, T.; Ramakrishnan, M.; Braczynski, A.K.; Richter, B.; Beli, P.; Hu, Y.; Saleh, M.; et al. IKKα controls ATG16L1 degradation to prevent ER stress during inflammation. *J. Exp. Med.* **2017**, *214*, 423–437. [CrossRef]

11. Hetz, C. The unfolded protein response: Controlling cell fate decisions under ER stress and beyond. *Nat. Rev. Mol. Cell. Biol.* **2012**, *13*, 89–102. [CrossRef] [PubMed]

12. Martindale, J.J.; Fernandez, R.; Thuerauf, D.; Whittaker, R.; Gude, N.; Sussman, M.A.; Glembotski, C.C. Endoplasmic reticulum stress gene induction and protection from ischemia/reperfusion injury in the hearts of transgenic mice with a tamoxifen-regulated form of ATF6. *Circ. Res.* **2006**, *98*, 1186–1193. [CrossRef] [PubMed]

13. Doroudgar, S.; Thuerauf, D.J.; Marcinko, M.C.; Belmont, P.J.; Glembotski, C.C. Ischemia activates the ATF6 branch of the endoplasmic reticulum stress response. *J. Biol. Chem.* **2009**, *284*, 29735–29745. [CrossRef] [PubMed]

14. Fratta Pasini, A.M.; Stranieri, C.; Rigoni, A.M.; De Marchi, S.; Peserico, D.; Mozzini, C.; Cominacini, L.; Garbin, U. Physical exercise reduces cytotoxicity and up-regulates Nrf2 and UPR expression in circulating cells of peripheral artery disease patients: An hypoxic adaptation? *J. Atheroscler. Thromb.* **2018**, *25*, 808–820. [CrossRef]

15. Ahmed, S.M.; Luo, L.; Namani, A.; Wang, X.J.; Tang, X. Nrf2 signaling pathway: Pivotal roles in inflammation. *Biochim. Biophys. Acta Mol. Basis Dis.* **2017**, *1863*, 585–597. [CrossRef]

16. Cullinan, S.B.; Diehl, J.A. PERK-dependent activation of Nrf2 contributes to redox homeostasis and cell survival following endoplasmic reticulum stress. *J. Biol. Chem.* **2004**, *279*, 20108–20117. [CrossRef]

17. Garbin, U.; Fratta Pasini, A.; Stranieri, C.; Cominacini, M.; Pasini, A.; Manfro, S.; Lugoboni, F.; Mozzini, C.; Guidi, G.; Faccini, G.; et al. Cigarette smoking blocks the protective expression of Nrf2/ARE pathway in peripheral mononuclear cells of young heavy smokers favouring inflammation. *PLoS ONE* **2009**, *4*, e8225. [CrossRef]

18. Mozzini, C.; Garbin, U.; Stranieri, C.; Pasini, A.; Solani, E.; Tinelli, I.A.; Cominacini, L.; Fratta Pasini, A.M. Endoplasmic reticulum stress and Nrf2 repression in circulating cells of type 2 diabetic patients without the recommended glycemic goals. *Free Radic. Res.* **2015**, *49*, 244–252. [CrossRef]

19. Mozzini, C.; Fratta Pasini, A.; Garbin, U.; Stranieri, C.; Pasini, A.; Vallerio, P.; Cominacini, L. Increased endoplasmic reticulum stress and Nrf2 repression in peripheral blood mononuclear cells of patients with stable coronary artery disease. *Free Radic. Biol. Med.* **2014**, *68*, 178–185. [CrossRef]

20. Jia, L.; Betters, J.L.; Yu, L. Niemann-pick C1-like1(NPC1L1) protein in intestinal and hepatic cholesterol transport. *Annu. Rev. Physiol.* **2011**, *73*, 239–259. [CrossRef]

21. Toth, P.P.; Davidson, M.H. Simvastatin and Ezetimibe: Combination therapy for the management of dyslipidaemia. *Expert Opin. Pharmacother.* **2005**, *6*, 131–139. [CrossRef] [PubMed]

22. Lee, D.H.; Han, D.H.; Nam, K.T.; Park, J.S.; Kim, S.H.; Lee, M.; Kim, G.; Min, B.S.; Cha, B.S.; Lee, Y.S.; et al. Ezetimibe, an NPC1L1 inhibitor, is a potent Nrf2 activator that protects mice from diet-induced nonalcoholic steatohepatitis. *Free Radic. Biol. Med.* **2016**, *99*, 520–532. [CrossRef] [PubMed]

23. Yu, J.; Li, X.; Matei, N.; McBride, D.; Tang, J.; Yan, M.; Zhang, J.H. Ezetimibe, a NPC1L1 inhibitor, attenuates neuronal apoptosis through AMPK dependent autophagy activation after MCAO in rats. *Exp. Neurol.* **2018**, *307*, 12–23. [CrossRef] [PubMed]

24. Yu, J.; Wang, W.N.; Matei, N.; Li, X.; Pang, J.W.; Mo, J.; Chen, S.P.; Tang, J.P.; Yan, M.; Zhang, J.H. Ezetimibe attenuates oxidative stress and neuroinflammation via the AMPK/Nrf2/TXNIP pathway after MCAO in rats. *Oxid. Med. Cell. Longev.* **2020**, 4717258. [CrossRef] [PubMed]

25. Nomura, M.; Ishii, H.; Kawakami, A.; Yoshida, M. Inhibition of hepatic Niemann-Pick C1-like1 improves hepatic insulin resistance. *Am. J. Physiol. Endocrinol. Metab.* **2009**, *297*, E1030–E1038. [CrossRef]

26. Trocha, M.; Merwid-Ląd, A.; Chlebda, E.; Sozański, T.; Pieśniewska, M.; Gliniak, H.; Szeląg, A. Influence of Ezetimibe on selected parameters of oxidative stress in rat liver subjected to ischemia/reperfusion. *Arch. Med. Sci.* **2014**, *10*, 817–824. [CrossRef]

27. Fratta Pasini, A.; Anselmi, M.; Garbin, U.; Franchi, E.; Stranieri, C.; Nava, M.C.; Boccioletti, V.; Vassanelli, C.; Cominacini, L. Enhanced levels of oxidized low-density lipoprotein prime monocytes to cytokine overproduction via upregulation of CD14 and toll-like receptor 4 in unstable angina. *Arterioscler. Thromb. Vasc. Biol.* **2007**, *27*, 1991–1997. [CrossRef]

28. Lin, X.L.; Xiao, W.J.; Xiao, L.L.; Liu, M.H. Molecular mechanisms of autophagy in cardiac ischemia/reperfusion injury (Review). *Mol. Med. Rep.* **2018**, *18*, 675–683. [CrossRef]

29. Tormos, A.M.; Taléns-Visconti, R.; Bonora-Centelles, A.; Pérez, S.; Sastre, J. Oxidative stress triggers cytokinesis failure in hepatocytes upon isolation. *Free Rad. Res.* **2015**, *49*, 927–934. [CrossRef]

30. Stone, L.L.H.; Chappuis, E.; Marquez, O.; McFalls, M.E.; Kelly, R.F.; Butterick, M. Mitochondrial respiratory capacity is restored in hibernating cardiomyocytes following co-culture with mesenchymal stem cells. *Cell Med.* **2019**, *11*, 1–7. [CrossRef]

31. Kemmerer, Z.A.; Ader, N.R.; Mulroy, S.S.; Eggler, A.L. Comparison of human Nrf2 antibodies: A tale of two proteins. *Toxicol. Lett.* **2015**, *238*, 83–89. [CrossRef] [PubMed]

32. Rasband, W.S.; Image, J. U.S. National Institutes of Health, Bethesda, Maryland, USA, 1997–2018. Available online: https://imagej.nih.gov/ij/ (accessed on 23 March 2020).

33. Kosoglou, T.; Statkevich, P.; Johnson-Levonas, A.O.; Paolini, J.F.; Bergman, A.J.; Alton, K.B. Ezetimibe. A Review of its metabolism, pharmacokinetics and drug interactions. *Clin. Pharmacokinet.* **2005**, *44*, 467–494. [CrossRef] [PubMed]

34. Hausenloy, D.J.; Yellon, D.M. Ischaemic conditioning and reperfusion injury. *Nat. Rev. Cardiol.* **2016**, *13*, 193–209. [CrossRef] [PubMed]

35. Cadenas, S. ROS and redox signaling in myocardial ischemia-reperfusion injury and cardioprotection. *Free Rad. Biol. Med.* **2018**, *117*, 76–89. [CrossRef] [PubMed]

36. Xu, B.; Zhang, J.; Strom, J.; Lee, S.; Chen, Q.M. Myocardial ischemic reperfusion induces de novo Nrf2 protein translation. *Biochim. Biophys. Acta* **2014**, *1842*, 1638–1647. [CrossRef]

37. Shen, Y.; Liu, X.; Shi, J.; Wu, X. Involvement of Nrf2 in myocardial ischemia and reperfusion injury. *Int. J. Biol. Macromol.* **2019**, *125*, 496–502. [CrossRef]

38. Jain, A.; Lamark, T.; Sjøttem, E.; Larsen, K.B.; Awuh, J.A.; Øvervatn, A.; McMahon, M.; Hayes, J.D.; Johansen, T. p62/SQSTM1is a target gene for transcription factor NRF2 and creates a positive feedback loop by inducing antioxidant response element-driven gene transcription. *J. Biol. Chem.* **2010**, *285*, 22576–22591. [CrossRef]

39. Hernandez-Mijares, A.; Bañuls, C.; Rovira-Llopis, S.; Diaz-Morales, N.; Escribano-Lopez, I.; de Pablo, C.; Alvarez, A.; Veses, S.; Rocha, M.; Victor, V.M. Effects of simvastatin, Ezetimibe and simvastatin/Ezetimibe on mitochondrial function and leukocyte/endothelial cell interactions in patients with hypercholesterolemia. *Atherosclerosis* **2016**, *247*, 40–47. [CrossRef]

40. Yuan, H.X.; Xiong, Y.; Guan, K.L. Nutrient sensing, metabolism, and cell growth control. *Mol. Cell.* **2013**, *49*, 379–387. [CrossRef]

41. Plaisance, V.; Brajkovic, S.; Tenenbaum, M.; Favre, D.; Ezanno, H.; Bonnefond, A.; Bonner, C.; Gmyr, V.; Kerr-Conte, J.; Gauthier, B.R.; et al. Endoplasmic reticulum stress links oxidative stress to impaired pancreatic beta-cell function caused by human oxidized LDL. *PLoS ONE* **2016**, *11*, e0163046. [CrossRef]

Article

Differential Effects of MitoVitE, α-Tocopherol and Trolox on Oxidative Stress, Mitochondrial Function and Inflammatory Signalling Pathways in Endothelial Cells Cultured under Conditions Mimicking Sepsis

Beverley E. Minter, Damon A. Lowes [†], Nigel R. Webster and Helen F. Galley *

Institute of Medical Sciences, University of Aberdeen, Aberdeen AB41 8TJ, UK;
beverley.minter1@abdn.ac.uk (B.E.M.); damon.lowes@rothamsted.ac.uk (D.A.L.);
n.r.webster@abdn.ac.uk (N.R.W.)
* Correspondence: h.f.galley@abdn.ac.uk; Tel.: +44-1224-437-363
† Present address: Rothamsted Research, Harpenden AL5 2JQ, UK.

Received: 9 January 2020; Accepted: 25 February 2020; Published: 26 February 2020

Abstract: Sepsis is a life-threatening response to infection associated with inflammation, oxidative stress and mitochondrial dysfunction. We investigated differential effects of three forms of vitamin E, which accumulate in different cellular compartments, on oxidative stress, mitochondrial function, mRNA and protein expression profiles associated with the human Toll-like receptor (TLR) -2 and -4 pathways. Human endothelial cells were exposed to lipopolysaccharide (LPS)/peptidoglycan G (PepG) to mimic sepsis, MitoVitE, α-tocopherol, or Trolox. Oxidative stress, mitochondrial function, mitochondrial membrane potential and metabolic activity were measured. NFκB-P65, total and phosphorylated inhibitor of NFκB alpha (NFκBIA), and STAT-3 in nuclear extracts, interleukin (IL)-6 and IL-8 production in culture supernatants and cellular mRNA expression of 32 genes involved in Toll-like receptor-2 and -4 pathways were measured. Exposure to LPS/PepG caused increased total radical production ($p = 0.022$), decreased glutathione ratio ($p = 0.016$), reduced membrane potential and metabolic activity (both $p < 0.0001$), increased nuclear NFκB-P65 expression ($p = 0.016$) and increased IL-6/8 secretion (both $p < 0.0001$). MitoVitE, α- tocopherol and Trolox were similar in reducing oxidative stress, NFκB activation and interleukin secretion. MitoVitE had widespread downregulatory effects on gene expression. Despite differences in site of actions, all forms of vitamin E were protective under conditions mimicking sepsis. These results challenge the concept that protection inside mitochondria provides better protection.

Keywords: sepsis; MitoVitE; antioxidant; mitochondria; gene expression; cytokines; mRNA

---

1. Introduction

Sepsis is a complex syndrome and is a leading cause of morbidity and mortality worldwide. It is characterized by a dysregulated immune response to infection, usually bacterial, leading to a systemic inflammatory response, oxidative stress, depletion of intracellular antioxidants, and ultimately organ failure [1,2]. Development of organ dysfunction associated with sepsis is now accepted to be due, at least in part, to mitochondrial dysfunction [3,4].

Vitamin E belongs to a group of compounds that includes both tocopherols and tocotrienols [5]. Vitamin E sequesters into the hydrophobic interior of membranes and α-tocopherol is the most biologically active form. Tocopherol is able to protect cell membranes from oxidation, reacting with lipid radicals produced during lipid peroxidation [5]. MitoVitE is essentially the chromanol moiety of vitamin E bound to a triphenyl phosphonium (TPP) cation and accumulates within mitochondria as a result of the large negative charge inside the mitochondrial inner membrane, with the vitamin moiety directed towards the matrix. MitoVitE has been shown to accumulate in all major organs of mice and rats after oral, intraperitoneal or intravenous administration and has potent antioxidant activity [6]. Trolox (6-hydroxy-2,5,7,8-tetra-methylchroman-2-carboxylic acid) is a synthetic, water soluble and cell-permeable derivative of vitamin E which accumulates in the cell cytosol. It is a potent antioxidant in several model systems [6–9].

Since mitochondria are both a major source of production and a target for damage of reactive oxygen species that contribute to oxidative stress in sepsis, antioxidants targeted to mitochondria have been proposed as a better approach than non-targeted forms for antioxidant protection in sepsis [10,11]. In this study, we investigated the relative effects of MitoVitE, α-tocopherol and Trolox on oxidative stress, mitochondrial function and expression of key genes and proteins involved in the toll-like receptor (TLR)-2 and -4 signalling pathways in human endothelial cells cultured in an environment mimicking acute bacterial sepsis. Despite differences in site of actions, we found that all three forms of vitamin E had protective effects in human endothelial cells under conditions mimicking sepsis.

2. Materials and Methods

2.1. Chemicals

Unless otherwise stated, all chemicals were obtained from Sigma-Aldrich (Poole, Dorset, UK)

2.2. Cell Culture

The human umbilical vein endothelial cell line (HUVEC-C, obtained from ATCC, Teddington, Middlesex, UK) was used from passages 7 to 10 as described previously in detail [12–14]. Microscopy images of the cells are shown in Supplementary Figure S1. For experimentation, cells were cultured in 96- or 6-well plates in the presence of 0.2 µg/mL lipopolysaccharide (LPS) plus 20 µg/mL peptidoglycan G (PepG) plus 5 µM MitoVitE, α-tocopherol acetate or Trolox or molecular grade ethanol as vehicle control. The duration of treatment was either 4 h, 24 h or 7 d based on expected expression profiles as detailed below. Cell viability was assessed using acid phosphatase activity [15].

2.3. Oxidative Stress

Total radical production was measured in intact cells as follows: following treatment for 24 h, cells were washed with PBS before being loaded with 50 µM of the oxidation sensitive dye 5-(-6)-carboxy-2′,7′-dichlorofluorescein diacetate (carboxy-DCFDA, molecular probes, Invitrogen, Paisley, UK) in Hank's balanced salt solution (HBSS) supplemented with 1g/L glucose, and incubated for 1 h in the dark at 37 °C. Following incubation, cells were washed with phosphate buffered saline (PBS, pH 7.4) and fluorescence was determined immediately over 3 h at 37 °C at an excitation wavelength of 485 nm/emission wavelength 530 nm.

For measurement of reduced glutathione (GSH), buffer containing 0.1% (*v/v*) Triton X-100 and 0.1 M potassium phosphate, pH 6.5, plus 20 µM monochlorobimane was added to cells after 24 h treatment, for 37 °C in the dark for 30 min. To measure oxidised glutathione (GSSG), buffer containing 0.1% (*v/v*) Triton X-100 and 0.1 M potassium phosphate, pH 6.5 and 2.2 mM diethylenetriamine-penta-acetic acid and 2 mM dithiothreitol were added to additional cells for 30 min at 37 °C followed by the addition of an equal volume of 40 µM monochlorobimane for 30 min at 37 °C. Fluorescence was determined at ambient room temperature (excitation 355 nm, emission 520 nm) and data were normalised to total cellular protein level, determined using the Bradford method [16].

2.4. Mitochondrial Function

Mitochondrial membrane potential was determined in intact cells using the fluorescent probe JC-1 (5,5,6,6-tetrachloro-1,1,3,3-tetraethylbenzimidazolcarbocyanine iodide (Invitrogen, Paisley, UK) [12–14]. Following treatments for 7 d, with a medium change after 3–4 days, cells were washed with PBS, incubated for 30 min with 10 µg/mL JC-1 in PBS at 37 °C in the dark, then washed with PBS and the red/green fluorescence ratio measured. A decrease in the ratio of red/green fluorescence indicates loss of mitochondrial membrane potential [12–14]. As a positive control, some cells were treated with 1 µM rotenone without LPS/PepG for 5 h prior to the addition of JC-1. Metabolic activity was analyzed by measuring the rate of reduction of Alamar Blue™ (Invitrogen, Paisley, UK) in intact cells after 7 d treatment as described above. Alamar Blue™ is a novel redox indicator that exhibits both fluorescent and colourimetric changes in response to changes in metabolic activity via oxidative metabolism [17]. Briefly, Alamar Blue™ was added to each well and fluorescence was measured every 15 min for 2 h at 37 °C at excitation 530 nm/emission 620 nm. Metabolic activity was determined as the rate of change in fluorescence over time [13,14].

2.5. Differential Gene Expression

The genes were selected based on their roles in the TLR2 and TLR4 pathways which ultimately result in activation of nuclear factor kappa B (NFκB). This transcription factor is known to be redox sensitive and has a crucial role in propagation of the inflammatory response to sepsis. The TLR2 and TLR4 signalling pathways are illustrated in Supplementary Figure S2 and the genes investigated are listed in Supplementary Table S1. Expression of 32 key genes was measured using pre-coated custom RT-PCR plates (Applied Biosystems, Warrington, UK).

Total RNA was isolated from treated cells after 4 h treatments using Trizol and further purified using the Qiagen RNeasy Mini Kit with on-column genomic DNA digestion. Total RNA (100ng) was used for cDNA synthesis using the Applied Biosystems high capacity RNA-to-cDNA TaqMan® Kit. For qPCR, cDNA was mixed with TaqMan® gene expression master mix (Applied Biosystems, Warrington, UK) and added to the pre-coated well plate. Targets were amplified and detected using the 7500 HT Fast Real-Time PCR System. Hypoxanthine-guanine phosphoribosyltransferase-1 (HPRT-1) was used as housekeeping gene. The PCR system run cycle: activation (95 °C, 10 min), melt (95 °C, 15 s), annealing and extension (60 °C, 1 min) for 40 cycles. Gene expression (as fold change) was determined using the Delta-Delta C_t ($2^{-\Delta\Delta CT}$) method using the SA Biosciences (Qiagen, Manchester, UK) web-based PCR array system from raw threshold cycle data (C_t). The $\Delta\Delta C_t$ algorithm is an approximation method to determine relative gene expression with quantitative real-time PCR (qPCR) experiments. Six independent experiments were performed and a minimum fold change of at least 2 compared to LPS/PepG treatment alone was pre-defined. p values were calculated using Student's T-tests of the replicate $2^{-\Delta\Delta CT}$ values for each gene and a p value of ≤ 0.05 was taken as significant.

2.6. Protein Expression

To determine NFκB activation, nuclear extracts from treated cells were prepared following 4 h treatments using the Novagens Nucbuster™ protein extraction kit (Merck Chemicals Ltd., Nottingham, UK). NFκB activation was measured as the amount of the p65 subunit present in the nucleus using the Novagen NoShift™ transcription factor assay kit (Merck, Nottingham, UK) [12,13]. To determine phosphorylated inhibitor of NFκB alpha (NFκBIA, also known as IκBα), and signal transducer and activator of transcription-3 (STAT-3) activation, after 4 h exposure to LPS/PepG, cells were lysed in TRIS Base buffer containing protease/ phosphatases inhibitors and adjusted to a protein concentration of 0.25 mg/mL. Commercially available enzyme immunoassay kits were used to quantify the total and phosphorylated proteins (InstantOne™ eBioscience Ltd., Hatfield, UK) according to the manufacturer's protocols. Commercially available enzyme immunoassay kits were used to quantify interleukin (IL)-6 and IL-8 secretion (R&D Systems, Oxford, UK) in culture supernatants of cells treated with LPS/PepG with and without the three forms of vitamin E for 24 h, as described in the manufacturer's protocol.

2.7. Statistical Analysis

For oxidative stress and mitochondrial function assays, six independent experiments were performed ($n = 6$). For protein expression, 3–6 independent experiments were performed. No assumptions were made about data distribution. Data were analysed using non-parametric

Kruskal Wallis testing with Mann Whitney post hoc testing where appropriate and are presented as median, interquartile and full range, or individual raw data points when $n \leq 6$. A p value of ≤ 0.05 was taken to be significant.

3. Results

3.1. Cell Viability

Acid phosphatase activity was similar regardless of cell treatment at both 24 h and 7 d, showing no detrimental effect on cell viability (Supplementary Figure S3).

3.1.1. Oxidative Stress

Exposure of endothelial cells to LPS/PepG resulted in a significant increase in total radical production compared to vehicle control ($p = 0.022$, Figure 1A). Co-treatment of cells with any of the forms of vitamin E plus LPS/PepG abrogated the increase in radical production (Figure 1A). The ratio of GSH:GSSG was significantly lower in LPS/PepG treated cells compared to vehicle control treated cells ($p = 0.016$, Figure 1B). Co-exposure to LPS/PepG in the presence of all of the forms of vitamin E prevented the LPS-PepG mediated decrease in the glutathione ratio (Figure 1B).

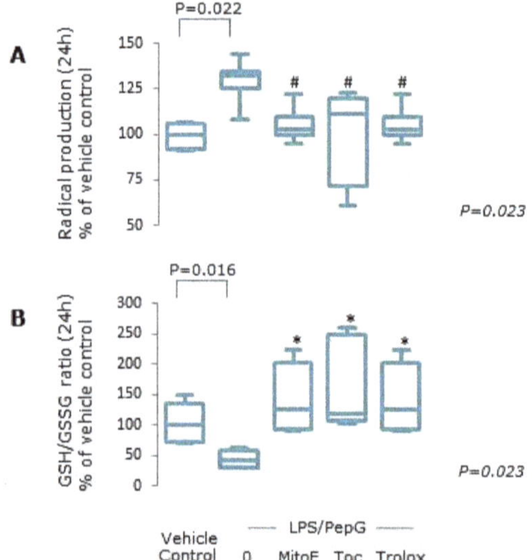

Figure 1. Oxidative stress. Endothelial cells were treated with vehicle control, lipopolysaccharide plus peptidoglycan (LPS/PepG) alone, or LPS/PepG plus 5 µM MitoVitE (MitoE), tocopherol (Toc) or Trolox for 24 h. (**A**) Total radical production, (**B**) reduced/oxidised glutathione. Box and whisker plots show median, interquartile and full range ($n = 6$). p value in italics refers to Kruskal–Wallis across LPS/PepG treated groups. # = significantly lower and * = significantly higher, than LPS/PepG alone ($p < 0.05$).

3.1.2. Mitochondrial Function

Cells exposed to LPS/PepG for 7 d had significantly lower mitochondrial membrane potential compared to vehicle control treated cells ($p < 0.0001$, Figure 2A). Membrane potential was significantly higher in cells exposed to LPS/PepG plus MitoVitE compared to LPS/PepG alone ($p = 0.003$) but not in those cells treated with Trolox or α-tocopherol; indeed, tocopherol worsened the loss of membrane potential (Figure 2A). Pre-treatment with rotenone resulted in around 50% loss of membrane potential in vehicle control treated cells (Figure 2A).

Metabolic activity was also significantly lower in cells exposed to LPS/PepG compared to vehicle control cells ($p < 0.0001$) and co-treatment with MitoVitE ($p = 0.03$), Trolox ($p = 0.002$) or α-tocopherol ($p = 0.002$) ameliorated the loss of metabolic activity (Figure 2B).

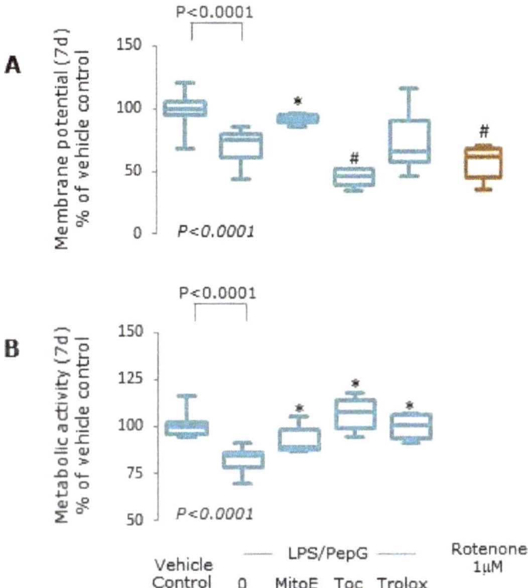

Figure 2. Oxidative stress. Endothelial cells were treated with vehicle control, lipopolysaccharide plus peptidoglycan (LPS/PepG) alone, or LPS/PepG plus 5 µM MitoVitE (MitoE), tocopherol (Toc) or Trolox for 7 d, or 1µM rotenone. (**A**) Mitochondrial membrane potential, (**B**) metabolic activity. Box and whisker plots show median, interquartile and full range ($n = 6$). p-value in italics refers to Kruskal–Wallis across LPS/PepG treated groups. # = significantly lower and * = significantly higher, than LPS/PepG alone ($p < 0.05$).

3.1.3. Gene Expression

Identities of the 32 genes analysed and their associated proteins and functions are summarized in Supplementary Table S1 and Supplementary Figure S2. Differential gene expression following LPS/PepG exposure for 4 h compared to vehicle control showed that expression of 7 genes were upregulated by at least 2–fold but only two of these were statistically significant: NFκBIA ($p = 0.02$) and NFκB1 ($p = 0.04$, Table 1, Figure 3A). Two genes were downregulated by >2 fold (XPO1 and IRAK4, both $p = 0.02$, Table 1, Figure 3A). In cells exposed to LPS/PepG plus MitoVitE, no genes were upregulated, but 12 genes were downregulated compared to LPS/PepG alone ($p < 0.05$, Table 1, Figure 3B). In contrast, only one gene, NFκB1, was downregulated in cells exposed to LPS/PepG plus α-tocopherol ($p = 0.002$, Table 1, Figure 3C) but none were downregulated by Trolox. PTGS2 was upregulated by α-tocopherol and Trolox ($p = 0.02$ and $p = 0.006$ respectively), a gene which was downregulated by MitoVitE ($p = 0.02$, Table 1, Figure 3D).

Table 1. Differential gene expression.

Gene Name.	LPS/PepG [a]	LPS/PepG + MitoVitE [b]	LPS/PepG + α-Tocopherol [b]	LPS/PepG +Trolox [b]
GUSB		−2.5 (0.27, 0.52) $p = 0.05$		
IkBKB		−2.5 0.25, 0.56 $p = 0.017$		
IRAK4	−2.2 (1.05, 3.42) $p = 0.02$			
MYD88		−2.4 (0.23, 0.61) $p = 0.05$		
NFκB1	+4.4 (0.06, 0.40) $p = 0.01$	−4.5 (0.17, 0.28) $p = 0.0003$	**−3.9** (0.18, 0.51) $p = 0.002$	
NFkBIA	+11.7 (0.00001, 0.18) 0.18) $p = 0.02$	-2.7 (0.18, 0.56) $p = 0.04$		
PPAR-α		−3.0 (0.17, 0.50) $p = 0.02$		
PTGS2		-3.5 (0.12, 0.45) $p = 0.02$	**+3.0** (1.29, 4.62) $p = 0.005$	**+2.9** (1.30, 4.50) $p = 0.006$
RIPK2		−5.4 (0.12, 0.25) $p = 0.001$		
STAT1		−3.7 (0.20, 0.34) $p = 0.0002$		
STAT3		−3.1 (0.25, 0.40) $p = 0.0002$		
TAB1		−2.8 (0.23, 0.49) $p = 0.009$		
TRAF6		−2.9 (0.24, 0.44) $p = 0.0003$		
XPO1	−2.7 (0.98, 4.33) $p = 0.02$			

[a] Mean fold change (red), 95% confidence interval (CI) and p-value compared to vehicle control or [b] to LPS/PepG alone. CI = indicateS 95% certainty that the mean value is the true mean of the population. For full gene names and functions see Supplementary Table S1.

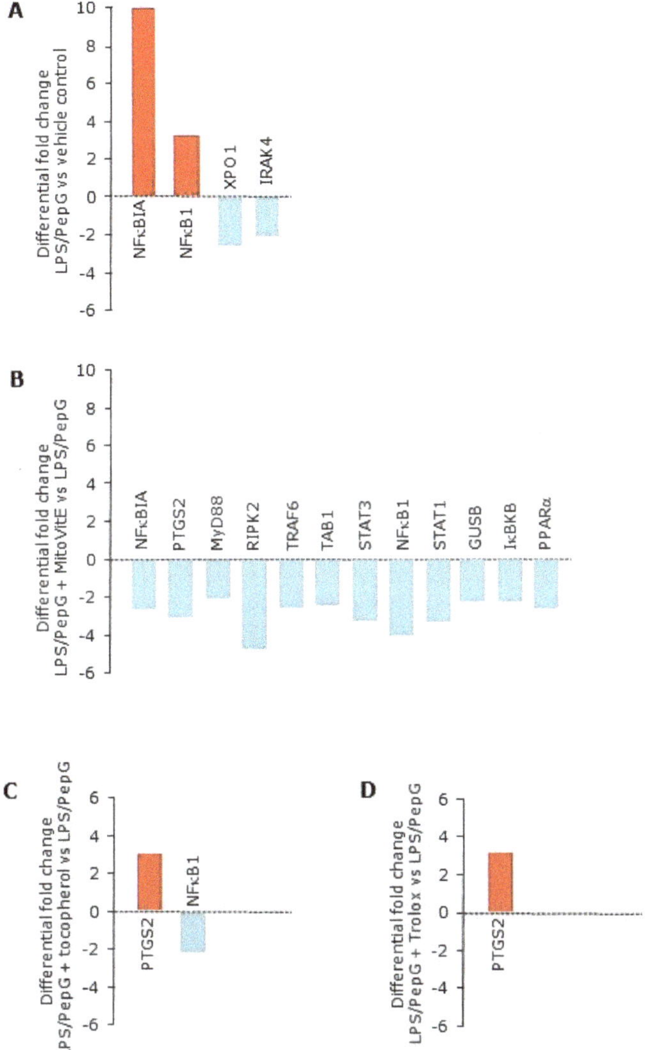

Figure 3. Gene expression. Differential gene expression as mean fold change in endothelial cells showing the effect of (**A**) Lipopolysaccharide plus peptidoglycan (LPS/PepG) compared to vehicle control (**B**) LPS/PepG plus 5 μM MitoVitE compared to LPS/PepG alone (**C**) LPS/PepG plus 5 μM tocopherol compared to LPS/PepG alone or (**D**) LPS/PepG plus 5 μM Trolox compared to LPS/PepG alone. Bars show mean fold changes where $p < 0.05$. Red = upregulation, blue = downregulation, $n = 6$. 95% confidence intervals and p values are given in Table 1.

3.1.4. Protein Expression

Nuclear NFκB-p65 protein was maximally expressed following 4 h LPS/PepG exposure (Supplementary Figure S4) and was higher than vehicle control treated cells ($p = 0.016$, Figure 4A). Co-treatment of cells with either of the three forms of vitamin E plus LPS/PepG resulted in decreased nuclear NFκB-p65 protein expression similar to that seen in vehicle control treated cells (Figure 4A). Significant increases in both IL-6 and IL-8 protein levels were seen in culture supernatants from cells exposed to LPS/PepG for 24 h compared to vehicle control treated cells (both $p < 0.0001$, Figure 4B,C).

All forms of vitamin E appeared to suppress IL-6 and IL-8 secretion when compared to LPS/PepG exposed cells. However, IL-6 levels were statistically significantly lower only in cells exposed to LPS/PepG plus α-tocopherol or Trolox compared to LPS/PepG alone (Figure 4B) and IL-8 levels were significantly lower only in Trolox treated cells (Figure 4C).

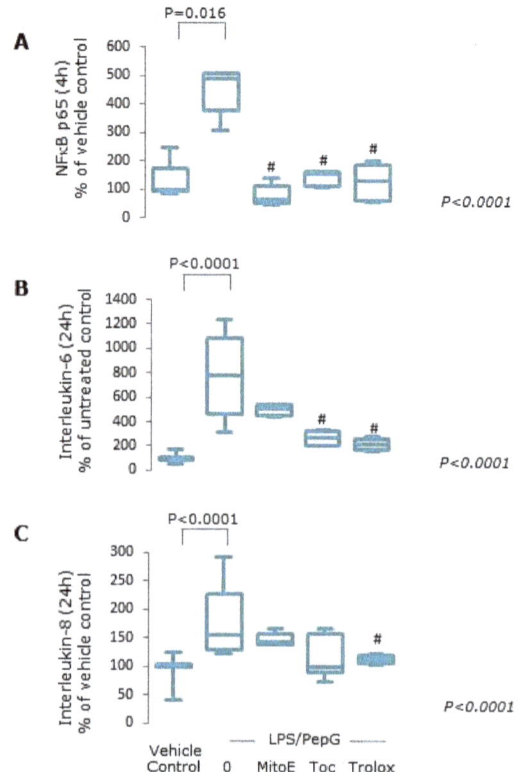

Figure 4. NFκB activation and cytokine expression. Endothelial cells treated with vehicle control, lipopolysaccharide plus peptidoglycan (LPS/PepG) alone, or LPS/PepG plus 5 μM MitoVitE (MitoE), tocopherol (Toc) or Trolox for 4 or 24 h. (**A**) Nuclear NFkB p65, (**B**) interleukin-6 (IL-6) and (**C**) interleukin-8. Box and whisker plots show median, interquartile and full range (*n* = 6). *p* value in italics refers to Kruskal–Wallis across LPS/PepG treated groups, # = significantly lower than LPS/PepG alone (*p* < 0.05).

Total and phosphorylated NFκBIA and STAT3 proteins were markedly increased in cells co-treated with LPS/PepG and α-tocopherol, whilst MitoVitE and Trolox had no effect (Figure 5).

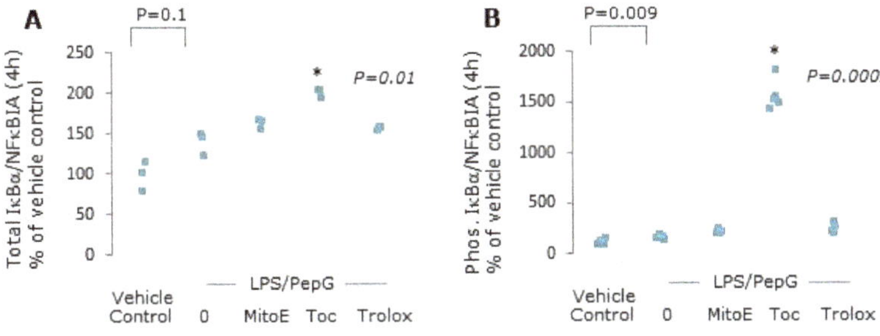

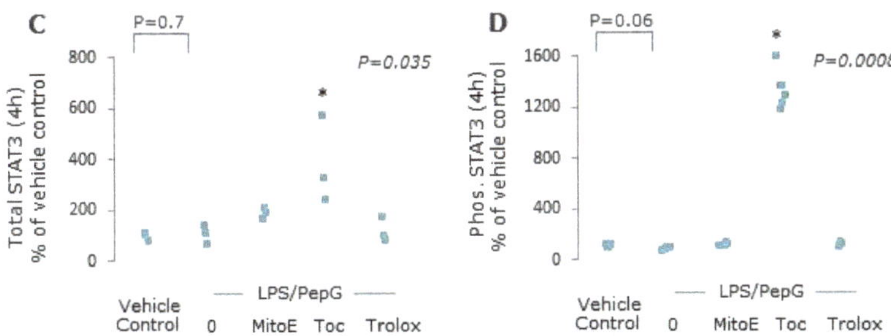

Figure 5. Total and phosphorylated IκBα and STAT3 expression. Endothelial cells were treated with vehicle control, lipopolysaccharide plus peptidoglycan (LPS/PepG) alone, or LPS/PepG plus 5 μM MitoVitE (MitoE), tocopherol (Toc) or Trolox for 4 h. (**A**) Total IκBα, (**B**) phosphorylated IκBα, (**C**) total STAT3 and (**D**) phosphorylated STAT3. Individual data points shown (*n* = 3–6). *p* value in italics refers to Kruskal–Wallis across LPS/PepG treated groups, * = significantly higher than LPS/PepG alone (*p* < 0.05).

4. Discussion

We found that exposure of human endothelial cells to LPS/PepG resulted in mitochondrial dysfunction and oxidative stress, associated with changes in expression of genes and selected proteins involved in TLR2 and TLR4 signalling and the inflammatory response to conditions of sepsis. Surprisingly, we found that although the effects of the three forms of vitamin E act in different cellular compartments, we found largely similar effects on mitochondrial function and oxidative stress. However, only the mitochondria targeted form of vitamin E, MitoVitE, had widespread downregulatory effects on genes involved in TLR pathways, resulting in a clear anti-inflammatory profile.

Oxidative stress and mitochondrial damage/dysfunction are consistent findings in sepsis and drive the dysregulated and prolonged inflammation seen in these patients [3,4]. TLRs are pattern recognition receptors which detect pathogen associated molecular patterns (PAMPs) that include LPS and PepG, and initiate signal transduction, culminating in the activation of NFκB and transcription of cytokines [18,19]. Exposure of endothelial cells to LPS/PepG resulted in maximal nuclear translocation of NFκB at 4 h. Since antioxidants acting in mitochondria have been advocated as a better protective strategy in sepsis than those that do not [10,11], and as NFκB has a fundamental and wide ranging role in propagation of immune responses in sepsis [20], we determined the effects of compartmentalized antioxidants on expression of genes involved in relevant pathways (see Supplementary Figure S2) at maximal NFκB protein expression and subsequent downstream effects on inflammatory mediator expression and mitochondrial function. Many studies report on the expression of genes known to be regulated by NFκB following PAMP-induced activation [18–21]. NFκB comprises DNA-binding subunits, principally NFκB1 and NFκB2, in heterodimers with transcriptional activators including P65, also known as RelA [18]. Inactivated NFκB complexes are retained in the cellular cytoplasm by inhibitors including NFκBIA. The process of NFκB activation involves translocation of NFκB heterodimers into the nucleus and activation of the transcriptional subunit, which is initiated by phosphorylation, resulting upregulation of a large number of genes required for host immune and inflammatory mediators, including chemokines, cytokines, enzymes and adhesion molecules. It is known that complete activation of NFκB requires a redox sensitive step and so antioxidants may have a role in its regulation [18,19]. However, effects on the signalling cascades leading to NFκB activation are less clear.

We and others have consistently reported low serum tocopherol levels and elevated oxidative stress biomarkers in patients with sepsis [22–25]. Although all tocopherols and tocotrienols have pronounced antioxidant activity in model systems in vitro, there are variable reports of the effect of administration of vitamin E on biomarkers of oxidative stress and there is little evidence of measurable metabolites of vitamin E after antioxidant reactions in vivo [26,27]. It has been shown that α-tocopherol, the most biologically relevant tocopherol, inhibits the signalling cascades initiated by LPS in neutrophils and microglial cells in vitro [28] and inhibits LPS- induced inflammatory responses in murine macrophages in vitro [29]. MitoVitE contains a 2-carbon chain which links the functional chromanol moiety of vitamin E to a TPP cation and is able to cross the outer mitochondrial membrane due to the positive charge of the TPP conjugate [6,30]. In Trolox, the chromanol is connected to a short-chain carboxylic acid making Trolox hydrophilic and cell-permeable such that it accumulates in the cell cytosol [6–9]. MitoVitE is more effective than Trolox both in vitro and in vivo [6,31]. Antioxidants which specifically accumulate within the mitochondrial matrix are suggested to offer better protection during sepsis-induced oxidative stress than non-targeted antioxidants, since mitochondria are the main site of radical production during inflammation [10,11]. We have shown that antioxidants which accumulate inside mitochondria are protective against mitochondrial dysfunction in human endothelial cells exposed to conditions which mimic those seen in sepsis [13,32] and have beneficial effects in an animal model of acute sepsis [32,33]. The present study was undertaken to determine the differential effects of compartmentalization of antioxidants using α-tocopherol, MitoVitE and Trolox, on mitochondrial function, oxidative stress and gene expression of proteins which are involved in the TLR2/4 signalling pathways, in an in vitro model relevant to sepsis. We have reported previously on this model [12–14,22,32].

Exposure of endothelial cells to LPS/PepG in vitro results in consumption of glutathione and initiation of inflammatory responses associated with activation of NFκB and mitochondrial oxidative stress [12–14,22,32]. We also found that following exposure of endothelial cells to LPS/PepG there was increased production of intracellular radicals, which was abrogated when the cells were cultured with all three forms of vitamin E tested. Alpha-tocopherol has been reported to reduce intracellular radical production in several other oxidative stress models [28,29] and Trolox is an effective intracellular radical scavenger in human cells [7,8]. We also showed that the three vitamin E derivatives ameliorated the reduction in the glutathione ratio caused by exposure to LPS/PepG. Only MitoVitE was able to prevent the loss of mitochondrial membrane potential, although all three compounds were able to maintain mitochondrial metabolic activity after LPS/PepG exposure of cells. This suggests that loss of membrane potential is an adaptive response to prevent an increase in radical formation by increasing uncoupling protein-2 to uncouple the electron transport chain [34]. Our results suggest that the loss of metabolic activity is either caused by damage to components of the tricarboxylic acid cycle and/or mitochondrial oxidative phosphorylation complexes with no substantial reduction of ATP synthesis since the resultant increase in AMP production would either drive an increase in metabolic activity, or again is an adaptive response in this model to limit damage. We have reported previously that other mitochondria targeted antioxidants are protective in this model [32,34]. We have also shown mitochondrial protective effects of MitoVitE in dorsal root ganglion cells exposed to the chemotherapy drug, paclitaxel, where MitoVitE but not Trolox was able to protect against mitochondrial dysfunction and loss of GSH [31]. Only MitoVitE is able to scavenge radicals at source and ameliorate mitochondrial uncoupling, as shown by maintenance of membrane potential.

NFκB-P65 subunit expression was maximal at 4 h following LPS/PepG exposure. We found significant upregulation, consistent with other studies, of NFκB-P50 mRNA in LPS/PepG exposed cells, whose protein is known to modulate NFκB activity, and NFκBIA, which encodes for an inhibitor subunit of NFκB; this may be a negative regulatory process to dampen the inflammatory response. Following phosphorylation, NFκBIA is ubiquinitated and degraded, which results in an increase in NFκB activity. However, the increase in NFκBIA phosphorylation was not reflected by p65, IL-6 or, IL-8 levels. Two genes were differentially downregulated by LPS/PepG exposure: XPO1 and IL-1 receptor-associated kinase 4 (IRAK4). XPO1 encodes for exportin-1 which is involved in trafficking of proteins including NFκBIA [35,36] whereas IRAK4 is a key mediator in TLR2 and -4 signalling [37]. Overexpression of XPO1 results in increased translocation of active NFκB [35].

The increase in LPS/PepG-induced NFκB-p65 translocation was abrogated by all forms of vitamin E tested, concomitant with lower IL-6 and IL-8 secretion. NFκB1 gene expression was also downregulated by MitoVitE and tocopherol, but not Trolox. Tocopherols and tocotrienols and a synthetic chroman carboxamide with a chemical structure similar to Trolox were previously shown to reduce LPS-induced NFκB activation and cytokine secretion [28,29,38,39]. In vitro, MitoVitE reduced NFκB nuclear translocation [40] and in vivo dampened cytokine responses and improved sepsis-induced organ dysfunction [33,41]. We also found higher total and phosphorylated STAT3 protein expression in cells treated with LPS/PepG plus tocopherol but not MitoVitE or Trolox. STAT3 is a transcription factor activated in response to cytokines by receptor associated Janus kinases and MAPK [42] which can localize within the mitochondria and is a regulator of the electron transport chain [43]. Suppression of STAT3 by tocopherols and tocotrienols has been reported (reviewed by Jiang in [5]) but the mechanism of changes in NFkBIA and STAT3 protein seen here with tocopherol are unclear. However, gene expression of STAT3 following tocopherol treatment remained similar to LPS/PepG and suggests modulation by tocopherol at the post-translational level.

Analysis of gene expression showed that endothelial cells exposed to either α-tocopherol or Trolox plus LPS/PepG resulted in upregulation of inducible prostaglandin endoperoxide synthase 2 (PTGS2, also known as COX-2). This protein is involved in prostanoid biosynthesis and is a mediator of the inflammatory response. In contrast, this gene was downregulated by MitoVitE compared along with several other crucial components of the TLR signalling cascade which. These include IκBKB, NFκB1, MyD88 and TRAF6 (see Supplementary Table S1 for functional roles). These genes encode for proteins that interact to initiate downstream signalling to NFκB (see Supplementary Figure S2). Downregulation of these genes indicate a clear anti-inflammatory profile for MitoVitE. In addition, STAT1 and STAT3 mRNA were also downregulated by MitoVitE. STAT1 has been shown to be involved in interferon signalling and cell survival. Many of these genes downregulated by MitoVitE have been shown to be upregulated in mononuclear cells from patients with sepsis [20]. It is intriguing that the functional effects of all three antioxidants on LPS/PepG-induced transcriptional activation, cytokine expression and mitochondrial function were broadly similar, yet the effects at the mRNA level were very different. It has been suggested that some of the anti-inflammatory effects of tocopherol are independent of its antioxidant activity [44]. Plasma membrane lipid rafts containing tocopherol can interact with TLRs and may be important in the response to LPS exposure and sepsis [45,46].

5. Conclusions

This was an in vitro study using human endothelial cells under conditions designed to mimic the changes seen in sepsis. As such is one of several tools for translational experimental approaches to investigate potential therapeutic strategies in patients with sepsis. Since the primary site of free radical production in sepsis is inside mitochondria, we expected that antioxidants that acting elsewhere in the cell may be less effective than those acting inside mitochondria. We did not expect to find that tocopherol and Trolox would be as functionally effective as MitoVitE at blunting the inflammatory consequences of key mediators involved in the response to LPS/PepG exposure, despite more marked anti-inflammatory effects of MitoVitE at the gene expression level. These results challenge current thinking and warrant further investigation.

Supplementary Materials: The following are available online at http://www.mdpi.com/2076-3921/9/3/195/s1, Figure S1: Microscopy images. Figure S2: Simplified Toll like receptor signaling pathway, Figure S3: Acid phosphatase activity, Figure S4: Western blot of nuclear NFkB P65 protein expression, Table S1: Gene ID and function.

Author Contributions: Conceptualization, N.R.W. and H.F.G.; methodology, D.A.L. and B.E.M.; formal analysis, B.E.M. and H.F.G.; investigation, B.E.M.; resources, B.E.M, D.A.L. and H.F.G; data curation, B.E.M.; writing—original draft preparation, B.E.M.; writing—review and editing, B.E.M. and H.F.G.; visualization, B.E.M. and H.F.G.; supervision, D.A.L, H.F.G. and N.R.W; project administration, B.E.M and D.A.L.; funding acquisition, D.A.L., H.F.G and N.R.W. All authors have read and agreed to the published version of the manuscript.

Funding: This research was funded by The British Journal of Anaesthesia/Royal College of Anaesthetists (PhD studentship to Beverley Minter).

Acknowledgments: We are very grateful to M.P. Murphy, MRC Mitochondrial Biology Unit, Wellcome Trust/MRC Building, Hills Road, Cambridge, UK for the generous gift of MitoVitE used in all the experiments, without which this work would not have been possible.

Conflicts of Interest: The authors declare no conflict of interest. The funders had no role in the design of the study; in the collection, analyses, or interpretation of data; in the writing of the manuscript, or in the decision to publish the results.

References

1. Singer, M.; Deutschman, C.S.; Seymour, C.W.; Shankar-Hari, M.; Annane, D.; Bauer, M.; Bellomo, R.; Bernard, G.R.; Chiche, J.D.; Coopersmith, C.M.; et al. The third international consensus definitions for sepsis and septic shock (Sepsis-3). *JAMA* **2016**, *315*, 801–810. [CrossRef] [PubMed]

2. Ramachandran, G. Gram-positive and gram-negative bacterial toxins in sepsis. *Virulence* **2014**, *5*, 213–218. [CrossRef] [PubMed]

3. Crouser, E.D. Mitochondrial dysfunction in septic shock and multiple organ dysfunction syndrome. *Mitochondrion* **2004**, *4*, 729–741. [CrossRef] [PubMed]

4. Singer, M. The role of mitochondrial dysfunction in sepsis-induced multi-organ failure. *Virulence* **2014**, *5*, 66–72. [CrossRef]

5. Jiang, Q. Natural forms of vitamin E: Metabolism, antioxidant, and anti-inflammatory activities and their role in disease prevention and therapy. *Free Radic. Biol. Med.* **2014**, *2*, 6–90. [CrossRef] [PubMed]

6. Jameson, V.J.; Cochemé, H.M.; Logan, A.; Hanton, L.R.; Smith, R.A.; Murphy, M.P. Synthesis of triphenylphosphonium vitamin E derivatives as mitochondria-targeted antioxidants. *Tetrahedron* **2015**, *71*, 8444–8453. [CrossRef]

7. Wu, T.W.; Hashimoto, N.; Wu, J.; Carey, D.; Li, R.K.; Mickle, D.A.; Weisel, R.D. The cytoprotective effect of Trolox demonstrated with three types of human cells. *Biochem. Cell Biol.* **1990**, *68*, 1189–1194. [CrossRef]

8. Guo, C.; He, Z.; Wen, L.; Zhu, L.; Lu, Y.; Deng, S.; Yang, Y.; Wei, Q.; Yuan, H. Cytoprotective effect of Trolox against oxidative damage and apoptosis in the NRK-52e cells induced by melamine. *Cell Biol. Int.* **2012**, *36*, 183–188. [CrossRef]

9. Hamad, I.; Arda, N.; Pekmez, M.; Karaer, S.; Temizkan, G. Intracellular scavenging activity of Trolox (6-hydroxy-2,5,7,8-tetramethylchromane-2-carboxylic acid) in the fission yeast, Schizosaccharomyces pombe. *J. Nat. Sci. Biol. Med.* **2010**, *1*, 16–21. [CrossRef]

10. Galley, H.F. Bench to bedside review: Targeting antioxidants to mitochondria in sepsis. *Crit. Care* **2010**, *14*, 230. [CrossRef]

11. Victor, V.M.; Espulgues, J.V.; Hernandez-Mijares, A.; Rocha, M. Oxidative stress and mitochondrial dysfunction in sepsis: A potential therapy with mitochondria-targeted antioxidants. *Infect. Disord. Drug Targets* **2009**, *9*, 376–389. [CrossRef] [PubMed]

12. Lowes, D.A.; Galley, H.F. The relative roles of mitochondrial thioredoxin and glutathione in protecting against mitochondrial dysfunction in an endothelial cell model of sepsis. *Biochem. J.* **2011**, *436*, 123–132. [CrossRef] [PubMed]

13. Lowes, D.A.; Almawash, A.M.; Webster, N.R.; Reid, V.; Galley, H.F. Melatonin and structurally similar compounds have differing effects on inflammation and mitochondrial function in endothelial cells under conditions mimicking sepsis. *Br. J. Anaesth.* **2011**, *107*, 193–201. [CrossRef] [PubMed]

14. McCreath, G.; Scullion, M.M.F.; Lowes, D.A.; Webster, N.R.; Galley, H.F. Pharmacological activation of endogenous protective pathways against oxidative stress under conditions of sepsis. *Br. J. Anaesth.* **2016**, *116*, 131–139. [CrossRef]

15. Yang, T.T.; Sinai, P.; Kain, S.R. An acid phosphatase assay for quantifying the growth of adherent and nonadherent cells. *Anal. Biochem.* **1996**, *241*, 103–108. [CrossRef]

16. Bradford, M.M. A rapid and sensitive method for the quantitation of microgram quantities of protein utilizing the principle of protein-dye binding. *Anal. Biochem.* **1976**, *72*, 248–254. [CrossRef]

17. Pagé, B.; Pagé, M.; Noël, L.C. A new fluorometric assay for cytotoxicity measurements in vitro. *Int. J. Oncol.* **2003**, *3*, 473–476. [CrossRef]

18. Zhang, C.; Jiang, H.; Wang, P.; Liu, H.; Sun, X. Transcription factor NF-kappa B represses ANT1 transcription and leads to mitochondrial dysfunctions. *Sci. Rep.* **2017**, *7*, 44708. [CrossRef]

19. Asehnoune, K.; Strassheim, D.; Mitra, S.; Kim, J.Y.; Abraham, E. Involvement of reactive oxygen species in Toll-like receptor 4-dependent activation of NF-kappB. *J. Immunol.* **2004**, *172*, 2522–2529. [CrossRef]

20. Severino, P.; Silva, E.; Baggio-Zappia, G.L.; Brunialti, M.K.; Nucci, L.A.; Rigato, O., Jr.; da Silva, I.D.; Machado, F.R.; Salomao, R. Patterns of gene expression in peripheral blood mononuclear cells and outcomes from patients with sepsis secondary to community acquired pneumonia. *PLoS ONE* **2014**, *9*, e91886. [CrossRef]

21. Talwar, S.; Munson, P.J.; Barb, J.; Fiuza, C.; Cintron, A.P.; Logun, C.; Tropea, M.; Khan, S.; Reda, D.; Shelhamer, J.H.; et al. Gene expression profiles of peripheral blood leukocytes after endotoxin challenge in humans. *Physiol. Genom.* **2006**, *25*, 203–215. [CrossRef] [PubMed]

22. Mertens, K.; Lowes, D.A.; Webster, N.R.; Tahib, J.; Hall, L.; Davies, M.; Beattie, J.H.; Galley, H.F. Low zinc and selenium levels in sepsis are associated with oxidative damage and inflammation. *Br. J. Anaesth.* **2015**, *114*, 990–999. [CrossRef] [PubMed]

23. Weber, S.U.; Lehmann, L.E.; Schewe, J.C.; Thiele, J.T.; Schröder, S.; Book, M.; Hoeft, A.; Stüber, F. Low serum alpha-tocopherol and selenium are associated with accelerated apoptosis in severe sepsis. *Biofactors* **2008**, *33*, 107–119. [CrossRef] [PubMed]

24. Lorente, L.; Martín, M.M.; Abreu-González, P.; Domínguez-Rodriguez, A.; Labarta, L.; Díaz, C.; Solé-Violán, J.; Ferreres, J.; Cabrera, J.; Igeño, J.C.; et al. Sustained high serum malondialdehyde levels are associated with severity and mortality in septic patients. *Crit. Care* **2013**, *17*, 290. [CrossRef]

25. Daga, M.K.; Khan, N.A.; Singh, H.; Chhoda, A.; Mattoo, S.; Gupta, B.K. Markers of oxidative stress and clinical outcome in critically ill septic patients: A preliminary study from North India. *J. Clin. Diagn. Res.* **2016**, *10*, OC35–OC38. [CrossRef]

26. Brigelius-Flohé, R.; Kelly, F.J.; Salonen, J.; Neuzil, J.; Zingg, J.M.; Azzi, S.A. The European perspective on vitamin E: Current knowledge and future research. *Am. J. Clin. Nutr.* **2002**, *76*, 703–716. [CrossRef]

27. Brigelius-Flohé, R. Vitamin E: The shrew waiting to be tamed. *Free Radic. Biol. Med.* **2009**, *46*, 543–554. [CrossRef]

28. Godbout, J.P.; Berg, B.M.; Kelley, K.W.; Johnson, R.W. α-Tocopherol reduces lipopolysaccharide-induced peroxide radical formation and interleukin-6 secretion in primary murine microglia and in brain. *J. Neuroimmunol.* **2004**, *149*, 101–109. [CrossRef]

29. Ng, L.T.; Ko, H.J. Comparative effects of tocotrienol-rich fraction, α-tocopherol and α-tocopheryl acetate on inflammatory mediators and nuclear factor kappa B expression in mouse peritoneal macrophages. *Food Chem.* **2012**, *134*, 920–925. [CrossRef]

30. Smith, R.A.J.; Porteous, C.M.; Coulter, C.V.; Murphy, M.P. Selective targeting of an antioxidant to mitochondria. *Eur. J. Biochem.* **1999**, *263*, 709–716. [CrossRef]

31. McCormick, B.; Lowes, D.A.; Colvin, L.; Torsney, C.; Galley, H.F. Mitochondria targeted vitamin E protects against paclitaxel-induced damage in dorsal root ganglion cells and neuropathic pain in vivo. *Br. J. Anaesth.* **2016**, *117*, 659–666. [CrossRef] [PubMed]

32. Lowes, D.A.; Thottakam, B.M.; Webster, N.R.; Murphy, M.P.; Galley, H.F. The mitochondria-targeted antioxidant MitoQ protects against organ damage in a lipopolysaccharide-peptidoglycan model of sepsis. *Free Radic. Biol. Med.* **2008**, *45*, 1559–1565. [CrossRef] [PubMed]

33. Lowes, D.A.; Webster, N.R.; Murphy, M.P.; Galley, H.F. Antioxidants that protect mitochondria reduce interleukin-6 and oxidative stress, improve mitochondrial function, and reduce biochemical markers of organ dysfunction in a rat model of acute sepsis. *Br. J. Anaesth.* **2013**, *110*, 472–480. [CrossRef] [PubMed]

34. Mailloux, R.J.; Harper, M. Uncoupling proteins and the control of mitochondrial reactive oxygen species production. *Free Radic. Biol. Med.* **2011**, *51*, 1106–1115. [CrossRef]

35. Xu, D.; Farmer, A.; Collett, G.; Grishin, N.V.; Chook, Y.M. Sequence and structural analyses of nuclear export signals in the NESdb database. *Mol. Biol. Cell* **2012**, *23*, 3677–3693. [CrossRef]

36. Kashyap, T.; Argueta, C.; Aboukameel, A.; Unger, T.J.; Klebanov, B.; Mohammad, R.M.; Muqbil, I.; Azmi, A.S.; Drolen, C.; Senapedis, W.; et al. Selinexor, a Selective Inhibitor of Nuclear Export (SINE) compound, acts through NF-κB deactivation and combines with proteasome inhibitors to synergistically induce tumor cell death. *Oncotarget* **2016**, *7*, 78883–78895. [CrossRef]

37. Patra, M.C.; Choi, S. Recent progress in the molecular recognition and therapeutic Importance of interleukin-1 receptor-associated kinase 4. *Molecules* **2016**, *21*, 1529. [CrossRef]

38. Godbout, J.P.; Berg, B.M.; Krzyszton, C.; Johnson, R.W. Alpha-tocopherol attenuates NFkappaB activation and pro-inflammatory cytokine production in brain and improves recovery from lipopolysaccharide-induced sickness behavior. *J. Neuroimmunol.* **2005**, *169*, 97–105. [CrossRef]

39. Kim, B.H.; Lee, K.H.; Chung, E.Y.; Chang, Y.S.; Lee, H.; Lee, C.K.; Min, K.R.; Kim, Y. Inhibitory effect of chroman carboxamide on interleukin-6 expression in response to lipopolysaccharide by preventing nuclear factor-kappaB activation in macrophages. *Eur. J. Pharmacol.* **2006**, *543*, 158–165. [CrossRef]

40. Hughes, G.; Murphy, M.P.; Ledgerwood, E.C. Mitochondrial reactive oxygen species regulate the temporal activation of nuclear factor kappaB to modulate tumour necrosis factor-induced apoptosis: Evidence from mitochondria-targeted antioxidants. *Biochem. J.* **2005**, *389*, 83–89. [CrossRef]

41. Zang, Q.S.; Sadek, H.; Maass, D.L.; Martinez, B.; Ma, L.; Kilgore, J.A.; Williams, N.S.; Frantz, D.E.; Wigginton, J.G.; Nwariaku, F.E.; et al. Specific inhibition of mitochondrial oxidative stress suppresses inflammation and improves cardiac function in a rat pneumonia-related sepsis model. *Am. J. Physiol. Heart Circ. Physiol.* **2012**, *302*, H1847–H1859. [CrossRef] [PubMed]

42. Levy, D.E.; Lee, C.-K. What does Stat3 do? *J. Clin. Investig.* **2002**, *109*, 1143–1148. [CrossRef] [PubMed]

43. Meier, J.A.; Hyun, M.; Cantwell, M.; Raza, A.; Mertens, C.; Raje, V.; Sisler, J.; Tracy, E.; Torres-Odio, S.; Gispert, S.; et al. Stress-induced dynamic regulation of mitochondrial STAT3 and its association with cyclophilin D reduce mitochondrial ROS production. *Sci. Signal.* **2017**, *10*, 472. [CrossRef] [PubMed]

44. Zingg, J.M.; Azzi, A. Non-antioxidant activities of vitamin E. *Curr. Med. Chem.* **2004**, *11*, 1113–1133. [CrossRef]

45. Ruysschaert, J.M.; Lonez, C. Role of lipid microdomains in TLR-mediated signalling. *Biochim. Biophys. Acta* **2015**, *1848*, 1860–1867. [CrossRef]

46. Ciesielska, A.; Kwiatkowska, K. Modification of pro-inflammatory signaling by dietary components: The plasma membrane as a target. *BioEssays* **2015**, *37*, 789–801. [CrossRef]

 *antioxidants*

Article

The Role of Sirtuin 3 in Radiation-Induced Long-Term Persistent Liver Injury

Francesca V. LoBianco [1,2], Kimberly J. Krager [1,2], Gwendolyn S. Carter [1,2], Sinthia Alam [1], Youzhong Yuan [3], Elise G. Lavoie [4], Jonathan A. Dranoff [4] and Nukhet Aykin-Burns [1,2,*]

[1] Division of Radiation Health, University of Arkansas for Medical Sciences, 4301 W. Markham St., Little Rock, AR 72205, USA; FVLobianco@uams.edu (F.V.L.); KJKrager@uams.edu (K.J.K.); GSCarter@uams.edu (G.S.C.); SAlam2@uams.edu (S.A.)

[2] Department of Pharmaceutical Sciences, University of Arkansas for Medical Sciences, 4301 W. Markham St., Little Rock, AR 72205, USA

[3] Department of Pathology, University of Arkansas for Medical Sciences, 4301 W. Markham St., Little Rock, AR 72205, USA; YYuan@uams.edu

[4] Department of Gastroenterology and Hepatology, University of Arkansas for Medical Sciences, 4301 W. Markham St., Little Rock, AR 72205, USA; EGLavoie@uams.edu (E.G.L.); JADranoff@uams.edu (J.A.D.)

* Correspondence: NAykinBurns@uams.edu

Received: 11 April 2020; Accepted: 30 April 2020; Published: 11 May 2020

Abstract: In patients with abdominal region cancers, ionizing radiation (IR)-induced long-term liver injury is a major limiting factor in the use of radiotherapy. Previously, the major mitochondrial deacetylase, sirtuin 3 (SIRT3), has been implicated to play an important role in the development of acute liver injury after total body irradiation but no studies to date have examined the role of SIRT3 in liver's chronic response to radiation. In the current study, ten-month-old $Sirt3^{-/-}$ and $Sirt3^{+/+}$ male mice received 24 Gy radiation targeted to liver. Six months after exposure, irradiated $Sirt3^{-/-}$ mice livers demonstrated histopathological elevations in inflammatory infiltration, the loss of mature bile ducts and higher DNA damage (TUNEL) as well as protein oxidation (3-nitrotyrosine). In addition, increased expression of inflammatory chemokines (IL-6, IL-1β, TGF-β) and fibrotic factors (Procollagen 1, α-SMA) were also measured in $Sirt3^{-/-}$ mice following 24 Gy IR. The alterations measured in enzymatic activities of catalase, glutathione peroxidase, and glutathione reductase in the livers of irradiated $Sirt3^{-/-}$ mice also implied that hydrogen peroxide and hydroperoxide sensitive signaling cascades in the absence of SIRT3 might contribute to the IR-induced long-term liver injury.

Keywords: ionizing radiation; liver; sirtuin 3; hydroperoxide; inflammation

---

1. Introduction

The lifetime probability of males developing an invasive cancer is 42% and for females, 38%. Approximately 50% of all cancer cases benefit from radiation therapy [1]. More than 333,000 new cases of abdominal-region cancers are expected in the U.S. alone in 2020 [1], and a serious limitation to radiation therapy for these cancers is the possibility of irradiating the liver [2–4]. Ionizing radiation (IR)-induced liver toxicity remains a major challenge, especially for patients with underlying liver dysfunction, which is common with hepatic malignancies [3–9]. Clinical studies showed that IR-induced liver pathologies could be observed within 3 months of IR, even with IR doses as low as 30 Gy [10]. Previous clinical studies have shown patients can develop radiation damage within three months post-irradiation, which is caused by a veno-occlusive process. Fibrin accumulation of central veins and sinusoids followed by collagen deposition increases, resulting in increased portal pressure. These clinical manifestations have been termed "radiation-induced liver disease (RILD)", which could progress to late-term radiation-induced fibrotic injury and eventual liver failure [4,11]. Received dose and volume of the liver that has been irradiated are two major parameters correlated with increased risk of developing RILD as well as its severity [12–14].

Stereotactic body radiation therapy (SBRT) could deliver high doses of IR to smaller volumes and provide survival benefits to those with liver and hepatobiliary cancers. However, conventional or SBRT IR doses delivered to the abdominal area, especially in patients with underlying liver dysfunction, are still constrained by concerns about RILD and hepatobiliary track toxicities [12–16]. The confounding factors that contribute to these pathologies are poorly understood, so determining the foremost molecular mechanisms is crucial for developing strategies to prevent and/or mitigate the side effects of radiotherapy [3,17,18].

Ionizing radiation can lead to oxidative events and biological activity changes in cells by directly interacting with target molecules, such as DNA, or through radiolysis of water. The biological activity alterations throughout the liver following radiation exposure are paramount because of the high mitochondrial content and oxidative metabolism [19–21]. Oxidative injury can arise from days to months after the initial exposure from the indirect generation of reactive oxygen species (ROS) and reactive nitrogen species (RNS), which are associated with chronic inflammatory responses [22–24]. Increased DNA damage, apoptosis and inflammation have been shown to play significant roles in RILD [4,13,25]. Several studies have also reported IR-induced mitochondrial dysfunction and oxidative stress in liver tissue that are thought to contribute to RILD [26–28]. Tao et al. (2010) and Coleman et al. (2014) demonstrated a causal link between ROS-mediated acute liver injury and decreased activity of manganese superoxide dismutase (MnSOD) in mice lacking sirtuin 3 (SIRT3) [29,30].

SIRT3 is the major mitochondrial nicotinamide adenine dinucleotide (NAD+)-dependent deacetylase. Mitochondrial deacetylation targets of SIRT3, such as peroxisome proliferator-activated receptor-gamma co-activator-1α, mitochondrial electron transport chain Complex I subunit NDUFA9, or Complex II subunit A, mitochondrial ribosomal protein L10 or TCA cycle enzyme isocitrate dehydrogenase 2, have been demonstrated to control a wide range of biological processes including but not limited to gene expression, metabolism, cancer and aging [30–37]. SIRT3 also plays a significant function as a stress response protein in regulation of ROS levels via posttranslational modification of antioxidant enzymes such as MnSOD and increasing their enzymatic activity. Therefore, it is logical to hypothesize that SIRT3, which is a pivotal mitochondrial stress-response protein, could also play a critical role in the long-term progression of liver toxicity seen in radiotherapy patients [38–46].

In an effort to best represent the age demographic most impacted in humans for developing liver and intrahepatic bile duct cancers, in the current study we used 10-month-old male mice for our liver only irradiation design. We followed the irradiated homozygous wild-type or Sirt3$^{-/-}$ mice for an additional 6-month period for the development of long-term liver injury. Our results indicated significant increases in expression of inflammatory and profibrotic markers as well as decreased numbers of bile ducts in the irradiated areas of Sirt3$^{-/-}$ livers compared to their sham treated controls. These mice also have more inflammatory cell infiltration in the portal and perivenular space. The DNA double strand breaks indicated by TUNEL staining and levels of protein oxidation in liver tissue were still elevated at the end of 6 months in Sirt3$^{-/-}$ irradiated mice. Interestingly, there were no significant changes in MnSOD activity in sham versus irradiated groups in either Sirt3$^{-/-}$ or Sirt3$^{+/+}$ mice. However, activity peroxide-removing enzymes were significantly altered only in irradiated Sirt3$^{-/-}$ livers, suggesting a shift in the types of reactive species contributing to the progression of long-term liver pathology following radiation exposure.

2. Materials and Methods

2.1. Image Guided Irradiation of Liver Tissue of Sirt3$^{+/+}$ and Sirt3$^{-/-}$ Male Mice

Littermates of Sirt3$^{+/+}$ and Sirt3$^{-/-}$ male mice on B6/Sv129 background were housed in a temperature and humidity controlled environment in filter top cages with *ad libitum* access to food and water at the University of Arkansas for Medical Sciences Animal Care facility until 10 months of age. Irradiation of the liver tissue was performed using the Small Animal Radiation Research Platform (SARRP, Xstrahl Inc., Suwanee, GA, USA) (Figure 1A). The mice were anaesthetized with 1% isoflurane inhalation for the duration of the radiation treatment. Each mouse was place supine on the horizontal mouse bed in the SARRP. A cone beam computed tomography (CBCT) image of each mouse was obtained to provide image guided radiation targeted to the liver at 60 kVp and 0.8 mA and reconstruction using 720 projections. From the image, precision targeting of the upper right lobe was determined. The liver was irradiated with 2 fractions of 12 Gy from a 90° and 0° gantry angle with a 7 mm and 5 mm tissue depth respectively. The liver treatments were performed utilizing a 0.5 mm copper filter with a 5 × 5 mm collimator using 220 kVp and 13 mA (Figure 1B). After 6 months, mice were euthanized and blood was collected through cardiac puncture. Additionally, irradiated liver tissue sections were flash-frozen in liquid nitrogen and stored at −80 °C or were fixed in formalin and paraffin embedded for further analysis. All radiation sham mice were anaesthetized and placed in the SARRP for an equivalent time as the irradiated treated mice. All animal protocols and procedures used in this study were approved (AUP# 3750) by the Institutional Animal Care and Use Committees of the University of Arkansas for Medical Sciences.

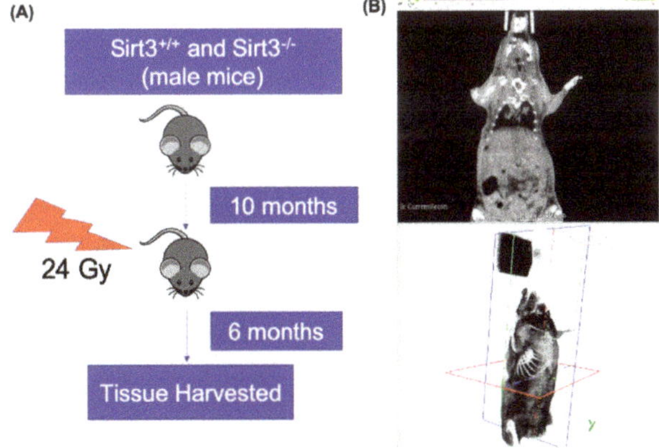

Figure 1. (**A**) Experimental timeline for the irradiation and tissue harvest from Sirt3$^{+/+}$ and Sirt3$^{-/-}$ male mice. (**B**) Image guided irradiation of the liver using Small Animal Radiation Research Platform (SARRP).

2.2. Immunohistochemistry and Histopathology Analysis

Sections were deparaffinized and rehydrated using decreasing concentrations of ethanol. One set of slides was stained with hematoxylin and eosin. These slides were then scored by a clinical pathologist to determine the level of liver injury in a double-blinded manner. Differences to the sham mice groups, when present, were noted in several categories including possible micro/macrovesicular steatosis, lymphoplasmacytic inflammation (i.e., portal, perivenular, and lobular regions), necrosis, fibrosis, angiectasis, and the presence of any regeneration nodules. Each liver section in each group was given a verbal score of "none", "mild" and "moderate" that was translated into the table as "−, +, and ++"; the table also includes how many animals out of the group presented the liver injury marker in the group.

Another set of slides was stained for DNA damage using a fluorescent Terminal deoxynucleotidyl transferase (TdT) dUTP Nick-End Labeling (TUNEL) assay. Analysis of DNA damage was determined by a double-blinded imaging and scoring of 10 random 40X fields per section for positive (green) compared to total hepatocyte nuclei (blue).

For the immunohistochemical staining for 3-nitrotyrosine and Cytokeratin-19, the tissue slides were deparaffinized and endogenous peroxidase was quenched followed by incubation in Dako protein-block to block nonspecific binding. Anti-3-nitrotyrosine rabbit polyclonal antibody (Millipore; #06284, 1:1000) was applied for 1 h in Dako antibody diluent buffer. Rat Anti-Cytokeratin-19 antibody (DSHB Hybridoma Product TROMA-III; #ab2133570, 1:300) was incubated for two hours in Dako diluent. All sets were then incubated in Vector Biotinylated Goat Anti–Rabbit–1:400 prepared in TBS-T for 30 min. Then slides were incubated in Vector ABC Elite for 30 min. Slides were developed with Dako diaminobenzidine (DAB). Slides were counterstained with hematoxylin and mounted. The negative control slides followed the same protocol but did not use the primary antibody. 3-Nitrotyrosine immunohistochemical staining was quantified by counting positive cells near similar sized central veins (cytoplasmic or nuclear staining) per 400× field with the following scoring system: 0 (0 positive cells), 1 (1–20 positive cells), 2 (21–30 positive cells), 3 (31–40 positive cells) and 4 (>41 positive cells). A total of $15 \times 400 \times$ fields were scored, and means of these scores were calculated. Bile ducts were scored and counted in Cytokeratin-19 stained slides by examining 8-10 regions containing at least one portal area per liver section.

2.3. Real Time Quantitative Reverse Transcription PCR (qRT-PCR)

Total RNA was extracted from flash-frozen liver using the DNA/RNA/Protein extraction kit (IBI Scientific IB47702, Peosta, IA, USA) according to the manufacturer's protocol. After extraction, the total RNA concentration was determined. cDNA was synthesized from 1000 ng RNA using the High-Capacity cDNA Reverse Transcription kit (Thermo Fisher Scientific, Waltham, MA, USA) as we previously described [47].

Fibrosis and inflammation marker levels of gene transcription were determined by single gene quantitative reverse transcription polymerase chain reaction (qRT-PCR) using Fast TaqMan Gene Expression Assays (Life Technologies, Carlsbad, CA, USA) and the Applied Biosystems 7500 Real Time PCR system. The genes of interest included: TGF-β, α-SMA, Procoll1, IL-6, and IL-1β; see Supplemental Table S1. The relative fold change of mRNA for each gene of interest was determined using the Comparative CT ($2^{-\Delta\Delta Ct}$) method. These results were normalized to house-keeping gene GAPDH values [47].

2.4. Antioxidant Enzyme Activity Measurements

Whole flash-frozen liver tissue was homogenized in 50 mM potassium phosphate buffer containing 1.34 mM diethylenetriaminepentaacetic acid (pH: 7.4) on ice. Protein concentrations were determined using a Lowry's Protein Assay as previously described in [48].

Catalase activity for liver homogenates was determined using a 1:10 dilution from the homogenate. Absorbance was read using the spectrophotometer at 240 nm for both the reference blank and the sample. 50 µL of the diluted sample was placed in a total of 4 mL of 50 mM (pH 7.0) phosphate buffer. Then 2 mL of this total solution was placed into a sample cuvette and a reference blank cuvette. The addition of 1 mL of 30 mM H_2O_2 into the sample cuvette began the reaction and absorbance change vs. time was measured for 2 min. The absorbance of the reference blank was subtracted from the sample cuvette to determine the level of activity per mg of protein as previously described [49].

Glutathione Peroxidase activity was calculated using an established protocol [50]. A 1:50 dilution of liver homogenate was added into a mixture of "working buffer" (containing 1.33 mM reduced glutathione (GSH), 1.33 U/mL glutathione reductase (GR) and 55.6 mM potassium phosphate buffer (pH: 7)) and 4 mM NADPH. This is incubated for 5 min at 30 °C, then the reaction is initiated when 15 mM cumene hydroperoxide is added. Absorption changes were measured at 340 nm for 3 min that was then compared to the blank absorbance for each sample cuvette. The activity is calculated in units per mg protein.

Glutathione Reductase activity was measured in a similar fashion to the Glutathione Peroxidase protocol. To initiate the reaction, a 1:10 dilution of liver homogenate was added to ddH$_2$O, PB/EDTA

(100 mM potassium phosphate buffer/3.4 mM EDTA), 30 mM oxidized glutathione (GSSG), 0.8 mM NADPH, and 1% bovine serum albumin (BSA). The decrease in absorbance at 340 nm was measured for 3 min and subtracted from the blank absorbance. The absorbance was used to calculate activity in units per mg protein as we previously described [51].

MnSOD activity was measured based on the protocol established by Spitz and Oberley [52]. It is determined by recording the rate of reduction of nitroblue tetrazolium (NBT) from the superoxide generated by xanthine oxidase. The superoxide dismutase (SOD) present in the samples will scavenge the superoxide generated. This is a competitive inhibition of the reduction of NBT. A unit of SOD is considered the amount of SOD required to inhibit 50% of the NBT reduction. This is determined using varying concentrations of sample to find the percent inhibition curve and determine the concentration at 50% inhibition. The MnSOD activity is determined by adding 0.33 M sodium cyanide buffer to inhibit the CuZnSOD enzyme activity.

2.5. Bilirubin Assay

Plasma bilirubin levels were measured following Bilirubin Assay Kit protocol (Sigma-Aldrich, St. Louis, MO, USA) using 50 µL of blood plasma in each well of 96 well plates to determine the total bilirubin, direct bilirubin and blank absorbance. Absorbance was measured at 530 nm. The indirect bilirubin levels were determined by subtracting the total bilirubin numbers from the direct bilirubin, which is also known as conjugated bilirubin.

2.6. Statistical Analysis

Statistical analysis was performed using GraphPad Prism 8.0 (GraphPad Software, San Diego, CA, USA). Data are expressed as mean ± 1SD, unless otherwise specified. One-way ANOVA analysis with Tukey's post-analysis was used to study the differences among the study group's means. Significance was determined at $p < 0.05$ and the 95% confidence interval. Statistical significance is expressed as * $p < 0.05$, ** $p < 0.01$, and *** $p < 0.005$.

3. Results

3.1. Exposure to 24 Gy IR Increased the Expression of Inflammatory Markers as Well as Lymphoplasmacytic Inflammation in the Livers of Sirt3$^{-/-}$ Mice

The histology of the liver sections was scored in a double-blinded manner to determine the level of injury after irradiation. The scoring demonstrated considerably increased inflammatory cells in the Sirt3$^{-/-}$ mice 6 months after 24 Gy IR, and while there were no severe cases yet, several animals displayed moderate inflammation in the portal and perivenular space of the murine livers (Figure 2).

We also measured the gene expression of inflammatory chemokines, IL-6, IL-1β, and TGF-β, in the livers of sham or 24 Gy treated mice from both Sirt3$^{+/+}$ and Sirt3$^{-/-}$ mice. Previous studies have demonstrated increased expression levels of these chemokines six months after liver-targeted irradiation. Additionally, IL-1β is an important chemokine in the formation of the liver's inflammasome that leads to increased liver inflammation and tissue damage. These increases in inflammatory chemokines and profibrotic factors could contribute to the removal of necrotic tissue after irradiation and stimulate extracellular matrix deposits that assist in regeneration. However, the chronic release of IL-6, IL-1β, and TGF-β would result in increased liver damage and fibrosis. Expression of IL-6, IL-1β, and TGF-β were measured in triplicates for each animal in a group. The values were averaged and normalized to GAPDH as the housekeeping gene. Fold change was determined by comparing each group to its own sham. Only Sirt3$^{-/-}$ mice, which received 24 Gy IR, showed a significant increased expression of these chemokines (Figure 3).

GROUPS	Lymphoplasmacytic inflammation					
	Portal		Perivenular		Lobular	
Sirt3 $^{-/-}$ Sham	+	(1/4)	−	(0/4)	+	(1/4)
Sirt3 $^{+/+}$ 24 Gy	+	(1/6)	+	(1/6)	−	(0/6)
Sirt3 $^{-/-}$ 24 Gy	++	(3/6)	++	(4/6)	+	(2/6)

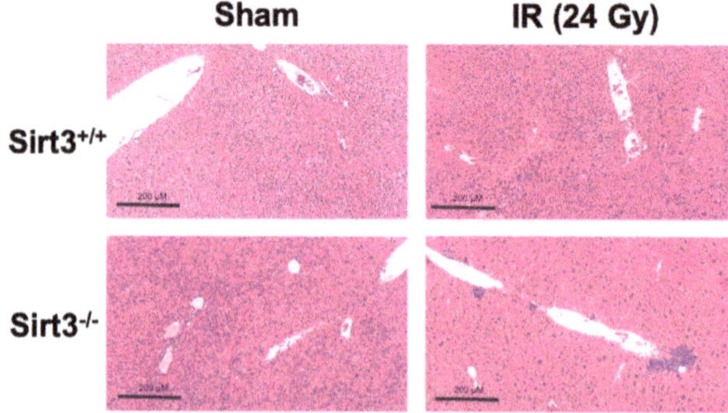

Figure 2. Increased numbers of inflammatory cells were seen in irradiated Sirt3$^{-/-}$ mice. H&E scoring: − (none), + (mild/focal), and ++ (moderate). All sections were normalized to the Sham Sirt3$^{+/+}$ group and presented as number of animals with marker per animals in each group (n = 4–6). Sham groups underwent the same procedures without receiving 24 Gy IR (irradiation).

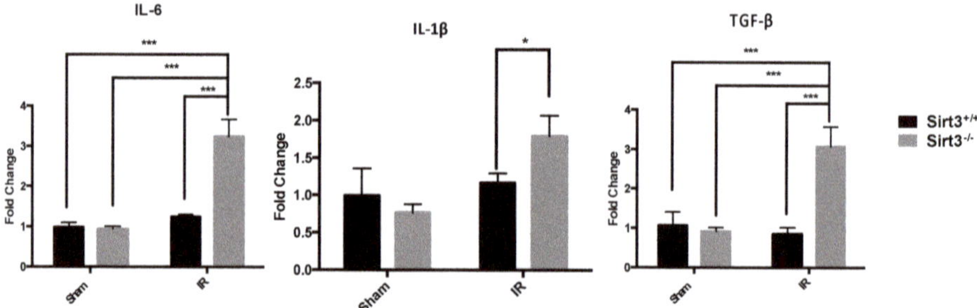

Figure 3. Expression of inflammatory markers determined by quantitative real-time PCR. IL6 and TGFβ were significantly increased in Sirt3$^{-/-}$ mice after irradiation compared to sham irradiated Sirt3$^{+/+}$ or Sirt3$^{-/-}$ as well as Sirt3$^{+/+}$ irradiated groups. IL1β was significantly increased in irradiated Sirt3$^{-/-}$ mice compared to irradiated Sirt3$^{+/+}$ livers (n = 4–6; * $p < 0.05$, *** $p < 0.001$).

3.2. The Number of Bile Ducts Was Decreased in Irradiated Livers of Sirt3$^{-/-}$ Mice

Liver tissue sections were stained with cytokeratin-19 to visualize and score the bile duct presence. Double-blind scoring showed a significant reduction of number of bile ducts per portal area in the Sirt3$^{-/-}$ irradiated mice (Figure 4), which could contribute to the development of cholestasis and eventually chronic fibrosis. Interestingly there were no changes in the plasma bilirubin levels (conjugated or unconjugated; Supplemental Figure S1) probably due to the small irradiation field that was used in the study.

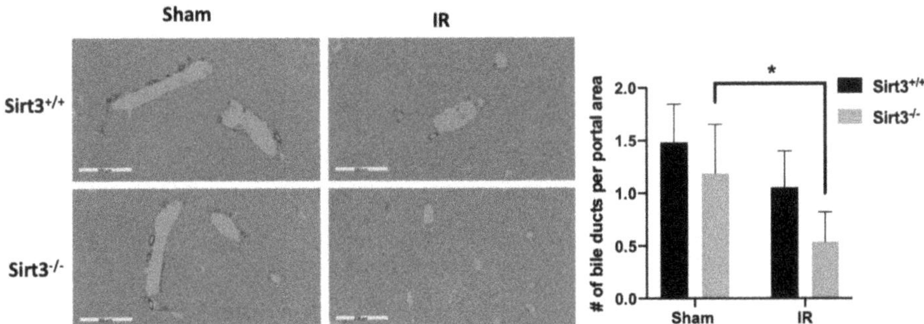

Figure 4. Sirt3$^{-/-}$ irradiated mice showed decreased number of bile ducts. Liver sections were stained and counted with anti-cytokeratin 19 to determine the presence of bile ducts. Scale bars: 200 µm. (8–10 fields were scored per mouse, $n = 6$; * $p < 0.01$).

3.3. Exposure to 24 Gy IR Increased the Expression of Profibrotic Markers in the Livers of Sirt3$^{-/-}$ Mice but Did Not Cause Fibrosis at 6 Months after IR Exposure

Because TGF-β is a cytokine growth factor that signals the release of additional inflammatory chemokines and fibrotic growth factors, next we measured gene expression of procollagen-1 and α-SMA in the sham and 24 Gy irradiated livers of both Sirt3$^{+/+}$ and Sirt3$^{-/-}$ mice. Expression of these two profibrotic marker proteins was measured in triplicates for each animal in a group. The values were averaged and normalized to GAPDH as the housekeeping gene. Only Sirt3$^{-/-}$ mice, which received 24 Gy IR, showed a significant increased expression of procollagen-1 and α-SMA (Figure 5). However, this increase did not translate into a bridging fibrosis (data not shown), which may develop in severe RILD patients.

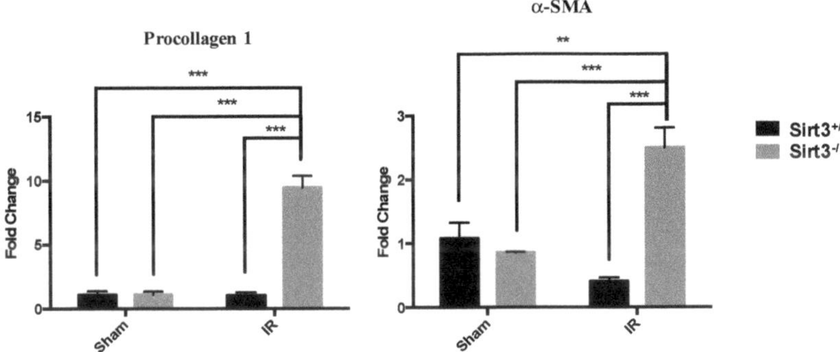

Figure 5. Expression of profibrotic markers were increased in all Sirt3$^{-/-}$ mice after irradiation compared to sham irradiated Sirt3$^{+/+}$ or Sirt3$^{-/-}$ as well as Sirt3$^{+/+}$ irradiated groups ($n = 4$–6; ** $p < 0.01$, *** $p < 0.001$).

3.4. Persistent Significant Increases in DNA Damage and Oxidative Stress in Liver Tissue of Sirt3$^{-/-}$ Mice Following 24 Gy IR

DNA damage is an early event during exposure to IR. However, chronic inflammation and oxidative/nitrosative stress have been suggested to contribute long-term detrimental effects of radiation exposures by creating a feed-forward mechanism for persistent injuries to macromolecules, such as DNA. Since we have seen increased expression of inflammatory markers 6 months post IR in liver tissue of Sirt3$^{-/-}$ mice, we determined the levels of protein oxidation using 3-nitrotyrosine immunohistochemistry. Blind scoring of liver sections were performed at 15 randomly selected central veins, and then given a score of 0 (0 positive cells), 1 (1–20 positive cells), 2 (21–30 positive cells), 3 (31–40 positive cells), or 4 (41 positive cells or greater). Our results showed that 24 Gy IR increased levels of protein oxidation in mouse livers, which were persistent months after the exposure (Figure 6).

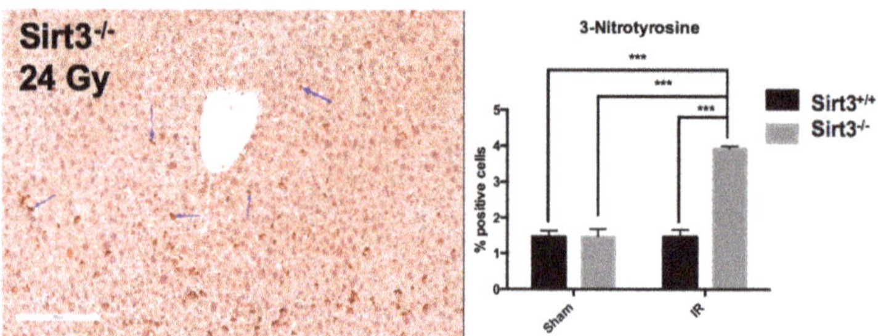

Figure 6. Protein oxidation, determined by 3-nitrotyrosine staining (indicated by the arrows), was increased in irradiated Sirt3$^{-/-}$ mice compared to sham irradiated Sirt3$^{+/+}$ or Sirt3$^{-/-}$ as well as Sirt3$^{+/+}$ irradiated groups. Scale bar: 200 μm (15 fields scored per mouse, $n = 4$–6; *** $p < 0.001$).

Therefore, it was not completely surprising when Sirt3$^{-/-}$ livers in irradiated group also exhibited a significant increase of TUNEL positive cells compared to sham irradiated mice, suggesting DNA fragmentation and possible apoptosis still exists in liver cells lacking SIRT3 as a stress response protein against IR-induced oxidative injury (Figure 7).

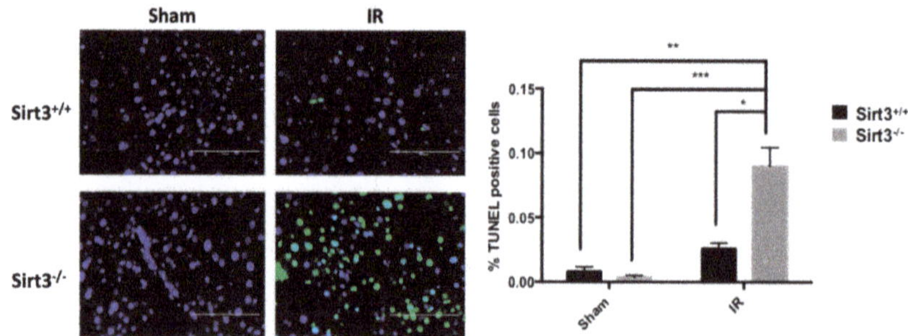

Figure 7. The number of TUNEL positive cells were significantly increased in Sirt3$^{-/-}$ mice after irradiation compared to sham irradiated Sirt3$^{+/+}$ or Sirt3$^{-/-}$ as well as Sirt3$^{+/+}$ irradiated groups. Scale bar: 100 μm ($n = 4$–6; * $p < 0.05$ ** $p < 0.01$, *** $p < 0.001$).

3.5. Mice Lacking SIRT3 Demonstrated Altered Activity of Antioxidant Enzymes, Which Are Responsible for Peroxide Removal in Liver Tissue Following 24 Gy IR

As a major deacetylation target of SIRT3, increased activity MnSOD has been shown to respond to exogenous cellular stressors including acute IR injury in liver tissue [29] and loss of SIRT3 has been associated with increased levels of superoxide [29,30,42]. Therefore, we first examined whether MnSOD activity has not been altered in irradiated Sirt3$^{-/-}$ mice but it was increased in Sirt3$^{+/+}$ mice, via deacetlyation of critical lysine residues at its active site. No significant changes were noted in liver MnSOD activity between the sham and irradiated groups in either genotype (Figure 8).

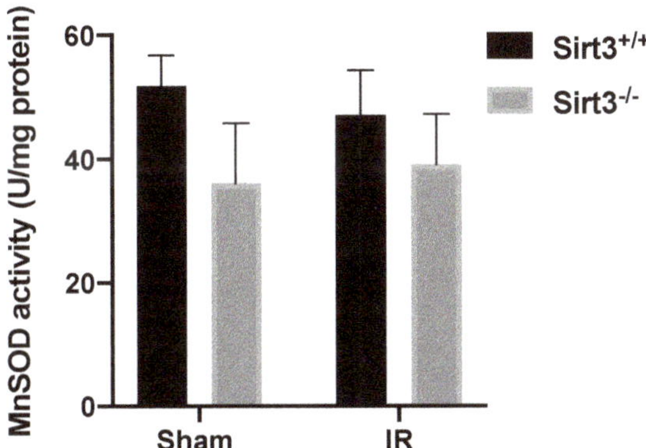

Figure 8. Exposure to 24 Gy liver only irradiation did not significantly alter enzymatic activity of MnSOD in Sirt3$^{+/+}$ or Sirt3$^{-/-}$ mice (n = 4–6).

Interestingly, the activity of antioxidant enzymes CAT and GPx were significantly higher in irradiated Sirt3$^{-/-}$ mice, while GR activity was significantly lower in this group compared to its sham irradiated counterparts (Figure 9).

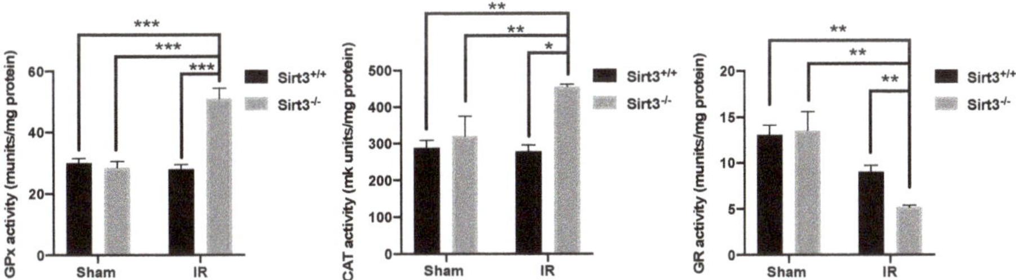

Figure 9. Glutathione Peroxidase (GPx) and catalase (CAT) activity were significantly increased while Glutathione Reductase (GR) enzymatic activity was decreased in irradiated Sirt3$^{-/-}$ mice compared to sham irradiated Sirt3$^{+/+}$ or Sirt3$^{-/-}$ as well as Sirt3$^{+/+}$ irradiated groups (n = 4–6; * $p < 0.05$, ** $p < 0.01$, *** $p < 0.001$).

This result suggested that peroxide mediated oxidative injury is more relevant in radiation-induced long-term liver toxicity.

4. Discussion

Almost half of all men and women have a lifetime probability of developing a new invasive cancer; therefore, radiation therapy and chemotherapy are being used more frequently [1]. The liver, located in the abdominal region, is a large organ that plays a critical role in metabolic homeostasis and detoxification. Because of this metabolic role, the liver contains about 25% of mitochondria in its cytosolic space [20]. This high mitochondrial content and oxygen requirement makes the liver especially susceptible to radiation-induced ROS that can lead to hepatic radiation-induced damage, even with doses as low as 30 Gy [10,19]. The typical clinical presentation of RILD develops 3 months after irradiation and is associated with increased abdominal girth, ascites, hepatomegaly, and veno-occlusive disease that can eventually progress to liver failure [4].

Although clinical observations are established and patient-oriented morphological characteristics of IR-induced pathologies are well described, we have limited understanding of molecular mechanisms that would explain dose-limiting toxicities in patients treated with high-dose IR [3,4,7,13,17,18,53–56].

In addition, no pharmacological interventions have been shown to be consistently efficacious [9,54]. Therefore, understanding the confounding factors and mechanisms involved in IR-induced long-term liver pathologies will particularly benefit the high-risk populations with underlying liver dysfunction. The lack of prevention and therapy post RILD diagnosis exacerbate the need for pre-clinical models that will allow the accurate study of radiation-induced liver damage development [2].

IR induces not only immediate production of free-radical mediated effects but also persistent increases in metabolic production of reactive species, which contribute to the long-term tissue effects of radiation [57]. The damage response cascades (i.e., inflammation, necrosis and fibrosis) are further stimulated from the consistent formation of reactive species in days or even months after initial irradiation exposure [19,24,58]. Similar damage response cascades were seen in our Sirt3$^{-/-}$ model (Figures 2–7) 6-months after 24 Gy radiation was delivered to a small (5×5 mm) field of the liver indicating persistent oxidative injury via the indirect generation of ROS and RNS. Superoxide anion, hydrogen peroxide, hydroxyl radical, and peroxynitrite are a few examples of the radiation-induced reactive species which can cause mitochondrial alterations that lead to oxidative and nitrosative stress signaling and genomic instability; all adding to the long-term injury and dysregulation of normal liver parenchymal function [59,60]. ROS and RNS can be removed through several different antioxidant mechanisms, including SODs, CAT, and the glutathione, thioredoxin, and peroxiredoxin systems, most of which are regulated via transcriptional and/or post-translational modifications [29–34]. SIRT3, a member of NAD+ dependent enzymes family, has been shown to increase the activity of antioxidant enzymes, including MnSOD [30–48,61]. In agreement with these studies, Coleman et al. demonstrated SIRT3's stress response function in an acute radiation-induced liver injury model, in which the authors provided evidence for $O_2^{\bullet-}$ mediated liver injury in the absence of SIRT3, 48 h after 4 Gy total body irradiation exposure [29]. Interestingly, in our current study of IR-induced long-term liver injury, the model suggested that $O_2^{\bullet-}$ may no longer be the major participant in the chronic liver damage, as we did not see any changes in enzymatic activity of MnSOD in the presence or absence of SIRT3 6-months following targeted irradiation (Figure 8).

After examination by a clinical pathologist, liver histology of the irradiated fields exhibited a moderate increase in lymphoplasmacytic inflammation in the perivenular space, implying an increased inflammation that can further signal of tissue necrosis or fibrosis. In an acute response to irradiation or other types of liver injury, increased inflammation is necessary for tissue regeneration. Inflammation facilitates a fibrotic network for regrowth and remodeling of liver tissue as well as vascular network using chemokines like TGF-β, IL-1β, and IL-6. However, continual elevations of inflammatory cells releasing, secreting and recruiting more cells using cytokines like TGF-β, can be associated with the development of chronic liver disease [62]. While IL-1β and IL-6 are initially released to protect the surrounding tissue from further damage in an acute response, its persistent and continued exposure causes recruitment of additional Kupffer cells within the liver that triggers a feed-forward cycle of continuous inflammation, damaging the hepatocytes [63]. In our targeted irradiation model, we found that there was a significant increase in TGF-β, IL-1β, and IL-6 mRNA expression, suggesting a persistent inflammatory response and continuous liver injury progression 6 months after treatment in the absence of SIRT3. This fold increase in the Sirt3$^{-/-}$ mice was significantly higher than in the Sirt3$^{+/+}$ irradiated group, suggesting that the release of IL-6 and IL-1β are no longer a beneficial injury response due to its new sustained nature. The elevation in inflammation cytokines and histochemistry corroborate the idea that the inflammation is chronic in our model and will eventually lead to further liver toxicity, demonstrating that functional SIRT3 is important for liver healing and the balance between inflammation and regeneration even 6 months after irradiation. Adding to this finding, the increased TUNEL staining confirms DNA degradation in the irradiated Sirt3$^{-/-}$ mice, further signifying the development of long-term hepatocyte injury from persistent inflammatory cascades. These results confirm the protective role of SIRT3, which was previously demonstrated in a sepsis model by describing its molecular links to DNA damage, apoptosis and inflammatory responses via

NLRP3 inflammasome upregulation as well as apoptosis-associated speck-like protein in the absence of functional SIRT3 [64].

Sinusoidal obstruction and hepatic fibrosis are classical features of radiation-induced liver injury in patients. Histology of the samples in the current study did not show any veno-occlusive events or an increase in fibrotic collagen staining. However, hepatobiliary tract toxicities including vanishment of bile ducts and elevated alkaline phosphatase are also common developments after radiation exposures, thus we used cytokeratin 19 staining for the visualization of bile ducts. The results demonstrate an interesting finding of possible bile ductopenia in the $Sirt3^{-/-}$ mice 6 months after IR exposure, which was not observed in wild-type irradiated mice. Although to date no direct mechanistic link has been established between SIRT3 function and bile duct injury, literature evidence suggests a strong correlation between increased oxidative stress and increased levels of apoptosis in biliary epithelial cells, which was mediated by glutathione in primary biliary cirrhosis [65]. Furthermore, increased inflammatory cytokine-induced oxidative stress has been shown to decrease the expression of biliary markers through miRNA506 regulation of DNA damage and apoptosis [66]. Corroborating with the findings of these studies, our results also revealed increased levels of 3-nitrotyrosine on irradiated $Sirt3^{-/-}$ livers (Figure 6), which evidenced the existence of sustained oxidative injury in irradiated $Sirt3^{-/-}$ livers. While the irradiated liver sections obtained from both genotypes did not demonstrate an increase in fibrotic collagen staining (data not shown), based on our findings with decreased number of functioning bile ducts in $Sirt3^{-/-}$ mice and elevated mRNA of fibrotic factors (α-SMA and procollagen 1), we believe these results indicate that these mice will eventually develop persistent fibrosis in the irradiated field [67].

The morphological evidence we presented in our current study strongly suggests a noteworthy role SIRT3 plays in the long-term radiation-induced liver toxicity. As an NAD+ dependent deacetylase, which resides primarily in mitochondria, SIRT3 functions to maintain redox homeostasis under stress. A total body irradiation study by Coleman et al., demonstrated that 48 h after exposure, there is an early increase in $O_2^{\bullet-}$ generation in $Sirt3^{-/-}$ mice livers, which mediated the increased acute liver injury in these animals [29]. We initially expected to see a similar response in MnSOD activity in the long-term progression of radiation-induced injury. However, there was no significant change in MnSOD activity between sham and irradiated groups at six months after irradiation (Figure 8). Because MnSOD, and thus $O_2^{\bullet-}$ were not likely to be involved in the IR-induced chronic liver injury at 6 months time point, we shifted our focus to other antioxidant networks that may now be attempting to overcome the oxidative stress from radiation.

As a number of SIRT3 deacetylation targets occur in mitochondria, which could indirectly contribute to the redox homeostasis by providing the necessary coenzyme NADPH (e.g., isocitrate dehydrogenase 2) [31,32], we examined alternate antioxidant enzymes. Our data indicate a possible switch from $O_2^{\bullet-}$ to hydroperoxide metabolism that contributes to the injury endpoints found in this study. In the absence of SIRT3, 6 months after radiation there was a significant decrease in GR. This is an important enzyme that requires NADPH for its enzymatic activity for glutathione recycling. Without SIRT3 there may have been a decrease in availability of the necessary coenzyme NADPH for the continued recycling of glutathione in response to the persistent elevated oxidative stress generated from radiation exposure.

Considering that peroxiredoxin 3, the principal peroxidase responsible for mitochondrial hydrogen peroxide, is also a SIRT3 deacetylation target [68], accumulation of hydroperoxides and hydrogen peroxide in irradiated livers resulted in the significant increases in GPx and CAT activities we measured in $Sirt3^{-/-}$ mice. However, with long-term elevations clearly still seen after 6 months, this antioxidant system also appears to be overwhelmed in the absence of SIRT3, leading to the liver toxicity seen in our study.

5. Conclusions

In conclusion, our results exhibited a vital function of SIRT3, a major mitochondrial deacetylase, in IR-induced long-term liver injury. Loss or decrease in SIRT3 levels could be an underlying factor and contributor to a damage-permissive phenotype in murine liver long after exposure to IR. Although our model did not fully represent the clinical indications of human RILD, we believe this was due to the small volume of irradiation chosen in the design. There were important signs that this murine model may relate to the clinical presentation of human RILD including ductopenia, lymphoplasmacytic inflammation, continued hepatocyte toxicity from DNA damage and elevations in the expression of fibrotic factors.

The data presented in the current study also strongly suggested that $O_2^{\bullet-}$ driven acute liver injury following IR exposure appears to shift towards a peroxide-mediated long-term injury and clearly is no longer limited to the mitochondria. Thus, specific SIRT3 target proteins and the reactive species contributing to the progression of RILD, merit investigation to establish causal mechanisms for IR-induced chronic liver toxicity and develop strategies to prevent RILD in cancer survivors.

Supplementary Materials: The following are available online at http://www.mdpi.com/2076-3921/9/5/409/s1, Figure S1: Conjugated and unconjugated plasma bilirubin levels did not show any significant changes between the sham and irradiated mice from both homozygous or SIRT3 knockout genotype, Table S1: Primers used in the single gene quantitative RT-PCR assays.

Author Contributions: Conceptualization, N.A.B.; investigation and data collection, F.V.L., K.J.K., S.A., G.S.C.; E.G.L.; formal analysis, N.A.B., F.V.L., J.A.D., Y.Y.; writing—original draft preparation, F.V.L.; writing—review and editing, N.A.-B., J.A.D., K.J.K.; supervision, N.A.B.; funding acquisition, N.A.B. All authors have read and agreed to the published version of the manuscript.

Funding: This research was funded by the following National Institutes of Health grants: NIGMS P20 GM109005; NIGMS T32GM106999; Arkansas Science and Technology Authority Award AWD50506 15-B-19.

Acknowledgments: The authors thank UAMS Experimental Pathology, DNA Damage and Toxicology, and Experimental Radiation Cores for their technical assistance.

Conflicts of Interest: The authors declare no conflict of interest. The funders had no role in the design of the study; in the collection, analyses, or interpretation of data; in the writing of the manuscript, or in the decision to publish the results.

References

1. Siegel, R.L.; Miller, K.D.; Jemal, A. Cancer statistics. *CA Cancer J. Clin.* **2020**. [CrossRef]
2. Kim, J.; Jung, Y. Radiation-induced liver disease: Current understanding and future perspectives. *Exp. Mol. Med.* **2017**. [CrossRef] [PubMed]
3. Benson, R.; Madan, R.; Kilambi, R.; Chander, S. Radiation induced liver disease: A clinical update. *J. Egypt Natl. Canc. Inst.* **2016**. [CrossRef] [PubMed]
4. Pan, C.C.; Kavanagh, B.D.; Dawson, L.A.; Li, X.A.; Das, S.K.; Miften, M.; Ten Haken, R.K. Radiation-Associated Liver Injury. *Int. J. Radiat. Oncol. Biol. Phys.* **2010**, *76*. [CrossRef]
5. Ben-Josef, E.; Normolle, D.; Ensminger, W.D.; Walker, S.; Tatro, D.; Ten Haken, R.K.; Knol, J.; Dawson, L.A.; Pan, C.; Lawrence, T.S. Phase II trial of high-dose conformal radiation therapy with concurrent hepatic artery floxuridine for unresectable intrahepatic malignancies. *J. Clin. Oncol.* **2005**. [CrossRef]
6. Feng, M.; Smith, D.E.; Normolle, D.P.; Knol, J.A.; Pan, C.C.; Ben-Josef, E.; Lu, Z.; Feng, M.R.; Chen, J.; Ensminger, W.; et al. A phase i clinical and pharmacology study using amifostine as a radioprotector in dose-escalated whole liver radiation therapy. *Int. J. Radiat. Oncol. Biol. Phys.* **2012**. [CrossRef]
7. Munoz-Schuffenegger, P.; Ng, S.; Dawson, L.A. Radiation-Induced Liver Toxicity. *Semin. Radiat. Oncol.* **2017**. [CrossRef]
8. Tanguturi, S.K.; Wo, J.Y.; Zhu, A.X.; Dawson, L.A.; Hong, T.S. Radiation Therapy for Liver Tumors: Ready for Inclusion in Guidelines? *Oncologist* **2014**. [CrossRef]
9. Toesca, D.A.S.; Ibragimov, B.; Koong, A.J.; Xing, L.; Koong, A.C.; Chang, D.T. Strategies for prediction and mitigation of radiation-induced liver toxicity. *J. Radiat. Res.* **2018**, *59*, i40–i49. [CrossRef]

10. Tai, A.; Erickson, B.; Li, X.A. Extrapolation of Normal Tissue Complication Probability for Different Fractionations in Liver Irradiation. *Int. J. Radiat. Onco.l Biol. Phys.* **2009**, *74*, 283–289. [CrossRef]

11. Lee, I.J.; Seong, J.; Shim, S.J.; Han, K.H. Radiotherapeutic Parameters Predictive of Liver Complications Induced by Liver Tumor Radiotherapy. *Int. J. Radiat. Oncol. Biol. Phys.* **2009**, *73*, 154–158. [CrossRef] [PubMed]

12. Fajardo, L.F.; Colby, T.V. Pathogenesis of veno-occlusive liver disease after radiation. *Arch. Pathol. Lab. Med.* **1980**, *104*, 584–588. [PubMed]

13. Guha, C.; Kavanagh, B.D. Hepatic Radiation Toxicity: Avoidance and Amelioration. *Semin. Radiat. Oncol.* **2011**. [CrossRef] [PubMed]

14. Coia, L.R.; Myerson, R.J.; Tepper, J.E. Late effects of radiation therapy on the gastrointestinal tract. *Int. J. Radiat. Oncol. Biol. Phys.* **1995**, *31*, 1213–1236. [CrossRef]

15. Hsieh, C.H.; Liu, C.Y.; Shueng, P.W.; Chong, N.S.; Chen, C.J.; Chen, M.J.; Lin, C.C.; Wang, T.E.; Lin, S.C.; Tai, H.C.; et al. Comparison of coplanar and noncoplanar intensity-modulated radiation therapy and helical tomotherapy for hepatocellular carcinoma. *Radiat. Oncol.* **2010**. [CrossRef]

16. Yamashita, H.; Onishi, H.; Murakami, N.; Matsumoto, Y.; Matsuo, Y.; Nomiya, T.; Nakagawa, K. Survival outcomes after stereotactic body radiotherapy for 79 Japanese patients with hepatocellular carcinoma. *J. Radiat. Res.* **2014**. [CrossRef]

17. Huang, Y.; Chen, S.W.; Fan, C.C.; Ting, L.L.; Kuo, C.C.; Chiou, J.F. Clinical parameters for predicting radiation-induced liver disease after intrahepatic reirradiation for hepatocellular carcinoma. *Radiat. Oncol.* **2016**. [CrossRef]

18. Jung, J.; Yoon, S.M.; Kim, S.Y.; Cho, B.; Park, J.H.; Kim, S.S.; Song, S.Y.; Lee, S.W.; Do Ahn, S.; Choi, E.K.; et al. Radiation-induced liver disease after stereotactic body radiotherapy for small hepatocellular carcinoma: Clinical and dose-volumetric parameters. *Radiat. Oncol.* **2013**. [CrossRef]

19. Fry, R.J.M.; Hall, E.J. Radiobiology for the Radiologist. *Radiat Res.* **1995**. [CrossRef]

20. Lautt, W.W. *Hepatic Circulation Physiology and Pathophysiology*; Morgan & Claypool Publishers: San Rafael, CA, USA, 2009. [CrossRef]

21. Sharp, G.B. The Relationship between Internally Deposited Alpha-Particle Radiation and Subsite-Specific Liver Cancer and Liver Cirrhosis: An Analysis of Published Data. *J. Radiat. Res.* **2002**, *43*, 371–380. Available online: https://academic.oup.com/jrr/article-abstract/43/4/371/986795 (accessed on 30 March 2020). [CrossRef]

22. Formenti, S.C.; Demaria, S. Local control by radiotherapy: Is that all there is? *Breast Cancer Res.* **2008**. [CrossRef] [PubMed]

23. Robbins, M.E.C.; Zhao, W. Chronic oxidative stress and radiation-induced late normal tissue injury: A review. *Int. J. Radiat. Biol.* **2004**. [CrossRef] [PubMed]

24. Kryston, T.B.; Georgiev, A.B.; Pissis, P.; Georgakilas, A.G. Role of oxidative stress and DNA damage in human carcinogenesis. *Mutat. Res. Fundam. Mol. Mech. Mutagen.* **2011**. [CrossRef] [PubMed]

25. Castilla, A.; Prieto, J.; Fausto, N. Transforming Growth Factors β1 and α in Chronic Liver Disease Effects of Interferon Alfa Therapy. *N. Engl. J. Med.* **1991**. [CrossRef] [PubMed]

26. Cai, X.; Hao, J.; Zhang, X.; Yu, B.; Ren, J.; Luo, C.; Li, Q.; Huang, Q.; Shi, X.; Li, W.; et al. The polyhydroxylated fullerene derivative C60(OH)24 protects mice from ionizing-radiation-induced immune and mitochondrial dysfunction. *Toxicol. Appl. Pharmacol.* **2010**. [CrossRef] [PubMed]

27. Kawamura, K.; Qi, F.; Kobayashi, J. Potential relationship between the biological effects of low-dose irradiation and mitochondrial ROS production. *J. Radiat. Res.* **2018**. [CrossRef]

28. Richardson, R.B.; Harper, M.E. Mitochondrial stress controls the radiosensitivity of the oxygen effect: Implications for radiotherapy. *Oncotarget* **2016**. [CrossRef]

29. Coleman, M.C.; Olivier, A.K.; Jacobus, J.A.; Mapuskar, K.A.; Mao, G.; Martin, S.M.; Riley, D.P.; Gius, D.; Spitz, D.R. Superoxide mediates acute liver injury in irradiated mice lacking Sirtuin 3. *Antioxid. Redox Signal.* **2014**. [CrossRef]

30. Tao, R.; Coleman, M.C.; Pennington, J.D.; Ozden, O.; Park, S.H.; Jiang, H.; Kim, H.S.; Flynn, C.R.; Hill, S.; McDonald, W.H.; et al. Sirt3-Mediated Deacetylation of Evolutionarily Conserved Lysine 122 Regulates MnSOD Activity in Response to Stress. *Mol. Cell.* **2010**. [CrossRef]

31. Zou, X.; Zhu, Y.; Park, S.H.; Liu, G.; O'Brien, J.; Jiang, H.; Gius, D. SIRT3-mediated dimerization of IDH2 directs cancer cell metabolism and tumor growth. *Cancer Res.* **2017**. [CrossRef]

32. Yu, W.; Dittenhafer-Reed, K.E.; Denu, J.M. SIRT3 protein deacetylates isocitrate dehydrogenase 2 (IDH2) and regulates mitochondrial redox status. *J. Biol. Chem.* **2012**. [CrossRef] [PubMed]

33. Jing, E.; O'Neill, B.T.; Rardin, M.J.; Kleinridders, A.; Ilkeyeva, O.R.; Ussar, S.; Bain, J.R.; Lee, K.Y.; Verdin, E.M.; Newgard, C.B.; et al. Sirt3 regulates metabolic flexibility of skeletal muscle through reversible enzymatic deacetylation. *Diabetes* **2013**. [CrossRef] [PubMed]

34. Kendrick, A.A.; Choudhury, M.; Rahman, S.M.; McCURDY, C.E.; Friederich, M.; Van Hove, J.L.; Watson, P.A.; Birdsey, N.; Bao, J.; Gius, D.; et al. Fatty liver is associated with reduced SIRT3 activity and mitochondrial protein hyperacetylation. *Biochem. J.* **2011**. [CrossRef] [PubMed]

35. Kincaid, B.; Bossy-Wetzel, E. Forever young: SIRT3 a shield against mitochondrial meltdown, aging, and neurodegeneration. *Front. Aging. Neurosci.* **2013**. [CrossRef]

36. Hirschey, M.D.; Shimazu, T.; Goetzman, E.; Jing, E.; Schwer, B.; Lombard, D.B.; Grueter, C.A.; Harris, C.; Biddinger, S.; Ilkayeva, O.R.; et al. SIRT3 regulates mitochondrial fatty-acid oxidation by reversible enzyme deacetylation. *Nature* **2010**. [CrossRef]

37. Brown, K.; Xie, S.; Qiu, X.; Mohrin, M.; Shin, J.; Liu, Y.; Zhang, D.; Scadden, D.T.; Chen, D. SIRT3 Reverses Aging-Associated Degeneration. *Cell Rep.* **2013**. [CrossRef]

38. Liu, J.; Li, D.; Zhang, T.; Tong, Q.; Ye, R.D.; Lin, L. SIRT3 protects hepatocytes from oxidative injury by enhancing ROS scavenging and mitochondrial integrity. *Cell Death Dis.* **2017**. [CrossRef]

39. Sundaresan, N.R.; Samant, S.A.; Pillai, V.B.; Rajamohan, S.B.; Gupta, M.P. SIRT3 Is a Stress-Responsive Deacetylase in Cardiomyocytes That Protects Cells from Stress-Mediated Cell Death by Deacetylation of Ku70. *Mol. Cell Biol.* **2008**. [CrossRef]

40. Ansari, A.; Rahman, M.S.; Saha, S.K.; Saikot, F.K.; Deep, A.; Kim, K.H. unction of the SIRT3 mitochondrial deacetylase in cellular physiology, cancer, and neurodegenerative disease. *Aging Cell.* **2017**. [CrossRef]

41. Chen, Y.; Zhang, J.; Lin, Y.; Lei, Q.; Guan, K.L.; Zhao, S.; Xiong, Y. Tumour suppressor SIRT3 deacetylates and activates manganese superoxide dismutase to scavenge ROS. *EMBO Rep.* **2011**. [CrossRef]

42. Kim, H.S.; Patel, K.; Muldoon-Jacobs, K.; Bisht, K.S.; Aykin-Burns, N.; Pennington, J.D.; van der Meer, R.; Nguyen, P.; Savage, J.; Owens, K.M.; et al. SIRT3 Is a Mitochondria-Localized Tumor Suppressor Required for Maintenance of Mitochondrial Integrity and Metabolism during Stress. *Cancer Cell.* **2010**. [CrossRef] [PubMed]

43. Pi, H.; Xu, S.; Reiter, R.J.; Guo, P.; Zhang, L.; Li, Y.; Li, M.; Cao, Z.; Tian, L.; Xie, J.; et al. SIRT3-SOD2-mROS-dependent autophagy in cadmium-induced hepatotoxicity and salvage by melatonin. *Autophagy* **2015**. [CrossRef] [PubMed]

44. Park, S.H.; Ozden, O.; Jiang, H.; Cha, Y.I.; Pennington, J.D.; Aykin-Burns, N.; Spitz, D.R.; Gius, D.; Kim, H.S. Sirt3, mitochondrial ROS, ageing, and carcinogenesis. *Int. J. Mol. Sci.* **2011**, *12*, 6226–6239. [CrossRef] [PubMed]

45. Someya, S.; Yu, W.; Hallows, W.C.; Xu, J.; Vann, J.M.; Leeuwenburgh, C.; Tanokura, M.; Denu, J.M.; Prolla, T.A. Sirt3 mediates reduction of oxidative damage and prevention of age-related hearing loss under Caloric Restriction. *Cell* **2010**. [CrossRef]

46. Zeng, H.; Li, L.; Chen, J.X. Loss of Sirt3 limits bone marrow cell-mediated angiogenesis and cardiac repair in post-myocardial infarction. *PLoS ONE* **2014**. [CrossRef]

47. Alam, S.; Carter, G.S.; Krager, K.J.; Li, X.; Lehmler, H.-J.; Aykin-Burns, N. PCB11 Metabolite, 3,3'-Dichlorobiphenyl-4-ol, Exposure Alters the Expression of Genes Governing Fatty Acid Metabolism in the Absence of Functional Sirtuin 3: Examining the Contribution of MnSOD. *Antioxidants* **2018**, *7*, 121. [CrossRef]

48. Lowry, O.H.; Rosebrough, N.J.; Farr, A.L.; Randall, R.J. Protein measurement with the Folin phenol reagent. *J. Biol. Chem.* **1951**. [CrossRef]

49. Aebi, H. Catalase in Vitro. *Methods Enzymol.* **1984**. [CrossRef]

50. Lawrence, R.A.; Burk, R.F. Glutathione peroxidase activity in selenium-deficient rat liver. *Biochem. Biophys. Res. Commun.* **1976**. [CrossRef]

51. Pinto, R.E.; Bartley, W. The effect of age and sex on glutathione reductase and glutathione peroxidase activities and on aerobic glutathione oxidation in rat liver homogenates. *Biochem. J.* **1969**. [CrossRef]

52. Spitz, D.R.; Oberley, L.W. An assay for superoxide dismutase activity in mammalian tissue homogenates. *Anal. Biochem.* **1989**. [CrossRef]

53. Barshishat-Kupper, M.; Tipton, A.J.; McCart, E.A.; McCue, J.; Mueller, G.P.; Day, R.M. Effect of ionizing radiation on liver protein oxidation and metabolic function in C57BL/6J mice. *Int. J. Radiat. Biol.* **2014**. [CrossRef]

54. Maeda, M.; Ishikawa, H.; Yoshida, Y.; Takahashi, T.; Ohkubo, Y.; Musha, A.; Komachi, M.; Nakazato, Y.; Nakano, T. Long-term pathological and immunohistochemical features in the liver after intraoperative whole-liver irradiation in rats. *J. Radiat. Res.* **2014**. [CrossRef]

55. Osmundson, E.C.; Wu, Y.; Luxton, G.; Bazan, J.G.; Koong, A.C.; Chang, D.T. Predictors of toxicity associated with stereotactic body radiation therapy to the central hepatobiliary tract. *Int. J. Radiat. Oncol. Biol. Phys.* **2015**. [CrossRef] [PubMed]

56. Seidensticker, M.; Burak, M.; Kalinski, T.; Garlipp, B.; Koelble, K.; Wust, P.; Antweiler, K.; Seidensticker, R.; Mohnike, K.; Pech, M.; et al. Radiation-Induced Liver Damage: Correlation of Histopathology with Hepatobiliary Magnetic Resonance Imaging, a Feasibility Study. *Cardiovasc. Interven. Radiol.* **2015**. [CrossRef] [PubMed]

57. Spitz, D.R.; Hauer-Jensen, M. Ionizing radiation-induced responses: Where free radical chemistry meets redox biology and medicine. *Antioxidants Redox Signal.* **2014**. [CrossRef]

58. Tamminga, J.; Kovalchuk, O. Role of DNA Damage and Epigenetic DNA Methylation Changes in Radiation-Induced Genomic Instability and Bystander Effects in Germline In Vivo. *Curr. Mol. Pharmacol.* **2012**. [CrossRef]

59. Barjaktarovic, Z.; Shyla, A.; Azimzadeh, O.; Schulz, S.; Haagen, J.; Dörr, W.; Sarioglu, H.; Atkinson, M.J.; Zischka, H.; Tapio, S. Ionising radiation induces persistent alterations in the cardiac mitochondrial function of C57BL/6 mice 40 weeks after local heart exposure. *Radiother. Oncol.* **2013**. [CrossRef]

60. Morgan, W.F.; Day, J.P.; Kaplan, M.I.; McGhee, E.M.; Limoli, C.L. Genomic Instability Induced by Ionizing Radiation. *Radiat. Res.* **1996**. [CrossRef]

61. Qiu, X.; Brown, K.; Hirschey, M.D.; Verdin, E.; Chen, D. Calorie restriction reduces oxidative stress by SIRT3-mediated SOD2 activation. *Cell Metab.* **2010**. [CrossRef]

62. Fausto, N.; Mead, J.E.; Gruppuso, P.A.; Castilla, A.; Jakowlew, S.B. Effects of TGF-beta s in the liver: Cell proliferation and fibrogenesis. *Ciba Found. Symp.* **1991**. [CrossRef]

63. Jin, X.; Zimmers, T.A.; Perez, E.A.; Pierce, R.H.; Zhang, Z.; Koniaris, L.G. Paradoxical effects of short- and long-term interleukin-6 exposure on liver injury and repair. *Hepatology* **2006**. [CrossRef] [PubMed]

64. Zhao, W.Y.; Zhang, L.; Sui, M.X.; Zhu, Y.H.; Zeng, L. Protective Effects of Sirtuin 3 in a Murine Model of Sepsis-Induced Acute Kidney Injury. *Sci. Rep.* **2016**. [CrossRef] [PubMed]

65. Salunga, T.L.; Cui, Z.G.; Shimoda, S.; Zheng, H.C.; Nomoto, K.; Kondo, T.; Takano, Y.; Selmi, C.; Alpini, G.; Gershwin, M.E.; et al. Oxidative Stress-Induced Apoptosis of Bile Duct Cells in Primary Biliary Cirrhosis. *J. Autoimmun.* **2007**. [CrossRef] [PubMed]

66. Erice, O.; Munoz-Garrido, P.; Vaquero, J.; Perugorria, M.J.; Fernandez-Barrena, M.G.; Saez, E.; Santos-Laso, A.; Arbelaiz, A.; Jimenez-Agüero, R.; Fernandez-Irigoyen, J.; et al. MicroRNA-506 Promotes Primary Biliary Cholangitis–like Features in Cholangiocytes and Immune Activation. *Hepatology* **2018**. [CrossRef]

67. Degott, C.; Feldmann, G.; Larrey, D.; Durand-Schneider, A.M.; Grange, D.; Machayekhi, J.P.; Moreau, A.; Potet, F.; Benhamou, J.P. Drug-induced prolonged cholestasis in adults: A histological semiquantitative study demonstrating progressive ductopenia. *Hepatology* **1992**. [CrossRef]

68. Wang, Z.; Sun, R.; Wang, G.; Chen, Z.; Li, Y.; Zhao, Y.; Liu, D.; Zhao, H.; Zhang, F.; Yao, J.; et al. SIRT3-Mediated Deacetylation of PRDX3 Alleviates Mitochondrial Oxidative Damage and Apoptosis Induced by Intestinal Ischemia/Reperfusion Injury. *Redox Biol.* **2019**. [CrossRef]

 *antioxidants*

Review

The Influence of Light on Reactive Oxygen Species and NF-kB in Disease Progression

Naresh Kumar Rajendran, Blassan P. George, Rahul Chandran, Ivan Mfouo Tynga, Nicolette Houreld and Heidi Abrahamse *

Laser Research Centre, Faculty of Health Sciences, University of Johannesburg, P.O. Box 17011, Johannesburg 2028, South Africa; naresh.r84@outlook.com (N.K.R.); blassang@uj.ac.za (B.P.G.); rahulc@uj.ac.za (R.C.); ivant@uj.ac.za (I.M.T.); nhoureld@uj.ac.za (N.H.)
* Correspondence: habrahamse@uj.ac.za; Tel.: +27-11-559-6550

Received: 6 November 2019; Accepted: 2 December 2019; Published: 12 December 2019

Abstract: Reactive oxygen species (ROS) are important secondary metabolites that play major roles in signaling pathways, with their levels often used as analytical tools to investigate various cellular scenarios. They potentially damage genetic material and facilitate tumorigenesis by inhibiting certain tumor suppressors. In diabetic conditions, substantial levels of ROS stimulate oxidative stress through specialized precursors and enzymatic activity, while minimum levels are required for proper wound healing. Photobiomodulation (PBM) uses light to stimulate cellular mechanisms and facilitate the removal of oxidative stress. Photodynamic therapy (PDT) generates ROS to induce selective tumor destruction. The regulatory roles of PBM via crosstalk between ROS and nuclear factor kappa-light-chain-enhancer of activated B cells (NF-κB) are substantial for the appropriate management of various conditions.

Keywords: photobiomodulation; reactive oxygen species (ROS); nuclear factor kappa-light-chain-enhancer of activated B cells (NF-κB); cancer; diabetes; wound healing

---

1. Introduction

Reactive oxygen species (ROS) are formed by the fractional reduction of molecular oxygen and include, but are not limited to, superoxide anions, hydrogen peroxide, and hydroxyl radicals, all obtained from sequential oxidation–reduction processes involving nicotinamide adenine dinucleotide phosphate (NADPH) oxidase, lipoxygenases, or cyclooxygenases [1]. Unusually high levels of ROS can allegedly be used for cancer diagnoses, varying according to tumor type, and are potent signaling molecules in cancer, leading to nuclear damage, genetic instability, and tumorigenesis [2–4]. However, at non-cytotoxic levels, ROS act as secondary messengers with signaling roles in many physiological systems to activate programmed cell death, gene expression, and other cell signaling cascades [5]. Increased ROS production was observed in diabetes and diabetic complications, leading to oxidative stress. As a result, a series of cell death mechanisms were observed within the cell, finally leading to tissue and organ damage. Elevated levels of blood glucose appear to be the prime source of free radicals, unbalancing the pool of antioxidants and ROS. Therefore, the down-regulation of ROS production and targeting factors resulting in their increased generation may have a significant role in controlling diabetic complications [6,7].

ROS plays a pivotal role in the initial stages of wound healing by killing invading bacteria and other microorganisms. However, under chronic conditions, increased production of free radicals was observed, thereby inhibiting the proliferation and migration of key cell types and leading to delayed wound healing [8,9]. The regulation of certain redox transcription factors is dependent on the level of ROS. Nuclear factor kappa-light-chain-enhancer of activated B cells (NF-κB) was the first discovered redox-regulated transcription factor. NF-κB is a protein complex with multiple functions in immune, inflammation, cell growth, and survival responses. ROS are able to both activate and suppress NF-κB signaling pathways [10,11]. Photobiomodulation (PBM) is a modern therapeutic approach which results in beneficial outcomes and the modulation of various signaling pathways in the presence of light at a specific wavelength. Photodynamic therapy (PDT) uses a specific wavelength light to activate the photosensitizer to induce cell death in conjunction with molecular oxygen.

Even though PBM is well-known for its cell-stimulating properties both in vitro and in vivo, clinical studies have been very mixed, and some contradicted non-clinical studies [12–15]; as a result, some clinicians consider PBM a very controversial therapy [16–18]. It is important to realize that the underlying cellular mechanisms of PBM are not fully understood [19,20]. Additionally, PBM treatment parameters vary, such as the wavelength, fluence, power density, pulse structure, and irradiation time. These are factors that preclude efficient clinical transition of PBM [21–24]. However, some studies reported the role of cytochrome c oxidase as an important chromophore in the cellular response to PBM [25]. A similar problem exists with PDT, in that it is not clinically accepted by many clinicians. Although photodynamic therapy (PDT) has a long history, there is a minimal amount of proven clinical research, making it difficult for it to be recognized as a first-line treatment approach in modern medicine. This review focuses on the effect of ROS on NF-κB activity, how ROS are affected by PBM/PDT, and its role in diabetes, wound healing, and cancer, respectively.

2. Sources and Stimuli of ROS

ROS are oxygen intermediates with unpaired electrons; both superoxide and hydroxyl radicals are highly unstable oxygen radicals [26]. Experimentally, hydrogen peroxide is a simple peroxide radical involved in various signaling functions and is frequently used as a source of all oxygen-related free radicals [27]. Elevated levels of hydrogen peroxide effectively oxidize cysteine residues (Cys-SH) to cysteine sulfenic acid (Cys-SOH) or cysteine disulphide (Cys-S-S-Cys) in various proteins, such as kinases, phosphatases, and transcription factors. A well-established mechanism by which ROS regulate cellular functions is through the redox balance of cysteine residues [28].

Mitochondria and NADPH promote endogenous ROS formation in cancer and reports have shown crosstalk between these two producers [29]. The mitochondrial oxidative generation of adenosine triphosphate (ATP) is a major source of free radicals. During the Krebs's cycle, unpaired electrons are transferred to the electron transport chain (ETC), resulting in the production of superoxide anions [30]. About ten mitochondrial sites generate numerous superoxide anions through different mechanisms [31]. The reactions of the five complexes of the ETC (complex I to V) are involved in the production of ATP and free radicals as byproducts. The radicals produced from complexes I and III have various cellular signaling roles [32]. Both superoxide anions and hydrogen peroxides are constantly produced by complex III, with 80% released into the intermembrane space and the rest into the mitochondrial matrix. These ROS intermediate radicals are necessary for cell differentiation, proliferation, survival, and adaptive immunity responses [33]. In complex IV and during oxidative phosphorylation reactions, the movement of electrons results in the reduction of oxygen to water. Almost all of the generated superoxide anions are effectively quenched by manganese superoxide dismutase (MnSOD) to form hydrogen peroxide, which serves as an important precursor for other free radicals and acts as a secondary messenger with the ability to diffuse across the mitochondrial membrane, mediated by a specialized protein from the aquaporin family [33,34]. Other than the positive functions, mitochondrial ROS also have some deleterious ones, including the formation and progression of various cancer types like chronic lymphocytic leukemia, acute myelogenous leukemia, and breast cancer [35]. Figure 1 depicts the generation of free radicals and their fate in the ETC. Free radicals formed by the ETC are immediately removed and converted to water by a series of enzymatic reactions.

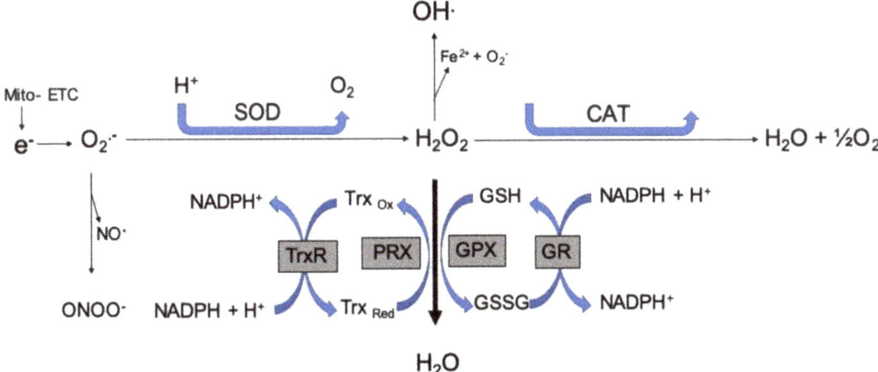

Figure 1. Free radical generation. Free radicals formed by the electron transport chain are immediately removed and converted to water through a series of enzymatic reactions. ETC: electron transport chain; SOD: superoxide dismutase; CAT: catalase; GPX: glutathione peroxidase; GR: glutathione reductase; PRX: peroxiredoxins; GSH: glutathione; GSSG: glutathione disulfide; Trx: thioredoxin; $O_2\bullet^-$: superoxide; NO•: nitric oxide; $ONOO^-$: peroxynitrite; $H_2O_2^-$: hydrogen peroxide.

2.1. Oxidative Stress and Cancer

Oxidative damage is linked to all phases of cancer, during which ROS act throughout its progression. Radicals generated in the mitochondria damage genetic material and produce mutations that initiate tumor progression due to the imbalance between ROS generation and antioxidant defence mechanisms [36]. Redox homeostasis is important to sustain normal and survival functions. Increased aerobic glycolysis and high levels of free radicals in cancer cells are linked to modifications in cell signaling pathways, which can be counteracted by antioxidant defence mechanisms [37]. ROS facilitate carcinogenesis by reversibly inhibiting certain tumor suppressors, such as phosphatase and tensin homolog (PTEN) and protein tyrosine phosphatases (PTPs), which enhance antioxidant expression and thus decrease ROS levels. However, high levels of ROS are beneficial during the developmental stages of tumors, promoting cancer vulnerability and death [38].

2.2. Oxidative Stress and Diabetes

Glucose metabolism is the mechanism behind the generation of oxidative stress. In a transition metal-dependent reaction, the enediol form of glucose is oxidized to enediol anion radicals, then converted to more reactive superoxide anions and ketoaldehydes. Through a dismutation reaction, superoxide anions are converted into hydrogen peroxides, which are precursors of highly reactive hydroxyl radicals. Under diabetic conditions, if these hydrogen peroxides are not degraded by glutathione peroxidase (GPx) or catalase (CAT), then the production of reactive hydroxyl radicals occurs in the presence of transition metals [39]. Superoxide radical formation during hyperglycaemia may also promote additional ROS formation through the oxidation of low-density lipoproteins [40].

Amadori products and advanced glycation end-products (AGEs) are emerging as major precursors of free radicals in diabetes. AGEs may fluoresce, produce ROS, and bind to specific cell surface receptors [41]. These generated AGEs bind to their receptors via the receptor for AGEs (RAGE) and deactivate enzymes, followed by the discharge of free radicals [42]. These enzymatic alterations may also quench and inhibit the anti-proliferative effect of nitric oxide (NO), which acts as a vital vasodilator in diabetic patients [43]. Altogether, the increased free radical production stimulates intracellular oxidative stress by AGEs, which upregulate NF-κB-controlled target genes [44]. In endothelial cells, ROS stimulates the hexosamine pathway and induces vascular complications in patients with high blood glucose levels [45]. The overexpression of the enzyme xanthine oxidase may contribute to the pathological condition, leading to type 2 diabetes. One of the mechanisms underlying this is the ability of xanthine oxidase to produce free radicals and oxidative stress [46].

2.3. Oxidative Stress and Wound Healing

The main processes involved in regulated wound healing are migration, adhesion, proliferation, neovascularization, remodeling, and apoptosis. Imbalance between antioxidants and free radicals with elevated ROS formation is often observed, followed by the induction of apoptosis. In chronic wounds, elevated free radicals coupled with reduced antioxidants results in the stimulation of pro-apoptotic transcription factors, caspase-3, and oxidative stress, leading to cell death and further delays in the healing process. The development of chronic impaired wound healing involves certain cell signaling reactions [47]. Fibroblast/keratinocyte cells with mutated mitochondrial DNA result in increased ROS generation, which affects nuclear transcriptional events by stimulating various signal transduction pathways and inducing cell cycle arrest, death, and oxidative stress processes [48,49]. In fibroblasts, mitochondria-generated ROS damages mitochondrial DNA and induces a persistent oxidative stress condition. Keratinocyte differentiation depends on the ETC for the accumulation of superoxide anions [50]. In contrast, in normal wound healing, minimal ROS levels mediate intracellular signaling for cell proliferation and collagen deposition. Low ROS levels and high antioxidant levels are essential for normal tissue repair [51].

2.4. Influence of Light on Oxidative Stress

Under two different conditions, light is used to either attenuate oxidative stress (as in the case of PBM) or generate excessive amounts of ROS (as in the case of PDT). PBM involves the use of non-ionizing radiation or light specifically in the visible and near-infrared regions of the electromagnetic spectrum for therapeutic applications and stimulation of various cellular mechanisms. When there is an imbalance between antioxidants and free radicals, PBM facilitates the displacement of free radicals and the reduction of the oxidative stress load on the organism [52]. The successful action of PBM depends on the activation of cytochrome c oxidase, which catalyzes the reduction of oxygen to water by increasing the mitochondrial membrane potential (MMP), ATP, cyclic adenosine monophosphate (cAMP), and NO [53,54]. NO is an essential signaling agent for the activation of certain cellular pathways. PBM increases the generation of NO by upregulating cytochrome c levels and/or by cleaving the metal complexes in cytochrome c [55]. The basic concepts and principles behind PBM are clearly described in the literature. It is understood that PBM acts on impaired or dysfunctional tissue through light and leads to mitochondria-mediated cellular processes, thereby inciting various mechanisms that result in a wide range of therapeutic functions [56]. Some studies proposed that light at wavelengths between 400 and 500 nm may generate free radicals through photosensitization, while ROS production by mitochondria was directly evident at longer wavelengths. PBM activates several transcription factors in the cytosol such as NF-kB, which trigger the transcription of genes and protect cells from oxidative stress [57].

PDT is a minimally invasive treatment that is becoming commonly accepted as a potential therapeutic option for localized cancers [58]. PDT depends on the successful accumulation of photosensitizers (PSs) in tumor cells to initiate the process of irradiation of targeted areas for activation of the PSs (Figure 2) [59,60]. The wavelength of light should match the absorption properties of the PSs for effective ROS generation, either through direct electron abstraction (type I) or transfer from a substrate to oxygen in a highly reactive state (type II), killing targeted tumors and leaving the neighboring healthy cells unaffected [61]. Due to light inaccessibility to all areas of the body, PDT was previously limited to superficial conditions, leaving deeper diseased organs less affected. Due to the advancement in science and the medical use of fiber optics, PDT now has the potential to treat both superficial lesions and deeper-seated tumors, which were previously inaccessible. The recent development of wireless photonic and contracted implantable devices for light delivery into deep regions, such as the brain and liver, rendered PDT treatment even more effective [62].

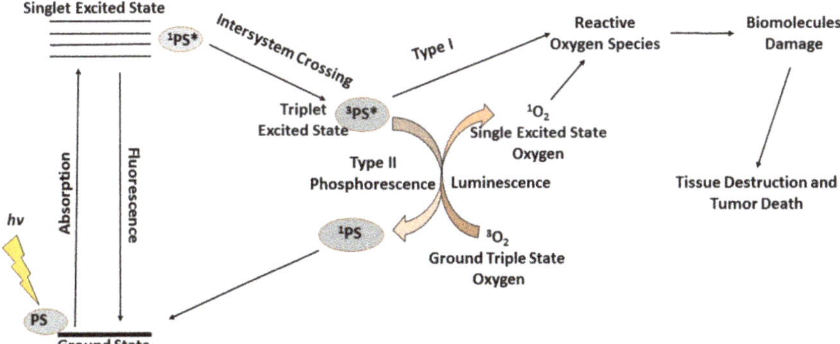

Figure 2. During photodynamic therapy (PDT), generated reactive oxygen species (ROS) kill the targeted tumor cells. PDT depends on the successful accumulation of photosensitizers (PSs) in tumor cells to initiate the process of irradiation of targeted areas for activation of the PSs. When irradiated with light of a specific wavelength, a photochemical reaction leads to the production of ROS in the presence of oxygen. When the PS absorbs the photon energy it is excited and jumps from the ground state to the excited singlet state and the energy is emitted back as fluorescence or heat via internal conversion. Should intersystem-crossing occur, the PS is converted to an excited triplet state, which can transfer electrons with cellular biomolecules, ultimately leading to the generation of ROS in a Type I reaction. Alternatively, the excited triplet state PS transfers its electrons to ground triplet state molecular oxygen ($^3O_2$), leading to the generation of an excited singlet oxygen ($^1O_2$) in a Type II reaction.

2.5. Effect of Light on NF-κB Activation and ROS Regulation

Activation of NF-κB depends on various parameters, like the cell compartment and dimer confinement. Cytoplasmic NF-κB dimers are maintained in a stationary and inactive state through their association with IκB (inhibitor of κB) proteins. When unconfined, NF-κB dimers translocate to nuclear regions and bind to targeted DNA sequences, promoting gene expression [63]. Other determinant factors include the synthesis of IκB proteins, activities of IKK (IκB kinase) complexes, and displacement of NF-κB family members, both transcriptional co-activators and DNA [64]. The N-terminal Rel homology domain (RHD) has affinities to κB sites and characterizes members of the NF-κB family, consisting of RelB, RelA/p65, c-Rel, p50 (NF-κB1), and p52 (NF-κB2) in mammals [65]. The binding affinities of their κB sites are of critical merit to exert positive and negative effects on transcription processes, dimerization formation, and maintenance [60]. Only p65, RelB, and c-Rel mediate transcription through their C-terminal transactivation domains (TADs). Though they have no TADs, through interactions with non-Rel proteins with transactivation ability and hetero-dimerization with TAD-containing NF-κB subunits, NF-κB1 and NF-κB2 upregulate transcription on the one hand and downregulate transcription on the other hand via competitive inhibition with TAD-containing dimers [66].

Activation of NF-κB can be observed by direct and indirect glutathionylation of NF-κB and I-κB, respectively. The p50 of cysteine-62 is sensitive to oxidation, which inhibits the translocation of NF-κB into the nucleus and prevents DNA binding [67]. However, as S-glutathionylation events occur, p50 is selectively reduced and DNA binding is restored [68,69]. ROS modification and glutathionylation of IκBα at cysteine 189 prevents phosphorylation events and degradation, thereby leading to NF-κB activation and targeted gene transcription (Figure 3) [70]. The IκB-mediated regulation characterizes the NF-κB pathway and IκB proteins (IκBα, IκBβ, IκBε, IκBz, B-cell lymphoma 3 (BCL-3), and IκBns), together with the precursor proteins NF-κB1 and NF-κB2, are well defined by the presence of multiple ankyrin repeat domains. NF-κB activation, via phosphorylation of IκBs on conserved serine residues, facilitates recognition by bTrCP proteins and K48-linked polyubiquitination by the Skp1–Culin–Roc1/Rbx1/Hrt-1–F-box family of E3 ligases acting alongside the E2 enzyme UbcH5 in a precise manner [63]. The NF-κB signaling pathways are classified into canonical and non-canonical routes. The canonical route is a classical representation of the generalized regulation of the NF-κB pathway. Upon ligand recognition, cytokine receptors, such as the antigen receptors, interleukin (IL)-1 receptor (IL-1R), tumor necrosis factor receptor (TNFR), and pattern recognition receptors (PRRs), stimulate and activate signaling cascades that culminate in the activation of IKKβ (IKK2). IκB proteins are phosphorylated by activated IKKβ, which exists in a complex with the closely related kinase IKKα (IKK1) and NF-κB essential modulator (NEMO, IKKγ). Many membrane-bound ligands, such as TNFR, TLR, and IL-1R, are involved in activation of the NF-κB pathway and act as upstream regulators. When activated, the IKK complex becomes the principal upstream part of all NF-κB pathways, thereby regulating the transcription of proinflammatory genes [71].

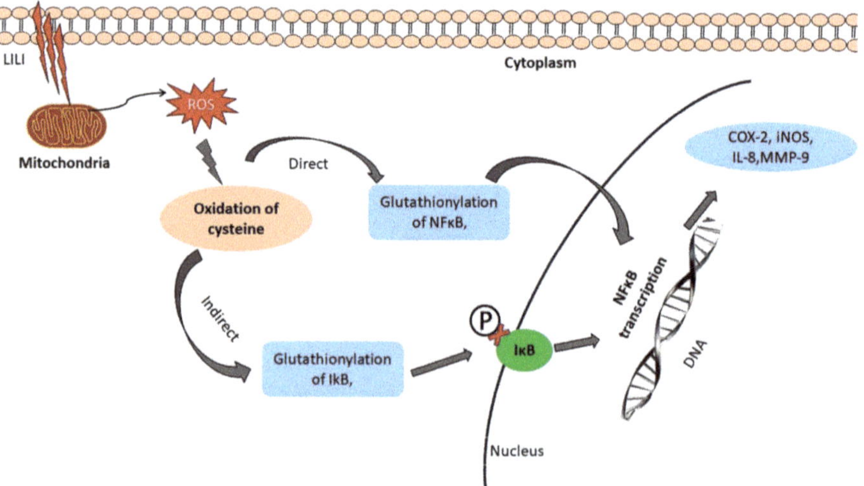

Figure 3. Mechanism of photobiomodulation (PBM). When low-powered light (LILI) enters the cell, it is immediately absorbed by cytochrome c oxidase, which is present in mitochondria. This results in an increased respiratory chain reaction and the overall redox reaction is altered. The increase in reactive oxygen species (ROS) stimulates the oxidation of cysteine molecules. The glutathionylation of IkB and S-glutathionylation of p50 activates NF-kB and transcription of specific genes. This further stimulates the activation of the IkB kinase complex, leading to phosphorylation, ubiquitination, and degradation of IkB proteins. Released NF-kB dimers then translocate into the nucleus, bind to specific DNA sequences, and promote the transcription of target genes such as interleukin (IL)-2, IL-6, IL-8, inducible nitric oxide synthase (iNOS), COX-2, and matrix metallopeptidases (MMP)-9.

The archetypical IκBα protein promptly degrades during the canonical activation route, causing the release of multiple NF-κB dimers. IκBα principally targets the p65:p50 heterodimer to form a IκBα: p65:p50 complex, which has translocating capabilities and masks the p65 nuclear localization signal. DNA binding is prevented by the IκBα nuclear export signal, which results in steady-state cytoplasmic localization of NF-κB dimers [72]. The degradation of IκBα leads to localization of NF-κB in the nucleus and removal of IkB from the NF-κB complex. Then, NF-κB dimers bind to specific DNA κB sites in the promoters and enhancers of target genes, leading to the formation of homodimers and heterodimers, and heterotypic interactions with other transcription factors, finally leading to regulation of transcriptional activity via post-translational modifications of targeted NF-κB subunits [66]. The termination of transcription depends on the synthesis of typical IκB proteins and the elimination of active NF-κB dimers from DNA. The difference between canonical and non-canonical routes remains whether they are NEMO-dependent or NEMO-independent. The non-canonical pathway is NEMO-independent and is determined by IKKα activation to induce the NF-kB/RelB activation complex, thereby leading to the generation of a p52/RelB complex and phosphorylation of p100 (Figure 4). IKKα proteins can activate the canonical route and expand to the non-canonical route through the induction of p100 expression [73].

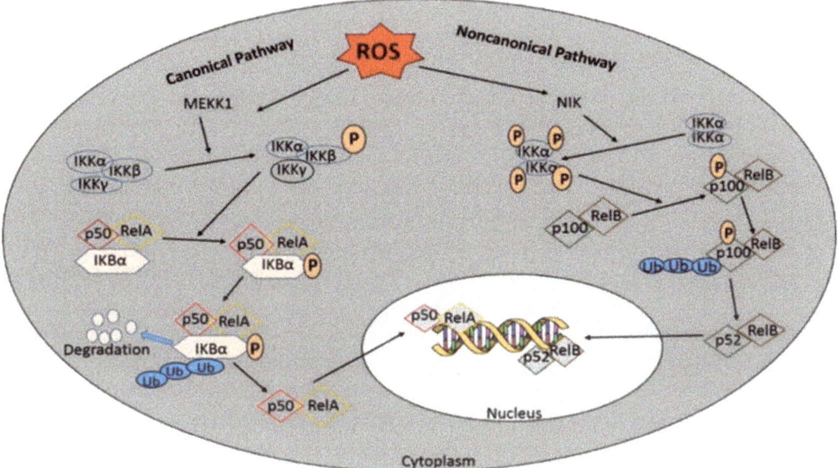

Figure 4. Activation of NF-κB by the canonical and/or the noncanonical pathway. The canonical NF-κB-activating pathway depends on the phosphorylation of IκB-kinase (IKK) β. The phosphorylation and ubiquitination of IκBα translocate NF-κB into the nucleus. where it helps in the transcription of target genes. Many membrane-bound ligands involved in activating the NF-κB pathway act as effective upstream regulators and the IKK complex is the common upstream component of all NF-κB pathways. In contrast, the non-canonical pathway is NF-kB essential modulator (NEMO)-independent and depends on IKKα activation to induce the NF-kB/RelB activation complex, leading to the phosphorylation of p100 and the generation of p52/RelB complexes.

2.6. Reciprocal Influence of ROS and NF-κB Activation

Hydrogen peroxide inhibits IKK activation by directly affecting IKKβ through its cysteine inhibitory ability in catalytic domains of tyrosine phosphatases. The interactions of ROS with cysteine residues are essential in order to affect the NF-κB pathways [74,75]. They facilitate the phosphorylation of serine residues in the activation loops of IKK in cells [75], and IKK triggers the ubiquitination and degradation of IkBα, which is usually phosphorylated on serine 32 and 36 [76]. The kinase upstream of IKK or mitogen-activated protein kinase 1 (MEKK1) is inactive when glutathionylated at C1238. ROS also target IKK and affect the NF-κB pathway, S-glutathionylation and IKKB functions via phosphorylation and degradation of IκBα on tyrosine residues and activation of the NF-κB pathway, as this may inhibit IκBα phosphorylation. They have the potential to influence the ubiquitination and degradation of IkB and activate NF-κB by inactivating Ubcl2 [77]. In the non-canonical route, NF-κB-inducing kinases (NIK), i.e., the upstream kinases, are activated by ROS through inhibition of phosphatase and oxidation of cysteine residues [78].

Moreover, ROS appear as vital bridging factors that mediate the crosstalk between JNK and NF-κB signaling pathways and interconnect ROS-mediated NF-κB and cell death, with the ability to inhibit JNK activation [79]. Furthermore, NF-κB downregulates JNK activation by suppressing TNFα-induced ROS accumulation. Due to their exerted oxidative stress, ROS directly participate in upregulation or downregulation of NF-κB through their interactions with cytoplasmic components. ROS are noxious at certain cellular levels and their inhibition results in the blockage of TNFα-induced programmed cell death facilitated by the NF-κB signaling pathway [80]. However, nuclear ROS might solely lead to a reduction of the binding ability of NF-κB to DNA [81]. The respective ROS accumulation and resulting oxidative stress generated in a specific subcellular localization becomes significant, as oxidative damage could be avoided and cell survival could be maintained [82]. Cell survival is the most probable outcome of the increased expression of NF-κB target genes, and only in a few exceptional cases does a cell death response prevail. Therefore, it is expected that, on one hand, ROS would modulate NF-κB responses, and on the other hand, NF-κB target genes would promote survival by ROS attenuation [83].

2.7. NF-κB in Cancer Progression

NF-κB is involved in various stages of oncogenesis and controls the expression of certain tumor-related genes. During initiation in pre-malignant cells, stimulated NF-κB promotes the expression of cytokines and chemokines, leading to the recruitment and activation of immune cells, which produce more chemokines and growth factors [71]. They act on both malignant and inflammatory cells in an autocrine or paracrine way, creating a complex inflammatory and pro-tumorigenic microenvironment. NF-κB mediated inflammation contributes to nuclear damage, oncogenic mutation, and tumor initiation and progression in pre-malignant cells [84]. Epidemiological studies revealed that over 20% of all cancer types are associated with chronic inflammation, which triggers carcinogenic events, leading to the malignant transformation of normal cells [85]. Abnormal NF-κB activity may promote the induction of tumors and apoptosis-resistant genes; this is one of the mechanisms by which resistance to radiotherapy and chemotherapy is acquired. Through cytidine deaminase enzymatic activity, activation-induced cytidine deaminases (AID) cause genetic alterations in DNA sequences, and their intrinsic mutagenic-enzyme forms can be induced in response to infectious agents and pro-inflammatory cytokines that facilitate NF-κB activation in epithelial cells [86]. While the major role of NF-κB in the regulation of AID expression is evident, NF-κB-mediated AID expression is accomplished via inflammatory mechanisms for the malignant transformation of epithelial cells during carcinogenesis. In normal cells, NF-κB is transiently activated, while in cancer cells, it exhibits sustained activation [87].

NF-κB is involved in tumorigenesis by upregulating the anti-apoptotic pathway and tumor cell survival. The canonical pathway of NF-κB is known to activate the transcription of a group of anti-apoptotic proteins, which is further subdivided into two categories, namely, inhibitors of apoptotic proteins and the Bcl-2 family members [88]. In addition to transcription, inhibition of NF-κB activity promotes JNK activity and apoptosis, suggesting that NF-κB inhibits apoptosis via inhibition of JNK activity [89]. Pro-survival functions of NF-κB are related to the expression of phosphoinositide 3-kinase (PI3K/Akt cascade), one of the key elements in promoting cell proliferation and growth [90] (Figure 5).

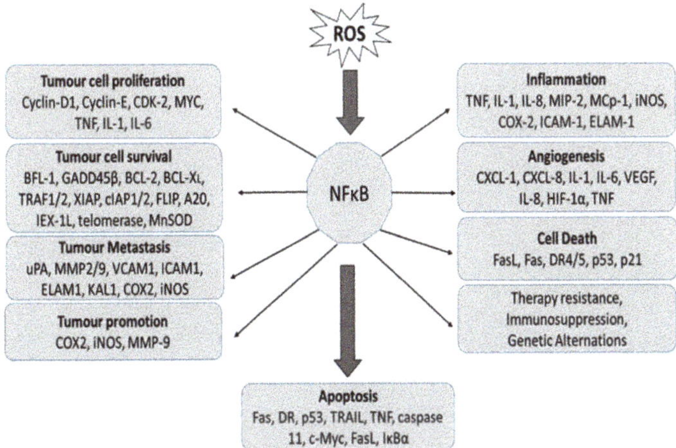

Figure 5. NF-κB activation in the progression of cancer by regulating genes involved in tumor cell proliferation, cell growth, survival, angiogenesis, tumor promotion, and metastasis. BCL2: B-cell lymphoma protein 2; BCL-XL, also known as BCL2-like 1; BFL1, also known as BCL2A1; CDK2: cyclin-dependent kinase 2; COX2: cyclooxygenase 2; CXCL: chemokine (C-X-C motif) ligand; DR: death receptor; ELAM1: endothelial adhesion molecule 1; FLIP, also known as CASP8; GADD45beta: growth arrest and DNA-damage-inducible protein beta; HIF1alpha: hypoxia-inducible factor 1 alpha; ICAM1: intracellular adhesion molecule 1; IEX-1L: radiation-inducible immediate early gene (also known as IER3); IL: interleukin; iNOS: inducible nitric oxide synthase; KAL1: Kallmann syndrome 1 sequence; MCP1: monocyte chemoattractant protein 1 (also known as CCL2); MIP2: macrophage inflammatory protein 2; MMP: matrix metalloproteinase; MnSOD: manganese superoxide dismutase (also known as SOD2); TNF: tumor necrosis factor; TRAF: TNF receptor-associated factor; uPA: urokinase plasminogen activator; VCAM1: vascular cell adhesion molecule 1; VEGF: vascular endothelial growth factor; XIAP: X-linked inhibitor of apoptosis protein.

2.8. NF-κB in the Pathogenesis of Diabetes

Insulin-dependent (type 1) and insulin-independent (type 2) diabetes are the two major variations of diabetes seen among humans. In type 1 diabetes, beta pancreatic cells are prone to autoimmune attack by cytokines, such as interferon and IL-1, and NF-κB is mostly inactive in resting cells. However, IL-1 can activate the translocation of NF-κB to the nucleus [91]. NF-κB regulates the expression of cytokine-induced genes, which play very significant roles in pro- or anti-apoptotic cascades (Figure 6). Cytokine-induced NF-κB activation can be reverted by a non-degradable mutant form of inhibitory κB (IκB) (S32A, S36A) in recombinant adenovirus (AdIB (SA) 2) in human pancreatic islet cells or by transfection with a stable inhibitor of NF-κB in purified rat cells [92,93]. Moreover, intravenous administration of a "dummy" NF-κB and mice deficient in NF-κB (p50) showed resistance to alloxan- and streptozotocin-induced islet cell death [94]. This could be considered a strategy to protect cytokine-induced in vitro and in vivo apoptosis, whereby inhibition of NF-κB is achieved. This highlighted the pro-apoptotic effect of NF-κB activation in β-cells of the pancreas. Blocking NF-κB for a prolonged period displayed impaired development of endocrine cells and expression of important genes involved in the insulin-secretion pathway [95]. Interestingly, resistance to NF-κB activation could decrease the time course of the development of diabetes. Hence, it is more important to consider the mechanisms that regulate NF-κB in cells and genes in diabetic conditions before building up inhibition strategies.

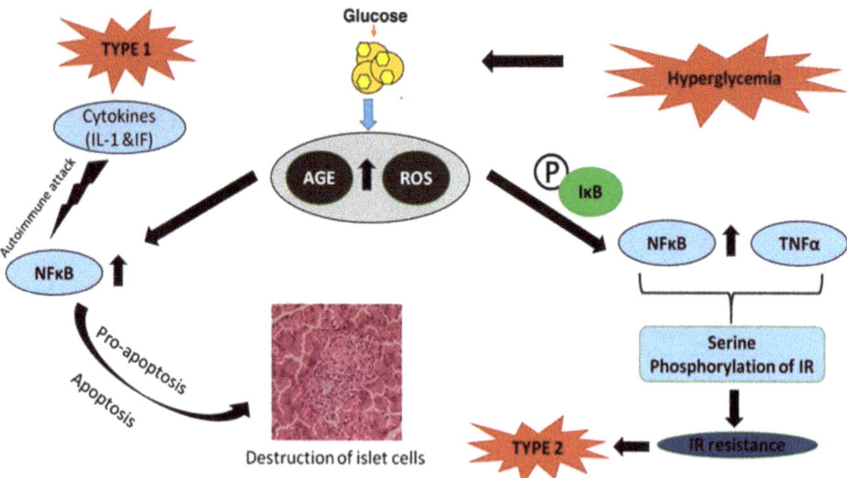

Figure 6. Role and activation of NF-κB in diabetes mellitus. NF-κB regulates the expression of cytokine-induced genes that affect pro- or anti-apoptotic cascades. NF-κB is one of the major causes in the development of insulin resistance. Tumor necrosis factor (TNF) predominantly induces insulin resistance by the serine phosphorylation of insulin receptor substrate-1 (IRS1), and NF-κB is a major cause of insulin resistance.

Type 2 diabetes is mainly characterized by insulin resistance. NF-κB is one of the chief components in the insulin resistance observed in type 2 diabetes. The anti-inflammatory agent aspirin prevents the degradation of NF-κB inhibitor (IB), whose expression in the liver attenuates the expression of NF-κB dependent genes, but also reduced the likelihood of type 2 diabetes development. Similarly, inflammation-induced hyperglycemia can be inhibited using specific inhibitors of interleukin-1 signaling [96]. Hence, it is obvious that the target genes of NF-κB, such as IL-1, TNF, and IL-6, and NF-κB itself, play vital roles in insulin resistance. NF-κB also antagonizes peroxisome proliferator-activated receptor (PPAR), which maintains glucose homeostasis in bone marrow-derived mesenchymal cells [97]. TNF predominantly induces insulin resistance by serine phosphorylation of insulin receptor substrate-1, which is also a potent activator of NF-κB (Figure 6). Evidence suggests that TNF is involved in insulin resistance in humans and animal adipose tissue and blocks reversion of the diabetic condition by masking the activity of NF-κB [98]. Hence, NF-κB is one of the major causes in the development of insulin resistance. Given the role of NF-κB in insulin regulation, it is important to understand its beneficial effects in insulin sensitivity and GLUT2 monitoring, a protein that controls glucose-stimulated insulin secretion by cells [95]. In certain cases, inhibition of the transcription factor NF-κB may inflict deleterious effects on GLUT 2, leading to diabetes progression. Although there are a lot of studies highlighting the role of NF-κB in type 1 and 2 diabetes, its specific trail in the pathogenesis of diabetes requires further investigation.

2.9. NF-κB in Wound Healing

NF-κB is essential for inflammatory and phagocytic cell migration as it strictly regulates genes involved in cell proliferation (granulocyte colony-stimulating factor and macrophage colony-stimulating factor), transformation, and survival. In wound healing, the classical NF-κB pathway is activated as an innate immune reaction and, as a result, activation of NF-κB is more important in the protection of cells from infections or microbes [63]. The effects of NF-κB on cell survival mainly depends on the type and number of stimulatory molecules [99]. In normal wound healing responses, NF-κB functions as a positive signaling molecule to enhance the proliferation phase, and both increase and regulate the migration of keratinocytes toward re-epithelization sites [100,101]. Under normal physiological conditions, NF-κB protein levels are standard, hence there is no defined physiological level to clear dead cells and microbes during responses to infections or wounds. Any defect in NF-κB activation inhibits the innate immune reaction, whereas the elevated production of NF-κB increases inflammatory cytokine levels and leads to tumor formation [102]. Every wound is different and requires integrated defence and repair mechanisms. In clinical settings, wound evaluation cannot solely depend upon Nrf2 and NF-κB but can indicate an inadequate defence against oxidative stress or innate immune reactions at large in non-healing wounds. For wounds with prolonged inflammation, compounds that decrease NF-κB levels or activate the Nrf2 pathway can be useful, and Nrf2 and NF-κB as markers are mainly utilized to assess healing processes. In immunosuppressed patients, NF-κB induction can help to initiate the correct immune reaction, but the multifaceted network of active molecules in the whole body can cause a different cell response than that observed in vivo [103].

3. Conclusions

Unusually high levels of ROS can allegedly be used for cancer diagnosis, varying according to tumor type, and are potent signaling molecules in cancer scenarios, thereby leading to nuclear damage, genetic instability, and tumorigenesis. The regulation of redox transcription factors is dependent on the generation of ROS levels. NF-κB was the first redox-regulated transcription factor to be discovered. It is a protein complex with multiple functions in immune, inflammation, cell growth, and survival responses. ROS have NF-κB stimulatory effects in the cytoplasm and inhibitory effects in the nucleus. Therefore, ROS have the ability to both activate and suppress NF-κB signaling pathways. PBM promotes photon availability and absorption by chromophores in the catalytic center of mitochondrial cytochrome c oxidase the and subsequent reduction of molecular oxygen. This disturbs the mitochondrial membrane potential and changes the fission–fusion homeostasis in the mitochondrial network, leading to increased levels of intermediates such as ATP, cAMP, and ROS. PBM-mediated ROS interact with certain family proteins that enclose highly reactive cysteine (Cys) residues and after the oxidation of Cys residues. The activity of NF-κB is regulated by direct activation or after disassembling from IkB, an inhibitor, then the released NF-κB translocate to the nucleus and promotes gene transcription. The regulation of crosstalk between ROS and NF-κB appears to be a key factor for better management of many conditions. PBM facilitates the reduction of oxidative stress and activation of mechanisms that upregulate or downregulate NF-κB, which are critical for the promotion of anti-apoptotic responses and tumor progression, insulin dependence or resistance in diabetes, cell proliferation and migration in wounded sites, or cell survival by ROS attenuation.

Author Contributions: N.K.R., B.P.G., R.C., and I.M.T. equally contributed to the writing of the manuscript. N.H. and H.A. contributed to the reviewing of the manuscript.

Funding: This work is based on the research supported by the South African Research Chairs Initiative of the Department of Science and Technology and National Research Foundation of South Africa (Grant No 98337).

Acknowledgments: The authors sincerely thank the University of Johannesburg, South Africa for their support.

References

1. Zorov, D.B.; Juhaszova, M.; Sollott, S.J. Mitochondrial reactive oxygen species (ROS) and ROS-induced ROS release. *Physiol. Rev.* **2014**, *94*, 909–950. [CrossRef] [PubMed]
2. Panieri, E.; Santoro, M.M. ROS homeostasis and metabolism: A dangerous liaison in cancer cells. *Cell Death Dis.* **2016**, *7*, e2253. [CrossRef] [PubMed]
3. Stanicka, J.; Russell, E.G.; Woolley, J.F.; Cotter, T.G. NADPH oxidase-generated hydrogen peroxide induces DNA damage in mutant FLT3-expressing leukemia cells. *J. Biol. Chem.* **2015**, *290*, 9348–9361. [CrossRef] [PubMed]

4. Roy, K.; Wu, Y.; Meitzler, J.L.; Juhasz, A.; Liu, H.; Jiang, G.; Lu, J.; Antony, S.; Doroshow, J.H. NADPH oxidases and cancer. *Clin. Sci.* **2015**, *128*, 863–875. [CrossRef]

5. Gorrini, C.; Harris, I.S.; Mak, T.W. Modulation of oxidative stress as an anticancer strategy. *Nat. Rev. Drug Discov.* **2013**, *12*, 931–947. [CrossRef]

6. Gonzalez, C.D.; Lee, M.S.; Marchetti, P.; Pietropaolo, M.; Towns, R.; Vaccaro, M.I.; Watada, H.; Wiley, J.W. The emerging role of autophagy in the pathophysiology of diabetes mellitus. *Autophagy* **2011**, *7*, 2–11. [CrossRef]

7. Krakauer, T. Inflammasome, mTORC1 activation, and metabolic derangement contribute to the susceptibility of diabetics to infections. *Med. Hypotheses* **2015**, *85*, 997–1001. [CrossRef]

8. O'Toole, E.A.; Goel, M.; Woodley, D.T. Hydrogen peroxide inhibits human keratinocyte migration. *Dermatol. Surg.* **1996**, *22*, 525–529. [CrossRef]

9. Auf Dem Keller, U.; Angelika, K.; Susanne, B.; Werner, S. Reactive oxygen species and their detoxification in healing skin wounds. *J. Investig. Dermatol. Symp. Proc.* **2006**, *11*, 106–111. [CrossRef]

10. Vander Heiden, M.G.; Cantley, L.C.; Thompson, C.B. Understanding the Warburg effect: The metabolic requirements of cell proliferation. *Science* **2009**, *324*, 1029–1033. [CrossRef]

11. Levine, A.J.; Puzio-Kuter, A.M. The control of the metabolic switch in cancers by oncogenes and tumor suppressor genes. *Science* **2010**, *330*, 1340–1344. [CrossRef] [PubMed]

12. Santos, L.; Olmo-Aguado, S.D.; Valenzuela, P.L.; Winge, K.; Iglesias-Soler, E.; Arguelles-Luis, J.; Alvarez-Valle, S.; Parcero-Iglesias, G.J.; Fernandez-Martinez, A.; Lucia, A. Photobiomodulation in Parkinson's disease: A randomized controlled trial. *Brain Stimul.* **2019**, *12*, 810–812. [CrossRef]

13. Chow, R.T.; Johnson, M.I.; Lopes-Martins, R.A.; Bjordal, J.M. Efficacy of low-level laser therapy in the management of neck pain: A systematic review and meta-analysis of randomised placebo or active-treatment controlled trials. *Lancet* **2009**, *374*, 1897–1908. [CrossRef]

14. Lavery, L.A.; Murdoch, D.P.; Williams, J.; Lavery, D.C. Does anodyne light therapy improve peripheral neuropathy in diabetes? A double-blind, sham-controlled, randomized trial to evaluate monochromatic infrared photoenergy. *Diabetes Care* **2008**, *31*, 316–321. [CrossRef] [PubMed]

15. Arnall, D.A.; Nelson, A.G.; Lopez, L.; Sanz, N.; Iversen, L.; Sanz, I.; Stambaugh, L.; Arnall, S.B. The restorative effects of pulsed infrared light therapy on significant loss of peripheral protective sensation in patients with long-term type 1 and type 2 diabetes mellitus. *Acta Diabetol.* **2006**, *43*, 26–33. [CrossRef] [PubMed]

16. Brosseau, L.; Robinson, V.; Wells, G.; Debie, R.; Gam, A.; Harman, K.; Morin, M.; Shea, B.; Tugwell, P. Low level laser therapy (Classes I, II and III) for treating rheumatoid arthritis. *Cochrane Database Syst. Rev.* **2005**, CD002049. [CrossRef]

17. Huang, Z.; Ma, J.; Chen, J.; Shen, B.; Pei, F.; Kraus, V.B. The effectiveness of low-level laser therapy for nonspecific chronic low back pain: A systematic review and meta-analysis. *Arthritis Res. Ther.* **2015**, *17*, 360. [CrossRef]

18. Yousefi-Nooraie, R.; Schonstein, E.; Heidari, K.; Rashidian, A.; Pennick, V.; Akbari-Kamrani, M.; Irani, S.; Shakiba, B.; Mortaz Hejri, S.A.; Mortaz Hejri, S.O. Low level laser therapy for nonspecific low-back pain. *Cochrane Database Syst. Rev.* **2008**, CD005107. [CrossRef]

19. Chung, H.; Dai, T.; Sharma, S.K.; Huang, Y.Y.; Carroll, J.D.; Hamblin, M.R. The nuts and bolts of low-level laser (light) therapy. *Ann. Biomed. Eng.* **2012**, *40*, 516–533. [CrossRef]

20. Yun, S.H.; Kwok, S.J.J. Light in diagnosis, therapy and surgery. *Nat. Biomed. Eng.* **2017**, *1*, 0008. [CrossRef]

21. Passarella, S.; Casamassima, E.; Molinari, S.; Pastore, D.; Quagliariello, E.; Catalano, I.M.; Cingolani, A. Increase of proton electrochemical potential and ATP synthesis in rat liver mitochondria irradiated in vitro by helium-neon laser. *FEBS Lett.* **1984**, *175*, 95–99. [CrossRef]

22. Lynnyk, A.; Lunova, M.; Jirsa, M.; Egorova, D.; Kulikov, A.; Kubinova, S.; Lunov, O.; Dejneka, A. Manipulating the mitochondria activity in human hepatic cell line Huh7 by low-power laser irradiation. *Biomed. Opt. Express* **2018**, *9*, 1283–1300. [CrossRef] [PubMed]

23. Lunova, M.; Smolkova, B.; Uzhytchak, M.; Janouskova, K.Z.; Jirsa, M.; Egorova, D.; Kulikov, A.; Kubinova, S.; Dejneka, A.; Lunov, O. Light-induced modulation of the mitochondrial respiratory chain activity: Possibilities and limitations. *Cell. Mol. Life Sci.* **2019**. [CrossRef] [PubMed]

24. Monro, S.; Colon, K.L.; Yin, H.; Roque, J., 3rd; Konda, P.; Gujar, S.; Thummel, R.P.; Lilge, L.; Cameron, C.G.; McFarland, S.A. Transition metal complexes and photodynamic therapy from a tumor-centered approach: Challenges, opportunities, and highlights from the development of TLD1433. *Chem. Rev.* **2019**, *119*, 797–828. [CrossRef] [PubMed]

25. Huang, Z.; Xu, H.; Meyers, A.D.; Musani, A.I.; Wang, L.; Tagg, R.; Barqawi, A.B.; Chen, Y.K. Photodynamic therapy for treatment of solid tumors—Potential and technical challenges. *Technol. Cancer Res. Treat.* **2008**, *7*, 309–320. [CrossRef]

26. Jayavelu, A.K.; Moloney, J.N.; Bohmer, F.D.; Cotter, T.G. NOX-driven ROS formation in cell transformation of FLT3-ITD-positive AML. *Exp. Hematol.* **2016**, *44*, 1113–1122. [CrossRef]

27. Reczek, C.R.; Chandel, N.S. ROS-dependent signal transduction. *Curr. Opin. Cell Biol.* **2015**, *33*, 8–13. [CrossRef]

28. Groeger, G.; Quiney, C.; Cotter, T.G. Hydrogen peroxide as a cell-survival signaling molecule. *Antioxid. Redox Signal.* **2009**, *11*, 2655–2671. [CrossRef]

29. Kroller-Schon, S.; Steven, S.; Kossmann, S.; Scholz, A.; Daub, S.; Oelze, M.; Xia, N.; Hausding, M.; Mikhed, Y.; Zinssius, E.; et al. Molecular mechanisms of the crosstalk between mitochondria and NADPH oxidase through reactive oxygen species—Studies in white blood cells and in animal models. *Antioxid. Redox Signal.* **2014**, *20*, 247–266. [CrossRef]

30. Quinlan, C.L.; Terberg, J.R.; Perevoshchikova, I.V. Native rates of superoxide production from multiple sites in isolated mitochondria measured using endogenous reporters. *Free Radic. Biol. Med.* **2012**, *53*, 1807–1817. [CrossRef]

31. Sabharwal, S.S.; Schumacker, P.T. Mitochondrial ROS in cancer: Initiators, amplifiers or an Achilles' heel? *Nat. Rev. Cancer* **2014**, *11*, 709–721. [CrossRef] [PubMed]

32. Murphy, M.P. How mitochondria produce reactive oxygen species. *Biochem. J.* **2009**, *417*, 1–13. [CrossRef]

33. Sena, L.A.; Chandel, N.S. Physiological roles of mitochondrial reactive oxygenspecies. *Mol. Cell* **2012**, *48*, 158–167. [CrossRef] [PubMed]

34. Bienert, G.P.; Muller, A.L.; Kristiansen, K.A.; Schulz, A.; Møller, I.M.; Schjoerring, J.K.; Jahn, T.P. Specific aquaporins facilitate the diffusion of hydrogen peroxide across membranes. *J. Biol. Chem.* **2007**, *282*, 1183–1192. [CrossRef]

35. Hart, P.C.; Mao, M.; de Abreu, A.L.; Ansenberger-Fricano, K.; Ekoue, D.N.; Ganini, D.; Kajdacsy-Balla, A.; Diamond, A.M.; Minshall, R.D.; Consolaro, M.E.; et al. MnSOD upregulation sustains the Warburg effect via mitochondrial ROS and AMPK-dependent signaling in cancer. *Nat. Commun.* **2015**, *6*, 6053. [CrossRef]

36. Cairns, R.A.; Harris, I.S.; Mak, T.W. Regulation of cancer cell metabolism. *Nat. Rev. Cancer* **2011**, *11*, 85–95. [CrossRef]

37. Finkel, T. Signal transduction by mitochondrial oxidants. *J. Biol. Chem.* **2012**, *287*, 4434–4440. [CrossRef]

38. Schafer, Z.T.; Grassian, A.R.; Song, L.; Jiang, Z.; Gerhart-Hines, Z.; Irie, H.Y.; Gao, S.; Puigserver, P.; Brugge, J.S. Antioxidant and oncogene rescue of metabolic defects caused by loss of matrix attachment. *Nature* **2009**, *461*, 109–113. [CrossRef]

39. Jiang, Z.Y.; Woollard, A.C.; Wolff, S.P. Hydrogen peroxide production during experimental protein glycation. *FEBS Lett.* **1990**, *268*, 69–71. [CrossRef]

40. Kawamura, M.; Heinecke, J.W.; Chait, A. Pathophysiological concentrations of glucose promote oxidative modification of low density lipoprotein by a superoxide dependent pathway. *J. Clin. Investig.* **1994**, *94*, 771–778. [CrossRef]

41. Goldin, A.; Beckman, J.A.; Schmidt, A.M.; Creage, A.M. Advanced Glycation End Products Sparking the Development of Diabetic Vascular Injury. *Basic Sci. Clin.* **2006**, *114*, 597–605.

42. McCarthy, A.D.; Etcheverry, S.B.; Cortizo, A.M. Effect of advanced glycation end products on the secretion of insulin-like growth factor-I and its binding proteins: Role in osteoblast development. *Acta Diabetol.* **2001**, *38*, 113–122. [CrossRef] [PubMed]

43. Vlassara, H. Recent progress in advanced glycation end products and diabetic complications. *Diabetes* **1997**, *46*, 19–25. [CrossRef] [PubMed]

44. Mohamed, A.K.; Bierhaus, A.; Schiekofer, S.; Tritschler, H.; Ziegler, R.; Nawroth, P.P. The role of oxidative stress and NF-kB activation in late diabetic complications. *BioFactors* **1999**, *10*, 157–167. [CrossRef]

45. Ceriello, A. Cardiovascular effects of acute hyperglycemia: Pathophysiological under pinnings. *Diabetes Vasc. Dis. Res.* **2008**, *5*, 260–268. [CrossRef]

46. Butler, R.; Morris, A.D.; Belch, J.J.; Hill, A.; Struthers, A.D. Allopurinol normalizes endothelial dysfunction in type 2 diabetics with mild hypertension. *Hypertension* **2000**, *35*, 746–751. [CrossRef]

47. Guo, S.; Dipietro, L.A. Factors affecting wound healing. *J. Dent. Res.* **2010**, *89*, 219–229. [CrossRef]

48. Kauppila, J.H.; Stewart, J.B. Mitochondrial DNA: Radically free of free-radical driven mutations. *Biochim. Biophys. Acta* **2015**, *1847*, 1354–1361. [CrossRef]

49. Greaves, L.C.; Reeve, A.K.; Taylor, R.W.; Turnbull, D.M. Mitochondrial DNA and disease. *J. Pathol.* **2012**, *226*, 274–286. [CrossRef]

50. Hornig-Do, H.T.; von Kleist-Retzow, J.C.; Lanz, K.; Wickenhauser, C.; Kudin, A.P.; Kunz, W.S.; Wiesner, R.J.; Schauen, M. Human epidermal keratinocytes accumulate superoxide due to low activity of Mn-SOD, leading to mitochondrial functional impairment. *J. Investig. Dermatol.* **2007**, *127*, 1084–1093. [CrossRef]

51. Sen, C.K.; Roy, S. Redox signals in wound healing. *Biochim. Biophys. Acta* **2008**, *1780*, 1348–1361. [CrossRef] [PubMed]

52. Hamblin, M.R. Shining light on the head: Photobiomodulation for brain disorders. *BBA. Clin.* **2016**, *6*, 113–124. [CrossRef] [PubMed]

53. De Freitas, L.F.; Hamblin, M.R. Proposed mechanisms of photobiomodulation or low-level light therapy. *IEEE J. Sel. Top Quant. Electron* **2016**, *22*, 4–17. [CrossRef] [PubMed]

54. Waypa, G.B.; Smith, K.A.; Schumacker, P.T. O_2 sensing, mitochondria and ROS signaling: The fog is lifting. *Mol. Asp. Med.* **2016**, *47–48*, 76–89. [CrossRef] [PubMed]

55. Zhao, Y.; Vanhoutte, P.M.; Leung, S.W. Vascular nitric oxide: Beyond eNOS. *J. Pharmacol. Sci.* **2015**, *129*, 83–94. [CrossRef] [PubMed]

56. Anders, J.J.; Lanzafame, R.J.; Arany, P.R. Low-level light/laser therapy versus photobiomodulation therapy. *Photomed. Laser Surg.* **2015**, *33*, 183–184. [CrossRef]

57. Chen, H.; Aaron, C.H.; Arany, P.R.; Hamblin, M.R. Role of reactive oxygen species in low level light theraphy. In *Mechanisms for Low-Light Theraphy IV*; Michael, R.H., Ronald, W.W., Juanita, A., Eds.; SPIE: San Jose, CA, USA, 2009; pp. 716502–716511.

58. Saini, R.; Lee, N.V.; Liu, K.Y.P.; Poh, C.F. Prospects in the Application of Photodynamic Therapy in Oral Cancer and Premalignant Lesions. *Cancers* **2016**, *8*, 83. [CrossRef]

59. Mitton, D.; Ackroyd, R. A brief overview of photodynamic therapy in Europe. *Photodiagn. Photodyn. Ther.* **2008**, *5*, 103–111. [CrossRef]

60. Castano, A.P.; Demidova, T.N.; Hamblin, M.R. Mechanisms in photodynamic therapy: Part two-cellular signaling, cell metabolism and modes of cell death. *Photodiagn. Photodyn. Ther.* **2005**, *2*, 1–23. [CrossRef]

61. Singleton, D.A.; Hang, C.; Szymanski, M.J.; Meyer, M.P.; Leach, A.G.; Kuwata, K.T.; Chen, J.S.; Greer, A.; Foote, C.S.; Houk, K.N. Mechanism of ene reactions of singlet oxygen. A two-step no-intermediate mechanism. *J. Am. Chem. Soc.* **2003**, *125*, 1319–1328. [CrossRef]

62. Bansal, A.; Yang, F.; Xi, T.; Zhang, Y.; Ho, J.S. In vivo wireless photonic photodynamic therapy. *Proc. Natl. Acad. Sci. USA* **2018**, *115*. [CrossRef] [PubMed]

63. Hayden, M.S.; Ghosh, S. NF-kB, the first quarter-century: Remarkable progress and outstanding questions. *Genes Dev.* **2012**, *26*, 203–234. [CrossRef] [PubMed]

64. Hoesel, B.; Schmid, J.A. The complexity of NF-kappa B signaling in inflammation and cancer. *Mol. Cancer* **2013**, *12*, 86–93. [CrossRef] [PubMed]

65. Mercurio, F.; DiDonato, J.A.; Rosette, C.; Karin, M. p105 and p98 precursor proteins play an active role in NF-kappa B-mediated signal transduction. *Genes Dev.* **1993**, *7*, 705–718. [CrossRef]

66. Pahl, H.L. Activators and target genes of Rel/NF-kappaB transcription factors. *Oncogene* **1999**, *18*, 6853–6866. [CrossRef]

67. Matthews, J.R.; Kaszubska, W.; Turcatti, G.; Wells, T.N.; Hay, R.T. Role of cysteine62 in DNA recognition by the P50 subunit of NF-kappa B. *Nucleic Acids Res.* **1993**, *21*, 1727–1734. [CrossRef]

68. Klatt, P.; Lamas, S. Regulation of protein function by S-glutathiolation in response to oxidative and nitrosative stress. *Eur. J. Biochem.* **2000**, *267*, 4928–4944. [CrossRef]

69. Pineda-Molinam, E.; Klatt, P.; Vazquez, J.; Mariana, A.; Garcia de Lacabo, M.; Parez-Sala, D.; Lamas, S. Glutathionylation of the p50 subunit of NF-kappaB: A mechanism for redoxinduced inhibition of DNA binding. *Biochemistry* **2001**, *40*, 14134–14142. [CrossRef]

70. Kil, I.S.; Kim, S.Y.; Park, J.W. Glutathionylation regulates IkappaB. *Biochem. Biophys. Res. Commun.* **2008**, *373*, 169–173. [CrossRef]

71. Karin, M. NF-kappaB as a critical link between inflammation and cancer. *Cold Spring Harb. Perspect. Biol.* **2009**, *1*, a000141. [CrossRef]

72. Saccani, S.; Pantano, S.; Natoli, G. Modulation of NF-kappa-B activity by exchange of dimers. *Mol. Cell* **2003**, *11*, 1563–1574. [CrossRef]

73. Oeckinghaus, A.; Ghosh, S. The NF-κB family of transcription factors and its regulation. *Cold Spring Harb. Perspect. Biol.* **2009**, *1*, a000034. [CrossRef] [PubMed]

74. Oliveira-Marques, V.; Marinho, H.S.; Cyrne, L.; Antunes, F. Role of hydrogen peroxide in NF-κB activation: From inducer to modulator. *Antioxid. Redox Signal.* **2009**, *11*, 2223–2243. [CrossRef] [PubMed]

75. Kamata, H.; Manabe, T.; Oka, S.; Kamata, K.; Hirata, H. Hydrogen peroxide activates IκB kinases through phosphorylation of serine residues in the activation loops. *FEBS Lett.* **2002**, *519*, 231–237. [CrossRef]

76. Takada, Y.; Mukhopadhyay, A.; Kundu, G.C.; Mahabeleshwar, G.H.; Singh, S.; Aggarwal, B.B. Hydrogen peroxide activates NF-κB through tyrosine phosphorylation of IκBα and serine phosphorylation of p65: Evidence for the involvement of IκBα kinase and Syk protein-tyrosine kinase. *J. Biol. Chem.* **2003**, *26*, 24233–24241. [CrossRef] [PubMed]

77. Reynaert, N.L.; van der Vliet, A.; Guala, A.S.; McGovern, T.; Hristova, M.; Pantano, C.; Heintz, N.H.; Heim, J.; Ho, Y.S.; Matthews, D.E.; et al. Dynamic redox control of NF-κB through glutaredoxin-regulated S-glutathionylation of inhibitory κB kinase β. *Proc. Natl. Acad. Sci. USA* **2006**, *103*, 13086–13091. [CrossRef]

78. Kim, J.H.; Na, H.J.; Kim, C.K.; Kim, J.Y.; Ha, K.S.; Lee, H.; Chung, H.T.; Kwon, H.J.; Kwon, Y.G.; Kim, Y.M. The non-provitamin a carotenoid, lutein, inhibits NF-κB-dependent gene expression through redox-based regulation of the phosphatidylinositol 3-kinase/PTEN/Akt and NF-κB-inducing kinase pathways: Role of H2O2 in NF-κB activation. *Free Radic. Biol. Med.* **2008**, *45*, 885–896. [CrossRef]

79. Nakano, H.; Nakajima, A.; Sakon-Komazawa, S.; Piao, J.H.; Xue, X.; Okumura, K. Reactive oxygen species mediate crosstalk between NF-κB and JNK. *Cell Death Differ.* **2006**, *13*, 730–737. [CrossRef]

80. Ventura, J.J.; Cogswell, P.; Flavell, R.A.; Baldwin, A.S., Jr.; Davis, R.J. JNK potentiates TNF-stimulated necrosis by increasing the production of cytotoxic reactive oxygen species. *Genes Dev.* **2004**, *18*, 2905–2915. [CrossRef]

81. Kabe, Y.; Ando, K.; Hirao, S.; Yoshida, M.; Handa, H. Redox regulation of NF-κB activation: Distinct redox regulation between the cytoplasm and the nucleus. *Antioxid. Redox Signal.* **2005**, *7*, 395–403. [CrossRef]

82. Saito, Y.; Nishio, K.; Ogawa, Y.; Kimata, J.; Kinumi, T.; Yoshida, Y.; Noguchi, N.; Niki, E. Turning point in apoptosis/necrosis induced by hydrogen peroxide. *Free Radic. Res.* **2006**, *40*, 619–630. [CrossRef] [PubMed]

83. Perkins, N.D.; Gilmore, T.D. Good cop, bad cop: The different faces of NF-kappa B. *Cell Death Differ.* **2006**, *13*, 759–772. [CrossRef] [PubMed]

84. Ak, P.; Levine, A.J. p53 and NF-kappa B: Different strategies for responding to stress lead to a functional antagonism. *FASEB J.* **2010**, *24*, 3643–3652. [CrossRef] [PubMed]

85. Zhang, R.; Kang, K.A.; Kim, K.C.; Na, S.Y.; Chang, W.Y.; Kim, G.Y.; Kim, H.S.; Hyun, J.W. Oxidative stress causes epigenetic alteration of CDX1 expression in colorectal cancer cells. *Gene* **2013**, *524*, 214–219. [CrossRef] [PubMed]

86. Shimizu, T.; Marusawa, H.; Endo, Y.; Chiba, T. Inflammation- mediated genomic instability: Roles of activation-induced cytidine deaminase in carcinogenesis. *Cancer Sci.* **2012**, *103*, 1201–1206. [CrossRef] [PubMed]

87. Takai, A.; Marusawa, H.; Minaki, Y.; Watanabe, T.; Nakase, H.; Kinoshita, K.; Tsujimoto, G.; Chiba, T. Targeting activation-induced cytidine deaminase prevents colon cancer development despite persistent colonic inflammation. *Oncogene* **2012**, *31*, 1733–1742. [CrossRef]

88. Luo, J.L.; Kamata, H.; Karin, M. IKK/NF-kappa B signaling: Balancing life and death–a new approach to cancer therapy. *J. Clin. Investig.* **2005**, *115*, 2625–2632. [CrossRef]

89. De Smaele, E.; Zazzeroni, F.; Papa, S.; Nguyen, D.U.; Jin, R.; Jones, J.; Cong, R.; Franzoso, G. Induction of gadd45beta by NF-kappa B downregulates pro-apoptotic JNK signaling. *Nature* **2001**, *414*, 308–313. [CrossRef]

90. Chen, C.; Edelstein, L.C.; Gélinas, C. The Rel/NFkappaB family directly activates expression of the apoptosis inhibitor Bcl-x (L). *Mol. Cell Biol.* **2000**, *20*, 2687–2695. [CrossRef]

91. Cardozo, A.K.; Heimberg, H.; Heremans, Y.; Leeman, R.; Kutlu, B.; Kruhøffer, M.; Ørntoft, T.; Eizirik, D.L. A comprehensive analysis of cytokine-induced and nuclear factor-B dependent genes in primary rat pancreatic beta cells. *J. Biol. Chem.* **2001**, *276*, 879–886. [CrossRef]

92. Heimberg, H.; Heremans, Y.; Jobin, C.; Leemans, R.; Cardozo, A.K.; Darville, M.; Eizirik, D.L. Inhibition of cytokine induced NF-B activation by adenovirus-mediated expression of a NF- B super-repressor prevents beta cell apoptosis. *Diabetes* **2001**, *50*, 2219–2224. [CrossRef] [PubMed]

93. Giannoukakis, N.; Rudert, W.A.; Trucco, M.; Robbins, P.D. Protection of human islets from the effects of interleukin- by adenoviral gene transfer of an IB repressor. *J. Biol. Chem.* **2000**, *47*, 36509–36513. [CrossRef] [PubMed]

94. Mabley, J.G.; Haskó, G.; Liaudet, L.; Soriano, F.G.; Southan, G.J.; Salzman, A.L.; Szabó, C. NFB1 (p50)-deficient mice are not susceptible to multiple low-dose streptozotocin induced diabetes. *J. Endocrinol.* **2002**, *173*, 457–464. [CrossRef] [PubMed]

95. Norlin, S.; Ahlgren, U.; Edlund, H. Nuclear factor-kB activity in k-cells is required for glucose-stimulated insulin secretion. *Diabetes* **2005**, *54*, 125–132. [CrossRef] [PubMed]

96. Arkan, M.C.; Hevener, A.L.; Greten, F.R.; Maeda, S.; Li, Z.W.; Long, J.M.; Wynshaw-Boris, A.; Poli, G.; Olefsky, J.; Karin, M. IKK- links inflammation to obesity-induced insulin resistance. *Nat. Med.* **2005**, *11*, 191–198. [CrossRef]

97. Suzawa, M.; Takada, I.; Yanagisawa, J.; Ohtake, F.; Ogawa, S.; Yamauchi, T.; Kadowaki, T.; Takeuchi, Y.; Shibuya, H.; Gotoh, Y.; et al. Cytokines suppress adipogenesis and PPAR-gamma function through the TAK1/TAB1/NIK cascade. *Nat. Cell Biol.* **2003**, *5*, 224–230. [CrossRef]

98. Hotamisligil, G.S.; Arner, P.; Caro, J.F.; Atkinson, R.L.; Spiegelman, B.M. Increased adipose tissue expression of tumor necrosis factor-alpha in human obesity and insulin resistance. *J. Clin. Investig.* **1995**, *95*, 2409–2415. [CrossRef]

99. Jobin, C.; Sartor, R.B. The IκB/NF-κB system: A key determinant of mucosal inflammation and protection. *Am. J. Physiol. Cell Physiol.* **2000**, *278*, 451–462. [CrossRef]

100. Na, J.; Lee, K.; Na, W.; Shin, J.Y.; Lee, M.J.; Yune, T.Y.; Lee, H.K.; Jung, H.S.; Kim, W.S.; Ju, B.G. Histone H3K27 demethylase JMJD3 in cooperation with NF-κB regulates keratinocyte wound healing. *J. Investig. Dermatol.* **2016**, *136*, 847–858. [CrossRef]

101. Schreml, S.; Szeimies, R.M.; Karrer, S.; Heinlin, J.; Landthaler, M.; Babilas, P. Wound healing in 21st century. *J. Am. Acad. Dermatol.* **2010**, *63*, 866–881. [CrossRef]

102. Lizzul, P.F.; Aphale, A.; Malaviya, R.; Sun, Y.; Masud, S.; Dombrovskiy, V.; Gottlieb, A.B. Differential expression of phosphorylated NF-κB/ RelA in normal and psoriatic epidermis and downregulation of NF-κB in response to treatment with etanercept. *J. Investig. Dermatol.* **2005**, *124*, 1275–1283. [CrossRef] [PubMed]

103. Ambrozovaa, N.; Ulrichova, J.; Galandakova, A. Models for the study of skin wound healing. The role of Nrf2 and NF-κB. *Biomed. Pap. Med. Fac. Palacky Univ. Olomouc Czech Repub.* **2017**, *161*, 1–13. [CrossRef] [PubMed]

Review

Potential Applications of NRF2 Modulators in Cancer Therapy

Emiliano Panieri [1,*], Aleksandra Buha [2], Pelin Telkoparan-Akillilar [3], Dilek Cevik [3], Demetrios Kouretas [4], Aristidis Veskoukis [4], Zoi Skaperda [4], Aristidis Tsatsakis [5], David Wallace [6], Sibel Suzen [7] and Luciano Saso [1]

[1] Department of Physiology and Pharmacology "Vittorio Erspamer", Sapienza University of Rome, 00185 Rome, Italy; luciano.saso@uniroma1.it

[2] Department of Toxicology "Akademik Danilo Soldatović", University of Belgrade-Faculty of Pharmacy, 11000 Belgrade, Serbia; aleksandra.buha@pharmacy.bg.ac.rs

[3] Department of Medical Biology, Faculty of Medicine, Yuksek Ihtisas University, 06520 Balgat, Ankara, Turkey; pelinta@yiu.edu.tr (P.T.-A.); dilekcevik@yiu.edu.tr (D.C.)

[4] Department of Biochemistry-Biotechnology University of Thessaly Viopolis, Mezourlo, 41500 Larissa, Greece; dkouret@uth.gr (D.K.); veskoukis@uth.gr (A.V.); skaperda@uth.gr (Z.S.)

[5] Laboratory of Toxicology Science and Research, Medical School, University of Crete, 71003 Heraklion, Crete, Greece; tsatsaka@uoc.gr

[6] School of Biomedical Science, Oklahoma State University Center for Health Sciences, Tulsa, OK 74107-1898, USA; david.wallace@okstate.edu

[7] Department of Pharmaceutical Chemistry, Faculty of Pharmacy, Ankara University, 06100 Tandogan, Ankara, Turkey; Sibel.Suzen@pharmacy.ankara.edu.tr

* Correspondence: emiliano.panieri@hotmail.it

Received: 30 January 2020; Accepted: 21 February 2020; Published: 25 February 2020

Abstract: The nuclear factor erythroid 2-related factor 2 (NRF2)–Kelch-like ECH-associated protein 1 (KEAP1) regulatory pathway plays an essential role in protecting cells and tissues from oxidative, electrophilic, and xenobiotic stress. By controlling the transactivation of over 500 cytoprotective genes, the NRF2 transcription factor has been implicated in the physiopathology of several human diseases, including cancer. In this respect, accumulating evidence indicates that NRF2 can act as a double-edged sword, being able to mediate tumor suppressive or pro-oncogenic functions, depending on the specific biological context of its activation. Thus, a better understanding of the mechanisms that control NRF2 functions and the most appropriate context of its activation is a prerequisite for the development of effective therapeutic strategies based on NRF2 modulation. In line of principle, the controlled activation of NRF2 might reduce the risk of cancer initiation and development in normal cells by scavenging reactive-oxygen species (ROS) and by preventing genomic instability through decreased DNA damage. In contrast however, already transformed cells with constitutive or prolonged activation of NRF2 signaling might represent a major clinical hurdle and exhibit an aggressive phenotype characterized by therapy resistance and unfavorable prognosis, requiring the use of NRF2 inhibitors. In this review, we will focus on the dual roles of the NRF2-KEAP1 pathway in cancer promotion and inhibition, describing the mechanisms of its activation and potential therapeutic strategies based on the use of context-specific modulation of NRF2.

Keywords: NRF2-KEAP1; ROS; cancer metabolism; antioxidant; oxidative stress; cancer therapy; chemoresistance; radioresistance

---

1. Introduction

Nuclear factor erythroid 2-related factor 2 (NRF2) is a key transcription factor and a key modulator of cellular antioxidant responses that regulates the expression of genes encoding antioxidant enzymes with a protective role against different types of oxidative changes. NRF2 in combination with its own negative regulator, Kelch-like ECH-associated protein 1 (KEAP1), has become the center of a debate regarding whether NRF2 suppresses the tumor promotion or, conversely, exerts pro-oncogenic functions. Based on this, the present review will describe the role of NRF2 in cancer prevention and promotion, discussing potential advantages and disadvantages derived from its therapeutic modulation in cancer prevention and treatment. As it is known, under non-stressed conditions, NRF2 is constitutively poly-ubiquitinated by the CUL3-KEAP1 E3 ubiquitin ligase complex and subjected to degradation through the proteasome pathway [1]. After exposure to several redox altering stimuli, highly reactive thiols of KEAP1 are subjected to instant modification, leading to NRF2 stabilization caused by its decreased affinity for the CUL3-KEAP1 complex. Subsequently, NRF2 translocates into the nucleus and binds to the antioxidant response element (ARE) located within the promoter region of specific target genes, inducing the expression of a large number of cytoprotective proteins with antioxidant and detoxifying roles [2–5]. The importance of NRF2 function has been demonstrated by several studies using NRF2-deficient mice showing an increased susceptibility to redox disturbances and xenobiotic stress [6–8]. It has also been shown that tissue oxidative damage after ischemia and reperfusion is efficiently counteracted by NRF2 induction [9]. In line with the protective roles of the KEAP1-NRF2 pathway, its activation seems to effectively prevent carcinogenesis by promoting a number of antioxidant mechanisms [10,11]. Thus, NRF2 activation may have beneficial role as a result of its suppressive effect on carcinogenesis. On the other hand, increasing evidence show that constitutive NRF2 activation contributes to the progression of various cancer types. Specifically, many studies have shown that the increased activation of NRF2 in cancer cells leads to its augmented transcriptional activity and promotes tumor progression [12–14], metastasis formation [15], resistance to chemo-radiotherapy [16–19], and is clinically associated with poor prognosis [20]. During the last decade, several mechanisms through which NRF2 signaling pathway is persistently activated in different types of cancers have been discussed. Regarding NRF2, its oncogenic activity promotes cancer cell growth and proliferation, suppression of cancer cell apoptosis, self-renewal of cancer stem cells, therapy resistance, increased angiogenesis and anti-inflammatory activities [21]. As the pro-tumorigenic role of this factor has gained interest, pharmacological modulation of the NRF2 pathway offers pioneering therapeutic opportunities against several diseases. Recent studies have brought to light a few small molecules displaying promising properties in NRF2 inhibition, but their applicability still needs to be further investigated [22–25]. Most of these molecules lack specificity and have off-target toxic effects since they easily react with cysteine residues of different molecules. Metabolic instability, low bioavailability, and poor membrane permeability are some of the basic drawbacks in the administration of many NRF2 inhibitors [26]. To date, dimethyl fumarate is the only NRF2 activator approved by the Food and Drug Administration (FDA) but its function in cancer prevention has not been examined yet.

Structure and Function of NRF2 and KEAP1

Nuclear factor erythroid 2-related factor 2 (NRF2) is a transcriptional factor encoded by the *NFE2L2* gene that belongs to "Cap'N'Collar" type of basic region leucine zipper factor family (CNC-bZIP) [27]. Human NRF2 protein is 605 amino acids long and contains seven conserved NRF2-ECH homology domains known as Neh1-Neh7 [27,28]. Neh2 is a major regulatory domain located to N-terminus of NRF2 and it has two binding sites known as DLG and ETGE. These sites help to regulate NRF2 stability by interacting with the Kelch domains of E3 ubiquitin ligase Kelch-like ECH-associated protein 1 (KEAP1), a substrate of Cullin 3-based ubiquitin E3 ligase complex that ubiquitinates and targets NRF2 for proteasomal degradation [29–32]. The Neh1 and Neh6 domains have also been shown to control NRF2 stability. The Neh1 contains a basic leucine zipper motif that is also known as DNA binding domain and it enhances NRF2 transcriptional activation [27,33]. The Neh6 domain is a serine-rich domain containing two motifs (DSGIS and DSAPGS) that negatively modulate NRF2 stability through beta-TrCP dependent but KEAP1 independent regulation [34]. The Neh3, Neh4, and Neh5 domains are known as trans-activation domains of NRF2. The carboxy-terminal Neh3 domain binds to CHD6 (a chromo-ATPase/helicase DNA-binding protein) that is the transcriptional co-activator of NRF2 [35]. The Neh4 and Neh5 domains interact with the CH3 domains of CBP (CREB-binding protein) that facilitates transactivation of NRF2 target genes [36,37]. In addition, a seventh domain of NRF2 known as Neh7 has been shown to interact with a nuclear receptor retinoic X receptor alpha (RXRa) that inhibits NRF2 target genes transcription [38]. A schematic representation of NRF2 structure is shown in Figure 1A.

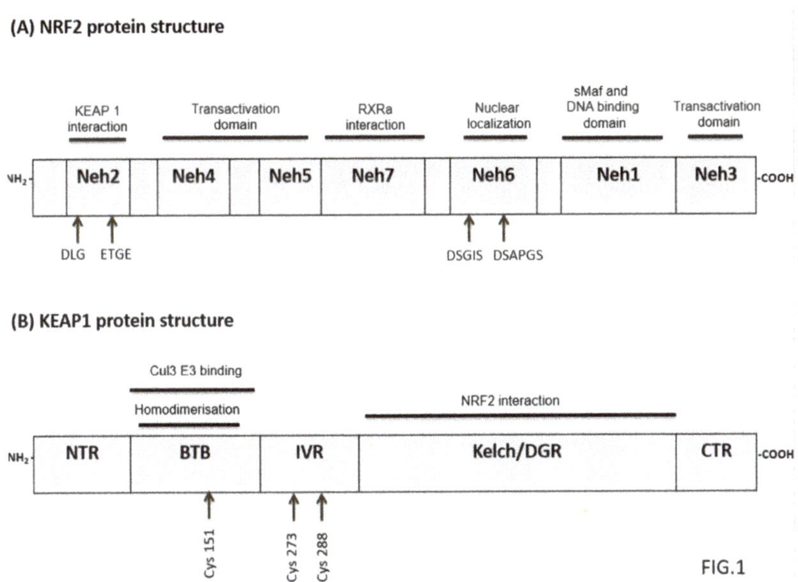

Figure 1. Domain architectures of Kelch-like ECH-associated protein 1 (KEAP1) and nuclear factor erythroid 2-related factor 2 (NRF2). (**A**) Human NRF2 protein is 605 amino acids long and contains seven Neh domains. The Neh1 contains a basic leucine zipper motif that is responsible for dimerization with sMaf protein and ARE sequence binding in DNA. Neh2 has two binding sites known as DLG and ETGE that control KEAP1 interaction. The Neh6 domain is a serine-rich domain containing two motifs (DSGIS and DSAPGS) that negatively regulate NRF2 stability. The Neh7 domain interacts with a nuclear receptor RXRα. The Neh3, Neh4, and Neh5 domains are known as trans-activation domains of NRF2. (**B**) KEAP1 is a 69-kDa protein and contains five domains. The BTB domain is critical for KEAP1 dimerization and recruitment of Cul3-based E3-ligase. The IVR domain has hypercritical cysteine residues, Cys273 and Cys288 that are essential for controlling NRF2 activity. Kelch/DGR domain negatively regulates NRF2 activation by interacting with conserved carboxyl terminus of Neh2 domain. BTB, broad complex, tram-track and bric-a-brac; CTR, C-terminal region; Cul3, Cullin3; IVR, intervening region; KEAP1, kelch-like ECH-associated protein 1; sMaf, musculoaponeurotic fibrosarcoma oncogene; Neh, NRF2-ECH homologous structure; NRF2, nuclear factor erythroid-2-related factor-2; NTR, N-terminal region; RXRα,.

KEAP1 is a 69-kDa protein, which belongs to the BTB-Kelch family of proteins [39]. All the members of this family assemble with Cullin-RING ligases that catalyze general protein ubiquitylation [39]. KEAP1 contains five domains including N-terminal region, the Cullin3 binding broad complex, tramtrack and broad complex/tramtrack/bric-a-brac (BTB) homodimerisation domain, the intervening region (IVR), the Kelch/double glycine repeat (DGR) domain and C-terminal domain [40,41]. The BTB domain is critical for KEAP1 dimerization and CUL3 assembly requires a BTB protein motif for ubiquitination and proteasomal degradation of NRF2 [41]. In addition, the BTB domain also contains a critical cysteine residue (Cys151) that has an important role in the activation of NRF2 [42]. The IVR/BACK domain contains highly reactive cysteines, namely Cys273 and Cys288 that function as a sensor for NRF2 inducers and are essential for controlling NRF2 activity [43]. Kelch/DGR domain negatively regulates NRF2 activation by interacting with conserved carboxyl terminus of Neh2 domain [44]. NRF2 controls antioxidant and cytoprotective genes expression and regulates cellular defense [23]. Under physiological conditions, NRF2 is kept at low levels in normal tissues by KEAP1-dependent ubiquitination and proteasomal degradation [29,30]. NRF2 localizes in the cytosol where ETGE and DLG motifs on its Neh2 domain associate with KEAP1 Kelch domain [45]. A schematic representation of KEAP1 structure is shown in Figure 1B. KEAP1 acts as an adaptor for NRF2 binding to the KEAP1-CUL3-E3 ubiquitin ligase complex, an event followed by rapid NRF2 proteasomal degradation [28]. Under oxidative stress (OS) conditions or in the presence of other stressors including reactive-oxygen species (ROS) or electrophiles, the activity of KEAP1 is decreased by direct modification of reactive cysteine residues in the IVR domain. These redox changes induced on KEAP1 thiols alter its structure and prevent KEAP1-mediated NRF2 ubiquitination [46,47]. Subsequently, NRF2 accumulates in the nucleus where it induces the expression of its target genes. In the nucleus, NRF2 heterodimerizes with small Maf proteins through Neh1 domain and promotes the antioxidant responsive elements (AREs)-dependent expression of cytoprotective genes [46] (see Figure 2). In this regard, over 500 NRF2 target genes have been so far identified and this number is expected to increase in the next coming future [48,49]. The genetic products of NRF2 activation can be categorized in functional categories covering multiple cellular processes including phase I, II, III drug/xenobiotic metabolism and detoxification, antioxidant response and redox homeostasis regulation, iron homeostasis, cellular metabolism, DNA repair, transcriptional activation, cell proliferation, and apoptosis [46,50]. A selected list of NRF2 target genes is presented in Table 1.

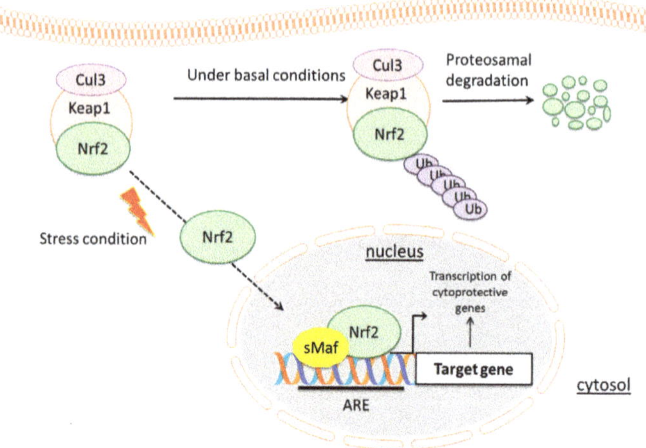

Figure 2. NRF2 regulation by KEAP1. Under basal conditions, NRF2 is constantly ubiquitinated through KEAP1 and degraded in the proteasome in cytosol. Under stress conditions, KEAP1-NRF2 interaction is stopped and free NRF2 translocates into nucleus. Then, NRF2 forms heterodimers with sMaf and binds to ARE sites within regulatory sites of antioxidant and detoxification genes. ARE, antioxidant response element; KEAP1, Kelch-like ECH-associated protein1; NRF2, nuclear erythroid-2 like factor-2; Retinoic X receptor alpha sMafs, small musculoaponeurotic fibrosarcoma oncogene family.

Table 1. Selected list of NRF2 target genes.

Gene	Coded Protein	Functional Category	Biological Role	Ref.
GCLC	Glutamate-Cysteine Ligase Catalytic Subunit	GSH Synthesis & Regeneration	GSH Synthesis	[51]
GCLM	Glutamate-Cysteine Ligase Modulatory Subunit	GSH Synthesis & Regeneration	GSH Synthesis	[51]
GSR1	Glutathione Reductase 1	GSH Synthesis & Regeneration	GSH Regeneration	[51]
SLC7A11	xCT, Light Subunit of Xc-Antiporter	GSH Synthesis & Regeneration	Cystine Uptake	[51]
PHGDH	Phosphoglycerate Dehydrogenase	GSH Synthesis & Regeneration	Serine/Glycine Synthesis	[21]
PSAT1	Phosphoserine Aminotransferase 1	GSH Synthesis & Regeneration	Serine/Glycine Synthesis	[21]
PSPH	Phosphoserine Phosphatase	GSH Synthesis & Regeneration	Serine/Glycine Synthesis	[21]
SHMT 1,2	Serine Hydroxymethyltransferase 1,2	GSH Synthesis & Regeneration	Serine/Glycine Synthesis	[21]
GPX1,2,4	Glutathione Peroxidase 1,2,4	ROS & Phase-II Xenobiotic Detoxification	ROS Scavenging	[21,51]
PRDX1,6	Peroxiredoxin 1,6	ROS & Phase-II Xenobiotic Detoxification	ROS Scavenging	[52]
TXN1	Thioredoxin 1	Thioredoxin-linked Antioxidant Role	Reduction of Sulfenylated-Proteins	[51]
TXNRD1	Thioredoxin Reductase-1	Thioredoxin-linked Antioxidant Role	Reduction of Thioredoxin	[51]
NQO1	NAD(P)H dehydrogenase Quinone 1	ROS & Phase-I Xenobiotic Detoxification	Reduction of quinones	[51,53]
AKR1B1	Aldo-Keto Reductase Family 1 Member B1	Phase-I Xenobiotic Detoxification	Reduction of aldehydes and ketones	[52]
AKR1B10	Aldo-Keto Reductase Family 1 Member B10	Phase-I Xenobiotic Detoxification	Reduction of aldehydes and ketones	[52]
AKR1C1	Aldo-Keto Reductase Family 1 Member C1	Phase-I Xenobiotic Detoxification	Reduction of aldehydes and ketones	[52]
ALDH1A1	Aldehyde Dehydrogenase 1 Family Member A1	Phase-I Xenobiotic Detoxification	Conversion of aldehydes to carboxylic acids	[54,55]
ALDH3A1	Aldehyde Dehydrogenase 3 Family Member A1	Phase-I Xenobiotic Detoxification	Conversion of aldehydes to carboxylic acids	[52]
GSTA 1,2,3,5	Glutathione-S Transferase A1,2,3,5	ROS & Phase-II Xenobiotic Detoxification	Conjugation of Glutathione to electrophiles	[51]
GSTM 1,2,3	Glutathione-S Transferase M1,2,3	ROS & Phase-II Xenobiotic Detoxification	Conjugation of Glutathione to electrophiles	[51]
UGT1A1,5	UDP Glucuronosyltransferase 1 A1,5	Phase-II Xenobiotic Detoxification	Conjugation of Glucuronic acid to electrophiles	[52]
ABCC1	ATP Binding Cassette Subfamily C Member 1/Multidrug resistance associated protein 1 (MRP1)	Phase-III Xenobiotic Detoxification	Transmembrane translocation of xenobiotics	[56]
ABCG2	ATP Binding Cassette Subfamily G Member 2	Phase-III Xenobiotic Detoxification	Transmembrane xenobiotic transporter	[57]
ABCB6	ATP Binding Cassette Subfamily B Member 6	Phase-III Xenobiotic Detoxification/Heme Synthesis	Transmembrane transport of xenobiotics and porphyrins	[52]
ABCC2	ATP Binding Cassette Subfamily C Member 2	Phase-III Xenobiotic Detoxification	Transmembrane transport of xenobiotics	[52]
SRXN1	Sulfiredoxin-1	Thioredoxin-linked Antioxidant Role	Reduction of Sulfinylated-Peroxiredoxins	[51]
G6PD	Glucose-6-Phosphate Dehydrogenase	NADPH Generation	Pentose Phosphate Pathway/Glucose to Glucose 6-Phosphate Conversion	[53]
PGD	6-Phosphogluconate Dehydrogenase	NADPH Generation	Pentose Phosphate Pathway/6-Phosphogluconate to Ribulose 5-Phosphate Conversion	[53]
ME1	Malic Enzyme 1	NADPH Generation	Malate to Pyruvate Conversion	[53]
IDH1	Isocitrate Dehydrogenase 1	NADPH Generation	Isocitrate to α-Ketoglutarate Conversion/TCA Cycle	[53]
TKT	Transketolase	NADPH Generation	Pentose Phosphate Pathway/Conversion of Xylulose 5-Phosphate and Ribose 5-Phosphate into Glyceraldehyde 3-Phosphate and Sedoheptulose 7-Phosphate	[53]
TALDO1	TransAldolase 1	NADPH Generation	Pentose Phosphate Pathway/Conversion of Glyceraldehyde 3-Phosphate and Sedoheptulose 7-Phosphate into Erythrose 4-Phosphate and Fructose 4-Phosphate	[53]
MTHFD2	Methylenetetrahydrofolate Dehydrogenase 2	NADPH Generation	Serine/Glycine Metabolism	[53]
MTHFDL1	Methylenetetrahydrofolate Dehydrogenase 1-like	NADPH Generation	Mitochondrial Tetrahydrofolate Synthesis	[58]
HMOX1	Heme Oxygenase 1	Heme & Iron Metabolism	Heme to Biliverdin Conversion	[51]
FTL	Ferritin Light Chain	Heme & Iron Metabolism	Iron Storage	[51]

2. NRF2 in Cancer Prevention and Therapeutic Implications

2.1. Therapeutic Modulation of NRF2-KEAP1 Pathway for Cancer Prevention

Historically, the NRF2-KEAP1 pathway has been the focus of extensive research aimed at assessing its potential role in human chronic diseases characterized by alterations of the redox homeostasis such as diabetes, cardiovascular diseases, neurodegenerative diseases and cancer. Among them, effort have been made to explore the chemopreventive properties of naturally occurring as well as synthetic compounds functioning as NRF2 activators or KEAP1 inhibitors in vitro and in vivo. The identification of the molecular mechanisms underlying NRF2 modulation has driven a renovated interest in the field of basic and clinical cancer research, fostering a growing number of studies. However, it is now increasingly recognized that NRF2 can exert oncogenic as well as oncosuppressive functions, so that the development of effective therapeutic approaches based on NRF2 modulation requires a careful evaluation of the specific context of its activation including not only the histotype, stage, and genetic background of a specific tumor but also the therapeutic scheme of administration and the target population that might benefit from treatment. In the following section, we will describe some of the most relevant NRF2 activators and their use in cancer treatment.

2.2. Activators of NRF2

Activation of the NRF2 system is complex and can follow canonical and non-canonical pathways. A difficulty in identifying activators and inhibitors of NRF2 or KEAP1 as modulators of inflammation and potential protectors against oxidative stress and carcinogenesis, is the dual nature of the NRF2-KEAP1 protein-protein interaction [26,59–62]. Generally, activation of NRF2 has been viewed as therapeutic, but recent evidence has suggested that this event can be pro-oncogenic as well, depending on the context of NRF2 activation [62]. For example, an increased NRF2 activity under "normal" conditions will lead to improved cell defense against carcinogenesis. By contrast, unrestrained NRF2 activation in a tumorigenic condition, can be protective against stressful conditions and promote therapy resistance [62]. In cases where NRF2 activation exerts pro-tumoral effects, the therapeutic applicability of NRF2 inhibitors has been explored. Since the NRF2-KEAP1 pathway can sense electrophiles as potential stressors, the use of electrophilic drugs to induce its activation would be a reasonable starting point [41,63]. There is however a substantial concern over potential side effects associated with the use of electrophilic compounds. This concern has stimulated the development of other "modulators" of NRF2 activity, considering that the optimal compound would not be a strong NRF2 activator since the strength of its activation is proportional to the electrophilic effects [64]. More work is needed to determine the potential of NRF2 activators as therapeutic drugs, but first, the pathways involved must be better elucidated [65].

2.2.1. Electrophilic and Non-Electrophilic NRF2 Activators

Collectively, NRF2-inducers fall into two classes—electrophilic and non-electrophilic [66]—with the majority of the currently identified inducers belonging to the former class. The mechanism of action for electrophilic NRF2 activators involves the interaction with the cysteine residues on KEAP1, resulting in a conformational change that releases NRF2 to its active conformational state. In the following subsections we will describe both types of molecules, but emphasizing that electrophilic activators are inherently toxic and when used at sufficiently high doses, will cause electrophilic cellular damage beyond NRF2 activation [66]. Despite the high risk for side effects, the quest for additional electrophilic covalent NRF2 activators remains [67] since the advantage of these molecules is due in part to the extremely high binding energy elicited by covalent binding compared to non-covalent and the discovery of a low-molecular weight compound able to maintain a high binding affinity is more likely to occur with covalent activators.

Electrophilic/Covalent NRF2 Activators

The largest class of chemicals so far identified with the ability to activate NRF2 through KEAP1 inhibition is represented by triterpenoids [68]. These molecules can bind to KEAP1 and induce a conformational change that prevents its association with NRF2, promoting its target genes transactivation. Among the others, 2-cyano-3,12-dioxooleana-1,9(11)-diene-28-oic acid (CDDO) is a synthetic derivative of the natural triterpenoid oleanolic acid with a very potent (low nanomolar doses) activating capacity on various NRF2-regulated proteins [66]. CDDO has received much attention due to its ability to hamper the development of certain tumors [66,69]. Despite KEAP1 contains 15 cysteine residues susceptible of modification by electrophilic compounds, each electrophile targets its unique series of residues, a phenomenon referred to as the "cysteine code" [70]. A key cysteine in the binding of triterpenoids to KEAP1 is C151 [70–72]. However, in order to improve potency, specificity and reduce potential side effects of CDDO, Methyl ester (CDDO-Me) and imidazole (CDDO-Im) derivatives have been subsequently developed. Both compounds have shown promising results, as they are able to activate NRF2 and stimulate ARE-expression at low doses [73,74]. Of note, the ability of the triterpenoids electrophilic region to react with thiol groups of many proteins and other thiol-containing molecules underscores the potential side effects associated with their use. For example CCDDO-Im has been shown to bind to mitochondrial glutathione (GSH), resulting in GSH depletion, increased oxidative stress, and increased NRF2 activation [75]. Although triterpenoids have therapeutic action at relatively low concentrations, the risk of serious side effects cannot be excluded.

The compound D3T (3*H*-1,2-dithiole-3-thione) has been shown to increase the nuclear accumulation of NRF2, an effect mediated in part by the activation of the ERK 1/2 pathway [76]. ERK 1/2 inhibition blocked the activation of NRF2 and the effects observed on other ARE-induced gene expression were similar for oltipraz, another dithiolethione, as well as the natural NRF2-activator, sulforaphane [76]. NRF2 also controls the expression of several isoforms of the multidrug resistance-associated protein (MDR), a molecular pump that extrudes chemicals outside the cells. Indeed, it has been show that the administration of oltipraz or butylated hydroxyanisole resulted in a clear upregulation in the MDRs expression as a consequence of NRF2 activation [77].

Non-Electrophilic/Non-Covalent NRF2 Activators

Recent studies have examined the therapeutic benefits provided by non-covalently bound non-electrophiles as NRF2 activators, since electrophilic compounds can have side effects and reduce the activity of certain proteins [78]. Binding studies revealed that cysteine 151 (C151) was an important target since non-covalent binding to this site promoted cell protection, while in contrast covalent binding to this residue enhanced cell toxicity [78]. Zhang et al., recently undertook a comprehensive analysis of nearly 200 chemicals to isolate potential non-electrophilic activators of NRF2. The list was initially shortened to 86 candidates and subsequently further reduced to only 22 molecules, after the exclusion of compounds having an electrophilic reactive site. Based on their structures, the chemicals were placed into one of seven groups consisting of: I—phenothiazine; II—tricyclics; III—trihexyphenidyl; IV—phenyl pyridine; V—quinolin-8-substituted; VI—tamoxifen substituted; and VII—hexetidine. The results indicated that each class was able to induce some level of NRF2-mediated increase in ARE-dependent protein expression [79]. These changes varied between class and protein, but the systematic approach used to identify biologically active non-electrophilic NRF2 activators opened many prospects for future development. Another non-covalent small molecule activator of NRF2, RA839, was shown to activate several pathways related to NRF2 signaling in bone marrow macrophages [80]. Investigations aimed to identify, develop, and then marketing a new non-covalent NRF2 activator have been hindered by the low affinity and low potency of existing compounds compared to the covalent agents. However, one potential activator with low toxicity, but still therapeutic utility is monomethylfumerate (MMF) [59]. Another fumarate-based compound that has demonstrated promise is dimethyl fumarate (DMF) [81,82]. In addition to screening strategies of existing chemicals based on their structure and their potential use as non-electrophilic NRF2 activators, other studies are underway to find novel compounds. Chemicals that have a naphthalene moiety have shown promise as non-electrophilic NRF2 activators. Exposure of lung epithelial cells to various naphthalene-based chemicals led to a marked increase in the expression of several antioxidant enzymes, such as quinone oxidoreductase (NQO1) and heme oxygenase 1 (HO-1), elicited by NRF2 activation [83]. Few of these new compounds were of similar potency to sulforaphane, an electrophilic NRF2 activator, in stimulating the expression of these antioxidant proteins.

2.2.2. Natural Compounds

A 2016 review systematically examined the activation of NRF2 by dietary factors, but also extended these findings to discuss how diet changes may restrict the nutritional utility of some activators [84]. A natural marine-based compound, honaucin A, obtained from cyanobacteria, has been reported to have anti-inflammatory properties both in vivo and in vitro. Only recently have investigators begun to focus on the pathways activated by this compound. Honaucin A forms a covalent bond with the sulfhydryl groups on KEAP1, resulting in the activation of NRF2 [85]. Many food-based NRF2 activators have been shown to work in multiple organs such as liver [86], lung [87], kidney [88], brain [89] and gastrointestinal tract [90]. Most of these NRF2 activating compounds are phenol, polyphenol, or triterpenoid. The use of compounds found in ordinary foods and spices has been an area of great interest in the last 20–25 years. For example, the spice curcumin, from turmeric and ginger family, has been reportedly used to treat a variety of disorders. Yet, little work has been done to fully understand the pathways associated with the therapeutic benefits induced by these chemicals. Recently, an entire book was dedicated to the beneficial actions of curcumin (*The Molecular Targets and Therapeutic uses of Curcumin in Health and Disease*, 2007). Several investigators focused on the molecular targets of curcumin, including NRF2, on curcumin action, and its uses as a neuroprotective agent against toxicants inducing oxidative stress but also as an antitumor agent [91–93]. The molecular action is non-specific with multiple systems and pathways being affected, including the involvement of NRF2 [94]. Other compounds have been examined with mixed results including silibinin, the active chemical from the milk thistle, and resveratrol, a biologically active compound from the skins of various grapes and berries. Both of these molecules have been limited in their usefulness in humans, due to mostly equivocal findings. Resveratrol has been reported to attenuate oxidative damage in the liver by increasing the expression and activity of antioxidant enzymes [95,96]. Silibinin has been shown to decrease metastasis by decreasing the activation of the PI3K-Akt and MAPK (mitogen-activated protein kinase) pathways in the lung [97], and to enhance apoptosis in colon cancer [98]. From previous reports, the effects of resveratrol and silibinin on the NRF2 pathway appear to be debatable and might depend on direct as well as indirect effects.

Foods containing compounds with positive effects on human health are sometimes referred to as "nutraceuticals." One chemical class found in cruciferous vegetables is represented by the isothiocyanates, that include compounds such as sulforaphane, isolated from broccoli, cauliflower, cabbage, and Brussel sprouts. The role of sulforaphane in ameliorating various health-related disorders has received a lot of attention, and some of its biological effects were described in 1992 [99]. The utility of a given compound in a clinical setting is dictated by several properties such as its bioavailability, potency, and interactions with other drugs. Some aspects of the sulforaphane effects within the body suggest that there could be some level of clinical utility under certain circumstances [100]. A recent review describes the multiple pathways affected by sulforaphane administration in reducing tumor growth, indicating that the NRF2-KEAP1 pathway was a critical targeted [101]. Sulforaphane is highly electrophilic due to a reactive carbon in the isothiocyanate group that readily reacts with many nucleophiles containing a sulfur, nitrogen, or oxygen center. By targeting the sulfhydryl groups of KEAP1, sulforaphane can non-covalently bind to these reactive groups, resulting in NRF2 activation [101]. Sulforaphane can also activate antioxidant response elements (AREs) associated with NRF2 [102]. Some of the activated redox regulators include glutathione S-transferase, catalase, glutathione peroxidase, and peroxiredoxins. To improve the functionality of food-based chemicals, it is common to use the parent compound as the "backbone" and then to develop modified versions of the parent endowed with greater efficacy, potency, and possibly also reduced side effects. An example is a modified sulforaphane, 6-methylsulfinylhexyl isothiocyanate (6-HITC), which was more potent than sulforaphane in increasing the expression and activity of various AREs in lung epithelial cells [103]. Curiously, not all the effects of sulforaphane are mediated by the NRF2-KEAP1-ARE pathway. Indeed, sulforaphane can inhibit multiple inflammasomes, sensor systems that activate pro-inflammatory mediators, as it was recently shown in macrophages and fibroblasts [104]. The mechanisms of action for sulforaphane and isothiocyanate is complex and may involve multiple mediators. A critical consideration is also the cell/organ type taken into consideration. Additional research is needed to help to clarify some of the nutraceutical pathways to better assess their in vivo benefits.

2.3. Potential Use of NRF2 Activators in Cancer Therapy

Whereas new findings indicate that NRF2 plays a dual role in cancer ([26,105,106], the potential use of NRF2 activators in cancer prevention and therapy needs to be further elucidated. It is widely accepted that effective chemoprevention should encompass induction of cytoprotective and detoxifying enzymes. Therefore, the use of compounds able to activate NRF2-KEAP1 pathway and induce genes involved in antioxidant defense appears to be a possible strategy in both cancer prevention and therapy [64]. As explained above, NRF2 activators fall into the class of electrophilic and non-electrophilic compounds [66] and can be of natural origin or semisynthetic/synthetic analogs [64]. As reviewed by Sanders et al. [107], phenolic and sulfur-containing compounds are the most promising agents in cancer prevention. Phenolic compounds such as curcumin and resveratrol exert their chemopreventive effect via activation of NRF2-KEAP1 signaling that induces phase-II detoxifying and antioxidant enzymes (20638930). Curcumin, a common spice obtained from the rhizomes of *Curcuma longa* (turmeric), was shown to induce the expression of antioxidant enzymes such as glutathione S-transferase, aldose reductase, and HO-1 through NRF2-KEAP1 signaling [108]. Apart from covalent modification of KEAP1 [109], activation of upstream kinases such as MAPK seems to be an additional mechanism of NRF2 activation [110,111]. However, the therapeutic success of this compound has been hampered by its limited bioavailability and rapid metabolism, the poor pharmacokinetic properties [112] and the lack of conclusive toxicity data [113]. Some studies reporting therapeutic uses of curcumin in various diseases including cancer are illustrated in Table 2. Resveratrol is a natural compound contained in edible plants and fruits such as grapes, peanuts, berries, and soy with the reported capacity to increase the NRF2 levels and promote its nuclear translocation. Resveratrol monomer has been shown to induce phase-II detoxifying enzymes by activating NRF2 signaling in several human cancer cell lines [114,115] and to protect from carcinogenicity derived from bioactivated carcinogens [116]. In contrast, the effects induced by its dimers are poorly understood although the monomer and the dimers have been reported to act differently in terms of NRF2/ARE induction [117]. Table 2 illustrates some recent in vitro and in vivo studies reporting the chemotherapeutic use of resveratrol. However, similarly to curcumin, its poor bioavailability and rapid clearance, made it necessary to develop analogs with improved pharmacokinetic properties and higher potency [64]. On the other hand, green tea polyphenols such as (-)-epigallocatechin-3-gallate (EGCG) and (-)-epicatechin-3-gallate (ECG) are known NRF2 activators showing potent induction of ARE-mediated luciferase activity [118]. EGCG potentiates cellular defense capacity against chemical carcinogens, UV, and oxidative stress via NRF2-mediated induction of genes codying for antioxidant or phase-II detoxifying enzymes, modulators of inflammation, cell growth, apoptosis, cell adhesion etc. [119]. However, it has been shown that EGCG has dual effects on NRF2-mediated ARE activation depending on its concentration, with higher doses producing down-regulation and lower doses enhancing HO-1 expression [118,120]. Table 2 summarizes the results of some studies exploring the potential role of EGCG in the treatment of different pathological conditions. Other potential chemopreventive agents that exert their properties through NRF2/ARE pathway are sulfur-containing compounds, such as sulforaphane, contained in cruciferous vegetables and diallyl sulfide derived from garlic [121,122]. In comparison with curcumin and resveratrol, sulforaphane exhibits more potent activation of NRF2 and significantly better bioavailability due to its lipophilic nature and low molecular weight [123–125]. Preclinical and clinical evaluation of sulforaphane in breast chemoprevention revealed the presence of its metabolites in the rat mammary gland after a single oral administration at concentrations known to alter gene expression and also in human breast tissue after a single dose of broccoli sprout in healthy women undergoing reduction mammoplasty [126]. These findings provided a strong rationale for evaluating the protective effects of a broccoli sprout preparation, claiming sulforaphane as a good candidate in the adjuvant therapy based on natural molecules against several types of cancer [127]. However, some studies indicate that this compound might exert pro-survival effects in cancer cells [128] and potentially interfere with the successful application of immunotherapy [129]. Table 2 contains some of the studies conducted in this field. In a recent review [130] discussing potential combinations of a conventional anticancer drug (cisplatin or doxorubicin) and a known antioxidant (sulforaphane or curcumin), it has emerged the necessity of preclinical evidence confirming that the natural compounds can potentiate the anti-cancer effect of traditional drugs but also reduce the side toxicity in normal tissues. A review by Robledinos-Anton et al. [65] provides the information on current clinical trials in progress based on NRF2 activators and their potential clinical use in various chronic disorders, including cancer. Namely, sulforaphane is in the phase-II clinical trial for the use in treatment of prostate cancer [131] (NCT01228084) and for breast cancer [132] (NCT00843167), while it has entered the phase-II for lung cancer prevention (NCT 03232138). Also, curcumin has entered phase-III for the treatment of prostate cancer [133] (NCT02064673) and resveratrol is in the phase-I for the treatment of colon cancer [134] (NCT00256334). In an attempt to improve their biological activity, many semisynthetic and synthetic NRF2 activators have been synthesized and these substances are preferentially used in clinical practice compared to the natural counterpart. For example, Bardoxolone methyl (CDDO-Me), a semisynthetic triterpenoid with the ability to protect cells and tissues from oxidative stress by increasing the NRF2 transcriptional activity, has demonstrated its efficacy as an anticancer drug in different mouse model [135]. So far, three clinical trials focusing on the use of CDDO-Me in cancer treatment have been registered. In a phase-I clinical trial investigating the tolerability, safety, efficacy and pharmacokinetics of CDDO-Me in advanced solid tumors and lymphoid malignancies, complete response was observed in a patient with mantle cell lymphoma and partial response in a patient with anaplastic thyroid carcinoma [136]. However, phase-III of the clinical trial designed to investigate the efficacy of CDDO-Me in patients with stage 4 chronic kidney disease and type 2 diabetes revealed significant increase of heart failure within four weeks of treatment [137]. In any case, since no evidence of direct cardiotoxicity was found [138,139], trials on this compound are still ongoing. Another triterpenoid analog, omaveloxolone, has been selected for cancer treatment in three clinical trials, out of which one is ongoing. Early clinical trials for the treatment of melanoma and NSCLC have been giving promising results [140]. Dimethyl fumarate (DMF) is a synthetic NRF2 activator which has already been used for the treatment of multiple sclerosis [141] and is currently tested in a number of clinical trials, mainly investigating its efficacy in lymphoma, leukemia and melanoma. However, it should be emphasized that DMF can also exert NRF2-independent effects, suggesting that its activity might also rely on alternative pathways as evidenced by a recent study [142]. Finally, Oltipraz is an organosulfur compound which has also entered clinical trials, namely phase-I trial to study its efficacy in preventing lung cancer in smokers (NCT00006457). However, further clinical trials are needed to confirm or challenge its possible use as a chemopreventive agent. In conclusion, while the list of natural, semisynthetic and synthetic NRF2 activators is steadily increasing, it is evident that the drug development is moving slowly due to the pleiotropic effects of NRF2 activators. More studies on detailed molecular mechanisms are necessary for their possible application in cancer chemoprevention, especially in consideration of the possible oncogenic role of NRF2 in cancer cells.

3. NRF2 in Cancer Promotion and Therapeutic Implications

3.1. Pro-Oncogenic Roles of the NRF2-KEAP1 Pathway

Given that NRF2 promotes cell survival in stress conditions [143], it is consequently accepted that enhanced NRF2 activity can be tumor promoting through several molecular mechanisms that protect cancer cells (Figure 3). This is a way by which cancer cells gain advantages over the normal cells, such as enhanced tumorigenic capacity, resistance to therapeutic agents and increased antioxidant activity leading to "NRF2 addiction" that turns this cellular guardian into a cancer driver [3]. Several important studies suggest that common oncogenes, such as *KRAS*, *BRAF*, and *MYC*, can directly promote the transcription of NRF2 through the modulation of signaling pathways such as the Raf-MEK-ERK-Jun cascade [14,53,144]. This overactivation of NRF2 leads to enhanced cytoprotective activity and, remarkably, to a decrease in ROS levels. Thus, probably oncogenes partly boost cancer development through an NRF2-dependent creation of a more favorable intracellular milieu for tumor cells selection. Constitutive activation of NRF2 in cancer promotion and the mechanisms that lead to this condition, are under debate. Many researchers have spotted cancer-associated mutations that activate NRF2 [145–150]. Mutations in NRF2 that lead to gain-of-function can be detected mainly in squamous cell carcinomas of the oesophagus, lung, larynx, and skin [151]. Additionally, aberrant NRF2 activation in cancer cells leads to remarkably increased expression of TKT and G6PD metabolic enzymes that contribute to metabolic reprogramming and cell proliferation [53]. Other evidence indicates that deficiencies in autophagy, and therefore activation of NRF2 and overexpression of p62, might promote induction of hepatic tumors [152]. As for the hormone related cancers, it has been reported that specific hormones lead to significant upregulation of NRF2 in ovarian cancer cell lines [153]. Epigenetic modifications seem to control the expression of KEAP1/NRF2 system and therefore it is important to investigate this type of regulation in cancer [154–156]. Another major topic is the enhancement of chemoresistance and radioresistance in cancer cells. For example, overactivation of NRF2 by pretreatment with a synthetic antioxidant was found to increase the survival of neuroblastoma cells treated with three chemotherapeutic drugs [157]. Finally, it has been demonstrated that radiation therapy leads to generation of ROS and depletion of GSH, frequently causing enhanced synthesis of antioxidant enzymes such as GCLC, HO-1 and TXRD1 by NRF2 activation, as reported in a recent study on prostate cancer cells [158]. One question that needs to be addressed, however, is whether the increase in NRF2 levels is a key step in cancer development. The existing body of evidence suggests that *KRAS* and *BRAF* increase the levels of JUN that in turn binds to well-known transcription starting sites of NRF2 promoting its induction. This finding suggests that some of these effects can be direct [14]. Moreover, many findings indicate that ROS levels should be suppressed in order to prevent cancer development due to their involvement in promoting and sustaining carcinogenesis [62,159]. It has also been proposed that utilizing drugs that boost ROS production can be an effective way of killing cancer cells [160]. Concomitantly, it is of utmost importance to determine the specific pathways and the equilibrium between ROS and NRF2 so as to elucidate the paradoxical role of KEAP1/NRF2 pathway in cancer. A schematic illustration of the pro-oncogenic functions of NRF2 is presented in Figure 3. In the following sections we will describe more in detail some of the most relevant hallmarks of cancer cells that are regulated by NRF2 activation.

3.1.1. Sustained Proliferation

The KEAP1-NRF2 pathway is a vital defense system against oxidative and electrophilic stress in normal cells as well as cancer cells that use it to foster their unrestricted growth [3]. Temporary activation of NRF2 is critical for the survival of non-cancerous cells and protection against carcinogenesis [161]. However, constant activation of this pathway is detrimental especially in a cancerous context since NRF2 exerts a pro-tumoral function by supporting sustained cancer cells proliferation through various mechanisms [3]. Studies with lung cancer, pancreatic cancer, and hepatocellular carcinoma cell lines showed that NRF2-KEAP1 status directly affects their proliferation rates since NRF2-negative cells proliferate slower, while KEAP1-deleted cells proliferate faster than their wild type counterparts [162–164]. Moreover, NRF2 was seen to promote oncogenic *K-RasG12D*-initiated tumor formation and proliferation of pancreatic and lung cancers in vivo and *K-RasG12D*, *BRAF V619E*, and *MYC*-induced NRF2 transcription [14,165]. The role of NRF2 on cancer cell proliferation relies on the functions of the genes regulated by its own transcriptional activation [166]. The genes that regulate the proliferative capacity of normal cells such as *NOTCH1*, *NPNT*, *BMPR1A*, *IGF1*, *ITGB2*, *PDGFC*, *VEGFC*, and *JAG1* are known NRF2 targets and contribute to cancer cells survival [51]. On top of these genes, NRF2 also regulates the expression of genes needed to fulfill the constant demand of protein synthesis of cancer cells such as *PHGDH*, *PSAT1*, *PSPH*, *SHMT1*, and *SHMT2* by activating the critical effector ATF4 [51,167]. Apart from increased protein synthesis, highly proliferating cells also require energy and small building blocks to synthesize other macromolecules [168]. In this context, NRF2 regulates the expression of enzymes such as *G6PD*,*TKT*, *TALDO1*, *PPAT*, *MTHFD2*, *IDH1*, and *ME1* in lung cancer cells [53]. Furthermore, NRF2 is involved in the regulation of genes required for the synthesis of GSH (reduced glutathione) and NADPH (reduced Nicotinamide adenine dinucleotide phosphate), two crucial molecules for cell proliferation [169]. Besides glucose metabolism, NRF2 also regulates genes involved in fatty acid and lipid metabolism [170]. There is also evidence of indirect involvement of NRF2 on cancer cell proliferation by regulating several non- coding microRNAs, such as mir-1 and miR-206. These miRNAs normally inhibit *TKDT* and *G6PD* genes, and their repression by HDAC4 through NRF2 supports cancer cells growth [171]. In summary, it is clear that the increased activation of NRF2 allows cancer cells to proliferate faster as a consequence of cytoprotective genes induction and metabolic reprogramming [168]. Due to the advantages granted by its activation, the cancer cells acquire a phenotype of "NRF2 addiction" which is characterized by aberrant NRF2 accumulation in both murine and human cancers [3]. Thus, the impairment of NRF2 pathway is expected to repress tumor growth, and this is the basis of developing drugs against NRF2 in a context-dependent manner for the targeted therapy of various cancers [172].

3.1.2. Angiogenesis Induction

The presence of constantly growing cells in tumor microenvironment causes depletion of oxygen and nutrients and creates an urgent need for a continuous supply of blood flow to fulfill the increased metabolic demand and to remove wastes and carbon dioxide [173]. Hypoxic tumor microenvironment induces the expression of *VEGF* through the transcription factor HIF-1a for the generation of new blood vessels, a process known as angiogenesis [174]. Other than VEGF, PDGF, angiopoietin, angiogenin and extracellular matrix elements participate also to the regulation of angiogenesis [175]. NRF2 modulates angiogenesis by itself and through its targeted genes [176]. Depletion of NRF2 decreases the levels of HIF-1a, which in turn causes reduction of blood vessels formation through regulation of VEGF in rat gastric epithelial cells, glioblastoma and colon cancer xenograft models [177–179]. In recent studies, it was reported that regulation of VEGF by NRF2 also depends on PI3K/AKT/mTOR pathways in endothelial cells [180,181]. These data suggest that both proliferative and pro-angiogenic effects of PI3K/AKT/mTOR and NRF2 pathways cooperate in favor of cancer cells. NRF2 also indirectly regulates HIF-1a levels by preventing its proteasomal degradation by the aid of NQO1 [161,182]. Angiogenesis is also regulated by common effectors of NRF2 and HIF1a including HO-1, platelet-derived growth factor (PDGFC), and fibroblast growth factor (FGF2) [183–185]. Among these genes, *HMOX1* was shown to promote angiogenesis in various cancers such as glioma, pancreatic cancer, and melanoma [176,186]. Intriguingly, there is an interplay between NRF2 and HIF1-a pathways since VEGF can in turn activate NRF2 via ERK1/2 signaling [187]. Another evidence supporting the crosstalk between NRF2 and HIF-1a pathway comes from the recent data showing that PIM kinases, upstream regulators of HIF-1a, promote NRF2 nuclear accumulation in response to hypoxia and in normoxia, leading to increased cancer cells survival in the hypoxic tumor microenvironment [188]. In conclusion, the induction of angiogenesis can be counted as one of the critical roles of NRF2 in promoting tumorigenesis.

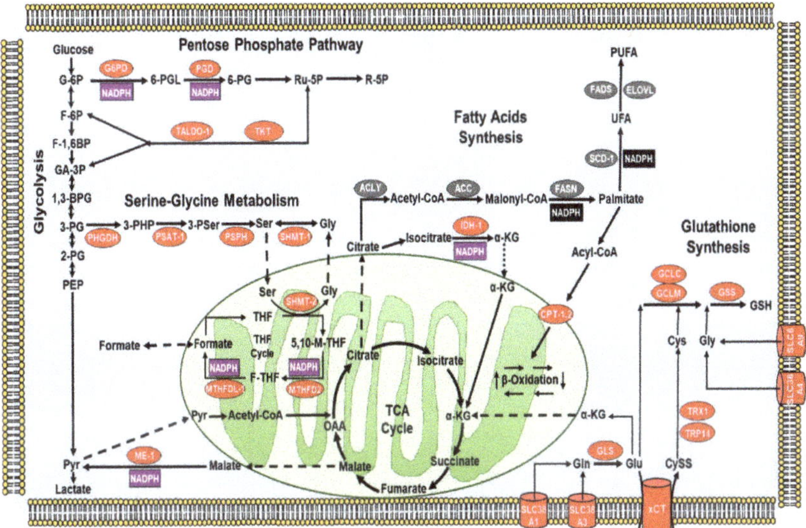

Figure 3. NRF2 rewires cancer cells metabolism to support the redox homeostasis. The enzymes marked in red are positively regulated while those in green are negatively modulated by NRF2. NADPH (reduced Nicotinamide adenine dinucleotide phosphate) production is indicated in violet while NADPH consumption in black. The abbreviations are: ACC1, Acetyl-CoA Carboxylase 1; ACL, ATP-Citrate Lyase; CPT, Carnitine PalmitoylTransferase; ELOVL, fatty acid Elongase; FADS, Fatty Acid Desaturase; FASN, Fatty Acid Synthase; G6PD, Glucose-6-Phosphate Dehydrogenase; GCLC, Glutamate-Cysteine Ligase, Catalytic subunit; GCLM, Glutamate-Cysteine Ligase, Modifier subunit; GLS, Glutaminase; GS, Glutathione Synthetase; IDH1, Isocitrate Dehydrogenase 1; ME1, Malic Enzyme 1; MTHFDL, MethyleneTetraHydroFolate Dehydrogenase 2; PGD, 6-phosphogluconate dehydrogenase; PHGDH, phosphoglycerate dehydrogenase; PPAT, phosphoribosyl pyrophosphate amidotransferase; PSAT1, phosphoserine aminotransferase; PSPH, phosphoserine phosphatase; SCD1, stearoyl CoA desaturase; SHMT 1-2, serine hydroxymethyltransferase 1 and 2; TALDO, transaldolase; TKT, transketolase; TXN, thioredoxin; UCP3, uncoupling protein 3; xCT, glutamate/cystine antiporter. G6P, glucose-6-phosphate; F6P, fructose-6-phosphate; F1,6BP, fructose-1,6-bisphosphate; GA3P, glyceraldehyde-3-phosphate; 3PG, 3-phosphoglycerate; PEP, phosphoenol pyruvate. PPP: 6PGL, 6-phosphoglucono-d-lactone; 6PG, 6-phosphogluconate. PRPP, 5-phospho-D-ribosyl-1 pyrophosphate; IMP, inosine monophosphate. Ser/Gly synthesis: 3PHP, 3 phosphohydroxypyruvate; 3PSer, 3-phosphoserine; THF, tetrahydrofolate; MTHF, methylenetetrahydrofolate; 5,10-FTHF, 5,10-methenyl-tetrahydrofolate. b-Oxidation: Acyl-CoA, acyl-coenzyme A; Ac-CoA, acetyl-coenzyme A. FA, fatty acid. GSH, glutathione, reduced.

3.1.3. Resistance to Apoptosis

NRF2 protects healthy cells from endogenous ROS and is a critical regulator of drug metabolism and antioxidant enzymes [21,46]. NRF2 leads to diminished apoptosis and increased drug resistance [189,190]. Inhibition of NRF2 signaling enhances apoptosis in response to oxidative insults [105,191]. Conversely, activation of NRF2 by chemopreventive agents decreases the number of apoptotic cells [192–194]. There are many studies showing elevated expression of NRF2 in various types of tumors such as non-small cells lung cancer (NSCLC), esophageal squamous cell cancer (ESCC), gastric cancer (GC), head and neck cancer (HNC), breast cancer (BC), ovarian cancer (OC), and endometrial cancer (EC) [46]. NRF2 signaling is activated during malignant transformation in response to radiotherapy/chemotherapy and it protects cancer cells from cell death by upregulating a number of ROS-scavenging enzymes that counterbalance the increased ROS production [195]. Moreover, NRF2 allows the cancer cells to escape death by cooperating with other pathways playing a role in apoptosis regulation. For instance, the tumor suppressor p53 inhibits NRF2 signaling by down regulating the expression of NRF2 target genes such as *x-CT*, *NQO1*, and *GST1* and triggers cell cycle arrest and apoptosis [196–198]. Under mild cellular stress conditions, the p21 protein, a major p53 target, binds to the DLG motif and prevents KEAP1-mediated NRF2 proteasomal degradation, activating the antioxidant response [199]. Additionally, other studies have shown that mutant p53 leads to constitutive NRF2 activation without affecting its expression and enhances cancer cells survival [200,201]. Glutathione-S-transferase pi 1 (GSTP1) is one of the downstream targets of NRF2 that inhibits proapoptotic c-Jun N-terminal kinases (JNKs) activity and promotes cell survival [202]. Importantly, p62 is another NRF2 target, which mediates autophagic degradation of KEAP1 and therefore enhances NRF2 stability, suppressing apoptosis [203]. Finally, Bcl-2 is a well-known anti-apoptotic protein that promotes increased cell survival and drug resistance [204,205]. NRF2 upregulates *Bcl-2* expression through direct binding of the ARE sequence on its promoter, which induces oncogenesis [206]. All of these studies indicate that NRF2 plays a critical role in tumor survival and drug resistance through the inhibition of apoptosis via different pathways.

3.1.4. NRF2 Signaling in Metastasis

Epithelial to mesenchymal transition (EMT) is a biological process that contributes to cancer metastasis and tissue invasion [207,208]. During EMT, the expression of E-cadherin and other epithelial phenotype-related genes is repressed while conversely the expression of mesenchymal phenotype-related genes such as vimentin and N-cadherin is activated by EMT regulators (Snail, Slug, Twist, Zeb, etc.) [209,210]. As a result, epithelial cells turn into mesenchymal cells by losing their cell–cell adhesion and cell polarity features and acquiring invasive and metastatic properties. Constitutively active NRF2 has been shown in human cancers with higher metastatic capacity [211]. In addition, the correlation of NRF2 expression with cancer progression, metastasis and drug resistance has been reported in many different studies [14,15,46,212]. NRF2 promotes EMT and invasion in pancreatic adenosquamous carcinoma cells through downregulation of E-cadherin gene expression [213]. NRF2 knockdown (KD) increases E-cadherin expression and downregulates N-cadherin and matrix metalloproteinase 2 and 9 (*MMP2*, *MMP9*) genes expression and reduces migration and invasion capacity of NSCLC cells [214]. Overexpression of NRF2 in BC cells promotes cell proliferation and metastasis via activating NRF2 target gene *NOTCH1* that in turn induces the expression of genes promoting EMT [215]. NRF2 also positively regulates the *RhoA* gene, which is a critical factor for growth and metastasis, while NRF2 down regulation inhibits proliferation of BC cells [216]. Furthermore, recent findings demonstrate that NRF2 expression is upregulated in human hepatocellular carcinoma (HCC) and that NRF2 promotes proliferation and tumor metastasis by regulating *Bcl-xL* and *MMP-9* genes expression [164]. In contrast, other studies have shown that low expression levels of NRF2 also play a critical role in cancer progression and metastasis formation. In human prostate cancers (PC), NRF2 and its target genes were shown to be significantly decreased during the metastatic process [217]. Additionally, it has also been reported that repression of NRF2 in HCC cell lines increased cell metastasis and invasiveness via TGF-β/Smad-dependent signaling [218]. Moreover, NRF2 deregulation was also shown in other cancers like OC, lung adenocarcinoma (LUAD), human head and neck squamous cell carcinoma (HNSCC) [217,219]. Taken together, all these studies demonstrate that NRF2 has both metastatic and anti-metastatic activity in different types of tumor and stages of cancer progression. It seems like cancer cells utilize both upregulation and downregulation of NRF2 signaling for their advantages. To interfere with cancer metastasis, it will be necessary to fully elucidate the role of NRF2 expression in the metastatic microenvironment.

3.1.5. Metabolic Reprogramming by NRF2: NADPH Links Tumor Growth and Redox Balance

It is well known that cancer cells reprogram their central metabolism to meet the energetic needs imposed by their uncontrolled growth. Initial studies mainly focused on the Warburg effect, while subsequent work also investigated changes in the one-carbon and fatty acids metabolism, pentose phosphate pathway (PPP), tricarboxylic acid cycle (TCA) and glutamine catabolism [173,220,221]. In general, NRF2 can control multiple metabolic routes by two different mechanisms, similarly to other oncogenes (e.g., *MYC* or *KRAS*): the first involves direct transactivation of key metabolic enzymes [222], while the second relies on the modulation of proteins controlling other signaling pathways such as PPARγ [223], Notch [215,224], AHR [225,226] and PI3K/AKT [53]. It is known that alterations of tumor metabolism are often paralleled by an increased antioxidant capacity, which is also part of the adaptive response mounted by cancer cells to face adverse conditions, including OS [173,227–230]. With this respect, certain metabolic reactions can play a dual role, providing intermediates for biosynthetic processes or essential cofactors used to modulate the intracellular redox balance [231–235]. Among them, NADPH represents a key player in supporting anabolic reactions and ROS-scavenging antioxidant systems. As first, NADPH is essential for the regulation of the glutathione/glutaredoxins (GRXs) system, that regenerates the reduced form of glutathione (GSH) once it is oxidized (GSSG) [236]. Secondly, NADPH is a key cofactor for the glutathione peroxidases (GPXs) that scavenge hydrogen peroxide or potentially harmful alkyl hydroperoxides. Lastly, NADPH is indispensable for the thioredoxin reductases (TRXRs), a class of enzymes that restore the reduced form of thioredoxins (TRXs) and indirectly contribute to the regulation of thiol groups in redox-sensitive proteins [237,238]. Importantly, NRF2 can enhance the expression of genes coding for NADPH-producing enzymes such as *G6PD* (Glucose-6-Phosphate Dehydrogenase) and *PGD* (Phospho Gluconate Dehydrogenase), or enzymes that regenerate glycolytic intermediates that can be diverted into the oxidative PPP branch, such as *TKT* (Transketolase) and *TALDO1* (Transaldolase-1) [53,182,215,239]. Mechanistically, either direct or indirect genes transactivation can occur, depending on the genetic and biological context of NRF2 activation. Indeed, by using H1437, A549 NSCLC and DU145 PC cells, Singh et al., reported that genetic induction of PPP genes was mediated by NRF2-dependent repression of miR-1 and miR-206, two negative regulators of *G6PD*, *PGD* and *TKT* expression, through yet unknown epigenetic mechanisms involving the histone deacetylase HDAC4 [240]. Also miR-1 inhibition was found to underlie NRF2-mediated upregulation of *G6PD* in HCC cells, an event that positively correlated with grading, metastasis number and poor prognosis in HCC patients [241]. Instead, a direct induction of *G6PD*, *PGD*, *TKT*, *TALDO1* and other NADPH-producing enzymes was reported in A549 NSCLC cells with sustained PI3K/AKT pathway activation [53], in agreement with previous ChIP-seq studies [48,182,242]. Also, Xu et al., demonstrated that NRF2 can bind to the AREs within intron 1 and 4 of the *TKT* gene, inducing its expression in HCC cells (SMMC and MHCC97/L) [239]. Consistently, the presence of functional NRF2 binding sites within the AREs of the *TKT* promoter was also reported in MEFs and A549 NSCLC cells, suggesting a direct transactivation mechanism [240]. Also, other data indicate that *NRF2* overexpression or *KEAP1* KD can upregulate the mRNA and protein levels of G6PD and TKT, enhancing tumor proliferation, migration and invasion of MCF7 and MDA-MB-231 BC cells by promoting EMT through G6PD/HIF-1α activation of Notch1 signaling, while *NRF2* silencing or *KEAP1* overexpression reverted these changes [215]. Interestingly, some studies suggest that NRF2 can control the expression of NADPH-producing enzymes involved in one-carbon metabolism such as *MTHFD2* (Methylenetetrahydrofolate Dehydrogenase 2) or in the TCA cycle, such as *IDH1* (Isocitrate Dehydrogenase 1) and *ME1* (Malic enzyme 1) [53,239,243], while others indicate that NRF2-dependent induction of the folate-cycle enzyme *MTHFDL1* (Methylenetetrahydrofolate Dehydrogenase 1-like) can increase the NADPH levels in HCC cells, supporting proliferation and redox homeostasis [58].

3.1.6. NRF2 Regulates Metabolic Processes Leading to GSH Synthesis and TCA Cycle Anaplerosis

In line of principle NRF2 can also regulate the redox balance of cancer cells via transcriptional induction of metabolic enzymes or membrane channels that control the availability of cysteine, glutamate and glycine, essential precursors in the GSH synthesis. In this regard, NRF2 was shown to enhance the expression of key enzymes for serine/glycine biosynthesis such as *PHGDH* (Phosphoglycerate Dehydrogenase), *PSAT1* (Phosphoserine Aminotransferase 1), *PSPH* (Phosphoserine

Phosphatase) and *SHMT1-2* (Serine Hydroxymethyltransferase 1-2) in a panel of NSCLC cells [167]. Also, a recent work underlined the importance of the NRF2-ATF4 pathway in the regulation of aminoacid metabolism in cancer. Indeed, NRF2 was seen to enhance the ATF4 transcriptional activity in autophagy-deficient HCT116 CRC cells, promoting the expression of genes (*SLC6A9*, *SLC36A4*, *SLC38A1*, and *SLC38A3*) coding for AATs (aminoacid transporters) involved in the uptake of glycine and glutamine. Notably, AATs inhibition sensitized autophagy-deficient CRC cells but not wild-type cells to apoptosis induced by glutamine withdrawal [244]. Other work has focused on xCT, an antiporter that couples the efflux of glutamate with the uptake of cystine (CySS), that is intracellularly reduced to cysteine (2x Cys) by GSH, TRX1 or TRP14 [245–249]. For example, Habib et al., showed that in MCF7 BC cells exposed to OS, an enhanced NRF2 nuclear translocation was responsible for the *SLC7A11* (solute carrier family 7 member 11) gene upregulation, leading to an increase in the xCT mRNA and protein levels and marked glutamate release. Of note, these changes were phenocopied by *NRF2* overexpression or *KEAP1* KD and reverted by its overexpression [250]. In a later study, *NRF2* and *SLC7A11* expression was found to be positively correlated across 947 cancer cell lines from the CCLE dataset [251], especially within 59 different BC cell lines. Here, *NRF2* KD markedly decreased *SLC7A11* expression and glutamate extrusion in Hs578T and MDA-MB-231 BC cells, improving cell viability upon glucose depletion while in the same conditions *NRF2* activation by DMF impaired cell viability. Thus, despite the enhanced activation of the NRF2-xCT axis might efficiently preserve the redox homeostasis of BC cells under OS, it might also decrease their metabolic flexibility, unveiling a specific vulnerability that can be therapeutically exploited [252]. In a study on human melanoma cells with constitutive *BRAF* activation (*BRAFV600E*), Khamari et al., explored potential metabolic changes promoting resistance (A375RIV1 cells) or sensitivity (A375-v cells) to the BRAF inhibitor Vemurafenib, by using an in vivo long-term treated xenograft mouse model. Here, A375RIV1 cells exhibited a strong activation of the NRF2 signaling, followed by an increased expression of genes involved in ROS scavenging (i.e., *GPX1*, *GPX2*), GSH synthesis (i.e., *GCLM*, and *xCT*) and NADPH-generation (i.e., *TKT*, *TALDO1*), compared to the A375-v counterpart. Of note, *NRF2* KD by siRNA decreased the protein content of its target genes in A375RIV1 cells, partly reversing their resistance to Vemurafenib [253]. Thus, the xCT system is key regulator of cancer cells redox balance, while its inactivation might sensitize malignant cells to OS inducers. Of note, xCT overexpression is expected to promote glutamine catabolism to support TCA cycle anaplerosis or GSH synthesis [254]. In this respect, a recent study from Sayin et al., reported that LOF mutations of the *KEAP1* gene can mediate glutamine addiction in both mouse (KPK) and human *KRAS*-driven LUAD cell lines. Here, NRF2 increased activation led to enhanced xCT/*SLC7A11* expression, causing an imbalance in the TCA cycle and sensitization of KPK cells to pharmacologic or metabolic glutamine depletion. Importantly, the glutaminase inhibitor CB-839, impaired cell growth in a panel of tumor cells including melanoma, colon, renal, bone, squamous and urinary-tract cancers with *KEAP1* LOF mutations while the use of KI696, a small-molecule activator of NRF2, conversely sensitized *KEAP1* WT cancer cells previously refractory to CB-839 [255]. Thus, oncogenic alterations in the NRF2-KEAP1 axis can induce defects in central carbon metabolism of cancer cells and reveal metabolic vulnerabilities that can be targeted. Notably, glutamine is the most abundant aminoacid in human serum and is essential for many cancer cells to generate TCA cycle intermediates and support the biosynthesis of nucelotides, N-acetyl glucosamines, fatty acids, GSH and other aminoacids [254]. Intriguingly, NRF2 can control different steps of glutamine fate, from its uptake to its metabolism. For example, early studies reported that NRF2 can induce the expression of the glutamine importer *SLC1A5* in HeLa cells through the ATF4 transcription factor [256] while a recent ChIP-Seq analysis on *KEAP1-/-* mice and human ESCC cells, revealed that NRF2 causes metabolic reprogramming by enhancing the expression of the *SLC1A4* glutamine transporter and other metabolic enzymes [257]. Lastly, the enzyme glutaminase, catalyzing the conversion of glutamine into glutamate, was found to be a direct NRF2 target gene in MCF7 and MCF10 BC cells treated with Sulforaphane or subdued to *KEAP1* KD by siRNA [258]. Therefore, NRF2 is profoundly implicated in the control of glutamine metabolism of cancer cells and most likely this regulatory node will be the focus of extensive research in the next future.

3.1.7. NRF2 in the Regulation of Fatty Acids Metabolism

Interestingly, NRF2 has been found to exert opposite changes in the metabolism of fatty acids, since a repression of FAS (fatty acid synthesis), but a stimulation of FAO (fatty acid oxidation) has been reported in isolated mitochondria, MEFs (mouse embryonic fibroblasts) and tissues of transgenic mice [259]. Despite the lack of data on malignant tumors a study on HEK-293T cells suggests that NRF2 might control the expression of *CPT1* and *CPT2*, two isoforms of the enzyme carnitine palmitoyltransferase (CPT) that catalyzes the rate-limiting step of FAO [260]. Another mechanism by which NRF2 can potentially support the redox balance of cancer cells is through the suppression of NADPH-consuming processes, including lipid biosynthesis. Indeed, by using murine models with variable *NRF2* expression, Wu et al., showed that *NRF2*-null mice exhibited increased hepatic mRNA levels of the enzymes fatty acid synthase (*FASN*), fatty acid desaturase (*FADS1*, *FADS2*), stearoyl-CoA desaturase (*SCD1*), fatty acid elongases (*ELOVL2,3,5,6* and *CYB5R3*), acetyl-CoA carboxylase-1 (*ACC1*) and ATP-citrate lyase (*ACLY*), while the opposite was seen in *KEAP1*-KO mice suggesting that NRF2 restrains lipogenesis and desaturation to prevent NADPH depletion [261]. On the other hand, NRF2 activation in mouse lung was conversely seen to induce the transcription of FAO genes and lipases, promoting degradation of damaged lipids and providing reducing equivalents in the form of NADPH [262]. In conclusion, the activity of NRF2 can significantly affect the efficiency of FAO and lipid biosynthesis, ultimately affecting bioenergetics as well as NADPH-linked antioxidant systems, underscoring the intimate connection between metabolic processes and redox homeostasis.

3.2. Strategies to Negatively Modulate NRF2 Signaling/Pathway

Tumors are complex entities composed by heterogeneous cell populations that dynamically adapt to their microenvironment driven by genetic/epigenetic alterations and metabolic rewiring [263–267]. Thus, cancer cell populations that become prevalent during certain phases of the malignant progression can frequently develop resistance to treatment [268–271]. Moreover, due to patients' individualities and tumors heterogeneity, variable response rates are often observed making the identification of more effective drugs very urgent [272,273]. In this regard, the concomitant inhibition of antioxidant circuits and metabolic pathways that support the redox balance of malignant cells, delineates a promising anti-cancer strategy [274–283]. It is known that most of the metabolic changes promoting cancer cells proliferation and tumor growth induce also an increased ROS generation, counterbalanced by an antioxidant response that prevents cell death [234,284,285]. Of note, since the cytotoxicity of conventional anti-cancer therapies largely relies on efficient ROS accumulation, the augmented antioxidant capacity of cancer cells constitutes a key determinant of therapy-resistance [286–288]. At the same time, given the strong dependency of malignant cells to their antioxidant systems, interfering with their redox control can induce OS-dependent cell death [289]. Since the NRF2 signaling plays a key role in tumorigenesis, malignant progression [290–292] and drug sensitivity [293–295], this transcription factor has emerged as a promising therapeutic target [16,282,296–298]. In this regard, several studies have tried to identify NRF2 activators to prevent ROS-dependent carcinogenesis while others have focused on the development of NRF2 inhibitors to overcome therapy resistance [299–301]. In the following sections, the most promising NRF2 inhibitors, described so far in the literature, will be presented.

3.2.1. Natural Compounds That Impair NRF2 Signaling by Interfering with Protein Synthesis

Despite the increasing demand for negative modulators of NRF2, selective inhibitors are neither currently available nor under clinical trial evaluation. Yet, alternative strategies have been explored based on the pharmacologic manipulation of various NRF2 modulators through natural and synthetic compounds [302,303]. Historically, many natural compounds have been used to treat human pathologies, including cancer. Importantly, these compounds have been often used to potentiate the efficacy of ROS-inducing agents against therapy-resistant cancers, by altering their redox homeostasis. Among the others, many studies focused on Brusatol, a quassinoid from the plant *Brucea javanica* [304–306], and its derivatives [307,308]. The antitumor effects of Brusatol have been ascribed both in solid and in hematologic tumors to the inhibition of proliferation and the disruption of antioxidant defenses caused by NRF2 depletion, due to global suppression of protein synthesis [309–311]. For instance, in A549 NSCLC cells resistant to radiation, Brusatol was found to dose-dependently decrease the NRF2 protein levels and to enhance the efficacy of ionizing radiations, by inducing ROS-dependent DNA damage and cell death [312]. Also, Brusatol was recently found to impair cell growth and proliferaton of human A375 melanoma cells both in vitro and in vivo (NOD/SCID xenografted mice) when used in combination with UVA treatment, leading to increased ROS generation and apoptosis due to AKT impairment [313]. Also, Karathedath et al., reported that NRF2 overexpression in AML cell lines and primary AML samples caused resistance to Cytarabine, Daunorubicin and Arsenic trioxide (ATO). Importantly, Brusatol was found to markedly decrease the NRF2 content and consequently the expression of its target genes *GCLC*, *GCLM*, *HMOX-1*, and *NQO1*, promoting ROS accumulation and apoptosis [314]. Finally, in a recent study from Xiang et al., the overactivation of NRF2 signaling was found to mediate Gemcitabine resistance in human PANC-1, PATU-8988 and BXPC3 pancreatic ductal adenocarcinoma cancer (PDAC) cell lines. Here, Brusatol induced ROS accumulation, growth inhibition and apoptosis by decreasing the protein levels of NRF2 and its target genes *HO-1* and *NQO-1*. Notably, Brusatol enhanced the antitumor activity of Gemcitabine in vivo by reducing the growth rate of PANC-1 xenografts implanted in BALB/c nude mice [315]. Similarly, other work has been directed to Halofuginone from the plant *Dichroa febrifuga*, which is rapidly emerging as one of the most promising NRF2 inhibitors. Indeed Halofuginone has been shown to impair proliferation, migration and invasion of HepG2 HCC cells [316] and MCF7 BC cells [317,318], to reverse the radioresistance of Lewis lung cancer cells (LLCC) [319], or to improve the drug delivery in PDAC cells [320]. In a recent study from Tsuchida K. et al., Halofuginone was seen to repress protein synthesis through prolyl-tRNA synthetase inhibition and to prevent NRF2 nuclear accumulation causing the impaired expression of enzymes involved in drug metabolism and transport, iron metabolism, GSH metabolism and ROS scavenging. As a result, KYSE70 human esophageal cancer (HEC) or A549 NSCLC cells, became more susceptible to Cisplatin based on in vitro and in vivo experiments, with limited side-effects [321]. It should be emphasized that Halofuginone has a very similar mechanism of action to that of Brusatol, confirming that the inhibition of protein synthesis remains a valid strategy to block NRF2 signaling.

3.2.2. Natural Compounds That Impair NRF2 Signaling by Acting on Functional Regulators

Other works have explored the potential use of Chrysin, a natural flavone found in many plant extracts such as honey, propolis, and blue passion flowers, endowed with anticancer effects against different tumors [322,323]. In an earlier study on BEL-7402 human HCC cells resistant to Doxorubicin, Chrysin significantly reduced NRF2 expression both at the mRNA and protein levels, by interfering with PI3K/AKT and ERK pathways. As a result, the expression of NRF2-related target genes *HMOX-1*, *AKR1B10* and *MRP5*, was significantly reduced and the chemoresistance attenuated [324]. In another work form Wang J. et al., Chrysin was seen to suppress the proliferation, migration and invasion of human T98, U251 and U87 glioblastoma (GBM) cells and to abrogate the in vivo tumorigenicity of U87 xenografts in BALB/c mice. Mechanistically, Chrysin impaired NRF2 nuclear translocation, by decreasing the protein levels of phospho-extracellular signal-regulated kinase-1/2 (ERK1/2) and two antioxidant enzymes HO-1 and NQO-1 [325].

3.2.3. Natural and Synthetic Compounds Blocking NRF2 Pathway by Still Unknown Mechanisms

Oridonin, a diterpenoid isolated from the herb *Rabdosia rubescens*, represents another promising compound whose potent antitumor effects have been described in leukemia [326], BC [327] and CRC cells [328]. Very recently, Oridonin was found to promote mitochondrial-dependent apoptosis by activating PPARγ and suppressing both NF-κB and NRF2 pathways in MG-63 and HOS osteosarcoma cells. Here, Oridonin prevented NRF2 nuclear translocation and decreased the expression of the *HMOX1, NQO1* genes and their encoded antioxidant proteins, leading to ROS-dependent apoptosis, in vitro and in vivo [329]. Also Plumbagin, a natural naphthoquinone from *Plumbago zeylanica L.*, has received substantial attention as a potent inducer of apoptosis in pancreatic, lung, breast and prostate cancer cells caused by ROS overproduction [114,330]. With this respect, by using human ovarian (OVCAR3), breast (SKOV3, MCF7), and endometrial (ECC1) cancer cells, a recent study has shown that Plumbagin promotes ROS generation via the mETC complexes I-III and synergizes with Brusatol to block NRF2 pathway, triggering cell death [331]. In another work, a SILAC-based quantitative proteomic approach was used to characterize the biological changes induced by Plumbagin in SCC25 tongue squamous cell carcinoma (TSCC) cells. Here, Plumbagin was seen to decrease the NRF2 nuclear translocation and to suppress the induction of its target genes. As a result, unbalanced ROS overproduction led to cell cycle arrest and stemness attenuation, triggering apoptosis [332]. Extensive research has also been pursued on the alkaloid Trigonelline, a coffee extract initially identified as a negative modulator of NRF2 signaling in HT29 human colorectal cancer cells (CRC), with the ability to decrease the nuclear and total content of NRF2 and to inhibit its target genes transactivation [333]. In a later study, the effects of Trigonelline on NRF2-mediated apoptosis evasion were studied in MiaPaca2, PANC-1, and Colo357 PDAC cell lines and in the human pancreatic duct cell line H6c7. Here, Trigonelline was shown to strongly suppress NRF2 activity by preventing its nuclear accumulation and to increase the efficacy of TRAIL and Etoposide both in vitro and in vivo, with few side-toxicity [334]. Importantly, three different studies from the group of Shin D. recently focused on the impact of Trigonelline on chemoresistance by using experimental models of HNC. First, using HNC cells resistant to Cisplatin, Trigonelline was found to restore the chemosensitivity of HN 2-10 and SNU cells both in vitro and in vivo when combined with inhibitors of GSH synthesis and TRX system [287]. Interestingly, in the same experimental model, NRF2 was seen to promote resistance to the ferroptosis inducer Artesunate, while conversely the combination with Trigonelline resulted in strong cytotoxicity due to ROS accumulation. Remarkably, the effect was restricted to Artesunate- or Cisplatin-resistant HNC cells, sparing normal oral keratinocytes and oral fibroblasts [301]. Lately, the same group reported that in HNC cells, suppression of NRF2 signaling by Trigonelline could reverse the resistance to ferroptosis both in vitro and in vivo [335]. Thus, Trigonelline is emerging as a promising molecule for combination regimens against tumors with widespread chemoresistance. Importantly, the use of high throughput screening (HTS) has played a major role in the discovery of NRF2 inhibitors. For example, AEM1, a benzodioxole substituted analog recently identified in A549 NSCLC cells, was seen to strongly suppress NRF2-mediated genes expression, without altering NRF2 protein stability, its phosphorylation or the KEAP1 content. Importantly, AEM1 reduced the growth rate of A549 NSCLC cells both in vitro and in vivo, enhancing also their sensitivity to Etoposide and 5-Fluorouracil. Of note, AEM1 was also able to decrease the NRF2-dependent induction of *HMOX1* in H838 and H460 NSCLC cells with LOF mutations of the *KEAP1* gene, suggesting that AEM1 might preferentially target tumor cells with constitutive NRF2 activation [336]. Similarly, Singh et al., conducted a quantitative high-throughput screening (qHTS) of the Molecular Libraries Small Molecule Repository (MLSMR) and identified a compound named ML385 that was able to reduce the transcriptional activity of NRF2 by preventing the binding of the NRF2-MAFG complex to the ARE sequence in the promoter of NSCLC cells. More in detail, ML385 attenuated NRF2 pathway by affecting the DNA binding activity of the NRF2-MAFG protein complex through direct interaction with NRF2. As a result, ML385 induced selective toxicity in both A549 and H460 NSCLC cells harboring *KEAP1* mutations, enhancing the cytotoxic effects of Doxorubicin, Carboplatin and Paclitaxel without affecting the non-tumorigenic BEAS2B cells. Importantly, the therapeutic efficacy of ML385 as a single agent and in combination with carboplatin was also confirmed in orthotopic lung cancer xenografts subcutaneously implanted in nude mice [25]. Moreover, in a very recent study from Hori R. et al., the K-563 compound was isolated from

Streptomyces sp. after HTS screening of almost 10000 culture broth samples. Of note, K-563 abrogated the expression of NRF2 target genes such as *GCLC, GCLM, AKR1C1* (reductase family 1 member C1), *ME1, NQO1* and *TXNRD1* in A549 NSCLC cells and in the human gallbladder cancer (GBC) cell line TGBC24TKB, without altering the NRF2 protein levels, its ARE-binding ability, or its nuclear localization. As a result, K-563 promoted ROS accumulation and synergized with either Cisplatin or Etoposide in A549 NSCLC cells, suppressing also cell proliferation in GBC cells (TGBC24TKB) without affecting normal human lung cells (BEAS-2B) [337]. In the context of hematologic tumors, Zhang et al., identified 4f, a pyrazolyl hydroxamic acid derivative with potent antineoplastic effects in human THP-1, HL60 and U937 AML (acute myeloid leukemia) cells. Here, 4f was found to decrease the NRF2 protein content and the mRNA levels of *HMOX1* and *GCLC* genes, promoting increased caspase-3 and PARP cleavage. Besides, 4f suppressed the growth of THP-1 xenografts seeded onto the CAMs of chicken eggs and also impaired blood vessels formation in vivo in a gelatin sponge assay, suggesting that 4f might be a promising treatment in advanced AML [338]. In summary, very promising results have been reported in case of repressors of protein synthesis, an evidence that might pave the way to the design of novel strategies to target the NRF2 pathway in cancers. Taken together, these studies provide a strong rationale for the identification and validation of compounds with the ability to disrupt the redox control exerted by NRF2 in solid as well as hematologic tumors. Moreover, it is expected that other experimental work as well as refined clinical trials will better define the molecular mechanisms, the specific context and the types of tumors wherein this approach might ensure the optimal efficacy in patients with advanced cancers.

3.2.4. Natural and Synthetic Compounds Targeting Functional Regulators of the NRF2-KEAP1 Pathway

An alternative strategy to hamper the pro-tumoral effects of NRF2 is through the modulation of functional regulators that ultimately converge on this signaling pathway. This approach offers at least two significant advantages. First, it relies on already established drugs with proven anticancer activity and, more importantly, it does not require selective NRF2 inhibitors. Of note, recent work supports the applicability of this strategy. For example, LGK-974, a specific inhibitor of the O-acyltransferase porcupine (PORCN) used to disrupt the WNT signaling, was recently found to prevent NRF2 nuclear translocation and its expression in HepG2 cells, presumably by impairing the WNT3A-dependent formation of the GSK-3/β-TrCP protein complex. As a result, the HepG2 cells were sensitized to otherwise non-toxic radiation doses, due to downregulation of NRF2 target genes such as *HMOX1* and *NQO1* and increased ROS production [339]. Also Wogonin, a flavonoid isolated from *Scutellaria baicalensis Georgi*, has emerged as a promising anticancer agent, due to its chemosensitizing ability in Doxocycline-resistant MCF7 BC cells [340]. Further investigations revealed that Wogonin reduced the NRF2 nuclear translocation and promoted increased ROS production, potentiating the cytotoxicity of Hydroxy-Camptothecin, Cisplatin and Etoposide also in HepG2 cells [341]. Other work was conducted on chronic myelogenous leukemia (CML) cells sensitive (K562) or resistant (K562/AO2) to Adriamycin (ADR). As first, Wogonin was found to decrease the NRF2 mRNA and protein levels in CML cells, causing a marked reduction in the HO-1, NQO1 and MRP1 proteins [342]. In a later work the same authors showed that Wogonin could decrease the binding of both p65 and p50 NF-kB subunits to the NRF2 promoter in K562/A02 cells, enhancing their sensitivity to ADR. In vivo, transplantation experiments of K562/A02 cells into NOD/SCID mice proved that the combination of ADR and Wogonin could reduce the nuclear content of NF-κB p65 and NRF2 [343]. Hence, Wogonin represents a promising inhibitor of NRF2 and a potent chemosensitizer in solid and hemathologic tumors. Also, Gao et al., explored the anti-cancer potential of Apigenin, a natural bioflavonoid found in many fruits and vegetables. As first, Apigenin was seen to reduce the expression of NRF2 and its targets HO-1, AKR1B10 and MRP5 both at the mRNA and protein levels in a KEAP1-independent way, by downregulating the PI3K/AKT pathway and to strongly sensitize BEL-7402/ADM cells to Doxorubicin both in vitro and in vivo, blocking tumor growth [344]. Additional work, led to the identification of the microRNA mir-101, as a modulator of NRF2 functions in HCC cells. Here, mir-101 was found to be downregulated in BEL-7402/ADM (resistant) and conversely highly expressed in BEL-7402 (sensitive) cells, while the NRF2 mRNA sequence analysis revealed the presence of a complementary site for mir-101 in the 3'-UTR. Importantly, mir-101 mimics markedly decreased the NRF2 protein levels in BEL-7402/ADM cells exerting chemo-sensitizing effects, while opposite changes were induced by antimir-101. Collectively, these data indicate that mir-101 re-expression might be a novel mechanism to blunt NRF2 signaling in HCC resistant cells [345]. Also, Retinoic acid, a Vitamin A metabolite, has emerged as a potential anticancer agent inducing differentiation, growth arrest, and apoptosis of cancer cells [346]. Initial studies on solid tumors have shown that All-*Trans* Retinoic Acid (ATRA) can induce RARα expression, which in turn forms a protein complex with NRF2 and antagonizes its transactivation [347,348], sensitizing chemoresistant neuroblastoma (NB) HTLA-230 cells to the proteasome inhibitor Bortezomib [349]. Other experimental work from Valenzuela M. et al., extended these observations to AML and APL cells treated with the ROS-inducer Arsenic Trioxide (ATO). Here, the combined use of ATRA and ATO was seen to prevent NRF2 nuclear translocation and to suppress the antioxidant response in HL-60 and THP-1 AML and in NB4 APL cells, inducing cytotoxicity. Additionally, the authors also showed that the ATRA-mediated inhibition of NRF2 depended on RARα, since ATRA was ineffective when parental NB4 cells were pre-treated with the RARα antagonist Ro-41-5253 or when applied to mutant NB4-R2 cells lacking RARα expression [350].

4. Conclusions and Future Perspectives

Since its discovery in 1994 [27] NRF2 has been connected to various cellular mechanisms, such as response to oxidative stress, mitochondrial respiration, stem cell quiescence, mRNA translation and autophagy [51]. The NRF2-KEAP1 pathway is a master regulator of cell protection mechanisms against exogenous or endogenous stress sources. Thus, in the past decade, NRF2 has emerged as an important target in cancer therapy. Many questions have arisen in relation to risks and benefits of negative modulators of NRF2 signaling. Recent research suggest that suppression of antioxidant mechanisms involving NRF2 might potentially induce a pro-oxidizing shift in tumor microenvironment and promote ROS-dependent cell death in many cancer types. Despite the absence of specific and selective NRF2 inhibitors, convincing indications show that the use of natural compounds with known therapeutic action might be effectively used in different types of tumors [351,352]. NRF2 has been recognized as one of the critical factors regulating an array of genes that protect cells against xenobiotics. NRF2-mediated transcriptional regulation is coordinated by a number of specific events within the cellular environment. Among them, the triggering stimulus, the cooperation with other activators and repressors, the interplay with different signaling pathways and the epigenetic landscape of the target gene promoters, can be regarded of utmost importance. Many approaches have been devised to target the NRF2 signaling pathway in cancer such as regulating NRF2 expression at the transcriptional level, controlling the NRF2 nuclear translocation, targeting the KEAP1-NRF2 binding for modulating NRF2 protein stability and regulating NRF2 binding to its target genes promoters. Several small-molecule NRF2 activators and inhibitors have been developed and successfully employed in cancer treatment. The important drawbacks of targeted therapy include resistance of cancer cells and difficulties of developing drugs to some tumor-specific targets [191,353]. Studies suggest that in tumors, high levels of NRF2 can occur in absence of genomic alterations in the NRF2 and KEAP1 genes. Additional research will help to increase the specificity of NRF2-based therapies. For example, crystallographic investigations of KEAP1 might provide chances to design and synthesize molecules that selectively interfere with the KEAP1-NRF2 binding [354]. Moreover, accumulating evidence suggests that NRF2 can interact with other pathways implicated in cell survival [355]. These important points offer great opportunities for pharmacological intervention to control the level of NRF2 and its therapeutic effects. NRF2 was firstly described as a tumor suppressor involved in the inhibition of tumor initiation and cancer metastasis. Recent evidence showed that it can also act as a pro-oncogenic factor. It is now well accepted that NRF2 has a dual role in carcinogenesis and that pharmacological induction of the NRF2 pathway might be chemopreventive in the early stages of tumorigenesis. However, adverse effects might occur in advanced stages of cancer by inducing therapy resistance [356,357]. Chemotherapy resistance is one of the crucial problems for the effective treatment of many cancers, and NRF2 inhibition might be a promising approach to overcome this phenomenon. Being located at the crossroad of many defensive pathways influencing cell life during chemical, oxidative, and metabolic stress, the NRF2-KEAP1 pathway has been the focus of broad research aimed at revealing its role in cancer [105]. Aberrant activation of the NRF2-KEAP1 pathway is often recognized in many tumors, promoting cancer growth, survival, metastasis formation and therapy resistance [358,359]. It is known that NRF2 activation in response to oxidative stress promotes cells survival. NRF2 also induces metabolic changes that sustain cell proliferation. Because of these benefits, cancer cells with sustained NRF2 activation frequently develop "NRF2 addiction" [3]. Disruption of NRF2 signaling is a promising therapeutic approach against NRF2-addicted cancers and some effective NRF2 inhibitors such as brusatol and halofuginone have been employed, while many others are under investigation. Nevertheless, administration of systemic NRF2 inhibitors may cause off-target effects on patients and this represents a key aspect for the identification of novel compounds that should ideally possess high specificity, bioactivity and limited side-toxicity [26]. To prevent these side effects, mechanistic insights should be revealed and new pharmacological targets besides NRF2 should be studied. Depending on metabolic fluxes of NRF2-addicted cancer cells, thorough metabolite examination might help to identify specific diagnostic markers. Research have suggested that the over-activation of NRF2 promotes tumor development, prevents cell apoptosis and enhances the therapy-resistance of cancer cells. We know that both NRF2 inducers and NRF2 inhibitors possess anticancer activity against different targets. NRF2 inducers protect normal cells from carcinogen effects, while NRF2 inhibitors halt cancer cells proliferation. The question is what we need exactly, NRF2 activators or inhibitors? According to the numerous recent studies related to the oncogenic activity of NRF2, the synthesis of novel selective inhibitors of NRF2-KEAP1 pathway could be the better strategy for cancer prevention and therapy [64]. The dilemma of whether NRF2 can be used as a pharmacological target needs to be harmonized with the research of clinical application and participation of mediators of oxidative stress in the anticancer therapy. In conclusion, the discovery, design, and synthesis of NRF2-centered methodologies are important and challenging tasks that might pave the way to novel therapeutic approaches in cancer treatment.

Table 2. Systematic presentation of studies on possible use of natural Nrf2 activators in therapy with possible mechanisms of action connected to Nrf2 activation pathways.

Type of Study	Experimental Model	Treatment Doses and Duration	Observed Mechanism of Action/Effects	Proposed Application in Therapy	Ref.
			Natural Products		
In vivo	male Albino rats Wistar strain	200 mg/kg dose of curcumin for four consecutive days oral administration	• enhanced nuclear translocation and ARE-binding of Nrf2 • curcumin protects against dimethylnitrosamine (DMN)-induced hepatic injury	chemopreventive agent	[360]
In vivo	male C57BL/6J mice	daily treated with curcumin at the dose of 50 mg/kg body weight by oral gavage	• mediated nuclear translocation of Nrf2, followed by induction of HO-1 • curcumin intervention dramatically reversed the defects in Nrf2 signaling induced by high fat diet	improving glucose intolerance	[361]
In vivo	male Sprague–Dawley rats	supplemented with curcumin (1 g/kg diet) for 16 weeks	• increased HO-1 expression was increased along with suppressed oxidative stress as well as reduced hepatic fat accumulation and fibrotic changes	attenuating the pathogenesis of fatty liver induced metabolic diseases	[362]
In vivo/In vitro	female specific pathogen-free BALB/c mice mouse macrophage RAW264.7 cells	on day 22, the mice were treated with curcumin (200 mg/kg) 1 h before ovalbumin challenge. cells treated with different concentrations (0, 5, 10, 25, and 50 μmol/L) of curcumin for 24 h or with 50-μmol/L curcumin for different lengths of time (0, 4, 8, 12 and 24 h)	• Nrf2 and HO-1 levels in lung tissues were significantly increased • heme oxygenase-1 and nuclear Nrf2 levels were enhanced in dose- and time-dependent manners in curcumin-treated RAW264.7 cells.	potentially effective drug in asthma treatment (alleviate airway inflammation in asthma through the Nrf2/HO-1 pathway)	[363]
In vitro	primary cultures of cerebellar granule neurons(CGNs) of rats	pretreated with 5–30 μM curcumin	• mediated neuro-protection against hemin induced damage via Nrf2 dependent HO-1 expression	neuroprotective agent	[364]
In vitro	human breast cancer cell line MCF-7	treatment with DMSO (vehicle) or various concentrations of curcumin for 12, 24 or 48 h	• inhibition of the proliferation of breast cancer cells through Nrf2-mediated down-regulation of Fen1 expression	chemotherapeutic agent	[365]
In vivo	female Sprague-Dawley rats	food supplemented with resveratrol equivalent to 50, 100, or 300 mg/kg body weight/d	• elevated protein and mRNA expression of hepatic Nrf2, attenuation of oxidative stress and suppression of inflammatory response mediated by Nrf2	prevention and intervention of human hepatocellular carcinoma	[366]
In vivo/In vitro	female August Copenhagen Irish rats non-tumorigenic human breast epithelial cell line MCF-10A	resveratrol given as a 50 mg subcutaneous pellet every other month during 8 months to animals subcutaneously treated with 3 mg E2 pellets prepared in cholesterol cells were treated with E2 (10 and 50 nM) and Res (50 μM) for up to 48 h	• resveratrol treatment alone or in combination with E2 significantly upregulated expression of nuclear factor erythroid 2-related factor 2 (NRF2) in mammary tissues; increased expression of NRF2-regulated genes codying for antioxidant enzymes (NQO1, SOD3 and OGG1) involved in protection against oxidative DNA damage • inhibition of suppression of FMO1 and AOX1 genes expression via regulation of NRF2	a potential chemopreventive agent in the development of therapeutic strategies for the prevention of estrogen-induced breast neoplasia	[367]

Table 2. *Cont.*

	Natural Products				
Type of Study	Experimental Model	Treatment Doses and Duration	Observed Mechanism of Action/Effects	Proposed Application in Therapy	Ref.
In vitro	MCF-10F and MCF-7 cells	retreated with 0.1 to 30 nmol/L TCDD with or without 25 μmol/L resveratrol for 72 h and then incubated with E2 (0.1–10 μmol/L) for 24 h	• induction of NAD(P)H quinone oxidoreductase 1 (NQO1) in MCF-10F cells exposed to resveratrol may involve the Nrf2-Keap1-ARE pathway (dissociation of Nrf2 from Keap 1 and translocation of Nrf2 to the nucleus, where it binds to the ARE to activate the transcription of NQO1 mRNA)	a potential chemopreventive agent against estrogen-initiated breast cancer	[368]
In vitro	primary rat hepatocytes were obtained from Sprague–Dawley male rats	cells were incubated in the presence of resveratrol for 24 and 48 h (at concentrations of 25, 50 and 75 μM)	• increase in the level of Nrf2 and induction of its translocation to the nucleus, and the increase in the concentration of the coding mRNA for Nrf2	protection of liver cells from oxidative stress induced damage (chemopreventive agents)	[115]
In vitro	human type II alveolar epithelial cell line, A549	cells were treated with various concentrations of native EGCG (5, 10, 20, 40, 60, 80 and 100 μM) and nano EGCG (1, 2, 4, 6, 8, 10 and 12 μM) and allowed to grow for 48 h	• increased expression of Nrf2 and Keap1 in native and nano EGCG treated A549 cells that could regulate apoptosis	chemotherapeutic in lung cancer	[369]
In vitro	bovine aortic endothelial cells (BAECs)	cells treated with various concentrations of EGCG	• upregulates Nrf2 levels in nuclear extracts and increases ARE-luciferase activity	therapeutic targets in a variety of oxidant- and inflammatory-mediated vascular diseases	[370]
In vivo	male albino Wistar rats; animal model of bleomycin-induced pulmonary fibrosis	intraperitoneally treated with EGCG at a dosage of 20 mg/kg body weight, once daily throughout 28 days	• EGCG enhances antioxidant activities and Phase II enzymes involving Nrf2–Keap1 signaling, with subsequent restraint inflammation during bleomycin-induced pulmonary fibrosis • no significant change in the expression of Keap1	treatment of diseases such as pulmonary fibrosis	[371]
In vitro	human hepatocytes (HHL5) and hepatoma (HepG2) cells	exposed to various concentrations of sulforaphane for different times with DMSO (0.1%) as control	• increased nuclear Nrf2 levels and intracellular GSH levels in both cell lines but with slightly different pattern indicating a potential risk of chemo-resistance of using sulforaphane for chemoprevention	possible induction of pro-survival effects in cancer cells	[128]
In vivo	male BALB/c mice (6 weeks)	5 μmol/animal sulforaphane plus different doses of microcystin-LR	• Nrf2 translocation to the nucleus in mouse livers • induction of Nrf2 downstream target genes, NQO1 and HO-1, two phase II enzymes	effective in cytoprotection against MC-LR-induced hepatotoxicity	[372]
In vitro	adult rat cardiomyocytes	exposed to 5μM sulforaphane with or without H_2O_2	• upregulation of Nrf2 by sulforaphane occurs at 1 h of incubation • increased protein expression of PGC-1α	protective action against oxidative damage, however, timeline of the sulforaphane actions needs to be established	[373]
In vivo	male BALB/c mice (6 weeks)	5 μmol/animal sulforaphane plus different doses of microcystin-LR	• Nrf2 translocation to the nucleus in mouse livers • induction of Nrf2 downstream target genes, NQO1 and HO-1, two phase II enzymes	effective in cytoprotection against MC-LR-induced hepatotoxicity	[372]

Table 2. *Cont.*

Natural Products					
Type of Study	Experimental Model	Treatment Doses and Duration	Observed Mechanism of Action/Effects	Proposed Application in Therapy	Ref.
			Electrophilic/Covalent		
		Triterpenoids			
In vitro	K562 myeloid leukemia cells	Exposed to 0.05–10 µM CDDO-Me for 24–48 h	• Down-regulated Na+/K+ ATPase • Arrested cells in G2/M and S phases • Increased mitochondria and death receptor-dependent and ER-mediated apoptosis • Triggered activation of autophagy	Activated and potentiated the effects of the apoptosis and autophagy pathways to kill K562 cancer cells	[374]
In vitro and In vivo	SKOV3, OVCAR3, A2780, A2780/CP70 and Hey2 ovarian cancer cells	0–50 µM CDDO-Me depending on cell assays 20 mg/kg CDDO-Me in xenograft model using nude mice	• Binds to USP7 cells, decreasing MDM2, MDMX and UHRF1 • Suppresses tumor growth in a xenograft model	Targets apoptosis-related substrates, increasing apoptosis and reducing growth of ovarian cancer cells	[375]
In vitro	MiaPaCa-2 and BxPC-3 cell lines 6 week old Scid/Ner mice	0.625–5µM CDDO-Me in cell culture CDDO-Me 7.5 mg/kg × 5 days/wk by oral gavage until day 40 (to treat primary tumor) or day 100 (to treat residual disease)	• Decreased proliferation and increased apoptosis in K-ras normal and mutant cells • Inhibits antiapoptotic pathways • Inhibited tumor growth in mice • Increased survival time of mice	Combination of in vitro and in vivo effects demonstrate that CDDO-Me will increase apoptosis in pancreatic ductal adenoma carcinoma cell lines and improve the survival of animals	[376]
In vitro	MDA-MB 435, MDA-MB 231, MDA-MB 468, BT-549, T47D and MCF-7 breast cancer cells	CDDO-Me 1.5 µM for 4 h	• Induces endoplasmic reticulum vacuolation • Induces apoptotic pathways • Increases intracellular calcium and generation of reactive oxygen species	CDDO-Me-induced c-FLIPL downregulation and relationship between Ca2+ influx and ROS generation are keys in controlling breast cancer growth.	[377]
In vitro	HO8910 and SKOV3 ovarian cancer cells	CDDO-Me concentration range of 0–100 µM 5 µM CDDO-Me in assays requiring a single concentration	• Upregulates Hsp70 • Degrades Hsp90-associated protein—ErbB2 and Akt • CDDO-Me reacts with nucleophiles on Hsp90 to form Michael adducts	May provide added insight to CDDO-Me action, with Hsp90 as a novel target	[378]
In vivo	C57BL/6 WT mice LSL-KrasG12D/+; Pdx-1-Cre (KC) mice Polyoma-middle T (PyMT) mice	CDDO-Im (100 mg/kg diet) fed 2 days prior to LPS injections	• Reduced the lethal effects of LPS injection in mutant mice • Increased migration of CD45+ cells • Increased percentage of Gr1+ myeloid-derived suppressor cells • CDDO-Im decreased IL-6, CCL-2, VEGF, and G-CSF in mutant mice	Postulating the use of CDDO-Im as prophylaxes in the development of pancreatic cancer within susceptible populations. This is due to the reduction in proinflammatory mediators.	[379]
In vitro	Human Jurkat E6-1 cells	CDDO-Im 1 nM and 10 nM for 30 min prior to activation with anti-CD3/anti-CD28	• Inhibited production of IL-2 • Suppression of CD25 in Nrf2-dependent manner with no effect on CD69	Nrf2 activation by CDDO-Im reduces IL-2 secretion and CD25 expression suggesting a role in potential anticancer therapy.	[380]

Table 2. *Cont.*

		Natural Products			
Type of Study	Experimental Model	Treatment Doses and Duration	Observed Mechanism of Action/Effects	Proposed Application in Therapy	Ref.
		Dithiolethiones			
In vitro	Mouse carcinoma Hepa-1c1c7 cells	Oltipraz 5–60 µM anetholedithione (ADT) 3–15 µM 1,2-dithiole-3-thione (D3T) 1–5 µM	• Dithiolethiones are reduced in Hepa-1c1c7 cells leading to generation of superoxide radicals • Superoxide dismutates to form hydrogen peroxide promoting translocation of Nrf2 to nucleus • Translocated Nrf2 results in the upregulation of various Phase II enzyme expression.	Use of D3T and members of this family may be able to modify KEAP-1 activity and upregulate the expression of Phase II enzymes	[381]
In vivo	Male Fischer 344 (100 g)	D3T administered by oral gavage 0.5 mmol/kg (in distilled water with 1% Cremaphor, and 25% glycerol	• Increased heme oxygenase (HO-1) activity • Increased expression of HO-1 • Increased levels of ferritin	Very early paper describing the potential utility of Dithiolethiones, like D3T, may offer protection against pro-carcinogenic compounds that increase oxidative stress	[382]
In vitro	HT29 colon adenocarcinoma cells	30µM D3T in DMSO vehicle (less than 0.1% final DMSO concentration)	• Increased expression and activity of multiple reductases—Thioredoxin reductase 1, prostaglandin reductase 1, NAD(P)Quinone oxidoreductase 1 • Potentiation of hydroxymethylacylfulvene action in alkylating DNA	D3T was a much stronger inducer of reductases compared to selenite and increased potency of the anticancer drug, hydroxymethylacylfulvene	[383]
In vivo	Male Fisher 344 rats (90–100 g)	Oral gavage of DST at 0.03 to 0.3 mmol/kg body wt at 3 days/week for 3 weeks Also 0.1 mmol/kg for measuring hepatic protein expression	• Reduced the pre-cancer potency of aflatoxin B by inhibiting the expression of glutathione S-transferase-placental isoform • Blocked aflatoxin-mediated increase in • Induced multiple hepatic genes associated with detoxifying aflatoxin—including glutathione • S-transferase A5 (GSTA5) and AFB1 aldehyde reductase	D3T is more potent than older Dithiolethiones like oltipraz and could be a probe for measuring anticancer potencies of this drug class	[384]
In vitro	HepG2 hepatic and LS180 colon cells	S-diclofenac and S-sulindac in range of 0–100 µM	• Inhibited activity and expression of CYP1A1, 1B1 and 1A2 • Regulates aryl hydrocarbon receptor pathway • Blocked binding of aryl hydrocarbon to the responsive element • Increased expression of anticancer enzymes, glutathione S transferase, glutamate cysteine ligase, and glutathione reductase	The NSAIDs with the dithiolethione group, S-diclofenac and S-sulindac, may function as effective anticancer agents	[385]
In vitro	HT29 and HCT116 colon adenocarcinoma	100 µM Oltipraz with 0.2% MeSO as a solvent control	• Induction of p65, IκB kinase α (IKKα), IκB kinase β (IKKβ), and NF-κB-inducing kinase (NIK) • IκBα phosphorylation was also induce • Induction of protein binding to a consensus NF-κB element • Transcriptional activation of QR was decreased • Both MEKK3 and NIK exert effects on IKKα/β activation via different pathways	This is a novel pathway involved in QR gene regulation and may provide insight to the actions of oltipraz as an anticancer agent	[386]

Table 2. *Cont.*

		Natural Products			
Type of Study	Experimental Model	Treatment Doses and Duration	Observed Mechanism of Action/Effects	Proposed Application in Therapy	Ref.
			Non-Electrophilic/Non-Covalent		
In vivo and in vitro	Male C57/Bl6 mice (22 g) and bone marrow-derived mouse macrophage cells	RA839 was dissolved in a vehicle of 95% (*v/v*) hydroxyethyl cellulose (0.5% (*w/v*))/5% (*v/v*) solutol—injected IP at 30 mg/kg. General RA839 interactions measured at a concentration of 10 μM	• Binds to KELCH domain with affinity of 6 μM • Blocks Nrf2-Keap-1 protein-protein interaction • Modified activity of multiple genes in mouse macrophage cells • Reduced the induction of nitric oxide release and inducible nitric oxide synthase activity • Induced Nrf2 gene expression in mouse liver	RA839 may be a useful tool in developing anticancer drugs that target the prevention of Nrf2-Keap1 protein interaction.	[80]
In vitro	THP-1 cells	Cells were exposed to TAT14 using a concentration range of 0–75 μM	• Activates heme oxygenase 1 expression and activity • Did not alter Nrf2 mRNA expression • Reduced LPS-induced TNF expression	The 14-mer TAT fragment interacts with Keap1, preventing Nrf2-Keap1 association, allowing Nrf2 to activate mediators downstream	[387]
In vitro	HepG2 hepatic and U2OS bone cell lines	ML334 and its isomers in a concentration range of 0–100 μM	• Binds with high affinity to KEAP1 • Strong inducer of ARE activity in both hepatic and bone cell lines	First description of ML334 as a potent inhibitor of Nrf2-Keap1 interaction. Highly potent.	[388]
In vitro	Immortalized baby mouse kidney epithelial cells (iBMK) and MDA-MB-231 breast cancer cells	Geopyxin F and other "geopyxin" isomers were used in a concentration range of 0–70 μM depending on assay	• Geopyxin F was the most potent of Geopyxin compounds at inducing Nrf2 activity • Geopyxin F displayed almost not toxicity/lethality in MDA-MB-231 cells • Geopyxin F increased autophagosome formation in iBMK cells • Geopyxin F prevents ubiquitination and stabilizes Nrf2 (KEAP1-dependent)—but is independent of interactions at the Cys151 in KEAP1	Geopyxin F, demonstrated a higher level of protection compared to electrophilic Nrf2 activators. Heightened potency suggests that Geopyxin F may be a useful anticancer compound.	[78]
In vitro	MCF-7 breast cancer cells	Multiple drugs were used as "off-label" activators of Nrf2 Astemizole 8 μM Tamoxifen 1 μM Trifluoperazine 10 μM	• Astemizole, Tamoxifen and Trifluoperazine increased expression of the NQO1, HO-1, and GCLM genes • After 24 h, the stimulated gene expression to basal was highest in Astemizole compared to sulforaphane • The genes for the detoxifying enzymes—NAD(P)H quinone oxidoreductase 1 (NQO1), heme oxygenase 1 (HO1) where most sensitive to upregulation	After large-scale screening, select drugs were chosen based on their ability to activate Nrf2. This shows that 'off-label' mechanisms may have benefit as anticancer drugs. Astemizole was the best candidate.	[79]

Author Contributions: Conceptualization, E.P. and L.S.; writing—original draft preparation, E.P.; writing—review and editing, E.P., A.B., P.T.-A., D.C., D.K., A.V., Z.S., A.T., D.W., S.S.; supervision, E.P.; funding acquisition, L.S. All authors have read and agreed to the published version of the manuscript.

Funding: This research received no external funding.

Conflicts of Interest: The authors declare no conflicts of interest.

References

1. Suzuki, T.; Muramatsu, A.; Saito, R.; Iso, T.; Shibata, T.; Kuwata, K.; Kawaguchi, S.I.; Iwawaki, T.; Adachi, S.; Suda, H.; et al. Molecular Mechanism of Cellular Oxidative Stress Sensing by Keap1. *Cell Rep.* **2019**, *28*, 746–758 e744. [CrossRef]

2. Kobayashi, A.; Kang, M.I.; Okawa, H.; Ohtsuji, M.; Zenke, Y.; Chiba, T.; Igarashi, K.; Yamamoto, M. Oxidative stress sensor Keap1 functions as an adaptor for Cul3-based E3 ligase to regulate proteasomal degradation of Nrf2. *Mol. Cell Biol.* **2004**, *24*, 7130–7139. [CrossRef] [PubMed]

3. Kitamura, H.; Motohashi, H. NRF2 addiction in cancer cells. *Cancer Sci.* **2018**, *109*, 900–911. [CrossRef] [PubMed]

4. Priftis, A.; Angeli-Terzidou, A.E.; Veskoukis, A.S.; Spandidos, D.A.; Kouretas, D. Cellspecific and roastingdependent regulation of the Keap1/Nrf2 pathway by coffee extracts. *Mol. Med. Rep.* **2018**, *17*, 8325–8331. [CrossRef] [PubMed]

5. Kerasioti, E.; Stagos, D.; Tzimi, A.; Kouretas, D. Increase in antioxidant activity by sheep/goat whey protein through nuclear factor-like 2 (Nrf2) is cell type dependent. *Food Chem. Toxicol.* **2016**, *97*, 47–56. [CrossRef]

6. Enomoto, A.; Itoh, K.; Nagayoshi, E.; Haruta, J.; Kimura, T.; O'Connor, T.; Harada, T.; Yamamoto, M. High sensitivity of Nrf2 knockout mice to acetaminophen hepatotoxicity associated with decreased expression of ARE-regulated drug metabolizing enzymes and antioxidant genes. *Toxicol. Sci.* **2001**, *59*, 169–177. [CrossRef]

7. Leung, L.; Kwong, M.; Hou, S.; Lee, C.; Chan, J.Y. Deficiency of the Nrf1 and Nrf2 transcription factors results in early embryonic lethality and severe oxidative stress. *J. Biol. Chem.* **2003**, *278*, 48021–48029. [CrossRef]

8. Antunes Dos Santos, A.; Ferrer, B.; Marques Goncalves, F.; Tsatsakis, A.M.; Renieri, E.A.; Skalny, A.V.; Farina, M.; Rocha, J.B.T.; Aschner, M. Oxidative Stress in Methylmercury-Induced Cell Toxicity. *Toxics* **2018**, *6*, 47. [CrossRef]

9. Honkura, Y.; Matsuo, H.; Murakami, S.; Sakiyama, M.; Mizutari, K.; Shiotani, A.; Yamamoto, M.; Morita, I.; Shinomiya, N.; Kawase, T.; et al. NRF2 Is a Key Target for Prevention of Noise-Induced Hearing Loss by Reducing Oxidative Damage of Cochlea. *Sci. Rep.* **2016**, *6*, 19329. [CrossRef]

10. Iida, K.; Itoh, K.; Kumagai, Y.; Oyasu, R.; Hattori, K.; Kawai, K.; Shimazui, T.; Akaza, H.; Yamamoto, M. Nrf2 is essential for the chemopreventive efficacy of oltipraz against urinary bladder carcinogenesis. *Cancer Res.* **2004**, *64*, 6424–6431. [CrossRef]

11. Satoh, H.; Moriguchi, T.; Saigusa, D.; Baird, L.; Yu, L.; Rokutan, H.; Igarashi, K.; Ebina, M.; Shibata, T.; Yamamoto, M. NRF2 Intensifies Host Defense Systems to Prevent Lung Carcinogenesis, but After Tumor Initiation Accelerates Malignant Cell Growth. *Cancer Res.* **2016**, *76*, 3088–3096. [CrossRef] [PubMed]

12. Satoh, H.; Moriguchi, T.; Takai, J.; Ebina, M.; Yamamoto, M. Nrf2 prevents initiation but accelerates progression through the Kras signaling pathway during lung carcinogenesis. *Cancer Res.* **2013**, *73*, 4158–4168. [CrossRef]

13. Tao, S.; de la Vega, M.R.; Chapman, E.; Ooi, A.; Zhang, D.D. The effects of NRF2 modulation on the initiation and progression of chemically and genetically induced lung cancer. *Mol. Carcinog.* **2018**, *57*, 182–192. [CrossRef] [PubMed]

14. DeNicola, G.M.; Karreth, F.A.; Humpton, T.J.; Gopinathan, A.; Wei, C.; Frese, K.; Mangal, D.; Yu, K.H.; Yeo, C.J.; Calhoun, E.S.; et al. Oncogene-induced Nrf2 transcription promotes ROS detoxification and tumorigenesis. *Nature* **2011**, *475*, 106–109. [CrossRef]

15. Wang, H.; Liu, X.; Long, M.; Huang, Y.; Zhang, L.; Zhang, R.; Zheng, Y.; Liao, X.; Wang, Y.; Liao, Q.; et al. NRF2 activation by antioxidant antidiabetic agents accelerates tumor metastasis. *Sci. Transl. Med.* **2016**, *8*, 334ra351. [CrossRef]

16. Ciamporcero, E.; Daga, M.; Pizzimenti, S.; Roetto, A.; Dianzani, C.; Compagnone, A.; Palmieri, A.; Ullio, C.; Cangemi, L.; Pili, R.; et al. Crosstalk between Nrf2 and YAP contributes to maintaining the antioxidant potential and chemoresistance in bladder cancer. *Free Radic. Biol. Med.* **2018**, *115*, 447–457. [CrossRef] [PubMed]

17. Mukhopadhyay, S.; Goswami, D.; Adiseshaiah, P.P.; Burgan, W.; Yi, M.; Guerin, T.M.; Kozlov, S.V.; Nissley, D.V.; McCormick, F. Undermining glutaminolysis bolsters chemotherapy while NRF2 promotes chemoresistance in KRAS-driven pancreatic cancers. *Cancer Res.* **2020**. [CrossRef] [PubMed]

18. Jeong, Y.; Hoang, N.T.; Lovejoy, A.; Stehr, H.; Newman, A.M.; Gentles, A.J.; Kong, W.; Truong, D.; Martin, S.; Chaudhuri, A.; et al. Role of KEAP1/NRF2 and TP53 Mutations in Lung Squamous Cell Carcinoma Development and Radiation Resistance. *Cancer Discov.* **2017**, *7*, 86–101. [CrossRef] [PubMed]

19. Lee, S.L.; Ryu, H.; Son, A.R.; Seo, B.; Kim, J.; Jung, S.Y.; Song, J.Y.; Hwang, S.G.; Ahn, J. TGF-beta and Hypoxia/Reoxygenation Promote Radioresistance of A549 Lung Cancer Cells through Activation of Nrf2 and EGFR. *Oxid. Med. Cell Longev.* **2016**, *2016*, 6823471. [CrossRef]

20. Solis, L.M.; Behrens, C.; Dong, W.; Suraokar, M.; Ozburn, N.C.; Moran, C.A.; Corvalan, A.H.; Biswal, S.; Swisher, S.G.; Bekele, B.N.; et al. Nrf2 and Keap1 abnormalities in non-small cell lung carcinoma and association with clinicopathologic features. *Clin. Cancer Res.* **2010**, *16*, 3743–3753. [CrossRef]

21. De la Vega, M.R.; Chapman, E.; Zhang, D.D. NRF2 and the Hallmarks of Cancer. *Cancer Cell* **2018**, *34*, 21–43. [CrossRef] [PubMed]

22. Gegotek, A.; Bielawska, K.; Biernacki, M.; Zareba, I.; Surazynski, A.; Skrzydlewska, E. Comparison of protective effect of ascorbic acid on redox and endocannabinoid systems interactions in in vitro cultured human skin fibroblasts exposed to UV radiation and hydrogen peroxide. *Arch. Dermatol. Res.* **2017**, *309*, 285–303. [CrossRef] [PubMed]

23. Ren, D.; Villeneuve, N.F.; Jiang, T.; Wu, T.; Lau, A.; Toppin, H.A.; Zhang, D.D. Brusatol enhances the efficacy of chemotherapy by inhibiting the Nrf2-mediated defense mechanism. *Proc. Natl. Acad. Sci. USA* **2011**, *108*, 1433–1438. [CrossRef] [PubMed]

24. Xia, C.; Bai, X.; Hou, X.; Gou, X.; Wang, Y.; Zeng, H.; Huang, M.; Jin, J. Cryptotanshinone Reverses Cisplatin Resistance of Human Lung Carcinoma A549 Cells through Down-Regulating Nrf2 Pathway. *Cell Physiol. Biochem.* **2015**, *37*, 816–824. [CrossRef] [PubMed]

25. Singh, A.; Venkannagari, S.; Oh, K.H.; Zhang, Y.Q.; Rohde, J.M.; Liu, L.; Nimmagadda, S.; Sudini, K.; Brimacombe, K.R.; Gajghate, S.; et al. Small Molecule Inhibitor of NRF2 Selectively Intervenes Therapeutic Resistance in KEAP1-Deficient NSCLC Tumors. *ACS Chem. Biol.* **2016**, *11*, 3214–3225. [CrossRef]

26. Wu, S.; Lu, H.; Bai, Y. Nrf2 in cancers: A double-edged sword. *Cancer Med.* **2019**, *8*, 2252–2267. [CrossRef]

27. Moi, P.; Chan, K.; Asunis, I.; Cao, A.; Kan, Y.W. Isolation of NF-E2-related factor 2 (Nrf2), a NF-E2-like basic leucine zipper transcriptional activator that binds to the tandem NF-E2/AP1 repeat of the beta-globin locus control region. *Proc. Natl. Acad. Sci. USA* **1994**, *91*, 9926–9930. [CrossRef]

28. Jaramillo, M.C.; Zhang, D.D. The emerging role of the Nrf2-Keap1 signaling pathway in cancer. *Genes Dev.* **2013**, *27*, 2179–2191. [CrossRef]

29. Tong, K.I.; Katoh, Y.; Kusunoki, H.; Itoh, K.; Tanaka, T.; Yamamoto, M. Keap1 recruits Neh2 through binding to ETGE and DLG motifs: Characterization of the two-site molecular recognition model. *Mol. Cell Biol.* **2006**, *26*, 2887–2900. [CrossRef]

30. Magesh, S.; Chen, Y.; Hu, L. Small molecule modulators of Keap1-Nrf2-ARE pathway as potential preventive and therapeutic agents. *Med. Res. Rev.* **2012**, *32*, 687–726. [CrossRef]

31. Zhang, D.D.; Lo, S.C.; Cross, J.V.; Templeton, D.J.; Hannink, M. Keap1 is a redox-regulated substrate adaptor protein for a Cul3-dependent ubiquitin ligase complex. *Mol. Cell Biol.* **2004**, *24*, 10941–10953. [CrossRef] [PubMed]

32. Itoh, K.; Wakabayashi, N.; Katoh, Y.; Ishii, T.; Igarashi, K.; Engel, J.D.; Yamamoto, M. Keap1 represses nuclear activation of antioxidant responsive elements by Nrf2 through binding to the amino-terminal Neh2 domain. *Genes Dev.* **1999**, *13*, 76–86. [CrossRef] [PubMed]

33. Itoh, K.; Igarashi, K.; Hayashi, N.; Nishizawa, M.; Yamamoto, M. Cloning and characterization of a novel erythroid cell-derived CNC family transcription factor heterodimerizing with the small Maf family proteins. *Mol. Cell Biol.* **1995**, *15*, 4184–4193. [CrossRef] [PubMed]

34. Chowdhry, S.; Zhang, Y.; McMahon, M.; Sutherland, C.; Cuadrado, A.; Hayes, J.D. Nrf2 is controlled by two distinct beta-TrCP recognition motifs in its Neh6 domain, one of which can be modulated by GSK-3 activity. *Oncogene* **2013**, *32*, 3765–3781. [CrossRef]

35. Nioi, P.; Nguyen, T.; Sherratt, P.J.; Pickett, C.B. The carboxy-terminal Neh3 domain of Nrf2 is required for transcriptional activation. *Mol. Cell Biol.* **2005**, *25*, 10895–10906. [CrossRef]

36. Katoh, Y.; Itoh, K.; Yoshida, E.; Miyagishi, M.; Fukamizu, A.; Yamamoto, M. Two domains of Nrf2 cooperatively bind CBP, a CREB binding protein, and synergistically activate transcription. *Genes Cells* **2001**, *6*, 857–868. [CrossRef]

37. Zhu, M.; Fahl, W.E. Functional characterization of transcription regulators that interact with the electrophile response element. *Biochem. Biophys. Res. Commun.* **2001**, *289*, 212–219. [CrossRef]

38. Wang, H.; Liu, K.; Geng, M.; Gao, P.; Wu, X.; Hai, Y.; Li, Y.; Li, Y.; Luo, L.; Hayes, J.D.; et al. RXRalpha inhibits the NRF2-ARE signaling pathway through a direct interaction with the Neh7 domain of NRF2. *Cancer Res.* **2013**, *73*, 3097–3108. [CrossRef]

39. Canning, P.; Cooper, C.D.; Krojer, T.; Murray, J.W.; Pike, A.C.; Chaikuad, A.; Keates, T.; Thangaratnarajah, C.; Hojzan, V.; Ayinampudi, V.; et al. Structural basis for Cul3 protein assembly with the BTB-Kelch family of E3 ubiquitin ligases. *J. Biol. Chem.* **2013**, *288*, 7803–7814. [CrossRef]

40. Ogura, T.; Tong, K.I.; Mio, K.; Maruyama, Y.; Kurokawa, H.; Sato, C.; Yamamoto, M. Keap1 is a forked-stem dimer structure with two large spheres enclosing the intervening, double glycine repeat, and C-terminal domains. *Proc. Natl. Acad. Sci. USA* **2010**, *107*, 2842–2847. [CrossRef]

41. Kansanen, E.; Kuosmanen, S.M.; Leinonen, H.; Levonen, A.L. The Keap1-Nrf2 pathway: Mechanisms of activation and dysregulation in cancer. *Redox Biol.* **2013**, *1*, 45–49. [CrossRef] [PubMed]

42. Cleasby, A.; Yon, J.; Day, P.J.; Richardson, C.; Tickle, I.J.; Williams, P.A.; Callahan, J.F.; Carr, R.; Concha, N.; Kerns, J.K.; et al. Structure of the BTB domain of Keap1 and its interaction with the triterpenoid antagonist CDDO. *PLoS ONE* **2014**, *9*, e98896. [CrossRef] [PubMed]

43. Wakabayashi, N.; Dinkova-Kostova, A.T.; Holtzclaw, W.D.; Kang, M.I.; Kobayashi, A.; Yamamoto, M.; Kensler, T.W.; Talalay, P. Protection against electrophile and oxidant stress by induction of the phase 2 response: Fate of cysteines of the Keap1 sensor modified by inducers. *Proc. Natl. Acad. Sci. USA* **2004**, *101*, 2040–2045. [CrossRef]

44. Canning, P.; Sorrell, F.J.; Bullock, A.N. Structural basis of Keap1 interactions with Nrf2. *Free Radic. Biol. Med.* **2015**, *88*, 101–107. [CrossRef]

45. Tian, W.; de la Vega, M.R.; Schmidlin, C.J.; Ooi, A.; Zhang, D.D. Kelch-like ECH-associated protein 1 (KEAP1) differentially regulates nuclear factor erythroid-2-related factors 1 and 2 (NRF1 and NRF2). *J. Biol. Chem.* **2018**, *293*, 2029–2040. [CrossRef]

46. Telkoparan-Akillilar, P.; Suzen, S.; Saso, L. Pharmacological Applications of Nrf2 Inhibitors as Potential Antineoplastic Drugs. *Int. J. Mol. Sci.* **2019**, *20*, 2025. [CrossRef]

47. Kobayashi, A.; Kang, M.I.; Watai, Y.; Tong, K.I.; Shibata, T.; Uchida, K.; Yamamoto, M. Oxidative and electrophilic stresses activate Nrf2 through inhibition of ubiquitination activity of Keap1. *Mol. Cell Biol.* **2006**, *26*, 221–229. [CrossRef]

48. Malhotra, D.; Portales-Casamar, E.; Singh, A.; Srivastava, S.; Arenillas, D.; Happel, C.; Shyr, C.; Wakabayashi, N.; Kensler, T.W.; Wasserman, W.W.; et al. Global mapping of binding sites for Nrf2 identifies novel targets in cell survival response through ChIP-Seq profiling and network analysis. *Nucleic Acids Res.* **2010**, *38*, 5718–5734. [CrossRef]

49. Pall, M.L.; Levine, S. Nrf2, a master regulator of detoxification and also antioxidant, anti-inflammatory and other cytoprotective mechanisms, is raised by health promoting factors. *Sheng Li Xue Bao* **2015**, *67*, 1–18.

50. Cuadrado, A.; Rojo, A.I.; Wells, G.; Hayes, J.D.; Cousin, S.P.; Rumsey, W.L.; Attucks, O.C.; Franklin, S.; Levonen, A.L.; Kensler, T.W.; et al. Therapeutic targeting of the NRF2 and KEAP1 partnership in chronic diseases. *Nat. Rev. Drug Discov.* **2019**, *18*, 295–317. [CrossRef]

51. Tonelli, C.; Chio, I.I.C.; Tuveson, D.A. Transcriptional Regulation by Nrf2. *Antioxid. Redox. Signal.* **2018**, *29*, 1727–1745. [CrossRef] [PubMed]

52. Hayes, J.D.; Dinkova-Kostova, A.T. The Nrf2 regulatory network provides an interface between redox and intermediary metabolism. *Trends. Biochem. Sci.* **2014**, *39*, 199–218. [CrossRef] [PubMed]

53. Mitsuishi, Y.; Taguchi, K.; Kawatani, Y.; Shibata, T.; Nukiwa, T.; Aburatani, H.; Yamamoto, M.; Motohashi, H. Nrf2 redirects glucose and glutamine into anabolic pathways in metabolic reprogramming. *Cancer Cell* **2012**, *22*, 66–79. [CrossRef] [PubMed]

54. Jung, K.A.; Choi, B.H.; Nam, C.W.; Song, M.; Kim, S.T.; Lee, J.Y.; Kwak, M.K. Identification of aldo-keto reductases as NRF2-target marker genes in human cells. *Toxicol. Lett.* **2013**, *218*, 39–49. [CrossRef]

55. Duong, H.Q.; You, K.S.; Oh, S.; Kwak, S.J.; Seong, Y.S. Silencing of NRF2 Reduces the Expression of ALDH1A1 and ALDH3A1 and Sensitizes to 5-FU in Pancreatic Cancer Cells. *Antioxidants (Basel)* **2017**, *6*, 52. [CrossRef]

56. Adachi, T.; Nakagawa, H.; Chung, I.; Hagiya, Y.; Hoshijima, K.; Noguchi, N.; Kuo, M.T.; Ishikawa, T. Nrf2-dependent and -independent induction of ABC transporters ABCC1, ABCC2, and ABCG2 in HepG2 cells under oxidative stress. *J. Exp. Ther. Oncol.* **2007**, *6*, 335–348.

57. Jia, Y.; Chen, J.; Zhu, H.; Jia, Z.H.; Cui, M.H. Aberrantly elevated redox sensing factor Nrf2 promotes cancer stem cell survival via enhanced transcriptional regulation of ABCG2 and Bcl-2/Bmi-1 genes. *Oncol. Rep.* **2015**, *34*, 2296–2304. [CrossRef]

58. Lee, D.; Xu, I.M.; Chiu, D.K.; Lai, R.K.; Tse, A.P.; Lan Li, L.; Law, C.T.; Tsang, F.H.; Wei, L.L.; Chan, C.Y.; et al. Folate cycle enzyme MTHFD1L confers metabolic advantages in hepatocellular carcinoma. *J. Clin. Investig.* **2017**, *127*, 1856–1872. [CrossRef]

59. Gazaryan, I.G.; Thomas, B. The status of Nrf2-based therapeutics: Current perspectives and future prospects. *Neural Regen. Res.* **2016**, *11*, 1708–1711. [CrossRef]

60. Pandey, P.; Singh, A.K.; Singh, M.; Tewari, M.; Shukla, H.S.; Gambhir, I.S. The see-saw of Keap1-Nrf2 pathway in cancer. *Crit. Rev. Oncol. Hematol.* **2017**, *116*, 89–98. [CrossRef]

61. Silva-Islas, C.A.; Maldonado, P.D. Canonical and non-canonical mechanisms of Nrf2 activation. *Pharmacol. Res.* **2018**, *134*, 92–99. [CrossRef] [PubMed]

62. Sporn, M.B.; Liby, K.T. NRF2 and cancer: The good, the bad and the importance of context. *Nat. Rev. Cancer* **2012**, *12*, 564–571. [CrossRef] [PubMed]

63. Kansanen, E.; Jyrkkanen, H.K.; Levonen, A.L. Activation of stress signaling pathways by electrophilic oxidized and nitrated lipids. *Free Radic Biol. Med.* **2012**, *52*, 973–982. [CrossRef]

64. Sova, M.; Saso, L. Design and development of Nrf2 modulators for cancer chemoprevention and therapy: A review. *Drug Des. Devel. Ther.* **2018**, *12*, 3181–3197. [CrossRef]

65. Robledinos-Anton, N.; Fernandez-Gines, R.; Manda, G.; Cuadrado, A. Activators and Inhibitors of NRF2: A Review of Their Potential for Clinical Development. *Oxid. Med. Cell Longev.* **2019**, *2019*, 9372182. [CrossRef]

66. Suzuki, T.; Yamamoto, M. Molecular basis of the Keap1-Nrf2 system. *Free Radic. Biol. Med.* **2015**, *88*, 93–100. [CrossRef]

67. Long, M.J.C.; Aye, Y. Privileged Electrophile Sensors: A Resource for Covalent Drug Development. *Cell Chem. Biol.* **2017**, *24*, 787–800. [CrossRef]

68. Liby, K.T.; Sporn, M.B. Synthetic oleanane triterpenoids: Multifunctional drugs with a broad range of applications for prevention and treatment of chronic disease. *Pharmacol. Rev.* **2012**, *64*, 972–1003. [CrossRef]

69. Taguchi, K.; Yamamoto, M. The KEAP1-NRF2 System in Cancer. *Front Oncol.* **2017**, *7*, 85. [CrossRef]

70. Taguchi, K.; Motohashi, H.; Yamamoto, M. Molecular mechanisms of the Keap1-Nrf2 pathway in stress response and cancer evolution. *Genes Cells* **2011**, *16*, 123–140. [CrossRef]

71. Huerta, C.; Jiang, X.; Trevino, I.; Bender, C.F.; Ferguson, D.A.; Probst, B.; Swinger, K.K.; Stoll, V.S.; Thomas, P.J.; Dulubova, I.; et al. Characterization of novel small-molecule NRF2 activators: Structural and biochemical validation of stereospecific KEAP1 binding. *Biochim. Biophys. Acta* **2016**, *1860*, 2537–2552. [CrossRef] [PubMed]

72. Shekh-Ahmad, T.; Eckel, R.; Dayalan Naidu, S.; Higgins, M.; Yamamoto, M.; Dinkova-Kostova, A.T.; Kovac, S.; Abramov, A.Y.; Walker, M.C. KEAP1 inhibition is neuroprotective and suppresses the development of epilepsy. *Brain* **2018**, *141*, 1390–1403. [CrossRef] [PubMed]

73. Reisman, S.A.; Buckley, D.B.; Tanaka, Y.; Klaassen, C.D. CDDO-Im protects from acetaminophen hepatotoxicity through induction of Nrf2-dependent genes. *Toxicol. Appl. Pharmacol.* **2009**, *236*, 109–114. [CrossRef] [PubMed]

74. Yates, M.S.; Tauchi, M.; Katsuoka, F.; Flanders, K.C.; Liby, K.T.; Honda, T.; Gribble, G.W.; Johnson, D.A.; Johnson, J.A.; Burton, N.C.; et al. Pharmacodynamic characterization of chemopreventive triterpenoids as exceptionally potent inducers of Nrf2-regulated genes. *Mol. Cancer Ther.* **2007**, *6*, 154–162. [CrossRef]

75. Samudio, I.; Konopleva, M.; Hail, N., Jr.; Shi, Y.X.; McQueen, T.; Hsu, T.; Evans, R.; Honda, T.; Gribble, G.W.; Sporn, M.; et al. 2-Cyano-3,12-dioxooleana-1,9-dien-28-imidazolide (CDDO-Im) directly targets mitochondrial glutathione to induce apoptosis in pancreatic cancer. *J. Biol. Chem.* **2005**, *280*, 36273–36282. [CrossRef]

76. Manandhar, S.; Cho, J.M.; Kim, J.A.; Kensler, T.W.; Kwak, M.K. Induction of Nrf2-regulated genes by 3H-1, 2-dithiole-3-thione through the ERK signaling pathway in murine keratinocytes. *Eur. J. Pharmacol.* **2007**, *577*, 17–27. [CrossRef]

77. Maher, J.M.; Dieter, M.Z.; Aleksunes, L.M.; Slitt, A.L.; Guo, G.; Tanaka, Y.; Scheffer, G.L.; Chan, J.Y.; Manautou, J.E.; Chen, Y.; et al. Oxidative and electrophilic stress induces multidrug resistance-associated protein transporters via the nuclear factor-E2-related factor-2 transcriptional pathway. *Hepatology* **2007**, *46*, 1597–1610. [CrossRef]

78. Liu, P.; Tian, W.; Tao, S.; Tillotson, J.; Wijeratne, E.M.K.; Gunatilaka, A.A.L.; Zhang, D.D.; Chapman, E. Non-covalent NRF2 Activation Confers Greater Cellular Protection than Covalent Activation. *Cell Chem. Biol.* **2019**, *26*, 1427–1435 e1425. [CrossRef]

79. Zhang, Q.Y.; Chu, X.Y.; Jiang, L.H.; Liu, M.Y.; Mei, Z.L.; Zhang, H.Y. Identification of Non-Electrophilic Nrf2 Activators from Approved Drugs. *Molecules* **2017**, *22*, 886. [CrossRef]

80. Winkel, A.F.; Engel, C.K.; Margerie, D.; Kannt, A.; Szillat, H.; Glombik, H.; Kallus, C.; Ruf, S.; Gussregen, S.; Riedel, J.; et al. Characterization of RA839, a Noncovalent Small Molecule Binder to Keap1 and Selective Activator of Nrf2 Signaling. *J. Biol. Chem.* **2015**, *290*, 28446–28455. [CrossRef]

81. Richardson, B.G.; Jain, A.D.; Speltz, T.E.; Moore, T.W. Non-electrophilic modulators of the canonical Keap1/Nrf2 pathway. *Bioorg. Med. Chem. Lett.* **2015**, *25*, 2261–2268. [CrossRef] [PubMed]

82. Satoh, T.; Lipton, S. Recent advances in understanding NRF2 as a druggable target: Development of pro-electrophilic and non-covalent NRF2 activators to overcome systemic side effects of electrophilic drugs like dimethyl fumarate. *F1000Res* **2017**, *6*, 2138. [CrossRef] [PubMed]

83. Jain, A.D.; Potteti, H.; Richardson, B.G.; Kingsley, L.; Luciano, J.P.; Ryuzoji, A.F.; Lee, H.; Krunic, A.; Mesecar, A.D.; Reddy, S.P.; et al. Probing the structural requirements of non-electrophilic naphthalene-based Nrf2 activators. *Eur. J. Med. Chem.* **2015**, *103*, 252–268. [CrossRef] [PubMed]

84. Senger, D.R.; Li, D.; Jaminet, S.C.; Cao, S. Activation of the Nrf2 Cell Defense Pathway by Ancient Foods: Disease Prevention by Important Molecules and Microbes Lost from the Modern Western Diet. *PLoS ONE* **2016**, *11*, e0148042. [CrossRef]

85. Mascuch, S.J.; Boudreau, P.D.; Carland, T.M.; Pierce, N.T.; Olson, J.; Hensler, M.E.; Choi, H.; Campanale, J.; Hamdoun, A.; Nizet, V.; et al. Marine Natural Product Honaucin A Attenuates Inflammation by Activating the Nrf2-ARE Pathway. *J. Nat. Prod.* **2018**, *81*, 506–514. [CrossRef] [PubMed]

86. Jadeja, R.N.; Upadhyay, K.K.; Devkar, R.V.; Khurana, S. Naturally Occurring Nrf2 Activators: Potential in Treatment of Liver Injury. *Oxid. Med. Cell Longev.* **2016**, *2016*, 3453926. [CrossRef] [PubMed]

87. Li, Y.R.; Li, G.H.; Zhou, M.X.; Xiang, L.; Ren, D.M.; Lou, H.X.; Wang, X.N.; Shen, T. Discovery of natural flavonoids as activators of Nrf2-mediated defense system: Structure-activity relationship and inhibition of intracellular oxidative insults. *Bioorg. Med. Chem.* **2018**, *26*, 5140–5150. [CrossRef]

88. Sun, W.; Liu, X.; Zhang, H.; Song, Y.; Li, T.; Liu, X.; Liu, Y.; Guo, L.; Wang, F.; Yang, T.; et al. Epigallocatechin gallate upregulates NRF2 to prevent diabetic nephropathy via disabling KEAP1. *Free Radic. Biol. Med.* **2017**, *108*, 840–857. [CrossRef]

89. Bao, F.; Tao, L.; Zhang, H. Neuroprotective Effect of Natural Alkaloid Fangchinoline against Oxidative Glutamate Toxicity: Involvement of Keap1-Nrf2 Axis Regulation. *Cell Mol. Neurobiol.* **2019**, *39*, 1177–1186. [CrossRef]

90. Kim, W.; Lee, H.; Kim, S.; Joo, S.; Jeong, S.; Yoo, J.W.; Jung, Y. Sofalcone, a gastroprotective drug, covalently binds to KEAP1 to activate Nrf2 resulting in anti-colitic activity. *Eur. J. Pharmacol.* **2019**, *865*, 172722. [CrossRef]

91. Cole, G.M.; Teter, B.; Frautschy, S.A. Neuroprotective effects of curcumin. *Adv. Exp. Med. Biol.* **2007**, *595*, 197–212. [CrossRef]

92. Kuttan, G.; Kumar, K.B.; Guruvayoorappan, C.; Kuttan, R. Antitumor, anti-invasion, and antimetastatic effects of curcumin. *Adv. Exp. Med. Biol.* **2007**, *595*, 173–184. [CrossRef] [PubMed]

93. Lin, J.K. Molecular targets of curcumin. *Adv. Exp. Med. Biol.* **2007**, *595*, 227–243. [CrossRef]

94. Aggarwal, B.B.; Sung, B. Pharmacological basis for the role of curcumin in chronic diseases: An age-old spice with modern targets. *Trends. Pharmacol. Sci.* **2009**, *30*, 85–94. [CrossRef] [PubMed]

95. Vendrely, V.; Peuchant, E.; Buscail, E.; Moranvillier, I.; Rousseau, B.; Bedel, A.; Brillac, A.; de Verneuil, H.; Moreau-Gaudry, F.; Dabernat, S. Resveratrol and capsaicin used together as food complements reduce tumor growth and rescue full efficiency of low dose gemcitabine in a pancreatic cancer model. *Cancer Lett.* **2017**, *390*, 91–102. [CrossRef] [PubMed]

96. Garcia-Nino, W.R.; Pedraza-Chaverri, J. Protective effect of curcumin against heavy metals-induced liver damage. *Food. Chem. Toxicol.* **2014**, *69*, 182–201. [CrossRef]

97. Chen, P.N.; Hsieh, Y.S.; Chiou, H.L.; Chu, S.C. Silibinin inhibits cell invasion through inactivation of both PI3K-Akt and MAPK signaling pathways. *Chem. Biol. Interact.* **2005**, *156*, 141–150. [CrossRef]

98. Agarwal, C.; Singh, R.P.; Dhanalakshmi, S.; Tyagi, A.K.; Tecklenburg, M.; Sclafani, R.A.; Agarwal, R. Silibinin upregulates the expression of cyclin-dependent kinase inhibitors and causes cell cycle arrest and apoptosis in human colon carcinoma HT-29 cells. *Oncogene* **2003**, *22*, 8271–8282. [CrossRef]

99. Zhang, Y.; Talalay, P.; Cho, C.G.; Posner, G.H. A major inducer of anticarcinogenic protective enzymes from broccoli: Isolation and elucidation of structure. *Proc. Natl. Acad. Sci. USA* **1992**, *89*, 2399–2403. [CrossRef]

100. Houghton, C.A.; Fassett, R.G.; Coombes, J.S. Sulforaphane and Other Nutrigenomic Nrf2 Activators: Can the Clinician's Expectation Be Matched by the Reality? *Oxid. Med. Cell Longev.* **2016**, *2016*, 7857186. [CrossRef]

101. Dinkova-Kostova, A.T.; Fahey, J.W.; Kostov, R.V.; Kensler, T.W. KEAP1 and Done? Targeting the NRF2 Pathway with Sulforaphane. *Trends. Food. Sci. Technol.* **2017**, *69*, 257–269. [CrossRef] [PubMed]

102. Kubo, E.; Chhunchha, B.; Singh, P.; Sasaki, H.; Singh, D.P. Sulforaphane reactivates cellular antioxidant defense by inducing Nrf2/ARE/Prdx6 activity during aging and oxidative stress. *Sci. Rep.* **2017**, *7*, 14130. [CrossRef] [PubMed]

103. Morimitsu, Y.; Nakagawa, Y.; Hayashi, K.; Fujii, H.; Kumagai, T.; Nakamura, Y.; Osawa, T.; Horio, F.; Itoh, K.; Iida, K.; et al. A sulforaphane analogue that potently activates the Nrf2-dependent detoxification pathway. *J. Biol. Chem.* **2002**, *277*, 3456–3463. [CrossRef] [PubMed]

104. Greaney, A.J.; Maier, N.K.; Leppla, S.H.; Moayeri, M. Sulforaphane inhibits multiple inflammasomes through an Nrf2-independent mechanism. *J. Leukoc. Biol.* **2016**, *99*, 189–199. [CrossRef]

105. Panieri, E.; Saso, L. Potential Applications of NRF2 Inhibitors in Cancer Therapy. *Oxid. Med. Cell Longev.* **2019**, *2019*, 8592348. [CrossRef]

106. Menegon, S.; Columbano, A.; Giordano, S. The Dual Roles of NRF2 in Cancer. *Trends. Mol. Med.* **2016**, *22*, 578–593. [CrossRef]

107. Sanders, K.; Moran, Z.; Shi, Z.; Paul, R.; Greenlee, H. Natural Products for Cancer Prevention: Clinical Update 2016. *Semin. Oncol. Nurs.* **2016**, *32*, 215–240. [CrossRef]

108. Kang, E.S.; Woo, I.S.; Kim, H.J.; Eun, S.Y.; Paek, K.S.; Kim, H.J.; Chang, K.C.; Lee, J.H.; Lee, H.T.; Kim, J.H.; et al. Up-regulation of aldose reductase expression mediated by phosphatidylinositol 3-kinase/Akt and Nrf2 is involved in the protective effect of curcumin against oxidative damage. *Free Radic. Biol. Med.* **2007**, *43*, 535–545. [CrossRef]

109. Balogun, E.; Hoque, M.; Gong, P.; Killeen, E.; Green, C.J.; Foresti, R.; Alam, J.; Motterlini, R. Curcumin activates the haem oxygenase-1 gene via regulation of Nrf2 and the antioxidant-responsive element. *Biochem. J.* **2003**, *371*, 887–895. [CrossRef]

110. McNally, S.J.; Harrison, E.M.; Ross, J.A.; Garden, O.J.; Wigmore, S.J. Curcumin induces heme oxygenase 1 through generation of reactive oxygen species, p38 activation and phosphatase inhibition. *Int. J. Mol. Med.* **2007**, *19*, 165–172. [CrossRef]

111. Kumar, G.; Mittal, S.; Sak, K.; Tuli, H.S. Molecular mechanisms underlying chemopreventive potential of curcumin: Current challenges and future perspectives. *Life Sci.* **2016**, *148*, 313–328. [CrossRef] [PubMed]

112. Adiwidjaja, J.; McLachlan, A.J.; Boddy, A.V. Curcumin as a clinically-promising anti-cancer agent: Pharmacokinetics and drug interactions. *Expert. Opin. Drug Metab. Toxicol.* **2017**, *13*, 953–972. [CrossRef] [PubMed]

113. Burgos-Moron, E.; Calderon-Montano, J.M.; Salvador, J.; Robles, A.; Lopez-Lazaro, M. The dark side of curcumin. *Int. J. Cancer* **2010**, *126*, 1771–1775. [CrossRef] [PubMed]

114. Hafeez, B.B.; Fischer, J.W.; Singh, A.; Zhong, W.; Mustafa, A.; Meske, L.; Sheikhani, M.O.; Verma, A.K. Plumbagin Inhibits Prostate Carcinogenesis in Intact and Castrated PTEN Knockout Mice via Targeting PKCepsilon, Stat3, and Epithelial-to-Mesenchymal Transition Markers. *Cancer Prev. Res. (Phila)* **2015**, *8*, 375–386. [CrossRef] [PubMed]

115. Rubiolo, J.A.; Mithieux, G.; Vega, F.V. Resveratrol protects primary rat hepatocytes against oxidative stress damage: Activation of the Nrf2 transcription factor and augmented activities of antioxidant enzymes. *Eur. J. Pharmacol.* **2008**, *591*, 66–72. [CrossRef] [PubMed]

116. Li, C.; Xu, X.; Wang, X.J.; Pan, Y. Imine resveratrol analogues: Molecular design, Nrf2 activation and SAR analysis. *PLoS ONE* **2014**, *9*, e101455. [CrossRef] [PubMed]

117. Li, C.; Xu, X.; Tao, Z.; Wang, X.J.; Pan, Y. Resveratrol dimers, nutritional components in grape wine, are selective ROS scavengers and weak Nrf2 activators. *Food. Chem.* **2015**, *173*, 218–223. [CrossRef] [PubMed]

118. Chen, C.; Yu, R.; Owuor, E.D.; Kong, A.N. Activation of antioxidant-response element (ARE), mitogen-activated protein kinases (MAPKs) and caspases by major green tea polyphenol components during cell survival and death. *Arch. Pharm. Res.* **2000**, *23*, 605–612. [CrossRef] [PubMed]

119. Na, H.K.; Surh, Y.J. Modulation of Nrf2-mediated antioxidant and detoxifying enzyme induction by the green tea polyphenol EGCG. *Food Chem. Toxicol.* **2008**, *46*, 1271–1278. [CrossRef]

120. Kweon, M.H.; Adhami, V.M.; Lee, J.S.; Mukhtar, H. Constitutive overexpression of Nrf2-dependent heme oxygenase-1 in A549 cells contributes to resistance to apoptosis induced by epigallocatechin 3-gallate. *J. Biol. Chem.* **2006**, *281*, 33761–33772. [CrossRef]

121. Kensler, T.W.; Curphey, T.J.; Maxiutenko, Y.; Roebuck, B.D. Chemoprotection by organosulfur inducers of phase 2 enzymes: Dithiolethiones and dithiins. *Drug Metabol. Drug Interact.* **2000**, *17*, 3–22. [CrossRef] [PubMed]

122. Gong, P.; Hu, B.; Cederbaum, A.I. Diallyl sulfide induces heme oxygenase-1 through MAPK pathway. *Arch. Biochem. Biophys.* **2004**, *432*, 252–260. [CrossRef]

123. Hanlon, N.; Coldham, N.; Gielbert, A.; Kuhnert, N.; Sauer, M.J.; King, L.J.; Ioannides, C. Absolute bioavailability and dose-dependent pharmacokinetic behaviour of dietary doses of the chemopreventive isothiocyanate sulforaphane in rat. *Br. J. Nutr.* **2008**, *99*, 559–564. [CrossRef] [PubMed]

124. Riedl, M.A.; Saxon, A.; Diaz-Sanchez, D. Oral sulforaphane increases Phase II antioxidant enzymes in the human upper airway. *Clin. Immunol.* **2009**, *130*, 244–251. [CrossRef] [PubMed]

125. Ye, L.; Dinkova-Kostova, A.T.; Wade, K.L.; Zhang, Y.; Shapiro, T.A.; Talalay, P. Quantitative determination of dithiocarbamates in human plasma, serum, erythrocytes and urine: Pharmacokinetics of broccoli sprout isothiocyanates in humans. *Clin. Chim. Acta* **2002**, *316*, 43–53. [CrossRef]

126. Cornblatt, B.S.; Ye, L.; Dinkova-Kostova, A.T.; Erb, M.; Fahey, J.W.; Singh, N.K.; Chen, M.S.; Stierer, T.; Garrett-Mayer, E.; Argani, P.; et al. Preclinical and clinical evaluation of sulforaphane for chemoprevention in the breast. *Carcinogenesis* **2007**, *28*, 1485–1490. [CrossRef]

127. Russo, M.; Spagnuolo, C.; Russo, G.L.; Skalicka-Wozniak, K.; Daglia, M.; Sobarzo-Sanchez, E.; Nabavi, S.F.; Nabavi, S.M. Nrf2 targeting by sulforaphane: A potential therapy for cancer treatment. *Crit. Rev. Food. Sci. Nutr.* **2018**, *58*, 1391–1405. [CrossRef]

128. Liu, P.; Wang, W.; Tang, J.; Bowater, R.P.; Bao, Y. Antioxidant effects of sulforaphane in human HepG2 cells and immortalised hepatocytes. *Food. Chem. Toxicol.* **2019**, *128*, 129–136. [CrossRef]

129. Liang, J.; Hansch, G.M.; Hubner, K.; Samstag, Y. Sulforaphane as anticancer agent: A double-edged sword? Tricky balance between effects on tumor cells and immune cells. *Adv. Biol. Regul.* **2019**, *71*, 79–87. [CrossRef]

130. Negrette-Guzman, M. Combinations of the antioxidants sulforaphane or curcumin and the conventional antineoplastics cisplatin or doxorubicin as prospects for anticancer chemotherapy. *Eur. J. Pharmacol.* **2019**, *859*, 172513. [CrossRef]

131. Joshi Alumkal, OHSU Knight Cancer Institute.

132. Jackie Shannon, OHSU Knight Cancer Institute.

133. Yair Lotan, MD, UT Southwestern Medical Center.

134. Chao Family Comprehensive Cancer Center, University of California, Irvine.

135. Borella, R.; Forti, L.; Gibellini, L.; De Gaetano, A.; De Biasi, S.; Nasi, M.; Cossarizza, A.; Pinti, M. Synthesis and Anticancer Activity of CDDO and CDDO-Me, Two Derivatives of Natural Triterpenoids. *Molecules* **2019**, *24*, 4097. [CrossRef] [PubMed]

136. Hong, D.S.; Kurzrock, R.; Supko, J.G.; He, X.; Naing, A.; Wheler, J.; Lawrence, D.; Eder, J.P.; Meyer, C.J.; Ferguson, D.A.; et al. A phase I first-in-human trial of bardoxolone methyl in patients with advanced solid tumors and lymphomas. *Clin. Cancer Res.* **2012**, *18*, 3396–3406. [CrossRef] [PubMed]

137. De Zeeuw, D.; Akizawa, T.; Audhya, P.; Bakris, G.L.; Chin, M.; Christ-Schmidt, H.; Goldsberry, A.; Houser, M.; Krauth, M.; Lambers Heerspink, H.J.; et al. Bardoxolone methyl in type 2 diabetes and stage 4 chronic kidney disease. *N. Engl. J. Med.* **2013**, *369*, 2492–2503. [CrossRef] [PubMed]

138. Chin, M.P.; Reisman, S.A.; Bakris, G.L.; O'Grady, M.; Linde, P.G.; McCullough, P.A.; Packham, D.; Vaziri, N.D.; Ward, K.W.; Warnock, D.G.; et al. Mechanisms contributing to adverse cardiovascular events in patients with type 2 diabetes mellitus and stage 4 chronic kidney disease treated with bardoxolone methyl. *Am. J. Nephrol.* **2014**, *39*, 499–508. [CrossRef]

139. Chin, M.P.; Rich, S.; Goldsberry, A.; O'Grady, M.; Meyer, C.J. Effects of Bardoxolone Methyl on QT Interval in Healthy Volunteers. *Cardiorenal. Med.* **2019**, *9*, 326–333. [CrossRef]

140. Creelan, B.C.; Gabrilovich, D.I.; Gray, J.E.; Williams, C.C.; Tanvetyanon, T.; Haura, E.B.; Weber, J.S.; Gibney, G.T.; Markowitz, J.; Proksch, J.W.; et al. Safety, pharmacokinetics, and pharmacodynamics of oral omaveloxolone (RTA 408), a synthetic triterpenoid, in a first-in-human trial of patients with advanced solid tumors. *Onco. Targets. Ther.* **2017**, *10*, 4239–4250. [CrossRef]

141. Linker, R.A.; Haghikia, A. Dimethyl fumarate in multiple sclerosis: Latest developments, evidence and place in therapy. *Ther. Adv. Chronic. Dis.* **2016**, *7*, 198–207. [CrossRef]

142. Schulze-Topphoff, U.; Varrin-Doyer, M.; Pekarek, K.; Spencer, C.M.; Shetty, A.; Sagan, S.A.; Cree, B.A.; Sobel, R.A.; Wipke, B.T.; Steinman, L.; et al. Dimethyl fumarate treatment induces adaptive and innate immune modulation independent of Nrf2. *Proc. Natl. Acad. Sci. USA* **2016**, *113*, 4777–4782. [CrossRef]

143. Niture, S.K.; Kaspar, J.W.; Shen, J.; Jaiswal, A.K. Nrf2 signaling and cell survival. *Toxicol. Appl. Pharmacol.* **2010**, *244*, 37–42. [CrossRef]

144. McCarthy, N. Tumorigenesis: Oncogene detox programme. *Nat. Rev. Cancer* **2011**, *11*, 622–623. [CrossRef]

145. Singh, A.; Misra, V.; Thimmulappa, R.K.; Lee, H.; Ames, S.; Hoque, M.O.; Herman, J.G.; Baylin, S.B.; Sidransky, D.; Gabrielson, E.; et al. Dysfunctional KEAP1-NRF2 interaction in non-small-cell lung cancer. *PLoS Med.* **2006**, *3*, e420. [CrossRef] [PubMed]

146. Shibata, T.; Ohta, T.; Tong, K.I.; Kokubu, A.; Odogawa, R.; Tsuta, K.; Asamura, H.; Yamamoto, M.; Hirohashi, S. Cancer related mutations in NRF2 impair its recognition by Keap1-Cul3 E3 ligase and promote malignancy. *Proc. Natl. Acad. Sci. USA* **2008**, *105*, 13568–13573. [CrossRef] [PubMed]

147. Ohta, T.; Iijima, K.; Miyamoto, M.; Nakahara, I.; Tanaka, H.; Ohtsuji, M.; Suzuki, T.; Kobayashi, A.; Yokota, J.; Sakiyama, T.; et al. Loss of Keap1 function activates Nrf2 and provides advantages for lung cancer cell growth. *Cancer Res.* **2008**, *68*, 1303–1309. [CrossRef]

148. Yoo, N.J.; Kim, H.R.; Kim, Y.R.; An, C.H.; Lee, S.H. Somatic mutations of the KEAP1 gene in common solid cancers. *Histopathology* **2012**, *60*, 943–952. [CrossRef] [PubMed]

149. Shibata, T.; Kokubu, A.; Gotoh, M.; Ojima, H.; Ohta, T.; Yamamoto, M.; Hirohashi, S. Genetic alteration of Keap1 confers constitutive Nrf2 activation and resistance to chemotherapy in gallbladder cancer. *Gastroenterology* **2008**, *135*, 1358–1368, 1368.e1–1368.e4. [CrossRef]

150. Konstantinopoulos, P.A.; Spentzos, D.; Fountzilas, E.; Francoeur, N.; Sanisetty, S.; Grammatikos, A.P.; Hecht, J.L.; Cannistra, S.A. Keap1 mutations and Nrf2 pathway activation in epithelial ovarian cancer. *Cancer Res.* **2011**, *71*, 5081–5089. [CrossRef]

151. Kim, Y.R.; Oh, J.E.; Kim, M.S.; Kang, M.R.; Park, S.W.; Han, J.Y.; Eom, H.S.; Yoo, N.J.; Lee, S.H. Oncogenic NRF2 mutations in squamous cell carcinomas of oesophagus and skin. *J. Pathol.* **2010**, *220*, 446–451. [CrossRef]

152. Komatsu, M. Potential role of p62 in tumor development. *Autophagy* **2011**, *7*, 1088–1090. [CrossRef]

153. Liao, H.; Zhou, Q.; Zhang, Z.; Wang, Q.; Sun, Y.; Yi, X.; Feng, Y. NRF2 is overexpressed in ovarian epithelial carcinoma and is regulated by gonadotrophin and sex-steroid hormones. *Oncol. Rep.* **2012**, *27*, 1918–1924. [CrossRef]

154. Wang, R.; An, J.; Ji, F.; Jiao, H.; Sun, H.; Zhou, D. Hypermethylation of the Keap1 gene in human lung cancer cell lines and lung cancer tissues. *Biochem. Biophys. Res. Commun.* **2008**, *373*, 151–154. [CrossRef]

155. Wang, L.; Zhang, C.; Guo, Y.; Su, Z.Y.; Yang, Y.; Shu, L.; Kong, A.N. Blocking of JB6 cell transformation by tanshinone IIA: Epigenetic reactivation of Nrf2 antioxidative stress pathway. *AAPS. J.* **2014**, *16*, 1214–1225. [CrossRef] [PubMed]

156. Paredes-Gonzalez, X.; Fuentes, F.; Su, Z.Y.; Kong, A.N. Apigenin reactivates Nrf2 anti-oxidative stress signaling in mouse skin epidermal JB6 P + cells through epigenetics modifications. *AAPS. J.* **2014**, *16*, 727–735. [CrossRef] [PubMed]

157. Wang, X.J.; Sun, Z.; Villeneuve, N.F.; Zhang, S.; Zhao, F.; Li, Y.; Chen, W.; Yi, X.; Zheng, W.; Wondrak, G.T.; et al. Nrf2 enhances resistance of cancer cells to chemotherapeutic drugs, the dark side of Nrf2. *Carcinogenesis* **2008**, *29*, 1235–1243. [CrossRef]

158. Jayakumar, S.; Kunwar, A.; Sandur, S.K.; Pandey, B.N.; Chaubey, R.C. Differential response of DU145 and PC3 prostate cancer cells to ionizing radiation: Role of reactive oxygen species, GSH and Nrf2 in radiosensitivity. *Biochim. Biophys. Acta* **2014**, *1840*, 485–494. [CrossRef] [PubMed]

159. Hayes, J.D.; McMahon, M.; Chowdhry, S.; Dinkova-Kostova, A.T. Cancer chemoprevention mechanisms mediated through the Keap1-Nrf2 pathway. *Antioxid. Redox. Signal.* **2010**, *13*, 1713–1748. [CrossRef] [PubMed]

160. Shaw, A.T.; Winslow, M.M.; Magendantz, M.; Ouyang, C.; Dowdle, J.; Subramanian, A.; Lewis, T.A.; Maglathin, R.L.; Tolliday, N.; Jacks, T. Selective killing of K-ras mutant cancer cells by small molecule inducers of oxidative stress. *Proc. Natl. Acad. Sci. USA* **2011**, *108*, 8773–8778. [CrossRef]

161. Oh, E.T.; Kim, J.W.; Kim, J.M.; Kim, S.J.; Lee, J.S.; Hong, S.S.; Goodwin, J.; Ruthenborg, R.J.; Jung, M.G.; Lee, H.J.; et al. NQO1 inhibits proteasome-mediated degradation of HIF-1alpha. *Nat. Commun.* **2016**, *7*, 13593. [CrossRef]

162. Homma, S.; Ishii, Y.; Morishima, Y.; Yamadori, T.; Matsuno, Y.; Haraguchi, N.; Kikuchi, N.; Satoh, H.; Sakamoto, T.; Hizawa, N.; et al. Nrf2 enhances cell proliferation and resistance to anticancer drugs in human lung cancer. *Clin. Cancer Res.* **2009**, *15*, 3423–3432. [CrossRef]

163. Lister, A.; Nedjadi, T.; Kitteringham, N.R.; Campbell, F.; Costello, E.; Lloyd, B.; Copple, I.M.; Williams, S.; Owen, A.; Neoptolemos, J.P.; et al. Nrf2 is overexpressed in pancreatic cancer: Implications for cell proliferation and therapy. *Mol. Cancer* **2011**, *10*, 37. [CrossRef] [PubMed]

164. Zhang, M.; Zhang, C.; Zhang, L.; Yang, Q.; Zhou, S.; Wen, Q.; Wang, J. Nrf2 is a potential prognostic marker and promotes proliferation and invasion in human hepatocellular carcinoma. *BMC Cancer* **2015**, *15*, 531. [CrossRef]

165. Tao, S.; Wang, S.; Moghaddam, S.J.; Ooi, A.; Chapman, E.; Wong, P.K.; Zhang, D.D. Oncogenic KRAS confers chemoresistance by upregulating NRF2. *Cancer Res.* **2014**, *74*, 7430–7441. [CrossRef] [PubMed]

166. Murakami, S.; Motohashi, H. Roles of Nrf2 in cell proliferation and differentiation. *Free Radic. Biol. Med.* **2015**, *88*, 168–178. [CrossRef]

167. DeNicola, G.M.; Chen, P.H.; Mullarky, E.; Sudderth, J.A.; Hu, Z.; Wu, D.; Tang, H.; Xie, Y.; Asara, J.M.; Huffman, K.E.; et al. NRF2 regulates serine biosynthesis in non-small cell lung cancer. *Nat. Genet.* **2015**, *47*, 1475–1481. [CrossRef] [PubMed]

168. Leinonen, H.M.; Kansanen, E.; Polonen, P.; Heinaniemi, M.; Levonen, A.L. Role of the Keap1-Nrf2 pathway in cancer. *Adv. Cancer Res.* **2014**, *122*, 281–320. [CrossRef]

169. Lu, S.C. Regulation of glutathione synthesis. *Mol. Aspects. Med.* **2009**, *30*, 42–59. [CrossRef] [PubMed]

170. Kitteringham, N.R.; Abdullah, A.; Walsh, J.; Randle, L.; Jenkins, R.E.; Sison, R.; Goldring, C.E.; Powell, H.; Sanderson, C.; Williams, S.; et al. Proteomic analysis of Nrf2 deficient transgenic mice reveals cellular defence and lipid metabolism as primary Nrf2-dependent pathways in the liver. *J. Proteom.* **2010**, *73*, 1612–1631. [CrossRef] [PubMed]

171. Shah, N.M.; Rushworth, S.A.; Murray, M.Y.; Bowles, K.M.; MacEwan, D.J. Understanding the role of NRF2-regulated miRNAs in human malignancies. *Oncotarget* **2013**, *4*, 1130–1142. [CrossRef] [PubMed]

172. Basak, P.; Sadhukhan, P.; Sarkar, P.; Sil, P.C. Perspectives of the Nrf-2 signaling pathway in cancer progression and therapy. *Toxicol. Rep.* **2017**, *4*, 306–318. [CrossRef] [PubMed]

173. Hanahan, D.; Weinberg, R.A. Hallmarks of cancer: The next generation. *Cell* **2011**, *144*, 646–674. [CrossRef] [PubMed]

174. Damert, A.; Ikeda, E.; Risau, W. Activator-protein-1 binding potentiates the hypoxia-induciblefactor-1-mediated hypoxia-induced transcriptional activation of vascular-endothelial growth factor expression in C6 glioma cells. *Biochem. J.* **1997**, *327*(Pt. 2), 419–423. [CrossRef]

175. Petrova, V.; Annicchiarico-Petruzzelli, M.; Melino, G.; Amelio, I. The hypoxic tumour microenvironment. *Oncogenesis* **2018**, *7*, 10. [CrossRef]

176. Leinonen, H.M.; Kansanen, E.; Polonen, P.; Heinaniemi, M.; Levonen, A.L. Dysregulation of the Keap1-Nrf2 pathway in cancer. *Biochem. Soc. Trans.* **2015**, *43*, 645–649. [CrossRef] [PubMed]

177. Shibuya, A.; Onda, K.; Kawahara, H.; Uchiyama, Y.; Nakayama, H.; Omi, T.; Nagaoka, M.; Matsui, H.; Hirano, T. Sofalcone, a gastric mucosa protective agent, increases vascular endothelial growth factor via the Nrf2-heme-oxygenase-1 dependent pathway in gastric epithelial cells. *Biochem. Biophys. Res. Commun.* **2010**, *398*, 581–584. [CrossRef] [PubMed]

178. Kim, T.H.; Hur, E.G.; Kang, S.J.; Kim, J.A.; Thapa, D.; Lee, Y.M.; Ku, S.K.; Jung, Y.; Kwak, M.K. NRF2 blockade suppresses colon tumor angiogenesis by inhibiting hypoxia-induced activation of HIF-1alpha. *Cancer Res.* **2011**, *71*, 2260–2275. [CrossRef] [PubMed]

179. Ji, X.; Wang, H.; Zhu, J.; Zhu, L.; Pan, H.; Li, W.; Zhou, Y.; Cong, Z.; Yan, F.; Chen, S. Knockdown of Nrf2 suppresses glioblastoma angiogenesis by inhibiting hypoxia-induced activation of HIF-1alpha. *Int. J. Cancer* **2014**, *135*, 574–584. [CrossRef]

180. Huang, Y.; Mao, Y.; Li, H.; Shen, G.; Nan, G. Knockdown of Nrf2 inhibits angiogenesis by downregulating VEGF expression through PI3K/Akt signaling pathway in cerebral microvascular endothelial cells under hypoxic conditions. *Biochem. Cell Biol.* **2018**, *96*, 475–482. [CrossRef]

181. Zhang, M.; Xu, Y.; Jiang, L. Irisin attenuates oxidized low-density lipoprotein impaired angiogenesis through AKT/mTOR/S6K1/Nrf2 pathway. *J. Cell Physiol.* **2019**, *234*, 18951–18962. [CrossRef]

182. Thimmulappa, R.K.; Mai, K.H.; Srisuma, S.; Kensler, T.W.; Yamamoto, M.; Biswal, S. Identification of Nrf2-regulated genes induced by the chemopreventive agent sulforaphane by oligonucleotide microarray. *Cancer Res.* **2002**, *62*, 5196–5203.

183. Kozakowska, M.; Dobrowolska-Glazar, B.; Okon, K.; Jozkowicz, A.; Dobrowolski, Z.; Dulak, J. Preliminary Analysis of the Expression of Selected Proangiogenic and Antioxidant Genes and MicroRNAs in Patients with Non-Muscle-Invasive Bladder Cancer. *J. Clin. Med.* **2016**, *5*, 29. [CrossRef]

184. Toth, R.K.; Warfel, N.A. Strange Bedfellows: Nuclear Factor, Erythroid 2-Like 2 (Nrf2) and Hypoxia-Inducible Factor 1 (HIF-1) in Tumor Hypoxia. *Antioxidants (Basel)* **2017**, *6*, 27. [CrossRef]

185. Xu, W.; Li, F.; Liu, Z.; Xu, Z.; Sun, B.; Cao, J.; Liu, Y. MicroRNA-27b inhibition promotes Nrf2/ARE pathway activation and alleviates intracerebral hemorrhage-induced brain injury. *Oncotarget* **2017**, *8*, 70669–70684. [CrossRef]

186. Yang, W.; Shen, Y.; Wei, J.; Liu, F. MicroRNA-153/Nrf-2/GPx1 pathway regulates radiosensitivity and stemness of glioma stem cells via reactive oxygen species. *Oncotarget* **2015**, *6*, 22006–22027. [CrossRef]

187. Li, L.; Pan, H.; Wang, H.; Li, X.; Bu, X.; Wang, Q.; Gao, Y.; Wen, G.; Zhou, Y.; Cong, Z.; et al. Interplay between VEGF and Nrf2 regulates angiogenesis due to intracranial venous hypertension. *Sci. Rep.* **2016**, *6*, 37338. [CrossRef] [PubMed]

188. Warfel, N.A.; Sainz, A.G.; Song, J.H.; Kraft, A.S. PIM Kinase Inhibitors Kill Hypoxic Tumor Cells by Reducing Nrf2 Signaling and Increasing Reactive Oxygen Species. *Mol. Cancer Ther.* **2016**, *15*, 1637–1647. [CrossRef] [PubMed]

189. Niture, S.K.; Jaiswal, A.K. Nrf2-induced antiapoptotic Bcl-xL protein enhances cell survival and drug resistance. *Free Radic. Biol. Med.* **2013**, *57*, 119–131. [CrossRef] [PubMed]

190. Kensler, T.W.; Wakabayashi, N.; Biswal, S. Cell survival responses to environmental stresses via the Keap1-Nrf2-ARE pathway. *Annu. Rev. Pharmacol. Toxicol.* **2007**, *47*, 89–116. [CrossRef] [PubMed]

191. Qin, J.J.; Cheng, X.D.; Zhang, J.; Zhang, W.D. Dual roles and therapeutic potential of Keap1-Nrf2 pathway in pancreatic cancer: A systematic review. *Cell Commun. Signal.* **2019**, *17*, 121. [CrossRef] [PubMed]

192. Li, J.; Lee, J.M.; Johnson, J.A. Microarray analysis reveals an antioxidant responsive element-driven gene set involved in conferring protection from an oxidative stress-induced apoptosis in IMR-32 cells. *J. Biol. Chem.* **2002**, *277*, 388–394. [CrossRef]

193. Niso-Santano, M.; Gonzalez-Polo, R.A.; Bravo-San Pedro, J.M.; Gomez-Sanchez, R.; Lastres-Becker, I.; Ortiz-Ortiz, M.A.; Soler, G.; Moran, J.M.; Cuadrado, A.; Fuentes, J.M.; et al. Activation of apoptosis signal-regulating kinase 1 is a key factor in paraquat-induced cell death: Modulation by the Nrf2/Trx axis. *Free Radic. Biol. Med.* **2010**, *48*, 1370–1381. [CrossRef]

194. Arlt, A.; Bauer, I.; Schafmayer, C.; Tepel, J.; Muerkoster, S.S.; Brosch, M.; Roder, C.; Kalthoff, H.; Hampe, J.; Moyer, M.P.; et al. Increased proteasome subunit protein expression and proteasome activity in colon cancer relate to an enhanced activation of nuclear factor E2-related factor 2 (Nrf2). *Oncogene* **2009**, *28*, 3983–3996. [CrossRef]

195. Wang, X.J.; Hayes, J.D.; Wolf, C.R. Generation of a stable antioxidant response element-driven reporter gene cell line and its use to show redox-dependent activation of nrf2 by cancer chemotherapeutic agents. *Cancer Res.* **2006**, *66*, 10983–10994. [CrossRef] [PubMed]

196. Skarby, T.V.; Hogestatt, E.D. Differential effects of calcium antagonists and Bay K 8644 on contractile responses to exogenous noradrenaline and adrenergic nerve stimulation in the rabbit ear artery. *Br. J. Pharmacol.* **1990**, *101*, 961–967. [CrossRef]

197. Polyak, K.; Xia, Y.; Zweier, J.L.; Kinzler, K.W.; Vogelstein, B. A model for p53-induced apoptosis. *Nature* **1997**, *389*, 300–305. [CrossRef]

198. Mendez-Garcia, L.A.; Martinez-Castillo, M.; Villegas-Sepulveda, N.; Orozco, L.; Cordova, E.J. Curcumin induces p53-independent inactivation of Nrf2 during oxidative stress-induced apoptosis. *Hum. Exp. Toxicol.* **2019**, *38*, 951–961. [CrossRef]

199. Chen, W.; Sun, Z.; Wang, X.J.; Jiang, T.; Huang, Z.; Fang, D.; Zhang, D.D. Direct interaction between Nrf2 and p21(Cip1/WAF1) upregulates the Nrf2-mediated antioxidant response. *Mol. Cell* **2009**, *34*, 663–673. [CrossRef]

200. Muller, P.A.; Vousden, K.H. p53 mutations in cancer. *Nat. Cell Biol.* **2013**, *15*, 2–8. [CrossRef] [PubMed]

201. Tung, M.C.; Lin, P.L.; Wang, Y.C.; He, T.Y.; Lee, M.C.; Yeh, S.D.; Chen, C.Y.; Lee, H. Mutant p53 confers chemoresistance in non-small cell lung cancer by upregulating Nrf2. *Oncotarget* **2015**, *6*, 41692–41705. [CrossRef] [PubMed]

202. Sau, A.; Filomeni, G.; Pezzola, S.; D'Aguanno, S.; Tregno, F.P.; Urbani, A.; Serra, M.; Pasello, M.; Picci, P.; Federici, G.; et al. Targeting GSTP1-1 induces JNK activation and leads to apoptosis in cisplatin-sensitive and -resistant human osteosarcoma cell lines. *Mol. Biosyst.* **2012**, *8*, 994–1006. [CrossRef]

203. Jain, A.; Lamark, T.; Sjottem, E.; Larsen, K.B.; Awuh, J.A.; Overvatn, A.; McMahon, M.; Hayes, J.D.; Johansen, T. p62/SQSTM1 is a target gene for transcription factor NRF2 and creates a positive feedback loop by inducing antioxidant response element-driven gene transcription. *J. Biol. Chem.* **2010**, *285*, 22576–22591. [CrossRef]

204. Danial, N.N.; Korsmeyer, S.J. Cell death: Critical control points. *Cell* **2004**, *116*, 205–219. [CrossRef]

205. Boise, L.H.; Gonzalez-Garcia, M.; Postema, C.E.; Ding, L.; Lindsten, T.; Turka, L.A.; Mao, X.; Nunez, G.; Thompson, C.B. bcl-x, a bcl-2-related gene that functions as a dominant regulator of apoptotic cell death. *Cell* **1993**, *74*, 597–608. [CrossRef]

206. Niture, S.K.; Jaiswal, A.K. Nrf2 protein up-regulates antiapoptotic protein Bcl-2 and prevents cellular apoptosis. *J. Biol. Chem.* **2012**, *287*, 9873–9886. [CrossRef]

207. Roche, J. The Epithelial-to-Mesenchymal Transition in Cancer. *Cancers (Basel)* **2018**, *10*, 52. [CrossRef] [PubMed]

208. Kalluri, R.; Neilson, E.G. Epithelial-mesenchymal transition and its implications for fibrosis. *J. Clin. Investig.* **2003**, *112*, 1776–1784. [CrossRef] [PubMed]

209. Wang, Y.; Shi, J.; Chai, K.; Ying, X.; Zhou, B.P. The Role of Snail in EMT and Tumorigenesis. *Curr. Cancer Drug Targets.* **2013**, *13*, 963–972. [CrossRef] [PubMed]

210. Lo, U.G.; Lee, C.F.; Lee, M.S.; Hsieh, J.T. The Role and Mechanism of Epithelial-to-Mesenchymal Transition in Prostate Cancer Progression. *Int. J. Mol. Sci.* **2017**, *18*, 79. [CrossRef] [PubMed]

211. Shibata, T.; Saito, S.; Kokubu, A.; Suzuki, T.; Yamamoto, M.; Hirohashi, S. Global downstream pathway analysis reveals a dependence of oncogenic NF-E2-related factor 2 mutation on the mTOR growth signaling pathway. *Cancer Res.* **2010**, *70*, 9095–9105. [CrossRef]

212. Hayden, A.; Douglas, J.; Sommerlad, M.; Andrews, L.; Gould, K.; Hussain, S.; Thomas, G.J.; Packham, G.; Crabb, S.J. The Nrf2 transcription factor contributes to resistance to cisplatin in bladder cancer. *Urol. Oncol.* **2014**, *32*, 806–814. [CrossRef]

213. Arfmann-Knubel, S.; Struck, B.; Genrich, G.; Helm, O.; Sipos, B.; Sebens, S.; Schafer, H. The Crosstalk between Nrf2 and TGF-beta1 in the Epithelial-Mesenchymal Transition of Pancreatic Duct Epithelial Cells. *PLoS ONE* **2015**, *10*, e0132978. [CrossRef]

214. Zhao, Q.; Mao, A.; Guo, R.; Zhang, L.; Yan, J.; Sun, C.; Tang, J.; Ye, Y.; Zhang, Y.; Zhang, H. Suppression of radiation-induced migration of non-small cell lung cancer through inhibition of Nrf2-Notch Axis. *Oncotarget* **2017**, *8*, 36603–36613. [CrossRef]

215. Zhang, H.S.; Zhang, Z.G.; Du, G.Y.; Sun, H.L.; Liu, H.Y.; Zhou, Z.; Gou, X.M.; Wu, X.H.; Yu, X.Y.; Huang, Y.H. Nrf2 promotes breast cancer cell migration via up-regulation of G6PD/HIF-1alpha/Notch1 axis. *J. Cell Mol. Med.* **2019**, *23*, 3451–3463. [CrossRef] [PubMed]

216. Zhang, C.; Wang, H.J.; Bao, Q.C.; Wang, L.; Guo, T.K.; Chen, W.L.; Xu, L.L.; Zhou, H.S.; Bian, J.L.; Yang, Y.R.; et al. NRF2 promotes breast cancer cell proliferation and metastasis by increasing RhoA/ROCK pathway signal transduction. *Oncotarget* **2016**, *7*, 73593–73606. [CrossRef]

217. Frohlich, D.A.; McCabe, M.T.; Arnold, R.S.; Day, M.L. The role of Nrf2 in increased reactive oxygen species and DNA damage in prostate tumorigenesis. *Oncogene* **2008**, *27*, 4353–4362. [CrossRef] [PubMed]

218. Rachakonda, G.; Sekhar, K.R.; Jowhar, D.; Samson, P.C.; Wikswo, J.P.; Beauchamp, R.D.; Datta, P.K.; Freeman, M.L. Increased cell migration and plasticity in Nrf2-deficient cancer cell lines. *Oncogene* **2010**, *29*, 3703–3714. [CrossRef] [PubMed]

219. Satoh, H.; Moriguchi, T.; Taguchi, K.; Takai, J.; Maher, J.M.; Suzuki, T.; Winnard, P.T., Jr.; Raman, V.; Ebina, M.; Nukiwa, T.; et al. Nrf2-deficiency creates a responsive microenvironment for metastasis to the lung. *Carcinogenesis* **2010**, *31*, 1833–1843. [CrossRef]

220. Pavlova, N.N.; Thompson, C.B. The Emerging Hallmarks of Cancer Metabolism. *Cell Metab.* **2016**, *23*, 27–47. [CrossRef] [PubMed]

221. Boroughs, L.K.; DeBerardinis, R.J. Metabolic pathways promoting cancer cell survival and growth. *Nat. Cell Biol.* **2015**, *17*, 351–359. [CrossRef] [PubMed]

222. Lee, S.B.; Sellers, B.N.; DeNicola, G.M. The Regulation of NRF2 by Nutrient-Responsive Signaling and Its Role in Anabolic Cancer Metabolism. *Antioxid. Redox. Signal.* **2018**, *29*, 1774–1791. [CrossRef]

223. Zhan, L.; Zhang, H.; Zhang, Q.; Woods, C.G.; Chen, Y.; Xue, P.; Dong, J.; Tokar, E.J.; Xu, Y.; Hou, Y.; et al. Regulatory role of KEAP1 and NRF2 in PPARgamma expression and chemoresistance in human non-small-cell lung carcinoma cells. *Free Radic. Biol. Med.* **2012**, *53*, 758–768. [CrossRef]

224. Fan, H.; Paiboonrungruan, C.; Zhang, X.; Prigge, J.R.; Schmidt, E.E.; Sun, Z.; Chen, X. Nrf2 regulates cellular behaviors and Notch signaling in oral squamous cell carcinoma cells. *Biochem. Biophys. Res. Commun.* **2017**, *493*, 833–839. [CrossRef]

225. Shin, S.; Wakabayashi, N.; Misra, V.; Biswal, S.; Lee, G.H.; Agoston, E.S.; Yamamoto, M.; Kensler, T.W. NRF2 modulates aryl hydrocarbon receptor signaling: Influence on adipogenesis. *Mol. Cell Biol.* **2007**, *27*, 7188–7197. [CrossRef] [PubMed]

226. Rajakumar, T.; Pugalendhi, P.; Thilagavathi, S.; Ananthakrishnan, D.; Gunasekaran, K. Allyl isothiocyanate, a potent chemopreventive agent targets AhR/Nrf2 signaling pathway in chemically induced mammary carcinogenesis. *Mol. Cell Biochem.* **2018**, *437*, 1–12. [CrossRef] [PubMed]

227. Hlouschek, J.; Ritter, V.; Wirsdorfer, F.; Klein, D.; Jendrossek, V.; Matschke, J. Targeting SLC25A10 alleviates improved antioxidant capacity and associated radioresistance of cancer cells induced by chronic-cycling hypoxia. *Cancer Lett.* **2018**, *439*, 24–38. [CrossRef] [PubMed]

228. Kawahara, B.; Moller, T.; Hu-Moore, K.; Carrington, S.; Faull, K.F.; Sen, S.; Mascharak, P.K. Attenuation of Antioxidant Capacity in Human Breast Cancer Cells by Carbon Monoxide through Inhibition of Cystathionine beta-Synthase Activity: Implications in Chemotherapeutic Drug Sensitivity. *J. Med. Chem.* **2017**, *60*, 8000–8010. [CrossRef]

229. Carter, D.R.; Sutton, S.K.; Pajic, M.; Murray, J.; Sekyere, E.O.; Fletcher, J.; Beckers, A.; De Preter, K.; Speleman, F.; George, R.E.; et al. Glutathione biosynthesis is upregulated at the initiation of MYCN-driven neuroblastoma tumorigenesis. *Mol. Oncol.* **2016**, *10*, 866–878. [CrossRef]

230. Kim, H.; Lee, G.R.; Kim, J.; Baek, J.Y.; Jo, Y.J.; Hong, S.E.; Kim, S.H.; Lee, J.; Lee, H.I.; Park, S.K.; et al. Sulfiredoxin inhibitor induces preferential death of cancer cells through reactive oxygen species-mediated mitochondrial damage. *Free Radic. Biol. Med.* **2016**, *91*, 264–274. [CrossRef]

231. Liao, J.; Liu, P.P.; Hou, G.; Shao, J.; Yang, J.; Liu, K.; Lu, W.; Wen, S.; Hu, Y.; Huang, P. Regulation of stem-like cancer cells by glutamine through beta-catenin pathway mediated by redox signaling. *Mol. Cancer* **2017**, *16*, 51. [CrossRef]

232. Amelio, I.; Markert, E.K.; Rufini, A.; Antonov, A.V.; Sayan, B.S.; Tucci, P.; Agostini, M.; Mineo, T.C.; Levine, A.J.; Melino, G. p73 regulates serine biosynthesis in cancer. *Oncogene* **2014**, *33*, 5039–5046. [CrossRef]

233. Ju, H.Q.; Lu, Y.X.; Wu, Q.N.; Liu, J.; Zeng, Z.L.; Mo, H.Y.; Chen, Y.; Tian, T.; Wang, Y.; Kang, T.B.; et al. Disrupting G6PD-mediated Redox homeostasis enhances chemosensitivity in colorectal cancer. *Oncogene* **2017**, *36*, 6282–6292. [CrossRef]

234. Panieri, E.; Santoro, M.M. ROS homeostasis and metabolism: A dangerous liason in cancer cells. *Cell Death Dis.* **2016**, *7*, e2253. [CrossRef]

235. Gentric, G.; Mieulet, V.; Mechta-Grigoriou, F. Heterogeneity in Cancer Metabolism: New Concepts in an Old Field. *Antioxid. Redox. Signal.* **2017**, *26*, 462–485. [CrossRef] [PubMed]

236. Hanschmann, E.M.; Godoy, J.R.; Berndt, C.; Hudemann, C.; Lillig, C.H. Thioredoxins, glutaredoxins, and peroxiredoxins–molecular mechanisms and health significance: From cofactors to antioxidants to redox signaling. *Antioxid. Redox. Signal.* **2013**, *19*, 1539–1605. [CrossRef] [PubMed]

237. Lu, J.; Holmgren, A. The thioredoxin antioxidant system. *Free Radic. Biol. Med.* **2014**, *66*, 75–87. [CrossRef] [PubMed]

238. Finkel, T. Signal transduction by reactive oxygen species. *J. Cell Biol.* **2011**, *194*, 7–15. [CrossRef] [PubMed]

239. Xu, I.M.; Lai, R.K.; Lin, S.H.; Tse, A.P.; Chiu, D.K.; Koh, H.Y.; Law, C.T.; Wong, C.M.; Cai, Z.; Wong, C.C.; et al. Transketolase counteracts oxidative stress to drive cancer development. *Proc. Natl. Acad. Sci. USA* **2016**, *113*, E725–E734. [CrossRef] [PubMed]

240. Singh, A.; Happel, C.; Manna, S.K.; Acquaah-Mensah, G.; Carrerero, J.; Kumar, S.; Nasipuri, P.; Krausz, K.W.; Wakabayashi, N.; Dewi, R.; et al. Transcription factor NRF2 regulates miR-1 and miR-206 to drive tumorigenesis. *J. Clin. Investig.* **2013**, *123*, 2921–2934. [CrossRef] [PubMed]

241. Kowalik, M.A.; Guzzo, G.; Morandi, A.; Perra, A.; Menegon, S.; Masgras, I.; Trevisan, E.; Angioni, M.M.; Fornari, F.; Quagliata, L.; et al. Metabolic reprogramming identifies the most aggressive lesions at early phases of hepatic carcinogenesis. *Oncotarget* **2016**, *7*, 32375–32393. [CrossRef]

242. MacLeod, A.K.; McMahon, M.; Plummer, S.M.; Higgins, L.G.; Penning, T.M.; Igarashi, K.; Hayes, J.D. Characterization of the cancer chemopreventive NRF2-dependent gene battery in human keratinocytes: Demonstration that the KEAP1-NRF2 pathway, and not the BACH1-NRF2 pathway, controls cytoprotection against electrophiles as well as redox-cycling compounds. *Carcinogenesis* **2009**, *30*, 1571–1580. [CrossRef]

243. Bai, M.; Yang, L.; Liao, H.; Liang, X.; Xie, B.; Xiong, J.; Tao, X.; Chen, X.; Cheng, Y.; Chen, X.; et al. Metformin sensitizes endometrial cancer cells to chemotherapy through IDH1-induced Nrf2 expression via an epigenetic mechanism. *Oncogene* **2018**, *37*, 5666–5681. [CrossRef]

244. Zhang, N.; Yang, X.; Yuan, F.; Zhang, L.; Wang, Y.; Wang, L.; Mao, Z.; Luo, J.; Zhang, H.; Zhu, W.G.; et al. Increased Amino Acid Uptake Supports Autophagy-Deficient Cell Survival upon Glutamine Deprivation. *Cell Rep.* **2018**, *23*, 3006–3020. [CrossRef]

245. Timmerman, L.A.; Holton, T.; Yuneva, M.; Louie, R.J.; Padro, M.; Daemen, A.; Hu, M.; Chan, D.A.; Ethier, S.P.; van 't Veer, L.J.; et al. Glutamine sensitivity analysis identifies the xCT antiporter as a common triple-negative breast tumor therapeutic target. *Cancer Cell* **2013**, *24*, 450–465. [CrossRef] [PubMed]

246. Lim, J.K.M.; Delaidelli, A.; Minaker, S.W.; Zhang, H.F.; Colovic, M.; Yang, H.; Negri, G.L.; von Karstedt, S.; Lockwood, W.W.; Schaffer, P.; et al. Cystine/glutamate antiporter xCT (SLC7A11) facilitates oncogenic RAS transformation by preserving intracellular redox balance. *Proc. Natl. Acad. Sci. USA* **2019**, *116*, 9433–9442. [CrossRef] [PubMed]

247. Ji, X.; Qian, J.; Rahman, S.M.J.; Siska, P.J.; Zou, Y.; Harris, B.K.; Hoeksema, M.D.; Trenary, I.A.; Heidi, C.; Eisenberg, R.; et al. xCT (SLC7A11)-mediated metabolic reprogramming promotes non-small cell lung cancer progression. *Oncogene* **2018**, *37*, 5007–5019. [CrossRef] [PubMed]

248. Koppula, P.; Zhang, Y.; Shi, J.; Li, W.; Gan, B. The glutamate/cystine antiporter SLC7A11/xCT enhances cancer cell dependency on glucose by exporting glutamate. *J. Biol. Chem.* **2017**, *292*, 14240–14249. [CrossRef] [PubMed]

249. Telorack, M.; Meyer, M.; Ingold, I.; Conrad, M.; Bloch, W.; Werner, S. A Glutathione-Nrf2-Thioredoxin Cross-Talk Ensures Keratinocyte Survival and Efficient Wound Repair. *PLoS Genet.* **2016**, *12*, e1005800. [CrossRef]

250. Habib, E.; Linher-Melville, K.; Lin, H.X.; Singh, G. Expression of xCT and activity of system xc(-) are regulated by NRF2 in human breast cancer cells in response to oxidative stress. *Redox. Biol.* **2015**, *5*, 33–42. [CrossRef]

251. Barretina, J.; Caponigro, G.; Stransky, N.; Venkatesan, K.; Margolin, A.A.; Kim, S.; Wilson, C.J.; Lehar, J.; Kryukov, G.V.; Sonkin, D.; et al. The Cancer Cell Line Encyclopedia enables predictive modelling of anticancer drug sensitivity. *Nature* **2012**, *483*, 603–607. [CrossRef]

252. Shin, C.S.; Mishra, P.; Watrous, J.D.; Carelli, V.; D'Aurelio, M.; Jain, M.; Chan, D.C. The glutamate/cystine xCT antiporter antagonizes glutamine metabolism and reduces nutrient flexibility. *Nat. Commun.* **2017**, *8*, 15074. [CrossRef]

253. Khamari, R.; Trinh, A.; Gabert, P.E.; Corazao-Rozas, P.; Riveros-Cruz, S.; Balayssac, S.; Malet-Martino, M.; Dekiouk, S.; Joncquel Chevalier Curt, M.; Maboudou, P.; et al. Glucose metabolism and NRF2 coordinate the antioxidant response in melanoma resistant to MAPK inhibitors. *Cell Death Dis.* **2018**, *9*, 325. [CrossRef]

254. Altman, B.J.; Stine, Z.E.; Dang, C.V. From Krebs to clinic: Glutamine metabolism to cancer therapy. *Nat. Rev. Cancer* **2016**, *16*, 773. [CrossRef]

255. Sayin, V.I.; LeBoeuf, S.E.; Singh, S.X.; Davidson, S.M.; Biancur, D.; Guzelhan, B.S.; Alvarez, S.W.; Wu, W.L.; Karakousi, T.R.; Zavitsanou, A.M.; et al. Activation of the NRF2 antioxidant program generates an imbalance in central carbon metabolism in cancer. *Elife* **2017**, *6*. [CrossRef] [PubMed]

256. Averous, J.; Bruhat, A.; Jousse, C.; Carraro, V.; Thiel, G.; Fafournoux, P. Induction of CHOP expression by amino acid limitation requires both ATF4 expression and ATF2 phosphorylation. *J. Biol. Chem.* **2004**, *279*, 5288–5297. [CrossRef] [PubMed]

257. Fu, J.; Xiong, Z.; Huang, C.; Li, J.; Yang, W.; Han, Y.; Paiboonrungruan, C.; Major, M.B.; Chen, K.N.; Kang, X.; et al. Hyperactivity of the transcription factor Nrf2 causes metabolic reprogramming in mouse esophagus. *J. Biol. Chem.* **2019**, *294*, 327–340. [CrossRef] [PubMed]

258. Agyeman, A.S.; Chaerkady, R.; Shaw, P.G.; Davidson, N.E.; Visvanathan, K.; Pandey, A.; Kensler, T.W. Transcriptomic and proteomic profiling of KEAP1 disrupted and sulforaphane-treated human breast epithelial cells reveals common expression profiles. *Breast Cancer Res. Treat.* **2012**, *132*, 175–187. [CrossRef]

259. Ludtmann, M.H.; Angelova, P.R.; Zhang, Y.; Abramov, A.Y.; Dinkova-Kostova, A.T. Nrf2 affects the efficiency of mitochondrial fatty acid oxidation. *Biochem. J.* **2014**, *457*, 415–424. [CrossRef] [PubMed]

260. Pang, S.; Lynn, D.A.; Lo, J.Y.; Paek, J.; Curran, S.P. SKN-1 and Nrf2 couples proline catabolism with lipid metabolism during nutrient deprivation. *Nat. Commun.* **2014**, *5*, 5048. [CrossRef]

261. Wu, K.C.; Cui, J.Y.; Klaassen, C.D. Beneficial role of Nrf2 in regulating NADPH generation and consumption. *Toxicol. Sci.* **2011**, *123*, 590–600. [CrossRef]

262. Lin, T.Y.; Cantley, L.C.; DeNicola, G.M. *NRF2 Rewires Cellular Metabolism to Support the Antioxidant Response*; IntechOpen: London, UK, 2016; Chapter 2; ISBN 978-953-51-2838-0. [CrossRef]

263. McGranahan, N.; Swanton, C. Clonal Heterogeneity and Tumor Evolution: Past, Present, and the Future. *Cell* **2017**, *168*, 613–628. [CrossRef]

264. Parker, J.J.; Canoll, P.; Niswander, L.; Kleinschmidt-DeMasters, B.K.; Foshay, K.; Waziri, A. Intratumoral heterogeneity of endogenous tumor cell invasive behavior in human glioblastoma. *Sci. Rep.* **2018**, *8*, 18002. [CrossRef]

265. Wang, D.C.; Wang, X. Systems heterogeneity: An integrative way to understand cancer heterogeneity. *Semin. Cell Dev. Biol.* **2017**, *64*, 1–4. [CrossRef]

266. Alderton, G.K. Tumour evolution: Epigenetic and genetic heterogeneity in metastasis. *Nat. Rev. Cancer* **2017**, *17*, 141. [CrossRef] [PubMed]

267. Mazor, T.; Pankov, A.; Song, J.S.; Costello, J.F. Intratumoral Heterogeneity of the Epigenome. *Cancer Cell* **2016**, *29*, 440–451. [CrossRef] [PubMed]

268. Jamal-Hanjani, M.; Quezada, S.A.; Larkin, J.; Swanton, C. Translational implications of tumor heterogeneity. *Clin. Cancer Res.* **2015**, *21*, 1258–1266. [CrossRef] [PubMed]

269. Robertson-Tessi, M.; Gillies, R.J.; Gatenby, R.A.; Anderson, A.R. Impact of metabolic heterogeneity on tumor growth, invasion, and treatment outcomes. *Cancer Res.* **2015**, *75*, 1567–1579. [CrossRef]

270. Dagogo-Jack, I.; Shaw, A.T. Tumour heterogeneity and resistance to cancer therapies. *Nat. Rev. Clin. Oncol.* **2018**, *15*, 81–94. [CrossRef]

271. Russo, M.; Siravegna, G.; Blaszkowsky, L.S.; Corti, G.; Crisafulli, G.; Ahronian, L.G.; Mussolin, B.; Kwak, E.L.; Buscarino, M.; Lazzari, L.; et al. Tumor Heterogeneity and Lesion-Specific Response to Targeted Therapy in Colorectal Cancer. *Cancer Discov.* **2016**, *6*, 147–153. [CrossRef]

272. Palmer, A.C.; Sorger, P.K. Combination Cancer Therapy Can Confer Benefit via Patient-to-Patient Variability without Drug Additivity or Synergy. *Cell* **2017**, *171*, 1678–1691 e1613. [CrossRef]

273. Wu, D.; Wang, D.C.; Cheng, Y.; Qian, M.; Zhang, M.; Shen, Q.; Wang, X. Roles of tumor heterogeneity in the development of drug resistance: A call for precision therapy. *Semin. Cancer Biol.* **2017**, *42*, 13–19. [CrossRef]

274. Anderton, B.; Camarda, R.; Balakrishnan, S.; Balakrishnan, A.; Kohnz, R.A.; Lim, L.; Evason, K.J.; Momcilovic, O.; Kruttwig, K.; Huang, Q.; et al. MYC-driven inhibition of the glutamate-cysteine ligase promotes glutathione depletion in liver cancer. *EMBO Rep.* **2017**, *18*, 569–585. [CrossRef]

275. Sun, L.; Yin, Y.; Clark, L.H.; Sun, W.; Sullivan, S.A.; Tran, A.Q.; Han, J.; Zhang, L.; Guo, H.; Madugu, E.; et al. Dual inhibition of glycolysis and glutaminolysis as a therapeutic strategy in the treatment of ovarian cancer. *Oncotarget* **2017**, *8*, 63551–63561. [CrossRef]

276. Goto, M.; Miwa, H.; Shikami, M.; Tsunekawa-Imai, N.; Suganuma, K.; Mizuno, S.; Takahashi, M.; Mizutani, M.; Hanamura, I.; Nitta, M. Importance of glutamine metabolism in leukemia cells by energy production through TCA cycle and by redox homeostasis. *Cancer Investig.* **2014**, *32*, 241–247. [CrossRef] [PubMed]

277. Zhong, W.; Weiss, H.L.; Jayswal, R.D.; Hensley, P.J.; Downes, L.M.; St Clair, D.K.; Chaiswing, L. Extracellular redox state shift: A novel approach to target prostate cancer invasion. *Free Radic. Biol. Med.* **2018**, *117*, 99–109. [CrossRef] [PubMed]

278. Habermann, K.J.; Grunewald, L.; van Wijk, S.; Fulda, S. Targeting redox homeostasis in rhabdomyosarcoma cells: GSH-depleting agents enhance auranofin-induced cell death. *Cell Death Dis.* **2017**, *8*, e3067. [CrossRef]

279. Jain, S.; Washington, A.; Leaf, R.K.; Bhargava, P.; Clark, R.A.; Kupper, T.S.; Stroopinsky, D.; Pyzer, A.; Cole, L.; Nahas, M.; et al. Decitabine Priming Enhances Mucin 1 Inhibition Mediated Disruption of Redox Homeostasis in Cutaneous T-Cell Lymphoma. *Mol. Cancer Ther.* **2017**, *16*, 2304–2314. [CrossRef]

280. Hass, C.; Belz, K.; Schoeneberger, H.; Fulda, S. Sensitization of acute lymphoblastic leukemia cells for LCL161-induced cell death by targeting redox homeostasis. *Biochem. Pharmacol.* **2016**, *105*, 14–22. [CrossRef] [PubMed]

281. Ma, M.Z.; Chen, G.; Wang, P.; Lu, W.H.; Zhu, C.F.; Song, M.; Yang, J.; Wen, S.; Xu, R.H.; Hu, Y.; et al. Xc-inhibitor sulfasalazine sensitizes colorectal cancer to cisplatin by a GSH-dependent mechanism. *Cancer Lett.* **2015**, *368*, 88–96. [CrossRef] [PubMed]

282. Luo, M.; Shang, L.; Brooks, M.D.; Jiagge, E.; Zhu, Y.; Buschhaus, J.M.; Conley, S.; Fath, M.A.; Davis, A.; Gheordunescu, E.; et al. Targeting Breast Cancer Stem Cell State Equilibrium through Modulation of Redox Signaling. *Cell Metab.* **2018**, *28*, 69–86 e66. [CrossRef]

283. Rashmi, R.; Huang, X.; Floberg, J.M.; Elhammali, A.E.; McCormick, M.L.; Patti, G.J.; Spitz, D.R.; Schwarz, J.K. Radioresistant Cervical Cancers Are Sensitive to Inhibition of Glycolysis and Redox Metabolism. *Cancer Res.* **2018**, *78*, 1392–1403. [CrossRef]

284. Wang, K.; Jiang, J.; Lei, Y.; Zhou, S.; Wei, Y.; Huang, C. Targeting Metabolic-Redox Circuits for Cancer Therapy. *Trends. Biochem. Sci.* **2019**, *44*, 401–414. [CrossRef]

285. Gorrini, C.; Harris, I.S.; Mak, T.W. Modulation of oxidative stress as an anticancer strategy. *Nat. Rev. Drug Discov.* **2013**, *12*, 931–947. [CrossRef]

286. Lu, B.C.; Li, J.; Yu, W.F.; Zhang, G.Z.; Wang, H.M.; Ma, H.M. Elevated expression of Nrf2 mediates multidrug resistance in CD133(+) head and neck squamous cell carcinoma stem cells. *Oncol. Lett.* **2016**, *12*, 4333–4338. [CrossRef] [PubMed]

287. Roh, J.L.; Jang, H.; Kim, E.H.; Shin, D. Targeting of the Glutathione, Thioredoxin, and Nrf2 Antioxidant Systems in Head and Neck Cancer. *Antioxid. Redox. Signal.* **2017**, *27*, 106–114. [CrossRef] [PubMed]

288. Ju, H.Q.; Gocho, T.; Aguilar, M.; Wu, M.; Zhuang, Z.N.; Fu, J.; Yanaga, K.; Huang, P.; Chiao, P.J. Mechanisms of Overcoming Intrinsic Resistance to Gemcitabine in Pancreatic Ductal Adenocarcinoma through the Redox Modulation. *Mol. Cancer Ther.* **2015**, *14*, 788–798. [CrossRef] [PubMed]

289. Gregory, M.A.; Nemkov, T.; Park, H.J.; Zaberezhnyy, V.; Gehrke, S.; Adane, B.; Jordan, C.T.; Hansen, K.C.; D'Alessandro, A.; DeGregori, J. Targeting Glutamine Metabolism and Redox State for Leukemia Therapy. *Clin. Cancer Res.* **2019**. [CrossRef]

290. Todoric, J.; Antonucci, L.; Di Caro, G.; Li, N.; Wu, X.; Lytle, N.K.; Dhar, D.; Banerjee, S.; Fagman, J.B.; Browne, C.D.; et al. Stress-Activated NRF2-MDM2 Cascade Controls Neoplastic Progression in Pancreas. *Cancer Cell* **2017**, *32*, 824–839 e828. [CrossRef]

291. Liu, D.; Zhang, Y.; Wei, Y.; Liu, G.; Liu, Y.; Gao, Q.; Zou, L.; Zeng, W.; Zhang, N. Activation of AKT pathway by Nrf2/PDGFA feedback loop contributes to HCC progression. *Oncotarget* **2016**, *7*, 65389–65402. [CrossRef]

292. Wang, J.; Liu, Z.; Hu, T.; Han, L.; Yu, S.; Yao, Y.; Ruan, Z.; Tian, T.; Huang, T.; Wang, M.; et al. Nrf2 promotes progression of non-small cell lung cancer through activating autophagy. *Cell Cycle* **2017**, *16*, 1053–1062. [CrossRef]

293. Gambardella, V.; Gimeno-Valiente, F.; Tarazona, N.; Martinez-Ciarpaglini, C.; Roda, D.; Fleitas, T.; Tolosa, P.; Cejalvo, J.M.; Huerta, M.; Rosello, S.; et al. NRF2 through RPS6 Activation Is Related to Anti-HER2 Drug Resistance in HER2-Amplified Gastric Cancer. *Clin. Cancer Res.* **2019**, *25*, 1639–1649. [CrossRef]

294. Kankia, I.H.; Khalil, H.S.; Langdon, S.P.; Moult, P.R.; Bown, J.L.; Deeni, Y.Y. NRF2 Regulates HER1 Signaling Pathway to Modulate the Sensitivity of Ovarian Cancer Cells to Lapatinib and Erlotinib. *Oxid. Med. Cell Longev.* **2017**, *2017*, 1864578. [CrossRef]

295. Wang, Q.; Ma, J.; Lu, Y.; Zhang, S.; Huang, J.; Chen, J.; Bei, J.X.; Yang, K.; Wu, G.; Huang, K.; et al. CDK20 interacts with KEAP1 to activate NRF2 and promotes radiochemoresistance in lung cancer cells. *Oncogene* **2017**, *36*, 5321–5330. [CrossRef]

296. Galan-Cobo, A.; Sitthideatphaiboon, P.; Qu, X.; Poteete, A.; Pisegna, M.A.; Tong, P.; Chen, P.H.; Boroughs, L.K.; Rodriguez, M.L.M.; Zhang, W.; et al. LKB1 and KEAP1/NRF2 pathways cooperatively promote metabolic reprogramming with enhanced glutamine dependence in KRAS-mutant lung adenocarcinoma. *Cancer Res.* **2019**. [CrossRef]

297. Marampon, F.; Codenotti, S.; Megiorni, F.; Del Fattore, A.; Camero, S.; Gravina, G.L.; Festuccia, C.; Musio, D.; De Felice, F.; Nardone, V.; et al. NRF2 orchestrates the redox regulation induced by radiation therapy, sustaining embryonal and alveolar rhabdomyosarcoma cells radioresistance. *J. Cancer Res. Clin. Oncol.* **2019**, *145*, 881–893. [CrossRef] [PubMed]

298. Benlloch, M.; Obrador, E.; Valles, S.L.; Rodriguez, M.L.; Sirerol, J.A.; Alcacer, J.; Pellicer, J.A.; Salvador, R.; Cerda, C.; Saez, G.T.; et al. Pterostilbene Decreases the Antioxidant Defenses of Aggressive Cancer Cells In Vivo: A Physiological Glucocorticoids- and Nrf2-Dependent Mechanism. *Antioxid. Redox. Signal.* **2016**, *24*, 974–990. [CrossRef] [PubMed]

299. Zhang, L.; Wang, H. FTY720 inhibits the Nrf2/ARE pathway in human glioblastoma cell lines and sensitizes glioblastoma cells to temozolomide. *Pharmacol. Rep.* **2017**, *69*, 1186–1193. [CrossRef] [PubMed]

300. Shin, D.; Kim, E.H.; Lee, J.; Roh, J.L. RITA plus 3-MA overcomes chemoresistance of head and neck cancer cells via dual inhibition of autophagy and antioxidant systems. *Redox. Biol.* **2017**, *13*, 219–227. [CrossRef]

301. Roh, J.L.; Kim, E.H.; Jang, H.; Shin, D. Nrf2 inhibition reverses the resistance of cisplatin-resistant head and neck cancer cells to artesunate-induced ferroptosis. *Redox. Biol.* **2017**, *11*, 254–262. [CrossRef] [PubMed]

302. Kim, E.H.; Baek, S.; Shin, D.; Lee, J.; Roh, J.L. Hederagenin Induces Apoptosis in Cisplatin-Resistant Head and Neck Cancer Cells by Inhibiting the Nrf2-ARE Antioxidant Pathway. *Oxid. Med. Cell Longev.* **2017**, *2017*, 5498908. [CrossRef] [PubMed]

303. Olayanju, A.; Copple, I.M.; Bryan, H.K.; Edge, G.T.; Sison, R.L.; Wong, M.W.; Lai, Z.Q.; Lin, Z.X.; Dunn, K.; Sanderson, C.M.; et al. Brusatol provokes a rapid and transient inhibition of Nrf2 signaling and sensitizes mammalian cells to chemical toxicity-implications for therapeutic targeting of Nrf2. *Free Radic. Biol. Med.* **2015**, *78*, 202–212. [CrossRef] [PubMed]

304. Lee, K.H.; Hayashi, N.; Okano, M.; Nozaki, H.; Ju-Ichi, M. Antitumor agents, 65. Brusatol and cleomiscosin-A, antileukemic principles from Brucea javanica. *J. Nat. Prod.* **1984**, *47*, 550–551. [CrossRef]

305. Eigebaly, S.A.; Hall, I.H.; Lee, K.H.; Sumida, Y.; Imakura, Y.; Wu, R.Y. Antitumor agents. XXXV: Effects of brusatol, bruceoside A, and bruceantin on P-388 lymphocytic leukemia cell respiration. *J. Pharm. Sci.* **1979**, *68*, 887–890. [CrossRef]

306. Hall, I.H.; Lee, K.H.; Eigebaly, S.A.; Imakura, Y.; Sumida, Y.; Wu, R.Y. Antitumor agents. XXXIV: Mechanism of action of bruceoside A and brusatol on nucleic acid metabolism of P-388 lymphocytic leukemia cells. *J. Pharm. Sci.* **1979**, *68*, 883–887. [CrossRef]

307. Zhao, M.; Lau, S.T.; Leung, P.S.; Che, C.T.; Lin, Z.X. Seven quassinoids from Fructus Bruceae with cytotoxic effects on pancreatic adenocarcinoma cell lines. *Phytother. Res.* **2011**, *25*, 1796–1800. [CrossRef] [PubMed]

308. Tang, W.; Xie, J.; Xu, S.; Lv, H.; Lin, M.; Yuan, S.; Bai, J.; Hou, Q.; Yu, S. Novel nitric oxide-releasing derivatives of brusatol as anti-inflammatory agents: Design, synthesis, biological evaluation, and nitric oxide release studies. *J. Med. Chem.* **2014**, *57*, 7600–7612. [CrossRef] [PubMed]

309. Willingham, W., Jr.; Stafford, E.A.; Reynolds, S.H.; Chaney, S.G.; Lee, K.H.; Okano, M.; Hall, I.H. Mechanism of eukaryotic protein synthesis inhibition by brusatol. *Biochim. Biophys. Acta* **1981**, *654*, 169–174. [CrossRef]

310. Harder, B.; Tian, W.; La Clair, J.J.; Tan, A.C.; Ooi, A.; Chapman, E.; Zhang, D.D. Brusatol overcomes chemoresistance through inhibition of protein translation. *Mol. Carcinog.* **2017**, *56*, 1493–1500. [CrossRef] [PubMed]

311. Vartanian, S.; Ma, T.P.; Lee, J.; Haverty, P.M.; Kirkpatrick, D.S.; Yu, K.; Stokoe, D. Application of Mass Spectrometry Profiling to Establish Brusatol as an Inhibitor of Global Protein Synthesis. *Mol. Cell Proteom.* **2016**, *15*, 1220–1231. [CrossRef] [PubMed]

312. Sun, X.; Wang, Q.; Wang, Y.; Du, L.; Xu, C.; Liu, Q. Brusatol Enhances the Radiosensitivity of A549 Cells by Promoting ROS Production and Enhancing DNA Damage. *Int. J. Mol. Sci.* **2016**, *17*, 997. [CrossRef]

313. Wang, M.; Shi, G.; Bian, C.; Nisar, M.F.; Guo, Y.; Wu, Y.; Li, W.; Huang, X.; Jiang, X.; Bartsch, J.W.; et al. UVA Irradiation Enhances Brusatol-Mediated Inhibition of Melanoma Growth by Downregulation of the Nrf2-Mediated Antioxidant Response. *Oxid. Med. Cell Longev.* **2018**, *2018*, 9742154. [CrossRef]

314. Karathedath, S.; Rajamani, B.M.; Musheer Aalam, S.M.; Abraham, A.; Varatharajan, S.; Krishnamurthy, P.; Mathews, V.; Velayudhan, S.R.; Balasubramanian, P. Role of NF-E2 related factor 2 (Nrf2) on chemotherapy resistance in acute myeloid leukemia (AML) and the effect of pharmacological inhibition of Nrf2. *PLoS ONE* **2017**, *12*, e0177227. [CrossRef]

315. Xiang, Y.; Ye, W.; Huang, C.; Yu, D.; Chen, H.; Deng, T.; Zhang, F.; Lou, B.; Zhang, J.; Shi, K.; et al. Brusatol Enhances the Chemotherapy Efficacy of Gemcitabine in Pancreatic Cancer via the Nrf2 Signalling Pathway. *Oxid. Med. Cell Longev.* **2018**, *2018*, 2360427. [CrossRef]

316. Huo, S.; Yu, H.; Li, C.; Zhang, J.; Liu, T. Effect of halofuginone on the inhibition of proliferation and invasion of hepatocellular carcinoma HepG2 cell line. *Int. J. Clin. Exp. Pathol.* **2015**, *8*, 15863–15870.

317. Xia, X.; Wang, L.; Zhang, X.; Wang, S.; Lei, L.; Cheng, L.; Xu, Y.; Sun, Y.; Hang, B.; Zhang, G.; et al. Halofuginone-induced autophagy suppresses the migration and invasion of MCF-7 cells via regulation of STMN1 and p53. *J. Cell Biochem.* **2018**, *119*, 4009–4020. [CrossRef] [PubMed]

318. Xia, X.; Wang, X.; Zhang, S.; Zheng, Y.; Wang, L.; Xu, Y.; Hang, B.; Sun, Y.; Lei, L.; Bai, Y.; et al. miR-31 shuttled by halofuginone-induced exosomes suppresses MFC-7 cell proliferation by modulating the HDAC2/cell cycle signaling axis. *J. Cell Physiol.* **2019**. [CrossRef] [PubMed]

319. Lin, R.; Yi, S.; Gong, L.; Liu, W.; Wang, P.; Liu, N.; Zhao, L.; Wang, P. Inhibition of TGF-beta signaling with halofuginone can enhance the antitumor effect of irradiation in Lewis lung cancer. *Onco. Targets. Ther.* **2015**, *8*, 3549–3559. [CrossRef] [PubMed]

320. Elahi-Gedwillo, K.Y.; Carlson, M.; Zettervall, J.; Provenzano, P.P. Antifibrotic Therapy Disrupts Stromal Barriers and Modulates the Immune Landscape in Pancreatic Ductal Adenocarcinoma. *Cancer Res.* **2019**, *79*, 372–386. [CrossRef] [PubMed]

321. Tsuchida, K.; Tsujita, T.; Hayashi, M.; Ojima, A.; Keleku-Lukwete, N.; Katsuoka, F.; Otsuki, A.; Kikuchi, H.; Oshima, Y.; Suzuki, M.; et al. Halofuginone enhances the chemo-sensitivity of cancer cells by suppressing NRF2 accumulation. *Free Radic. Biol. Med.* **2017**, *103*, 236–247. [CrossRef] [PubMed]

322. Davatgaran-Taghipour, Y.; Masoomzadeh, S.; Farzaei, M.H.; Bahramsoltani, R.; Karimi-Soureh, Z.; Rahimi, R.; Abdollahi, M. Polyphenol nanoformulations for cancer therapy: Experimental evidence and clinical perspective. *Int. J. Nanomedicine.* **2017**, *12*, 2689–2702. [CrossRef]

323. Li, X.; Huang, Q.; Ong, C.N.; Yang, X.F.; Shen, H.M. Chrysin sensitizes tumor necrosis factor-alpha-induced apoptosis in human tumor cells via suppression of nuclear factor-kappaB. *Cancer Lett.* **2010**, *293*, 109–116. [CrossRef]

324. Gao, A.M.; Ke, Z.P.; Shi, F.; Sun, G.C.; Chen, H. Chrysin enhances sensitivity of BEL-7402/ADM cells to doxorubicin by suppressing PI3K/Akt/Nrf2 and ERK/Nrf2 pathway. *Chem. Biol. Interact.* **2013**, *206*, 100–108. [CrossRef]

325. Wang, J.; Wang, H.; Sun, K.; Wang, X.; Pan, H.; Zhu, J.; Ji, X.; Li, X. Chrysin suppresses proliferation, migration, and invasion in glioblastoma cell lines via mediating the ERK/Nrf2 signaling pathway. *Drug Des. Devel. Ther.* **2018**, *12*, 721–733. [CrossRef]

326. Li, F.F.; Yi, S.; Wen, L.; He, J.; Yang, L.J.; Zhao, J.; Zhang, B.P.; Cui, G.H.; Chen, Y. Oridonin induces NPM mutant protein translocation and apoptosis in NPM1c+ acute myeloid leukemia cells in vitro. *Acta Pharmacol. Sin.* **2014**, *35*, 806–813. [CrossRef]

327. Wang, S.; Zhong, Z.; Wan, J.; Tan, W.; Wu, G.; Chen, M.; Wang, Y. Oridonin induces apoptosis, inhibits migration and invasion on highly-metastatic human breast cancer cells. *Am. J. Chin. Med.* **2013**, *41*, 177–196. [CrossRef] [PubMed]

328. Ren, C.M.; Li, Y.; Chen, Q.Z.; Zeng, Y.H.; Shao, Y.; Wu, Q.X.; Yuan, S.X.; Yang, J.Q.; Yu, Y.; Wu, K.; et al. Oridonin inhibits the proliferation of human colon cancer cells by upregulating BMP7 to activate p38 MAPK. *Oncol. Rep.* **2016**, *35*, 2691–2698. [CrossRef] [PubMed]

329. Lu, Y.; Sun, Y.; Zhu, J.; Yu, L.; Jiang, X.; Zhang, J.; Dong, X.; Ma, B.; Zhang, Q. Oridonin exerts anticancer effect on osteosarcoma by activating PPAR-gamma and inhibiting Nrf2 pathway. *Cell Death Dis.* **2018**, *9*, 15. [CrossRef] [PubMed]

330. Inbaraj, J.J.; Chignell, C.F. Cytotoxic action of juglone and plumbagin: A mechanistic study using HaCaT keratinocytes. *Chem. Res. Toxicol.* **2004**, *17*, 55–62. [CrossRef] [PubMed]

331. Kapur, A.; Beres, T.; Rathi, K.; Nayak, A.P.; Czarnecki, A.; Felder, M.; Gillette, A.; Ericksen, S.S.; Sampene, E.; Skala, M.C.; et al. Oxidative stress via inhibition of the mitochondrial electron transport and Nrf-2-mediated anti-oxidative response regulate the cytotoxic activity of plumbagin. *Sci. Rep.* **2018**, *8*, 1073. [CrossRef]

332. Pan, S.T.; Qin, Y.; Zhou, Z.W.; He, Z.X.; Zhang, X.; Yang, T.; Yang, Y.X.; Wang, D.; Zhou, S.F.; Qiu, J.X. Plumbagin suppresses epithelial to mesenchymal transition and stemness via inhibiting Nrf2-mediated signaling pathway in human tongue squamous cell carcinoma cells. *Drug Des. Devel. Ther.* **2015**, *9*, 5511–5551. [CrossRef]

333. Boettler, U.; Sommerfeld, K.; Volz, N.; Pahlke, G.; Teller, N.; Somoza, V.; Lang, R.; Hofmann, T.; Marko, D. Coffee constituents as modulators of Nrf2 nuclear translocation and ARE (EpRE)-dependent gene expression. *J. Nutr. Biochem.* **2011**, *22*, 426–440. [CrossRef]

334. Arlt, A.; Sebens, S.; Krebs, S.; Geismann, C.; Grossmann, M.; Kruse, M.L.; Schreiber, S.; Schafer, H. Inhibition of the Nrf2 transcription factor by the alkaloid trigonelline renders pancreatic cancer cells more susceptible to apoptosis through decreased proteasomal gene expression and proteasome activity. *Oncogene* **2013**, *32*, 4825–4835. [CrossRef]

335. Shin, D.; Kim, E.H.; Lee, J.; Roh, J.L. Nrf2 inhibition reverses resistance to GPX4 inhibitor-induced ferroptosis in head and neck cancer. *Free Radic. Biol. Med.* **2018**, *129*, 454–462. [CrossRef]

336. Bollong, M.J.; Yun, H.; Sherwood, L.; Woods, A.K.; Lairson, L.L.; Schultz, P.G. A Small Molecule Inhibits Deregulated NRF2 Transcriptional Activity in Cancer. *ACS Chem. Biol.* **2015**, *10*, 2193–2198. [CrossRef]

337. Hori, R.; Yamaguchi, K.; Sato, H.; Watanabe, M.; Tsutsumi, K.; Iwamoto, S.; Abe, M.; Onodera, H.; Nakamura, S.; Nakai, R. The discovery and characterization of K-563, a novel inhibitor of the Keap1/Nrf2 pathway produced by Streptomyces sp. *Cancer Med.* **2019**, *8*, 1157–1168. [CrossRef] [PubMed]

338. Zhang, J.; Su, L.; Ye, Q.; Zhang, S.; Kung, H.; Jiang, F.; Jiang, G.; Miao, J.; Zhao, B. Discovery of a novel Nrf2 inhibitor that induces apoptosis of human acute myeloid leukemia cells. *Oncotarget* **2017**, *8*, 7625–7636. [CrossRef] [PubMed]

339. Tian, D.; Shi, Y.; Chen, D.; Liu, Q.; Fan, F. The Wnt inhibitor LGK-974 enhances radiosensitivity of HepG2 cells by modulating Nrf2 signaling. *Int. J. Oncol.* **2017**, *51*, 545–554. [CrossRef] [PubMed]

340. Zhong, Y.; Zhang, F.; Sun, Z.; Zhou, W.; Li, Z.Y.; You, Q.D.; Guo, Q.L.; Hu, R. Drug resistance associates with activation of Nrf2 in MCF-7/DOX cells, and wogonin reverses it by down-regulating Nrf2-mediated cellular defense response. *Mol. Carcinog.* **2013**, *52*, 824–834. [CrossRef]

341. Qian, C.; Wang, Y.; Zhong, Y.; Tang, J.; Zhang, J.; Li, Z.; Wang, Q.; Hu, R. Wogonin-enhanced reactive oxygen species-induced apoptosis and potentiated cytotoxic effects of chemotherapeutic agents by suppression Nrf2-mediated signaling in HepG2 cells. *Free Radic. Res.* **2014**, *48*, 607–621. [CrossRef]

342. Xu, X.; Zhang, Y.; Li, W.; Miao, H.; Zhang, H.; Zhou, Y.; Li, Z.; You, Q.; Zhao, L.; Guo, Q. Wogonin reverses multi-drug resistance of human myelogenous leukemia K562/A02 cells via downregulation of MRP1 expression by inhibiting Nrf2/ARE signaling pathway. *Biochem. Pharmacol.* **2014**, *92*, 220–234. [CrossRef]

343. Xu, X.; Zhang, X.; Zhang, Y.; Yang, L.; Liu, Y.; Huang, S.; Lu, L.; Kong, L.; Li, Z.; Guo, Q.; et al. Wogonin reversed resistant human myelogenous leukemia cells via inhibiting Nrf2 signaling by Stat3/NF-kappaB inactivation. *Sci. Rep.* **2017**, *7*, 39950. [CrossRef]

344. Gao, A.M.; Ke, Z.P.; Wang, J.N.; Yang, J.Y.; Chen, S.Y.; Chen, H. Apigenin sensitizes doxorubicin-resistant hepatocellular carcinoma BEL-7402/ADM cells to doxorubicin via inhibiting PI3K/Akt/Nrf2 pathway. *Carcinogenesis* **2013**, *34*, 1806–1814. [CrossRef]

345. Gao, A.M.; Zhang, X.Y.; Ke, Z.P. Apigenin sensitizes BEL-7402/ADM cells to doxorubicin through inhibiting miR-101/Nrf2 pathway. *Oncotarget* **2017**, *8*, 82085–82091. [CrossRef]

346. Connolly, R.M.; Nguyen, N.K.; Sukumar, S. Molecular pathways: Current role and future directions of the retinoic acid pathway in cancer prevention and treatment. *Clin. Cancer Res.* **2013**, *19*, 1651–1659. [CrossRef]

347. Wang, X.J.; Hayes, J.D.; Henderson, C.J.; Wolf, C.R. Identification of retinoic acid as an inhibitor of transcription factor Nrf2 through activation of retinoic acid receptor alpha. *Proc. Natl. Acad. Sci. USA* **2007**, *104*, 19589–19594. [CrossRef] [PubMed]

348. Pickering, A.M.; Linder, R.A.; Zhang, H.; Forman, H.J.; Davies, K.J. Nrf2-dependent induction of proteasome and Pa28alphabeta regulator are required for adaptation to oxidative stress. *J. Biol. Chem.* **2012**, *287*, 10021–10031. [CrossRef] [PubMed]

349. Furfaro, A.L.; Piras, S.; Domenicotti, C.; Fenoglio, D.; De Luigi, A.; Salmona, M.; Moretta, L.; Marinari, U.M.; Pronzato, M.A.; Traverso, N.; et al. Role of Nrf2, HO-1 and GSH in Neuroblastoma Cell Resistance to Bortezomib. *PLoS ONE* **2016**, *11*, e0152465. [CrossRef]

350. Valenzuela, M.; Glorieux, C.; Stockis, J.; Sid, B.; Sandoval, J.M.; Felipe, K.B.; Kviecinski, M.R.; Verrax, J.; Buc Calderon, P. Retinoic acid synergizes ATO-mediated cytotoxicity by precluding Nrf2 activity in AML cells. *Br. J. Cancer* **2014**, *111*, 874–882. [CrossRef]

351. Matthews, J.H.; Liang, X.; Paul, V.J.; Luesch, H. A Complementary Chemical and Genomic Screening Approach for Druggable Targets in the Nrf2 Pathway and Small Molecule Inhibitors to Overcome Cancer Cell Drug Resistance. *ACS Chem. Biol.* **2018**, *13*, 1189–1199. [CrossRef] [PubMed]

352. Sompakdee, V.; Prawan, A.; Senggunprai, L.; Kukongviriyapan, U.; Samathiwat, P.; Wandee, J.; Kukongviriyapan, V. Suppression of Nrf2 confers chemosensitizing effect through enhanced oxidant-mediated mitochondrial dysfunction. *Biomed. Pharmacother.* **2018**, *101*, 627–634. [CrossRef] [PubMed]

353. Esteva, F.J.; Hubbard-Lucey, V.M.; Tang, J.; Pusztai, L. Immunotherapy and targeted therapy combinations in metastatic breast cancer. *Lancet. Oncol.* **2019**, *20*, e175–e186. [CrossRef]

354. McMahon, M.; Thomas, N.; Itoh, K.; Yamamoto, M.; Hayes, J.D. Dimerization of substrate adaptors can facilitate cullin-mediated ubiquitylation of proteins by a "tethering" mechanism: A two-site interaction model for the Nrf2-Keap1 complex. *J. Biol. Chem.* **2006**, *281*, 24756–24768. [CrossRef]

355. Wakabayashi, N.; Slocum, S.L.; Skoko, J.J.; Shin, S.; Kensler, T.W. When NRF2 talks, who's listening? *Antioxid. Redox. Signal.* **2010**, *13*, 1649–1663. [CrossRef]

356. Cort, A.; Ozben, T.; Saso, L.; De Luca, C.; Korkina, L. Redox Control of Multidrug Resistance and Its Possible Modulation by Antioxidants. *Oxid. Med. Cell Longev.* **2016**, *2016*, 4251912. [CrossRef]

357. Milkovic, L.; Siems, W.; Siems, R.; Zarkovic, N. Oxidative stress and antioxidants in carcinogenesis and integrative therapy of cancer. *Curr. Pharm. Des.* **2014**, *20*, 6529–6542. [CrossRef] [PubMed]

358. Rocha, C.R.; Kajitani, G.S.; Quinet, A.; Fortunato, R.S.; Menck, C.F. NRF2 and glutathione are key resistance mediators to temozolomide in glioma and melanoma cells. *Oncotarget* **2016**, *7*, 48081–48092. [CrossRef] [PubMed]

359. Lu, K.; Alcivar, A.L.; Ma, J.; Foo, T.K.; Zywea, S.; Mahdi, A.; Huo, Y.; Kensler, T.W.; Gatza, M.L.; Xia, B. NRF2 Induction Supporting Breast Cancer Cell Survival Is Enabled by Oxidative Stress-Induced DPP3-KEAP1 Interaction. *Cancer Res.* **2017**, *77*, 2881–2892. [CrossRef] [PubMed]

360. Zhang, J.; Bai, K.W.; He, J.; Niu, Y.; Lu, Y.; Zhang, L.; Wang, T. Curcumin attenuates hepatic mitochondrial dysfunction through the maintenance of thiol pool, inhibition of mtDNA damage, and stimulation of the mitochondrial thioredoxin system in heat-stressed broilers. *J. Anim. Sci.* **2018**, *96*, 867–879. [CrossRef] [PubMed]

361. He, H.J.; Wang, G.Y.; Gao, Y.; Ling, W.H.; Yu, Z.W.; Jin, T.R. Curcumin attenuates Nrf2 signaling defect, oxidative stress in muscle and glucose intolerance in high fat diet-fed mice. *World J. Diabetes* **2012**, *3*, 94–104. [CrossRef] [PubMed]

362. Oner-Iyidogan, Y.; Tanrikulu-Kucuk, S.; Seyithanoglu, M.; Kocak, H.; Dogru-Abbasoglu, S.; Aydin, A.F.; Beyhan-Ozdas, S.; Yapislar, H.; Kocak-Toker, N. Effect of curcumin on hepatic heme oxygenase 1 expression in high fat diet fed rats: is there a triangular relationship? *Can. J. Physiol. Pharmacol.* **2014**, *92*, 805–812. [CrossRef] [PubMed]

363. Liu, L.; Shang, Y.; Li, M.; Han, X.; Wang, J.; Wang, J. Curcumin ameliorates asthmatic airway inflammation by activating nuclear factor-E2-related factor 2/haem oxygenase (HO)-1 signalling pathway. *Clin. Exp. Pharmacol. Physiol.* **2015**, *42*, 520–529. [CrossRef]

364. Gonzalez-Reyes, S.; Guzman-Beltran, S.; Medina-Campos, O.N.; Pedraza-Chaverri, J. Curcumin pretreatment induces Nrf2 and an antioxidant response and prevents hemin-induced toxicity in primary cultures of cerebellar granule neurons of rats. *Oxid. Med. Cell. Longev.* **2013**, *2013*, 801418. [CrossRef]

365. Chen, B.; Zhang, Y.; Wang, Y.; Rao, J.; Jiang, X.; Xu, Z. Curcumin inhibits proliferation of breast cancer cells through Nrf2-mediated down-regulation of Fen1 expression. *J. Steroid Biochem. Mol. Biol.* **2014**, *143*, 11–18. [CrossRef]

366. Bishayee, A.; Bhatia, D.; Thoppil, R.J.; Darvesh, A.S.; Nevo, E.; Lansky, E.P. Pomegranate-mediated chemoprevention of experimental hepatocarcinogenesis involves Nrf2-regulated antioxidant mechanisms. *Carcinogenesis* **2011**, *32*, 888–896. [CrossRef]

367. Singh, B.; Shoulson, R.; Chatterjee, A.; Ronghe, A.; Bhat, N.K.; Dim, D.C.; Bhat, H.K. Resveratrol inhibits estrogen-induced breast carcinogenesis through induction of NRF2-mediated protective pathways. *Carcinogenesis* **2014**, *35*, 1872–1880. [CrossRef]

368. Lu, F.; Zahid, M.; Wang, C.; Saeed, M.; Cavalieri, E.L.; Rogan, E.G. Resveratrol prevents estrogen-DNA adduct formation and neoplastic transformation in MCF-10F cells. *Cancer Prev. Res. (Phila)* **2008**, *1*, 135–145. [CrossRef]

369. Velavan, B.; Divya, T.; Sureshkumar, A.; Sudhandiran, G. Nano-chemotherapeutic efficacy of (-)-epigallocatechin 3-gallate mediating apoptosis in A549cells: Involvement of reactive oxygen species mediated Nrf2/Keap1signaling. *Biochem. Biophys. Res. Commun.* **2018**, *503*, 1723–1731. [CrossRef]

370. Wu, C.C.; Hsu, M.C.; Hsieh, C.W.; Lin, J.B.; Lai, P.H.; Wung, B.S. Upregulation of heme oxygenase-1 by Epigallocatechin-3-gallate via the phosphatidylinositol 3-kinase/Akt and ERK pathways. *Life Sci.* **2006**, *78*, 2889–2897. [CrossRef] [PubMed]

371. Sriram, N.; Kalayarasan, S.; Sudhandiran, G. Epigallocatechin-3-gallate augments antioxidant activities and inhibits inflammation during bleomycin-induced experimental pulmonary fibrosis through Nrf2-Keap1 signaling. *Pulm. Pharmacol. Ther.* **2009**, *22*, 221–236. [CrossRef] [PubMed]

372. Sun, X.; Mi, L.; Liu, J.; Song, L.; Chung, F.L.; Gan, N. Sulforaphane prevents microcystin-LR-induced oxidative damage and apoptosis in BALB/c mice. *Toxicol. Appl. Pharmacol.* **2011**, *255*, 9–17. [CrossRef]

373. Corssac, G.B.; Campos-Carraro, C.; Hickmann, A.; da Rosa Araujo, A.S.; Fernandes, R.O.; Bello-Klein, A. Sulforaphane effects on oxidative stress parameters in culture of adult cardiomyocytes. *Biomed. Pharmacother.* **2018**, *104*, 165–171. [CrossRef] [PubMed]

374. Wang, X.Y.; Zhang, X.H.; Peng, L.; Liu, Z.; Yang, Y.X.; He, Z.X.; Dang, H.W.; Zhou, S.F. Bardoxolone methyl (CDDO-Me or RTA402) induces cell cycle arrest, apoptosis and autophagy via PI3K/Akt/mTOR and p38 MAPK/Erk1/2 signaling pathways in K562 cells. *Am. J. Transl. Res.* **2017**, *9*, 4652–4672.

375. Qin, D.; Wang, W.; Lei, H.; Luo, H.; Cai, H.; Tang, C.; Wu, Y.; Wang, Y.; Jin, J.; Xiao, W.; et al. CDDO-Me reveals USP7 as a novel target in ovarian cancer cells. *Oncotarget* **2016**, *7*, 77096–77109. [CrossRef]

376. Gao, X.; Deeb, D.; Liu, Y.; Liu, P.; Zhang, Y.; Shaw, J.; Gautam, S.C. CDDO-Me inhibits tumor growth and prevents recurrence of pancreatic ductal adenocarcinoma. *Int. J. Oncol.* **2015**, *47*, 2100–2106. [CrossRef]

377. Jeong, S.A.; Kim, I.Y.; Lee, A.R.; Yoon, M.J.; Cho, H.; Lee, J.S.; Choi, K.S. Ca2+ influx-mediated dilation of the endoplasmic reticulum and c-FLIPL downregulation trigger CDDO-Me-induced apoptosis in breast cancer cells. *Oncotarget* **2015**, *6*, 21173–21192. [CrossRef]

378. Qin, D.J.; Tang, C.X.; Yang, L.; Lei, H.; Wei, W.; Wang, Y.Y.; Ma, C.M.; Gao, F.H.; Xu, H.Z.; Wu, Y.L. Hsp90 Is a Novel Target Molecule of CDDO-Me in Inhibiting Proliferation of Ovarian Cancer Cells. *PLoS ONE* **2015**, *10*, e0132337. [CrossRef]

379. Leal, A.S.; Sporn, M.B.; Pioli, P.A.; Liby, K.T. The triterpenoid CDDO-imidazolide reduces immune cell infiltration and cytokine secretion in the KrasG12D;Pdx1-Cre (KC) mouse model of pancreatic cancer. *Carcinogenesis* **2016**, *37*, 1170–1179. [CrossRef]

380. Zagorski, J.W.; Maser, T.P.; Liby, K.T.; Rockwell, C.E. Nrf2-Dependent and -Independent Effects of tert-Butylhydroquinone, CDDO-Im, and H2O2 in Human Jurkat T Cells as Determined by CRISPR/Cas9 Gene Editing. *J. Pharmacol. Exp. Ther.* **2017**, *361*, 259–267. [CrossRef] [PubMed]

381. Holland, R.; Navamal, M.; Velayutham, M.; Zweier, J.L.; Kensler, T.W.; Fishbein, J.C. Hydrogen peroxide is a second messenger in phase 2 enzyme induction by cancer chemopreventive dithiolethiones. *Chem. Res. Toxicol.* **2009**, *22*, 1427–1434. [CrossRef] [PubMed]

382. Primiano, T.; Kensler, T.W.; Kuppusamy, P.; Zweier, J.L.; Sutter, T.R. Induction of hepatic heme oxygenase-1 and ferritin in rats by cancer chemopreventive dithiolethiones. *Carcinogenesis* **1996**, *17*, 2291–2296. [CrossRef] [PubMed]

383. Erzinger, M.M.; Bovet, C.; Uzozie, A.; Sturla, S.J. Induction of complementary function reductase enzymes in colon cancer cells by dithiole-3-thione versus sodium selenite. *J. Biochem. Mol. Toxicol.* **2015**, *29*, 10–20. [CrossRef]

384. Roebuck, B.D.; Curphey, T.J.; Li, Y.; Baumgartner, K.J.; Bodreddigari, S.; Yan, J.; Gange, S.J.; Kensler, T.W.; Sutter, T.R. Evaluation of the cancer chemopreventive potency of dithiolethione analogs of oltipraz. *Carcinogenesis* **2003**, *24*, 1919–1928. [CrossRef]

Antioxidants **2020**, *9*, 193

385. Bass, S.E.; Sienkiewicz, P.; Macdonald, C.J.; Cheng, R.Y.; Sparatore, A.; Del Soldato, P.; Roberts, D.D.; Moody, T.W.; Wink, D.A.; Yeh, G.C. Novel dithiolethione-modified nonsteroidal anti-inflammatory drugs in human hepatoma HepG2 and colon LS180 cells. *Clin. Cancer Res.* **2009**, *15*, 1964–1972. [CrossRef]
386. Nho, C.W.; O'Dwyer, P.J. NF-kappaB activation by the chemopreventive dithiolethione oltipraz is exerted through stimulation of MEKK3 signaling. *J. Biol. Chem.* **2004**, *279*, 26019–26027. [CrossRef]
387. Steel, R.; Cowan, J.; Payerne, E.; O'Connell, M.A.; Searcey, M. Anti-inflammatory Effect of a Cell-Penetrating Peptide Targeting the Nrf2/Keap1 Interaction. *ACS Med. Chem. Lett.* **2012**, *3*, 407–410. [CrossRef]
388. Hu, L.; Magesh, S.; Chen, L.; Wang, L.; Lewis, T.A.; Chen, Y.; Khodier, C.; Inoyama, D.; Beamer, L.J.; Emge, T.J.; et al. Discovery of a small-molecule inhibitor and cellular probe of Keap1-Nrf2 protein-protein interaction. *Bioorg. Med. Chem. Lett.* **2013**, *23*, 3039–3043. [CrossRef]

Review

Antioxidants as a Potential Target against Inflammation and Oxidative Stress in Attention-Deficit/Hyperactivity Disorder

Lourdes Alvarez-Arellano [1], **Nadia González-García** [2], **Marcela Salazar-García** [3] **and Juan Carlos Corona** [2,*]

[1] CONACYT-Hospital Infantil de México Federico Gómez, Mexico City 06720, Mexico; lourdes.alvareza@gmail.com
[2] Laboratory of Neurosciences, Hospital Infantil de México Federico Gómez, Mexico City 06720, Mexico; nadiag.him@gmail.com
[3] Laboratorio de Investigación en Biología del Desarrollo y Teratogénesis Experimental, Hospital Infantil de México Federico Gómez, Mexico City 06720, Mexico; msalazar.investigacion@gmail.com
* Correspondence: jcorona@himfg.edu.mx; Tel.: +52-(55)-52289917

Received: 12 December 2019; Accepted: 30 January 2020; Published: 21 February 2020

Abstract: Psychostimulants and non-psychostimulants are the medications prescribed for the treatment of attention-deficit/hyperactivity disorder (ADHD). However, several adverse results have been linked with an increased risk of substance use and side effects. The pathophysiology of ADHD is not completely known, although it has been associated with an increase in inflammation and oxidative stress. This review presents an overview of findings following antioxidant treatment for ADHD and describes the potential amelioration of inflammation and oxidative stress using antioxidants that might have a future as multi-target adjuvant therapy in ADHD. The use of antioxidants against inflammation and oxidative conditions is an emerging field in the management of several neurodegenerative and neuropsychiatric disorders. Thus, antioxidants could be promising as an adjuvant ADHD therapy.

Keywords: antioxidants; ADHD; inflammation; oxidative stress; Nrf2

---

1. Introduction

1.1. Attention-Deficit/Hyperactivity Disorder

Attention-deficit/hyperactivity disorder (ADHD) is the most common neurobehavioural disorder and is a chronic, often lifelong, condition [1,2]. The pooled worldwide prevalence of ADHD is 7.2% for children and adolescents and 3–5% for adults [3–5]. Boys are more than twice as likely as girls to receive a diagnosis of ADHD [6]. Approximately 50% of individuals diagnosed in childhood and adolescence persist with symptoms into adult life [1,7]. Comorbidity in ADHD is very common at roughly 70%, the main disorders being emotional or behavioural conditions, such as anxiety, oppositional defiant, depression and substance use disorders, and developmental conditions, such as learning and language disorders, autism spectrum disorders (ASD), and physical conditions (tics and sleep apnoea) [8–10]. The diagnosis of ADHD is based on the fifth edition of the Diagnostic and Statistical Manual of Mental Disorders (DSM-5) and the criteria include: inattention and/or hyperactive and impulsive symptoms for the last six months or more, onset before the age of 12 years old, and symptoms causing at least moderate psychological, social, and/or educational or occupational impairments based on interview and/or direct observation in multiple settings [6].

1.2. Pathophysiology

The pathophysiological mechanisms of ADHD are still not understood. However, biochemical, psychological, and environmental factors have generally been accepted as causes of the disorder. Several studies have suggested deregulation in catecholaminergic neurotransmission to be the cause [11,12]. Moreover, increasing evidence indicates a critical role for neuroinflammation [13,14]. Also, the involvement of oxidative stress is highlighted as a pathophysiological cause of ADHD [15,16].

1.3. Pharmacological Treatment

The management of ADHD comprises multimodal treatments that include psychosocial and educational interventions [6]. Nevertheless, pharmacological treatment is the first choice of therapy to improve symptoms, using psychostimulants such as methylphenidate (MPH) and amphetamines, which inhibit the reuptake of dopamine and norepinephrine, thus increasing catecholaminergic activity in the prefrontal cortex, striatum, and hippocampus, improving symptoms [6,17]. The second choice of therapy is with non-psychostimulants such as atomoxetine (ATX), which is a selective norepinephrine reuptake inhibitor, guanfacine, and clonidine, which are selective α-2 adrenergic receptor agonists [18,19].

However, psychostimulants have been associated with side effects such as appetite loss, headache, stomach pain, agitation, sleep disturbance, anxiety, and insomnia [2,20]. Moreover, these medications have a persistent effect on decreasing growth velocity and hallucinations or other psychotic symptoms [21]. Furthermore, non-psychostimulants are associated with changes in cardiovascular parameters, cardiovascular events, somnolence, gastrointestinal tract symptoms, nausea, diarrhoea, vomiting, decreased appetite, fatigue, and dizziness [2,22].

Neuroinflammation and oxidative stress play a role in the pathophysiology of ADHD due to genetic and environmental factors, catecholaminergic dysregulation, and medications used for treatment, all factors which could produce inflammation and oxidative stress, which increases the symptoms and, as a consequence, leads to establishing a vicious circle (Figure 1). Hence, antioxidants against inflammation and oxidative stress have been used to manage other disorders such as Alzheimer's disease, Parkinson's disease, Huntington's disease, autism, schizophrenia, and depression [23–28]. Accordingly, antioxidant modulators could be helpful as a multi-target adjuvant therapy in ADHD.

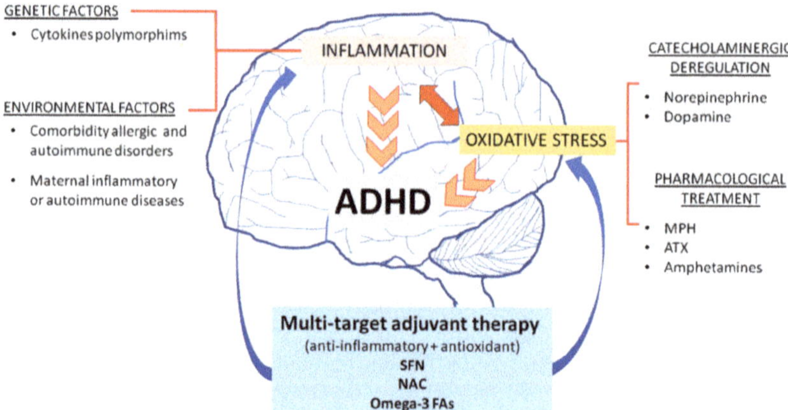

Figure 1. Role of inflammation and oxidative stress in the pathophysiology of ADHD and potential adjuvant therapy. Environmental and genetic factors, catecholaminergic dysregulation and pharmacological treatment can establish a vicious circle, producing inflammation and oxidative stress, therefore contributing to increase the symptoms. SFN, sulforaphane; NAC, N-Acetylcysteine; omega-3 FAs, omega-3 fatty acids; MPH, methylphenidate; ATX, atomoxetine.

2. Inflammation and the Relationship with ADHD

Increasing evidence supports the role of neuroinflammation in the pathophysiology of ADHD [14]. Inflammation in the brain is characterized by the activation of glial cells (oligodendrocytes, astrocytes, microglia and ependymal cells) and the production of cytokines, chemokines, prostaglandins, nitric oxide (NO), reactive oxygen species (ROS), and immune cell infiltration, including monocytes/macrophages, neutrophils, dendritic cells, T cells, and B cells. Glial cells, principally microglia (brain-resident immune cells), are responsible for the maintenance of homeostasis after brain injury [29,30].

Increased ADHD symptoms have been reported in patients with an uncontrolled inflammatory environment. Thus, elevated levels of interleukin-6 receptor (IL-6R), RANTES (regulated upon activation, normal t cell expressed and secreted) and tumour necrosis factor-α (TNF-α) in children with ADHD were associated with a major intensity of symptoms such as hyperactivity and inattention [31]. Moreover, serum levels of IL-6 and IL-10 were significantly higher in ADHD children compared with healthy controls. However, IL-6 levels did not correlate with the severity of ADHD symptoms [32,33]. In an animal model of ADHD, using spontaneously hypertensive rats (SHR), the serum levels of IL-1α, MCP-1, RANTES, and IP-10 were elevated in five-week-old compared to control rats. Interestingly, serum levels of IL-6 were similar in five-week-old animals of both strains, with elevated levels in 10-week-old SHR, which correlates with that found in children with ADHD [34]. The association between the dysregulation of the inflammatory response and the pathophysiology of ADHD is possible as a result of the role of inflammation in neurogenesis, differentiation, and neuronal function [30,35–37]. Furthermore, neuroinflammation can induce aggravating factors such as blood–brain barrier disruption, altered neurotransmitter metabolism, oxidative stress, and neurodegeneration [38].

The inflammatory mechanisms and the association of dysregulation in ADHD remain to be fully clarified but involve genetic and/or environmental factors. Several studies have reported an association between ADHD and the polymorphism of cytokines such as IL-2, IL-6 and TNF-α [39]. However, there are controversial results for interleukin-1 receptor antagonist (IL-1RA; also known as IL-1RN) gene variable number tandem repeat (VNTR) polymorphism: the four-repeat allele was associated with increased risk and the two-repeat allele with reduced risk for ADHD [40]. On the other hand, no association of this IL-1RN polymorphism with ADHD was found in a larger sample [41]. Comorbidity with allergic and autoimmune disorders such as atopic eczema, allergic rhinitis and asthma is another factor that could increase the risk for ADHD and future research may lead to a better understanding of the mechanisms underlying the observed comorbidity [38,42–44]. In a meta-analysis and a large-scale genome-wide association study (GWAS), associations between asthma and ADHD were found in both children and adults [45,46]. Also, patients with ADHD and asthma had similar brain region dysfunctions and higher levels of cytokines and IgE compared to healthy children, resulting in alterations to the regions in the brain associated with emotional and behavioural control [38,42,47]. The involvement of autoantibodies in ADHD has been suggested but this is an unknown causal association [33,48]. High antibody levels of anti-Purkinje cells (anti-Yo), anti-basal ganglia and anti-dopamine transporter were found in patients with ADHD [33,48–50]. Moreover, elevated anti-Yo antibody levels were correlated with IL-6 and IL-10 levels [33]. Further studies are needed to clarify the comorbidity or causal relationship between allergic and autoimmune disorders and ADHD. Maternal history of autoimmune disease has also been associated with an increased risk of ADHD. The inflammation caused during prenatal development by several maternal diseases, such as infections, asthma, diabetes, obesity, and autoimmune disease, was associated with ADHD in offspring [13,51].

3. Antioxidant Treatment Against Inflammation in ADHD

3.1. Sulforaphane Exerts Anti-Inflammatory Activity

Sulforaphane (SFN) is found in highest concentrations in vegetables such as cauliflower and broccoli sprouts, and it has been shown that SFN is an activator of nuclear factor erythroid 2-related factor 2 (Nrf2) [52]. Nrf2 is a transcription factor widely recognized as a master regulator of cellular redox homeostasis [53,54]. Regulation is carried out by binding Nrf2 to a specific DNA sequence known as the antioxidant response element (ARE) found in the promoter regions of genes that encode detoxification enzymes such as NADPH quinone oxidoreductase 1 (NQO1), haem oxygenase 1 (HO-1) and glutathione peroxidase 1 (GPx1), among others [55]. Nrf2 regulates enzymes responsible for GSH syntheses, such as the glutamate-cysteine catalytic subunit (GCLC) and glutamate-cysteine ligase modifier subunit (GCLM), and enzymes related to GSH utilization, such as glutathione S transferase (GST), glutathione peroxidase, and glutathione reductase [54]. Thus, SFN activates Nrf2 and stimulates transcription of genes involved in GSH synthesis.

SFN has been considered as a therapeutic target in several inflammation-associated diseases, including neurodevelopmental disorders such as psychosis and autism spectrum disorder [56–58]. Action mechanisms of antioxidants used against inflammation are shown in Figure 2.

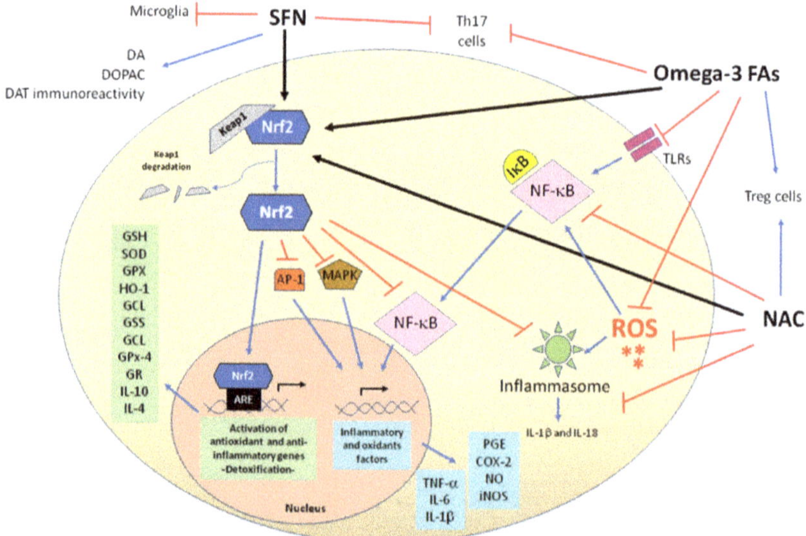

Figure 2. Overview of the mechanisms of SFN, NAC and Omega-3 FAs used as modulators against inflammation and oxidative stress. DA, dopamine; DOPAC, 3,4-dihydroxyphenylacetic acid; DAT, dopamine transporter; PGE, prostaglandins; ROS, reactive oxygen species; NO, nitric oxide; GCL, glutamate cystine ligase; GPx, glutathione peroxidase; GSH, glutathione; SOD, superoxide dismutase; GR, glutatione reductase; GSS, glutathione synthetase; glutathione peroxidase-4, GPx-4.

The molecular mechanism by which SFN exerts its anti-inflammatory function is by inducing Nrf2 pathway activation, which contributes to the anti-inflammatory process by regulating HO-1 gene expression [59–61]. Nrf2 leads to the inhibition of nuclear factor kappa B (NF-κB), activator protein-1 (AP-1) and mitogen-activated protein kinase (MAPK) classical inflammatory pathways, resulting in decreased expression of the inflammatory mediators (iNOS, COX-2, NO and prostaglandins) and pro-inflammatory cytokines (TNF-α, IL-6 and IL-1β). In addition, Nrf2/HO-1 activation increases anti-inflammatory cytokines (IL-10 and IL-4) [62–65]. Moreover, SFN has a prophylactic and a therapeutic effect by inhibiting both the inflammatory response and microglial activation [62].

Inflammasomes are multiprotein complexes necessary for the release of pro-inflammatory cytokines (IL-1α and IL-18). Interestingly, it has been reported that SFN inhibited the activation of NLRP1, NLRP3, and NLRC4 inflammasomes [66,67]. However, the precise mechanism by which it does so is controversial, inasmuch as the SFN-mediated inhibition of the inflammasomes was independent of caspase-1 activity, ROS modulation, and Nrf2 synthesis [66]. Contrary to these findings, other studies showed that ROS generation and Nrf2 were required for inflammasome activation [67,68]. SFN protected against neuroinflammation by preventing the increase of NF-κB activity, TNF-α, and depletion of the IL-10 level in the cortex and hippocampus of okadaic-acid-treated rats by the activation of the Nrf2/HO-1 pathway [69]. Recently, it was shown that treatment with SFN in ASD-induced improvement of social interaction and behavioural deficits may be due to the inhibition of STAT3 expression and suppression of Th17 cell response [27,70]. Although SFN supplementation in patients with ADHD has not been studied, this could constitute a promising approach against inflammation linked with ADHD. Nevertheless, more studies are needed to confirm safety, efficacy, therapeutic doses, effect in combination with conventional therapy and long-term side effects to be considered as adjuvant therapy in ADHD. The effects of antioxidants and the main outcomes are summarized in Table 1.

Table 1. Summary of the findings on the potential beneficial effects of antioxidants against inflammation and oxidative stress.

Against Inflammation	Type of Study	Outcome
SFN	Mouse model of atopic dermatitis [61]	Reduced inflammation, suppressed JAK1/STAT3 signaling and activated Nrf2/HO-1 pathway
SFN	Microglial cells [62]	Reduced inflammatory mediators (iNOS, COX-2, NO, and PGE2) and proinflammatory cytokines (TNF-α, IL-6, and IL-1β), increased anti-inflammatory cytokines (IL-10 and IL-4) and increased the expression of Nrf2 and HO-1.
SFN	Mouse model of acute lung injury and Macrophages [63,65]	Decreased lactate dehydrogenase, IL-6, TNF-α, NF-kB, PGE2 production, COX-2, MMP-9 and iNOS protein expression
SFN	Mouse model of peritonitis [66]	Inhibited inflammasome activation and IL-1β secretion and inhibited cell recruitment to peritoneum.
SFN	Mouse macrophages [67]	Blocked activation of NLRP3 and NLRC4 inflammasomes and IL-1β secretion
SFN	Rat [69]	Inhibited NF-kB activity and TNF-α secretion and prevent decreased IL-10
SFN	Mouse model of autism and Autism patients [27,70]	Reduced Th17 response and expression of NF-kB and iNOS Randomized double-blind study; decreased symptoms
NAC	Mild-stress rat model [71]	Inhibited pro-inflammatory cytokines (IL-1β, IL-6 and TNF-α)
NAC	Bipolar depression patients [72]	Randomized placebo-controlled trial; no effects on the biological parameters evaluated
NAC	Systemic lupus erythematosus patients [73,74]	Randomized double-blind placebo-controlled trial and randomized controlled trial; reduced the ADHD symptoms and also inhibited the autoimmune inflammatory process by suppression of the mammalian target of rapamycin (mTOR) and increased regulatory T cells
NAC	Human retinal pigment epithelial cell line [75]	Decreased IL-18, IL-1β mRNA, ROS and blocked inflammasome activation
NAC	Rat [76]	Improved brain oxidant/antioxidant status and reversed the overproduction of pro-inflammatory cytokines in brain and serum
Omega-3 FAs	Macrophage and mouse dendritic cell lines [77,78]	Inhibited dimerization and recruitment of TLR2 and TLR4 recruitment to lipid rafts and reduced T-cell proliferation and increased the proportion of T cells expressing FoxP3
Omega-3 FAs	Mouse [79,80]	Regulated CD4$^+$ T-cell function and reduced Th17cell polarazation
Omega-3 FAs	Children with ADHD [81]	Double-blind study; decreased plasma inflammatory mediators

Table 1. *Cont.*

Against Oxidative Stress	Type of Study	Outcome
SFN	Mouse [82]	Increased dopamine, DOPAC and dopamine transporter immunoreactivity in the striatum
SFN	Rat [69]	Activation of HO-1, glutamate-cysteine ligase catalytic subunit and Nrf2 and protected against memory impairment
SFN	Mouse model of autism [70]	Improved the autism-like symptoms and upregulated SOD, glutathione reductase and GPx
SFN and NAC	Rat with epilepsy [83]	Reduced oxidative stress, delayed the onset of epilepsy, blocked disease progression and reduced the frequency of spontaneous seizures
SFN	Healthy subjects [26]	Clinical pilot study; increased GSH
NAC	Rats [84]	Protected against amphetamine-induced damage
NAC	Paediatric Tourette's syndrome [85]	Randomized double-blind placebo-controlled trial; did not show a significant difference with placebo
NAC	A girl with ADHD [86]	A case-study; reduced the frequency of self-cutting and reduced the symptoms and depression
NAC	Mouse model of postoperative cognitive dysfunction [87]	Reduced oxidative stress and inflammation in the hippocampus and improved cognitive function by activation of the Nrf2/HO-1 pathway
Omega-3 FAs	Children with ADHD [81]	Double-blind study; decreased oxidative stress
Omega-3 FAs	Children with ADHD [88]	Randomized controlled trial; no significant differences among the treatments. One subgroup improved spelling, reading and attention and decreased hyperactivity
Omega-3 FAs	Children with ADHD [89,90]	Randomized pilot study and placebo-controlled trial; no significant improvement
Omega-3 FAs	Children with ADHD [91–94]	Pilot studies and randomized placebo-controlled trials; improved working memory function and improved symptoms and behaviour
Omega-3 FAs	Rat astrocytes [95]	Increased glutamate-cysteine ligase, Nrf2, glutathione synthetase and glutathione peroxidase-4 proteins

3.2. N-Acetylcysteine Decreases Inflammatory Response

N-Acetylcysteine (NAC), a precursor of L-cysteine and the antioxidant glutathione (GSH), is found in plants, especially the onion [96–98]. NAC has been used as an adjuvant therapy in many psychiatric disorders (e.g., Alzheimer's disease, schizophrenia, autism, addiction, substance abuse, obsessive-compulsive and mood disorders [24,99–104]), with promising results and no relevant side effects after its administration against inflammation [97]. The use of NAC in a chronic unpredictable mild-stress animal model inhibited the levels of pro-inflammatory cytokines (IL-1β, IL-6, and TNF-α) in the hippocampus and prefrontal cortex and exhibited antidepressant-like effects [71]. However, in a recent study, NAC treatment did not show any effect on the serum levels of IL-6, IL-8, TNF-α, IL-10, or C-reactive protein in a clinical trial for bipolar depression, possibly due to the small sample size used in this study [72]. On the other hand, a study derived from an ADHD self-report scale symptom checklist revealed that NAC reduced the ADHD symptoms in patients with systemic lupus erythematosus and also inhibited the autoimmune inflammatory process via suppression of the mammalian target of rapamycin (mTOR) and increased regulatory T cells [73,74]. Moreover, NAC has been reported to block inflammasome activation as well as IL-18 and IL-1β production [75]. Recently, NAC protected against cisplatin-induced toxicity in rat brain by modulation of inflammation and oxidative stress [76]. Thus, more studies are required to support the efficacy of NAC as a possible adjuvant treatment for ADHD.

3.3. Omega-3 Fatty Acids Prevent Inflammation

Omega-3 fatty acids (omega-3 FAs) are polyunsaturated fatty acids, whose primary sources are in oily fish. They are components of neuronal membranes and have a main role in neurotransmission, neuronal development and function [105]. Two of the main omega-3 FAs are docosahexaenoic acid (DHA) and eicosapentaenoic acid (EPA) and it has been demonstrated that supplementation with omega-3 FAs has beneficial effects in several neurodegenerative and neuropsychiatric disorders such as Parkinson´s and Alzheimer´s diseases, depression, bipolar disorder, anxiety, and schizophrenia [106,107]. These omega-3 FAs are also known to have anti-inflammatory effects [108,109]. It has been demonstrated that DHA decreased the TLR-dependent inflammatory signalling pathway by inhibiting dimerization and the recruitment of receptors to lipid rafts, which resulted in the reduced production of pro-inflammatory cytokines [77,110]. DHA also reduces T-cell activation, proliferation and promoted polarization into regulatory T cells (Treg; $CD4^+/CD25^+/FoxP3^+$) and interferes with the polarization of Th17 cells [78–80]. In a double-blind study, supplementation with omega-3 FAs for eight weeks was shown to decrease the plasma IL-6 level and any hyperactivity symptoms in children with ADHD [81]. However, a similar eight-week study found no effect on ADHD, but this could be due to the dose of omega-3 FAs [111]. Thus, further studies are needed to confirm the therapeutic dose, safety, and effectiveness of omega-3 FAs as a possible therapy against inflammation in ADHD.

4. Oxidative Stress and the Relationship with ADHD

There is increasing evidence for the involvement of oxidative stress in the pathophysiology of ADHD [15,16], but some studies have shown low levels of malondialdehyde (MDA) and the DNA damage indicator 8-hydroxy-2′-deoxyguanosine (8-OHdG) in ADHD children [112,113]. MDA is the degradation product of the main chain reactions that lead to the oxidation of polyunsaturated fatty acids, and therefore serves as an oxidative stress marker. Recently, it was reported that the total antioxidant capacity, catalase and GSH were significantly lower but that MDA was not significantly different in children with ADHD [114]. However, high levels of MDA have been observed in both adults [115,116] and children with ADHD [117], and increased plasma MDA and urinary 8-OHdG levels were found in children with ADHD compared to healthy children [118]. Another study has shown low total antioxidant levels in children with ADHD [119]. Furthermore, the SHR model showed an increase in ROS production in the hippocampus, cortex, and striatum [120]. It was shown that paraoxonase-1 (PON1) and arylesterase activity were decreased (the arylesterase activity linked to PON1 is known to protect lipoproteins from oxidation) and there was also a decrease in the total antioxidant status. Moreover, the total oxidant status and oxidative stress index were increased in children with ADHD, suggesting that there is significantly increased oxidative stress in ADHD [121]. It seems that nitrosative stress also has a role in ADHD because of increased oxidative and nitrosative stress and impaired oxidant–antioxidant balance has been demonstrated in children with ADHD [122]. The NO levels were significantly higher in children, adolescents and adults [117,123] and also significant increases in NO synthase activity were observed in children and adolescents with ADHD [124]. Although there are inconsistent results regarding the relationship of oxidative stress and ADHD, in a meta-analysis it was confirmed that ADHD is associated with increased oxidative stress in ADHD patients [16]. Taken together, the data suggest that both oxidative and nitrosative stress in ADHD have the potential to contribute to this condition [15].

On the other hand, it has been shown that dopamine and norepinephrine can easily undergo auto-oxidation, forming ROS [125,126], which could lead to cell damage and damage to DNA [127,128]. Also, it has been shown that ATX treatment increases extracellular norepinephrine and dopamine levels [19,129], which would produce an increase in oxidative stress and, as a consequence, cell damage and mitochondrial dysfunction [130]. The brain is particularly susceptible to oxidative stress because of its high lipid content and the high demand for energy consumption [131]. Neurons use mitochondria as the main producers of ATP. Mitochondria regulate ion homeostasis and the redox state, and they are also both producers of ROS and targets of ROS-induced damage, with such effects leading to the collapse of bioenergetic function and the initiation of cell death [132]. For this reason, oxidative stress could also be involved or interrelated with the catecholaminergic pathway in ADHD. However, the exact relationship between both processes and ADHD remains unclear.

Regarding ADHD medication, several reports have demonstrated that MPH has an impact on the generation of oxidative damage. Increases in DNA damage have been found mainly in the striatum of young and adult rats after MPH treatment [133]. In specific regions of the brain of young rats, chronic treatment with MPH increased oxidative damage, as assessed by the thiobarbituric acid reactive species and protein carbonyl assays, and this effect was dependent on the dose [134]. Furthermore, acute or chronic treatment with MPH altered the activity of catalase and superoxide dismutase (SOD) enzymes in the brain of young rats [135]. One study showed that acute and chronic treatment with MPH in the SHR model increased oxidative stress [136]. Finally, the acute administration of high doses of MPH can cause oxidative and inflammatory changes in brain cells and induce neurodegeneration in the hippocampus and cerebral cortex of adult rats [137].

Hence, the growing research in looking for alternative therapies for ADHD has focused on the neuroprotective effects of natural products as antioxidants because they may be high-efficiency alternative treatments with fewer side effects [2].

5. Antioxidant Treatment Against Oxidative Stress in ADHD

As a consequence of oxidative stress, cells have the capacity to increase their antioxidant defences through the activation of Nrf2. Thus, Nrf2 pathway activation takes place to act against the accumulation of ROS. Consequently, Nrf2 activators have been proposed as antioxidant targets in neurodegenerative and neuropsychiatric disorders (Alzheimer's disease, Parkinson's disease, Huntington's disease, autism, schizophrenia, and depression) to counteract the increase in oxidative stress [138–140]. The action mechanisms of the antioxidants used against oxidative stress are shown in Figure 2.

5.1. Sulforaphane Exerts Antioxidant Activity

In mice, the repeated administration of amphetamines induced decreases of dopamine and 3,4-dihydroxyphenylacetic acid DOPAC levels, as well as dopamine transporter immunoreactivity in the striatum, and such effects were significantly attenuated by the treatment with SFN [82]. Activation of the cellular antioxidant machinery (HO-1, glutamate-cysteine ligase catalytic subunit and Nrf2) resulted in SFN-mediated protection against memory impairment in rats treated with okadaic acid [69]. Moreover, in a mouse model of autism, SFN improved the autism-like symptoms and upregulated antioxidant defences such as SOD, glutathione reductase, and GPx [70]. The combination of NAC and SFN significantly reduced oxidative stress, delayed the onset of epilepsy, blocked disease progression, and reduced the frequency of spontaneous seizures in animals [83]. Also, in a clinical pilot study, treatment with SFN increased the antioxidant GSH, suggesting a need to explore possible correlations between GSH and clinical/neuropsychological measures and any positive influence that the treatment of SFN could have on neuropsychiatric disorders [26]. Even though the antioxidative effect of SFN supplementation in ADHD has not been studied, this could constitute a promising approach for oxidative imbalances linked with ADHD. Nevertheless, more research is needed to confirm the efficacy, therapeutic doses, and effect in combination with conventional therapy to be considered as adjuvant therapy in ADHD. The effects of antioxidants and the main outcomes are summarized in Table 1.

5.2. N-Acetylcysteine Exerts Antioxidant Activity

The NAC molecule scavenges ROS and there has been growing evidence of its role in attenuating psychiatric and neurological disorders and associated pathophysiological processes such as oxidative stress, mitochondrial dysfunction and glutamate and dopamine dysregulation [96,97,141]. In rats, amphetamine produces oxidative stress (by increasing hydroxyl radical formation and MDA) and dopaminergic neurotoxicity, but treatment with NAC protected against amphetamine-induced damage [84]. The treatment of paediatric Tourette´s syndrome with NAC in a randomized, double-blind placebo-controlled trial did not show a significant difference between NAC and placebo for reducing tic severity or any secondary outcomes such as depression, anxiety and ADHD [85]. In a case-study, a 17-year-old girl was successfully treated with NAC, reducing the frequency of self-cutting and the symptoms of ADHD and depression [86]. Finally, in a mouse model of postoperative cognitive dysfunction, NAC reduced oxidative stress and inflammation in the hippocampus and improved cognitive function by activation of the Nrf2/HO-1 pathway [87]. Thus, it seems that part of the protection produced by NAC is via the activation of antioxidant pathways that involve Nrf2. However, more research is required to support the efficacy of NAC as a possible treatment for ADHD.

5.3. Omega-3 Fatty Acids Exert Antioxidant Activity

The omega-3 FAs supplementation can be an enhancer factor in the antioxidant defence against ROS [142] and has been demonstrated to be effective against oxidative stress in the treatment of ADHD. A pilot study demonstrated that supplementation with alpha-linolenic acid, an omega-3 FAs, improved ADHD symptoms [91]. The findings of a small pilot study demonstrated that supplementation with high doses of EPA and DHA improved the behaviour of children with ADHD [92]. In a double-blind study, eight weeks of EPA and DHA supplementation decreased oxidative stress in children with ADHD [81]. Supplements with high EPA, DHA or omega-6 FAs as a control demonstrated no significant differences among the treatments. However, in one subgroup of children there was improvement in spelling, reading and attention and a decrease in hyperactivity [88]. On the other hand, in one study, EPA and DHA supplementation in children with ADHD demonstrated no significant improvements in outcome [89]. Furthermore, no beneficial results were observed in a randomized placebo-controlled clinical trial with EPA and DHA supplementation in children with ADHD [90]. There were increased EPA and DHA concentrations in erythrocyte membranes and improved working memory function on supplementation with a mix of omega-3 FAs [93]. The effects of dietary omega-3 FAs supplementation on ADHD showed a reduction in ADHD symptoms [94]. Finally, omega-3 FAs improved the antioxidant defence (by increasing glutamate-cysteine ligase, Nrf2, glutathione synthetase and glutathione peroxidase-4 proteins) in astrocytes treated with hydrogen peroxide, and Nrf2 activation was dependent on the proportion of DHA to EPA incorporated into the membrane phospholipids [95]. Thus, omega-3 FAs could be improving the antioxidant defences at least in part through the activation of the Nrf2 pathway.

6. Conclusions

The association of ADHD with an increase in inflammation and oxidative stress could play a role in the pathophysiological process. Nowadays, ADHD has no therapeutic option able to counteract the progression of the disorder, and therapy with MPH and amphetamines might be increasing oxidative stress. Thus, all evidence points towards inflammation and oxidative stress as factors which are influencing ADHD, whereas antioxidants may perhaps be able to ameliorate ADHD progression due to their anti-inflammatory and antioxidant properties. Consequently, some of the antioxidants discussed in this review might establish a new therapeutic approach for the treatment of ADHD. While the use of natural antioxidants for diverse disorders has been considered as a safe and healthier approach for patients, they are still far from being standard treatments, due to the lack of controlled clinical studies that may well corroborate both their high efficacy and safety. Accordingly, better designed and more rigorous research and clinical trials are required before they can be established as a therapeutic alternative and further studies would also be necessary to corroborate the use of these antioxidants administered as a co-treatment with the current medications. Nevertheless, antioxidants could be considered as a multi-target adjuvant therapy for ADHD.

Author Contributions: L.A.-A., N.G.-G., M.S.-G. and J.C.C. wrote a manuscript draft. J.C.C. finalized the manuscript with inputs from L.A.-A. All authors have read and agreed to the published version of the manuscript.

Funding: This work was supported by Fondos Federales HIM 2016/013 SSA 1258.

Conflicts of Interest: The authors declare no conflicts of interest.

References

1.   Faraone, S.V.; Asherson, P.; Banaschewski, T.; Biederman, J.; Buitelaar, J.K.; Ramos-Quiroga, J.A.; Rohde, L.A.; Sonuga-Barke, E.J.; Tannock, R.; Franke, B. Attention-Deficit/Hyperactivity Disorder. *Nat. Rev. Dis. Primers* **2015**, *1*, 15020. [CrossRef] [PubMed]
2.   Corona, J.C. Natural Compounds for the Management of Parkinson's Disease and Attention-Deficit/Hyperactivity Disorder. *BioMed Res. Int.* **2018**, *2018*, 4067597. [CrossRef] [PubMed]
3.   Polanczyk, G.; de Lima, M.S.; Horta, B.L.; Biederman, J.; Rohde, L.A. The Worldwide Prevalence of ADHD: A Systematic Review and Metaregression Analysis. *Am. J. Psychiatry* **2007**, *164*, 942–948. [CrossRef] [PubMed]
4.   Sayal, K.; Prasad, V.; Daley, D.; Ford, T.; Coghill, D. ADHD in Children and Young People: Prevalence, Care Pathways, and Service Provision. *Lancet Psychiatry* **2018**, *5*, 175–186. [CrossRef]
5.   Willcutt, E.G. The Prevalence of DSM-IV Attention-Deficit/Hyperactivity Disorder: A Meta-Analytic Review. *Neurotherapeutics* **2012**, *9*, 490–499. [CrossRef]
6.   Wolraich, M.L.; Hagan, J.F.; Allan, C.; Chan, E.; Davison, D.; Earls, M.; Evans, S.W.; Flinn, S.K.; Froehlich, T.; Frost, J.; et al. Clinical Practice Guideline for the Diagnosis, Evaluation, and Treatment of Attention-Deficit/Hyperactivity Disorder in Children and Adolescents. *Pediatrics* **2019**, *144*, e20192528. [CrossRef]
7.   Wilens, T.E.; Faraone, S.V.; Biederman, J. Attention-Deficit/Hyperactivity Disorder in Adults. *JAMA* **2004**, *292*, 619–623. [CrossRef]
8.   Jensen, P.S.; Hinshaw, S.P.; Kraemer, H.C.; Lenora, N.; Newcorn, J.H.; Abikoff, H.B.; March, J.S.; Arnold, L.E.; Cantwell, D.P.; Conners, C.K.; et al. ADHD Comorbidity Findings from the MTA Study: Comparing Comorbid Subgroups. *J. Am. Acad. Child Adolesc. Psychiatry* **2001**, *40*, 147–158. [CrossRef]
9.   Tejeda-Romero, C.; Kobashi-Margáin, R.A.; Alvarez-Arellano, L.; Corona, J.C.; González-García, N. Differences in Substance Use, Psychiatric Disorders and Social Factors between Mexican Adolescents and Young Adults. *Am. J. Addict.* **2018**, *27*, 625–631. [CrossRef]
10.   Newcorn, J.H.; Halperin, J.M.; Jensen, P.S.; Abikoff, H.B.; Arnold, L.E.; Cantwell, D.P.; Conners, C.K.; Elliott, G.R.; Epstein, J.N.; Greenhill, L.L.; et al. Symptom Profiles in Children with ADHD: Effects of Comorbidity and Gender. *J. Am. Acad. Child Adolesc. Psychiatry* **2001**, *40*, 137–146. [CrossRef]
11.   Del Campo, N.; Chamberlain, S.R.; Sahakian, B.J.; Robbins, T.W. The Roles of Dopamine and Noradrenaline in the Pathophysiology and Treatment of Attention-Deficit/Hyperactivity Disorder. *Boil. Psychiatry* **2011**, *69*, e145–e157. [CrossRef] [PubMed]
12.   Prince, J. Catecholamine Dysfunction in Attention-Deficit/Hyperactivity Disorder: An Update. *J. Clin. Psychopharmacol.* **2008**, *28*, S39–S45. [CrossRef] [PubMed]
13.   Instanes, J.T.; Halmøy, A.; Engeland, A.; Haavik, J.; Furu, K.; Klungsøyr, K. Attention-Deficit/Hyperactivity Disorder in Offspring of Mothers with Inflammatory and Immune System Diseases. *Boil. Psychiatry* **2017**, *81*, 452–459. [CrossRef] [PubMed]
14.   Dunn, G.A.; Nigg, J.T.; Sullivan, E.L. Neuroinflammation as a Risk Factor for Attention Deficit Hyperactivity Disorder. *Pharmacol. Biochem. Behav.* **2019**, *182*, 22–34. [CrossRef]
15.   Lopresti, A.L. Oxidative and Nitrosative Stress in ADHD: Possible Causes and the Potential of Antioxidant-Targeted Therapies. *ADHD Atten. Deficit Hyperact. Disord.* **2015**, *7*, 237–247. [CrossRef]
16.   Joseph, N.; Zhang-James, Y.; Perl, A.; Faraone, S.V. Oxidative Stress and ADHD: A Meta-Analysis. *J. Atten. Disord.* **2015**, *19*, 915–924. [CrossRef]
17.   Koda, K.; Ago, Y.; Cong, Y.; Kita, Y.; Takuma, K.; Matsuda, T. Effects of Acute and Chronic Administration of Atomoxetine and Methylphenidate on Extracellular Levels of Noradrenaline, Dopamine and Serotonin in the Prefrontal Cortex and Striatum of Mice. *J. Neurochem.* **2010**, *114*, 259–270. [CrossRef]
18.   Cinnamon Bidwell, L.; Dew, R.E.; Kollins, S.H. Alpha-2 Adrenergic Receptors and Attention-Deficit/Hyperactivity Disorder. *Curr. Psychiatry Rep.* **2010**, *12*, 366–373. [CrossRef]
19.   Bymaster, F.; Katner, J.S.; Nelson, D.L.; Hemrick-Luecke, S.K.; Threlkeld, P.G.; Heiligenstein, J.H.; Morin, S.M.; Gehlert, D.R.; Perry, K.W. Atomoxetine Increases Extracellular Levels of Norepinephrine and Dopamine in Prefrontal Cortex of Rat: A Potential Mechanism for Efficacy in Attention Deficit/Hyperactivity Disorder. *Neuropsychopharmacology* **2002**, *27*, 699–711. [CrossRef]
20.   Clemow, D.B. Misuse of Methylphenidate. *Curr. Top Behav. Neurosci.* **2017**, *34*, 99–124.

21. Swanson, J.M.; Elliott, G.R.; Greenhill, L.L.; Wigal, T.; Arnold, L.E.; Vitiello, B.; Hechtman, L.; Epstein, J.N.; Pelham, W.E.; Abikoff, H.B.; et al. Effects of Stimulant Medication on Growth Rates Across 3 Years in the MTA Follow-up. *J. Am. Acad. Child Adolesc. Psychiatry* **2007**, *46*, 1015–1027. [CrossRef] [PubMed]

22. Reed, V.A.; Buitelaar, J.K.; Anand, E.; Day, K.A.; Treuer, T.; Upadhyaya, H.P.; Coghill, D.R.; Kryzhanovskaya, L.A.; Savill, N.C. The Safety of Atomoxetine for the Treatment of Children and Adolescents with Attention-Deficit/Hyperactivity Disorder: A Comprehensive Review of Over a Decade of Research. *CNS Drugs* **2016**, *30*, 603–628. [CrossRef] [PubMed]

23. Taghizadeh, M.; Tamtaji, O.R.; Dadgostar, E.; Kakhaki, R.D.; Bahmani, F.; Abolhassani, J.; Aarabi, M.H.; Kouchaki, E.; Memarzadeh, M.R.; Asemi, Z. The Effects of Omega-3 Fatty Acids and Vitamin E Co-Supplementation on Clinical and Metabolic Status in Patients with Parkinson's disease: A randomized, double-blind, placebo-controlled trial. *Neurochem. Int.* **2017**, *108*, 183–189. [CrossRef]

24. Adair, J.C.; Knoefel, J.E.; Morgan, N. Controlled Trial of N-Acetylcysteine for Patients with Probable Alzheimer's disease. *Neurology* **2001**, *57*, 1515–1517. [CrossRef]

25. Puri, B.K.; Leavitt, B.R.; Hayden, M.; Ross, C.A.; Rosenblatt, A.; Greenamyre, J.T.; Hersch, S.; Vaddadi, K.S.; Sword, A.; Horrobin, D.F.; et al. Ethyl-EPA in Huntington Disease: A Double-Blind, Randomized, Placebo-Controlled Trial. *Neurology.* **2005**, *65*, 286–292. [CrossRef] [PubMed]

26. Sedlak, T.W.; Nucifora, L.G.; Koga, M.; Shaffer, L.S.; Higgs, C.; Tanaka, T.; Wang, A.M.; Coughlin, J.M.; Barker, P.B.; Fahey, J.W.; et al. Sulforaphane Augments Glutathione and Influences Brain Metabolites in Human Subjects: A Clinical Pilot Study. *Mol. Neuropsychiatry* **2018**, *3*, 214–222. [CrossRef]

27. Lynch, R.; Diggins, E.L.; Connors, S.L.; Zimmerman, A.W.; Singh, K.; Liu, H.; Talalay, P.; Fahey, J.W. Sulforaphane from Broccoli Reduces Symptoms of Autism: A Follow-up Case Series from a Randomized Double-blind Study. *Glob. Adv. Health Med.* **2017**, *6*. [CrossRef]

28. Ellegaard, P.K.; Licht, R.W.; Poulsen, H.E.; Nielsen, R.E.; Berk, M.; Dean, O.M.; Mohebbi, M.; Nielsen, C.T. Add-On Treatment with N-Acetylcysteine for Bipolar Depression: A 24-Week Randomized Double-Blind Parallel Group Placebo-Controlled Multicentre Trial (NACOS-Study Protocol). *Int. J. Bipolar Disord.* **2018**, *6*, 11. [CrossRef]

29. Shabab, T.; Khanabdali, R.; Moghadamtousi, S.Z.; Kadir, H.A.; Mohan, G. Neuroinflammation Pathways: A General Review. *Int. J. Neurosci.* **2017**, *127*, 624–633. [CrossRef]

30. Kohman, R.A.; Rhodes, J.S. Neurogenesis, Inflammation and Behavior. *Brain Behav. Immun.* **2013**, *27*, 22–32. [CrossRef]

31. Oades, R.D.; Myint, A.-M.; Dauvermann, M.R.; Schimmelmann, B.G.; Schwarz, M.J. Attention-Deficit Hyperactivity Disorder (ADHD) and Glial Integrity: An Exploration of Associations of Cytokines and Kynurenine Metabolites with Symptoms and Attention. *Behav. Brain Funct.* **2010**, *6*, 32. [CrossRef] [PubMed]

32. Darwish, A.H.; Elgohary, T.M.; Nosair, N.A. Serum Interleukin-6 Level in Children with Attention-Deficit Hyperactivity Disorder (ADHD). *J. Child Neurol.* **2019**, *34*, 61–67. [CrossRef] [PubMed]

33. Donfrancesco, R.; Nativio, P.; Di Benedetto, A.; Villa, M.P.; Andriola, E.; Melegari, M.G.; Cipriano, E.; Di Trani, M. Anti-Yo Antibodies in Children with ADHD: First Results About Serum Cytokines. *J. Atten. Disord.* **2016**. [CrossRef] [PubMed]

34. Kozłowska, A.; Wojtacha, P.; Równiak, M.; Kolenkiewicz, M.; Huang, A.C.W. ADHD Pathogenesis in the Immune, Endocrine and Nervous Systems of Juvenile and Maturating SHR and WKY Rats. *Psychopharmacology* **2019**, *236*, 2937–2958. [CrossRef] [PubMed]

35. Pawley, L.C.; Hueston, C.M.; O'Leary, J.D.; Kozareva, D.A.; Cryan, J.F.; O'Leary, O.F.; Nolan, Y.M. Chronic Intrahippocampal Interleukin-1β Overexpression in Adolescence Impairs Hippocampal Neurogenesis but not Neurogenesis-Associated Cognition. *Brain Behav. Immun.* **2020**, *83*, 172–179. [CrossRef]

36. Augusto-Oliveira, M.; Arrifano, G.P.; Lopes-Araujo, A.; Santos-Sacramento, L.; Takeda, P.Y.; Anthony, D.C.; Malva, J.O.; Crespo-Lopez, M.E. What Do Microglia Really Do in Healthy Adult Brain? *Cells* **2019**, *8*, 1293. [CrossRef]

37. Yirmiya, R.; Goshen, I. Immune Modulation of Learning, Memory, Neural Plasticity and Neurogenesis. *Brain Behav. Immun.* **2011**, *25*, 181–213. [CrossRef]

38. Buske-Kirschbaum, A.; Schmitt, J.; Plessow, F.; Romanos, M.; Weidinger, S.; Roessner, V. Psychoendocrine and Psychoneuroimmunological Mechanisms in the Comorbidity of Atopic Eczema and Attention Deficit/Hyperactivity Disorder. *Psychoneuroendocrinology* **2013**, *38*, 12–23. [CrossRef]

39. Drtilkova, I.; Sery, O.; Theiner, P.; Uhrova, A.; Zackova, M.; Balastikova, B.; Znojil, V. Clinical and Molecular-Genetic Markers of ADHD in Children. *Neuroendocrinol. Lett.* **2008**, *29*, 320–327.

40. Segman, R.H.; Meltzer, A.; Gross-Tsur, V.; Kosov, A.; Frisch, A.; Inbar, E.; Darvasi, A.; Levy, S.; Goltser, T.; Weizman, A.; et al. Preferential Transmission of Interleukin-1 Receptor Antagonist Alleles in Attention Deficit Hyperactivity Disorder. *Mol. Psychiatry* **2002**, *7*, 72–74. [CrossRef]

41. Misener, V.L.; Schachar, R.; Ickowicz, A.; Malone, M.; Roberts, W.; Tannock, R.; Kennedy, J.L.; Pathare, T.; Barr, C.L. Replication Test for Association of the IL-1 Receptor Antagonist Gene, IL1RN, with Attention-Deficit/Hyperactivity Disorder. *Neuropsychobiology* **2004**, *50*, 231–234. [CrossRef] [PubMed]

42. Wang, L.-J.; Yu, Y.-H.; Fu, M.-L.; Yeh, W.-T.; Hsu, J.-L.; Yang, Y.-H.; Chen, W.J.; Chiang, B.-L.; Pan, W.-H. Attention Deficit-Hyperactivity Disorder is Associated with Allergic Symptoms and Low Levels of Hemoglobin and Serotonin. *Sci. Rep.* **2018**, *8*, 10229. [CrossRef] [PubMed]

43. Yang, C.-F.; Yang, C.-C.; Wang, I.-J. Association between Allergic Diseases, Allergic Sensitization and Attention-Deficit/Hyperactivity Disorder in Children: A Large-Scale, Population-Based Study. *J. Chin. Med. Assoc.* **2018**, *81*, 277–283. [CrossRef] [PubMed]

44. Nielsen, P.R.; Benros, M.E.; Dalsgaard, S. Associations Between Autoimmune Diseases and Attention-Deficit/Hyperactivity Disorder: A Nationwide Study. *J. Am. Acad. Child Adolesc. Psychiatry* **2017**, *56*, 234–240. [CrossRef] [PubMed]

45. Cortese, S.; Sun, S.; Zhang, J.; Sharma, E.; Chang, Z.; Kuja-Halkola, R.; Almqvist, C.; Larsson, H.; Faraone, S.V. Association Between Attention Deficit Hyperactivity Disorder and Asthma: A Systematic Review and Meta-Analysis and a Swedish Population-Based Study. *Lancet Psychiatry* **2018**, *5*, 717–726. [CrossRef]

46. Zhu, Z.; Zhu, X.; Liu, C.-L.; Shi, H.; Shen, S.; Yang, Y.; Hasegawa, K.; Camargo, C.A.; Liang, L. Shared Genetics of Asthma and Mental Health Disorders: A Large-Scale Genome-Wide Cross-Trait Analysis. *Eur. Respir. J.* **2019**, *54*, 1901507. [CrossRef]

47. Rosenkranz, M.A.; Busse, W.W.; Johnstone, T.; Swenson, C.A.; Crisafi, G.M.; Jackson, M.M.; Bosch, J.A.; Sheridan, J.F.; Davidson, R.J. Neural Circuitry Underlying the Interaction between Emotion and Asthma Symptom Exacerbation. *Proc. Natl. Acad. Sci. USA* **2005**, *102*, 13319–13324. [CrossRef]

48. Passarelli, F.; Donfrancesco, R.; Nativio, P.; Pascale, E.; Di Trani, M.; Patti, A.M.; Vulcano, A.; Gozzo, P.; Villa, M.P. Anti-Purkinje Cell Antibody as a Biological Marker in Attention Deficit/Hyperactivity Disorder: A Pilot Study. *J. Neuroimmunol.* **2013**, *258*, 67–70. [CrossRef]

49. Giana, G.; Romano, E.; Porfirio, M.C.; D'Ambrosio, R.; Giovinazzo, S.; Troianiello, M.; Barlocci, E.; Travaglini, D.; Granstrem, O.; Pascale, E.; et al. Detection of Auto-Antibodies to DAT in the Serum: Interactions with DAT Genotype and Psycho-Stimulant Therapy for ADHD. *J. Neuroimmunol.* **2015**, *278*, 212–222. [CrossRef]

50. Toto, M.; Margari, F.; Simone, M.; Craig, F.; Petruzzelli, M.G.; Tafuri, S.; Margari, L. Antibasal Ganglia Antibodies and Antistreptolysin O in Noncomorbid ADHD. *J. Atten. Disord.* **2015**, *19*, 965–970. [CrossRef]

51. Rivera, H.M.; Christiansen, K.J.; Sullivan, E.L. The Role of Maternal Obesity in the Risk of Neuropsychiatric Disorders. *Front. Mol. Neurosci.* **2015**, *9*, 194. [CrossRef] [PubMed]

52. Ding, Y.; A Klomparens, E. The Neuroprotective Mechanisms and Effects of Sulforaphane. *Brain Circ.* **2019**, *5*, 74–83. [CrossRef] [PubMed]

53. Sova, M.; Saso, L. Design and Development of Nrf2 Modulators for Cancer Chemoprevention and Therapy: A Review. *Drug Des. Dev. Ther.* **2018**, *12*, 3181–3197. [CrossRef] [PubMed]

54. Dinkova-Kostova, A.T.; Abramov, A.Y. The Emerging Role of Nrf2 in Mitochondrial Function. *Free Radic. Boil. Med.* **2015**, *88*, 179–188. [CrossRef] [PubMed]

55. Smith, R.E. The Effects of Dietary Supplements That Overactivate the Nrf2/ARE System. *Curr. Med. Chem.* **2019**, *26*, 1. [CrossRef]

56. Hashimoto, K. Recent Advances in the Early Intervention in Schizophrenia: Future Direction from Preclinical Findings. *Curr. Psychiatry Rep.* **2019**, *21*, 75. [CrossRef]

57. Singh, K.; Connors, S.L.; Macklin, E.A.; Smith, K.D.; Fahey, J.W.; Talalay, P.; Zimmerman, A.W. Sulforaphane Treatment of Autism Spectrum Disorder (ASD). *Proc. Natl. Acad. Sci. USA* **2014**, *111*, 15550–15555. [CrossRef]

58. Myzak, M.C.; Dashwood, R.H. Chemoprotection by Sulforaphane: Keep One Eye beyond Keap1. *Cancer Lett.* **2006**, *233*, 208–218. [CrossRef]

59. Ahmed, S.M.U.; Luo, L.; Namani, A.; Wang, X.J.; Tang, X. Nrf2 Signaling Pathway: Pivotal Roles in Inflammation. *Biochim. Biophys. Acta (BBA)-Mol. Basis Dis.* **2017**, *1863*, 585–597. [CrossRef]

60. Aladaileh, S.H.; Hussein, O.E.; Abukhalil, M.H.; Saghir, S.A.M.; Bin-Jumah, M.; Alfwuaires, M.A.; Germoush, M.O.; Almaiman, A.A.; Mahmoud, A.M. Formononetin Upregulates Nrf2/HO-1 Signaling and Prevents Oxidative Stress, Inflammation and Kidney Injury in Methotrexate-Induced Rats. *Antioxidants* **2019**, *8*, 430. [CrossRef]

61. Wu, W.; Peng, G.; Yang, F.; Zhang, Y.; Mu, Z.; Han, X. Sulforaphane Has a Therapeutic Effect in an Atopic Dermatitis Murine Model and Activates the Nrf2/HO-1 axis. *Mol. Med. Rep.* **2019**, *20*, 1761–1771. [CrossRef]

62. Subedi, L.; Lee, J.H.; Yumnam, S.; Ji, E.; Kim, S.Y. Anti-Inflammatory Effect of Sulforaphane on LPS-Activated Microglia Potentially through JNK/AP-1/NF-kappaB Inhibition and Nrf2/HO-1 Activation. *Cells* **2019**, *8*, 194. [CrossRef] [PubMed]

63. Qi, T.; Xu, F.; Yan, X.; Li, S.; Li, H. Sulforaphane Exerts Anti-Inflammatory Effects against Lipopolysaccharide-Induced Acute Lung Injury in Mice through the Nrf2/ARE Pathway. *Int. J. Mol. Med.* **2016**, *37*, 182–188. [CrossRef] [PubMed]

64. Luo, J.-F.; Shen, X.-Y.; Lio, C.K.; Dai, Y.; Cheng, C.-S.; Liu, J.-X.; Yao, Y.-D.; Yu, Y.; Xie, Y.; Luo, P.; et al. Activation of Nrf2/HO-1 Pathway by Nardochinoid C Inhibits Inflammation and Oxidative Stress in Lipopolysaccharide-Stimulated Macrophages. *Front. Pharmacol.* **2018**, *9*, 911. [CrossRef] [PubMed]

65. Rakariyatham, K.; Wu, X.; Tang, Z.; Han, Y.; Wang, Q.; Xiao, H. Synergism between Luteolin and Sulforaphane in Anti-Inflammation. *Food Funct.* **2018**, *9*, 5115–5123. [CrossRef]

66. Greaney, A.J.; Maier, N.K.; Leppla, S.H.; Moayeri, M. Sulforaphane Inhibits Multiple Inflammasomes through an Nrf2-Independent Mechanism. *J. Leukoc. Biol.* **2016**, *99*, 189–199. [CrossRef]

67. Lee, J.; Ahn, H.; Hong, E.-J.; An, B.-S.; Jeung, E.-B.; Lee, G.-S. Sulforaphane Attenuates Activation of NLRP3 and NLRC4 Inflammasomes but not AIM2 Inflammasome. *Cell. Immunol.* **2016**, *306*, 53–60. [CrossRef]

68. Zhao, C.; Gillette, D.D.; Li, X.; Zhang, Z.; Wen, H. Nuclear Factor E2-Related Factor-2 (Nrf2) Is Required for NLRP3 and AIM2 Inflammasome Activation. *J. Boil. Chem.* **2014**, *289*, 17020–17029. [CrossRef]

69. Dwivedi, S.; Rajasekar, N.; Hanif, K.; Nath, C.; Shukla, R. Sulforaphane Ameliorates Okadaic Acid-Induced Memory Impairment in Rats by Activating the Nrf2/HO-1 Antioxidant Pathway. *Mol. Neurobiol.* **2016**, *53*, 5310–5323. [CrossRef]

70. Nadeem, A.; Ahmad, S.F.; Al-Harbi, N.O.; Attia, S.M.; Bakheet, S.A.; Ibrahim, K.E.; Alqahtani, F.; Alqinyah, M. Nrf2 Activator, Sulforaphane Ameliorates Autism-Like Symptoms through Suppression of Th17 Related Signaling and Rectification of Oxidant-Antioxidant Imbalance in Periphery and Brain of BTBR T+tf/J mice. *Behav. Brain Res.* **2019**, *364*, 213–224. [CrossRef]

71. Fernandes, J.; Gupta, G.L. N-acetylcysteine Attenuates Neuroinflammation Associated Depressive Behavior Induced by Chronic Unpredictable Mild Stress in Rat. *Behav. Brain Res.* **2019**, *364*, 356–365. [CrossRef] [PubMed]

72. Panizzutti, B.; Bortolasci, C.; Hasebe, K.; Kidnapillai, S.; Gray, L.; Walder, K.; Berk, M.; Mohebbi, M.; Dodd, S.; Gama, C.; et al. Mediator Effects of Parameters of Inflammation and Neurogenesis from a N-Acetyl Cysteine Clinical-Trial for Bipolar Depression. *Acta Neuropsychiatr.* **2018**, *30*, 334–341. [CrossRef] [PubMed]

73. Garcia, R.J.; Francis, L.; Dawood, M.; Lai, Z.-W.; Faraone, S.V.; Perl, A. Attention Deficit and Hyperactivity Disorder Scores Are Elevated and Respond to N-Acetylcysteine Treatment in Patients with Systemic Lupus Erythematosus. *Arthritis Rheum.* **2013**, *65*, 1313–1318. [CrossRef] [PubMed]

74. Lai, Z.-W.; Hanczko, R.; Bonilla, E.; Caza, T.N.; Clair, B.; Bartos, A.; Miklóssy, G.; Jimah, J.; Doherty, E.; Tily, H.; et al. N-Acetylcysteine Reduces Disease Activity by Blocking Mammalian Target of Rapamycin in T Cells from Systemic Lupus Erythematosus Patients: A Randomized, Double-Blind, Placebo-Controlled Trial. *Arthritis Rheum.* **2012**, *64*, 2937–2946. [CrossRef]

75. Wang, P.; Chen, F.; Wang, W.; Zhang, X.-D. Hydrogen Sulfide Attenuates High Glucose-Induced Human Retinal Pigment Epithelial Cell Inflammation by Inhibiting ROS Formation and NLRP3 Inflammasome Activation. *Mediat. Inflamm.* **2019**, *2019*, 8908960. [CrossRef]

76. Abdel-Wahab, W.M.; I Moussa, F. Neuroprotective Effect of N-Acetylcysteine against Cisplatin-Induced Toxicity in Rat Brain by Modulation of Oxidative Stress and Inflammation. *Drug Des. Dev. Ther.* **2019**, *13*, 1155–1162. [CrossRef]

77. Wong, S.W.; Kwon, M.-J.; Choi, A.M.K.; Kim, H.-P.; Nakahira, K.; Hwang, D.H. Fatty Acids Modulate Toll-like Receptor 4 Activation through Regulation of Receptor Dimerization and Recruitment into Lipid Rafts in a Reactive Oxygen Species-dependent Manner. *J. Boil. Chem.* **2009**, *284*, 27384–27392. [CrossRef]

78. Carlsson, J.A.; Wold, A.E.; Sandberg, A.S.; Ostman, S.M. The Polyunsaturated Fatty Acids Arachidonic Acid and Docosahexaenoic Acid Induce Mouse Dendritic Cells Maturation but Reduce T-Cell Responses in Vitro. *PLoS ONE* **2015**, *10*, e0143741. [CrossRef]

79. Kim, W.; Barhoumi, R.; McMurray, D.N.; Chapkin, R.S. Dietary Fish Oil and DHA Down-Regulate Antigen-Activated CD4+ T-Cells While Promoting the Formation of Liquid-Ordered Mesodomains. *Br. J. Nutr.* **2014**, *111*, 254–260. [CrossRef]

80. Monk, J.M.; Hou, T.Y.; Turk, H.F.; McMurray, D.N.; Chapkin, R.S. n3 PUFAs Reduce Mouse CD4+ T-cell Ex Vivo Polarization into Th17 Cells. *J. Nutr.* **2013**, *143*, 1501–1508. [CrossRef]

81. Hariri, M.; Djazayery, A.; Djalali, M.; Saedisomeolia, A.; Rahimi, A.; Abdolahian, E. Effect of n-3 Supplementation on Hyperactivity, Oxidative Stress and Inflammatory Mediators in Children with Attention-Deficit-Hyperactivity Disorder. *Malays. J. Nutr.* **2012**, *18*, 329–335. [PubMed]

82. Chen, H.; Wu, J.; Zhang, J.; Fujita, Y.; Ishima, T.; Iyo, M.; Hashimoto, K. Protective Effects of the Antioxidant Sulforaphane on Behavioral Changes and Neurotoxicity in Mice after the Administration of Methamphetamine. *Psychopharmacology* **2012**, *222*, 37–45. [CrossRef] [PubMed]

83. Pauletti, A.; Terrone, G.; Shekh-Ahmad, T.; Salamone, A.; Ravizza, T.; Rizzi, M.; Pastore, A.; Pascente, R.; Liang, L.-P.; Villa, B.R.; et al. Targeting Oxidative Stress Improves Disease Outcomes in a Rat Model of Acquired Epilepsy. *Brain* **2019**, *142*, e39. [CrossRef] [PubMed]

84. Wan, F.J.; Tung, C.S.; Shiah, I.S.; Lin, H.C. Effects of Alpha-phenyl-N-tert-butyl Nitrone and N-Acetylcysteine on Hydroxyl Radical Formation and Dopamine Depletion in the Rat Striatum Produced by d-Amphetamine. *Eur. Neuropsychopharmacol.* **2006**, *16*, 147–153. [CrossRef] [PubMed]

85. Bloch, M.H.; Panza, K.E.; Yaffa, A.; Alvarenga, P.G.; Jakubovski, E.; Mulqueen, J.M.; Landeros-Weisenberger, A.; Leckman, J.F. N-Acetylcysteine in the Treatment of Pediatric Tourette Syndrome: Randomized, Double-Blind, Placebo-Controlled Add-On Trial. *J. Child Adolesc. Psychopharmacol.* **2016**, *26*, 327–334. [CrossRef] [PubMed]

86. Rus, C.P. [A Girl with Self-Harm Treated with N-Acetylcysteine (NAC)]. *Tijdschr. Psychiatr.* **2017**, *59*, 181–184.

87. Yan, Y. [Effect of N-Acetylcysteine on Cognitive Function and Nuclear Factor Erythroid 2 Related Factor 2/Heme Oxygenase-1 Pathway in Mouse Models of Postoperative Cognitive Dysfunction]. *Zhongguo Yi Xue Ke Xue Yuan Xue Bao* **2019**, *41*, 529–535.

88. Milte, C.M.; Parletta, N.; Buckley, J.D.; Coates, A.M.; Young, R.M.; Howe, P.R. Eicosapentaenoic and Docosahexaenoic Acids, Cognition, and Behavior in Children with Attention-Deficit/Hyperactivity Disorder: A Randomized Controlled Trial. *Nutrition* **2012**, *28*, 670–677. [CrossRef]

89. Stevens, L.; Zhang, W.; Peck, L.; Kuczek, T.; Grevstad, N.; Mahon, A.; Zentall, S.S.; Arnold, L.E.; Burgess, J.R. EFA Supplementation in Children with Inattention, Hyperactivity, and Other Disruptive Behaviors. *Lipids* **2003**, *38*, 1007–1021. [CrossRef]

90. Matsudaira, T.; Gow, R.V.; Kelly, J.; Murphy, C.; Potts, L.; Sumich, A.; Ghebremeskel, K.; Crawford, M.A.; Taylor, E. Biochemical and Psychological Effects of Omega-3/6 Supplements in Male Adolescents with Attention-Deficit/Hyperactivity Disorder: A Randomized, Placebo-Controlled, Clinical Trial. *J. Child Adolesc. Psychopharmacol.* **2015**, *25*, 775–782. [CrossRef]

91. Joshi, K.; Lad, S.; Kale, M.; Patwardhan, B.; Mahadik, S.P.; Patni, B.; Chaudhary, A.; Bhave, S.; Pandit, A. Supplementation with Flax Oil and Vitamin C Improves the Outcome of Attention Deficit Hyperactivity Disorder (ADHD). *Prostaglandins Leukot. Essent. Fat. Acids* **2006**, *74*, 17–21. [CrossRef]

92. Sorgi, P.J.; Hallowell, E.M.; Hutchins, H.L.; Sears, B. Effects of an Open-Label Pilot Study with High-Dose EPA/DHA Concentrates on Plasma Phospholipids and Behavior in Children with Attention Deficit Hyperactivity Disorder. *Nutr. J.* **2007**, *6*, 16. [CrossRef] [PubMed]

93. Widenhorn-Muller, K.; Schwanda, S.; Scholz, E.; Spitzer, M.; Bode, H. Effect of Supplementation with Long-Chain Omega-3 Polyunsaturated Fatty Acids on Behavior and Cognition in Children with Attention Deficit/Hyperactivity Disorder (ADHD): A Randomized Placebo-Controlled Intervention Trial. *Prostaglandins Leukot. Essent. Fat. Acids* **2014**, *91*, 49–60. [CrossRef] [PubMed]

94. Bos, D.J.; Oranje, B.; Veerhoek, E.S.; Van Diepen, R.M.; Weusten, J.M.; Demmelmair, H.; Koletzko, B.; Velden, M.G.D.S.-V.D.; Eilander, A.; Hoeksma, M.; et al. Reduced Symptoms of Inattention after Dietary Omega-3 Fatty Acid Supplementation in Boys with and without Attention Deficit/Hyperactivity Disorder. *Neuropsychopharmacology* **2015**, *40*, 2298–2306. [CrossRef] [PubMed]

95. Zgórzyńska, E.; Dziedzic, B.; Gorzkiewicz, A.; Stulczewski, D.; Bielawska, K.; Su, K.-P.; Walczewska, A. Omega-3 Polyunsaturated Fatty Acids Improve the Antioxidative Defense in Rat Astrocytes via an Nrf2-Dependent Mechanism. *Pharmacol. Rep.* **2017**, *69*, 935–942. [CrossRef]

96. Šalamon, Š.; Kramar, B.; Marolt, T.P.; Poljšak, B.; Milisav, I. Medical and Dietary Uses of N-Acetylcysteine. *Antioxidants* **2019**, *8*, 111. [CrossRef]

97. Ooi, S.L.; Green, R.; Pak, S.C. N-Acetylcysteine for the Treatment of Psychiatric Disorders: A Review of Current Evidence. *BioMed Res. Int.* **2018**, *2018*, 2469486. [CrossRef] [PubMed]

98. Slattery, J.; Kumar, N.; Delhey, L.; Berk, M.; Dean, O.; Spielholz, C.; Frye, R. Clinical trials of N-acetylcysteine in psychiatry and neurology: A systematic review. *Neurosci. Biobehav. Rev.* **2015**, *55*, 294–321.

99. Duailibi, M.S.; Cordeiro, Q.; Brietzke, E.; Ribeiro, M.; LaRowe, S.; Berk, M.; Trevizol, A.P. N-acetylcysteine in the treatment of craving in substance use disorders: Systematic review and meta-analysis. *Am. J. Addict.* **2017**, *26*, 660–666. [CrossRef]

100. Chen, A.T.; Chibnall, J.T.; A Nasrallah, H. Placebo-Controlled Augmentation Trials of the Antioxidant NAC in Schizophrenia: A Review. *Ann. Clin. Psychiatry* **2016**, *28*, 190–196.

101. Zheng, W.; Qiu, Y.; Berk, M.; Zhang, Q.-E.; Cai, D.-B.; Yang, X.-H.; Ungvari, G.S.; Ng, C.H.; Ning, Y.-P.; Xiang, Y.-T. N-acetylcysteine for Major Mental Disorders: A Systematic Review and Meta-Analysis of Randomized Controlled Trials. *Acta Psychiatr. Scand.* **2018**, *137*, 391–400. [CrossRef] [PubMed]

102. Oliver, G.; Dean, O.; Camfield, D.; Blair-West, S.; Ng, C.; Berk, M.; Sarris, J. N-Acetyl Cysteine in the Treatment of Obsessive Compulsive and Related Disorders: A Systematic Review. *Clin. Psychopharmacol. Neurosci.* **2015**, *13*, 12–24. [CrossRef] [PubMed]

103. Nikoo, M.; Radnia, H.; Farokhnia, M.; Mohammadi, M.R.; Akhondzadeh, S. N-acetylcysteine as an Adjunctive Therapy to Risperidone for Treatment of Irritability in Autism: A Randomized, Double-Blind, Placebo-Controlled Clinical Trial of Efficacy and Safety. *Clin. Neuropharmacol.* **2015**, *38*, 11–17. [CrossRef] [PubMed]

104. Garcia-Keller, C.; Smiley, C.; Monforton, C.; Melton, S.; Kalivas, P.W.; Gass, J. N-Acetylcysteine Treatment during Acute Stress Prevents Stress-Induced Augmentation of Addictive Drug Use and Relapse. *Addict. Boil.* **2019**, e12798. [CrossRef]

105. Healy-Stoffel, M.; Levant, B. N-3 (Omega-3) Fatty Acids: Effects on Brain Dopamine Systems and Potential Role in the Etiology and Treatment of Neuropsychiatric Disorders. *CNS Neurol. Disord. Drug Targets* **2018**, *17*, 216–232. [CrossRef]

106. Avallone, R.; Vitale, G.; Bertolotti, M. Omega-3 Fatty Acids and Neurodegenerative Diseases: New Evidence in Clinical Trials. *Int. J. Mol. Sci.* **2019**, *20*, 4256. [CrossRef]

107. Nasir, M.; Bloch, M.H. Trim the Fat: The Role of Omega-3 Fatty Acids in Psychopharmacology. *Ther. Adv. Psychopharmacol.* **2019**, *9*, 2045125319869791. [CrossRef]

108. Calder, P.C. Omega-3 Fatty Acids and Inflammatory Processes: From Molecules to Man. *Biochem. Soc. Trans.* **2017**, *45*, 1105–1115. [CrossRef]

109. Endres, S.; Ghorbani, R.; Kelley, V.E.; Georgilis, K.; Lonnemann, G.; Van Der Meer, J.W.M.; Cannon, J.G.; Rogers, T.S.; Klempner, M.S.; Weber, P.C.; et al. The Effect of Dietary Supplementation with n-3 Polyunsaturated Fatty Acids on the Synthesis of Interleukin-1 and Tumor Necrosis Factor by Mononuclear Cells. *N. Engl. J. Med.* **1989**, *320*, 265–271. [CrossRef]

110. Lee, J.Y.; Zhao, L.; Youn, H.S.; Weatherill, A.R.; Tapping, R.; Feng, L.; Lee, W.H.; Fitzgerald, K.A.; Hwang, D.H. Saturated Fatty Acid Activates but Polyunsaturated Fatty Acid Inhibits Toll-Like Receptor 2 Dimerized with Toll-like Receptor 6 or 1. *J. Boil. Chem.* **2004**, *279*, 16971–16979. [CrossRef]

111. Mohammadzadeh, S.; Baghi, N.; Yousefi, F.; Yousefzamani, B. Effect of Omega-3 plus Methylphenidate as an Alternative Therapy to Reduce Attention Deficit-Hyperactivity Disorder in Children. *Korean J. Pediatr.* **2019**, *62*, 360–366. [CrossRef] [PubMed]

112. Öztop, D.; Altun, H.; Baskol, G.; Ozsoy, S. Oxidative Stress in Children with Attention Deficit Hyperactivity Disorder. *Clin. Biochem.* **2012**, *45*, 745–748. [CrossRef] [PubMed]

113. Spahis, S.; Vanasse, M.; Bélanger, S.A.; Ghadirian, P.; Grenier, E.; Levy, E. Lipid Profile, Fatty Acid Composition and Pro- and Anti-Oxidant Status in Pediatric Patients with Attention-Deficit/Hyperactivity Disorder. *Prostaglandins Leukot. Essent. Fat. Acids* **2008**, *79*, 47–53. [CrossRef] [PubMed]

114. Naeini, A.A.; Nasim, S.; Najafi, M.; Ghazvini, M.; Hassanzadeh, A. Relationship between Antioxidant Status and Attention Deficit Hyperactivity Disorder among Children. *Int. J. Prev. Med.* **2019**, *10*, 41. [CrossRef] [PubMed]

115. Bulut, M.; Selek, S.; Gergerlioglu, H.S.; Savas, H.A.; Yilmaz, H.R.; Yuce, M.; Ekici, G. Malondialdehyde Levels in Adult Attention-Deficit Hyperactivity Disorder. *J. Psychiatry Neurosci.* **2007**, *32*, 435–438.

116. Bulut, M.; Selek, S.; Bez, Y.; Kaya, M.C.; Günes, M.; Karababa, F.; Çelik, H.; Savaş, H.A. Lipid Peroxidation Markers in Adult Attention Deficit Hyperactivity Disorder: New Findings for Oxidative Stress. *Psychiatry Res. Neuroimaging* **2013**, *209*, 638–642. [CrossRef]

117. Ceylan, M.; Sener, S.; Bayraktar, A.C.; Kavutcu, M.; Bayraktar, A.C. Oxidative Imbalance in Child and Adolescent Patients with Attention-Deficit/Hyperactivity Disorder. *Prog. Neuro-Psychopharmacol. Boil. Psychiatry* **2010**, *34*, 1491–1494. [CrossRef]

118. Verlaet, A.A.J.; Breynaert, A.; Ceulemans, B.; De Bruyne, T.; Fransen, E.; Pieters, L.; Savelkoul, H.F.J.; Hermans, N. Oxidative Stress and Immune Aberrancies in Attention-Deficit/Hyperactivity Disorder (ADHD): A Case-Control Comparison. *Eur. Child Adolesc. Psychiatry* **2019**, *28*, 719–729. [CrossRef]

119. Dvorakova, M.; Sivoňová, M.; Trebatická, J.; Škodáček, I.; Waczulikova, I.; Muchová, J.; Ďuračková, Z. The Effect of Polyphenolic Extract from Pine Bark, Pycnogenol® on the Level of Glutathione in Children Suffering from Attention Deficit Hyperactivity Disorder (ADHD). *Redox Rep.* **2006**, *11*, 163–172. [CrossRef]

120. Leffa, D.T.; Bellaver, B.; De Oliveira, C.; De Macedo, I.C.; De Freitas, J.S.; Grevet, E.H.; Caumo, W.; Rohde, L.A.; Quincozes-Santos, A.; Torres, I.L.S. Increased Oxidative Parameters and Decreased Cytokine Levels in an Animal Model of Attention-Deficit/Hyperactivity Disorder. *Neurochem. Res.* **2017**, *42*, 3084–3092. [CrossRef]

121. Sezen, H.; Kandemir, H.; Savik, E.; Kandemir, S.B.; Kilicaslan, F.; Bilinc, H.; Aksoy, N. Increased Oxidative Stress in Children with Attention Deficit Hyperactivity Disorder. *Redox Rep.* **2016**, *21*, 248–253. [CrossRef] [PubMed]

122. Avcil, S.; Uysal, P.; Yenisey, Ç.; Abas, B.I. Elevated Melatonin Levels in Children with Attention Deficit Hyperactivity Disorder: Relationship to Oxidative and Nitrosative Stress. *J. Atten. Disord.* **2019**, 1087054719829816. [CrossRef] [PubMed]

123. Selek, S.; Savaş, H.A.; Gergerlioglu, H.S.; Bulut, M.; Yilmaz, H.R.; Yılmaz, H.R. Oxidative Imbalance in Adult Attention deficit/hyperactivity disorder. *Boil. Psychol.* **2008**, *79*, 256–259. [CrossRef]

124. Ceylan, M.F.; Sener, S.; Bayraktar, A.C.; Kavutcu, M. Changes in Oxidative Stress and Cellular Immunity Serum Markers in Attention-Deficit/Hyperactivity Disorder. *Psychiatry Clin. Neurosci.* **2012**, *66*, 220–226. [CrossRef] [PubMed]

125. Goldstein, D.S.; Kopin, I.J.; Sharabi, Y. Catecholamine autotoxicity. Implications for Pharmacology and Therapeutics of Parkinson Disease and Related Disorders. *Pharmacol. Ther.* **2014**, *144*, 268–282. [CrossRef] [PubMed]

126. Napolitano, A.; Manini, P.; D'Ischia, M. Oxidation Chemistry of Catecholamines and Neuronal Degeneration: An Update. *Curr. Med. Chem.* **2011**, *18*, 1832–1845. [CrossRef]

127. Neri, M.; Cerretani, D.; Fiaschi, A.I.; Laghi, P.F.; Lazzerini, P.E.; Maffione, A.B.; Micheli, L.; Bruni, G.; Nencini, C.; Giorgi, G.; et al. Correlation between Cardiac Oxidative Stress and Myocardial Pathology due to Acute and Chronic Norepinephrine Administration in Rats. *J. Cell. Mol. Med.* **2007**, *11*, 156–170. [CrossRef]

128. Spencer, W.A.; Jeyabalan, J.; Kichambre, S.; Gupta, R.C. Oxidatively Generated DNA Damage after Cu(II) Catalysis of Dopamine and Related Catecholamine Neurotransmitters and Neurotoxins: Role of Reactive Oxygen Species. *Free Radic. Biol. Med.* **2011**, *50*, 139–147. [CrossRef]

129. Swanson, C.J.; Perry, K.W.; Koch-Krueger, S.; Katner, J.; Svensson, K.A.; Bymaster, F.P. Effect of the Attention Deficit/Hyperactivity Disorder Drug Atomoxetine on Extracellular Concentrations of Norepinephrine and Dopamine in Several Brain Regions of the Rat. *Neuropharmacology.* **2006**, *50*, 755–760. [CrossRef]

130. Corona, J.C.; Carreón-Trujillo, S.; González-Pérez, R.; Gómez-Bautista, D.; Vázquez-González, D.; Salazar-García, M. Atomoxetine Produces Oxidative Stress and Alters Mitochondrial Function in Human Neuron-Like Cells. *Sci. Rep.* **2019**, *9*, 13011–13019. [CrossRef]

131. Cobley, J.N.; Fiorello, M.L.; Bailey, D.M. 13 Reasons Why the Brain is Susceptible to Oxidative Stress. *Redox Boil.* **2018**, *15*, 490–503. [CrossRef] [PubMed]

132. Corona, J.C.; Duchen, M.R. Impaired Mitochondrial Homeostasis and Neurodegeneration: Towards New Therapeutic Targets? *J. Bioenerg. Biomembr.* **2015**, *47*, 89–99. [CrossRef] [PubMed]

133. Andreazza, A.C.; Frey, B.N.; Valvassori, S.S.; Zanotto, C.; Gomes, K.M.; Comim, C.M.; Cassini, C.; Stertz, L.; Ribeiro, L.C.; Quevedo, J.; et al. DNA Damage in Rats after Treatment with Methylphenidate. *Prog. Neuro-Psychopharmacol. Boil. Psychiatry* **2007**, *31*, 1282–1288. [CrossRef] [PubMed]

134. Martins, M.R.; Reinke, A.; Petronilho, F.C.; Gomes, K.M.; Dal-Pizzol, F.; Quevedo, J. Methylphenidate Treatment Induces Oxidative Stress in Young Rat Brain. *Brain Res.* **2006**, *1078*, 189–197. [CrossRef] [PubMed]

135. Gomes, K.M.; Petronilho, F.C.; Mantovani, M.; Garbelotto, T.; Boeck, C.R.; Dal-Pizzol, F.; Quevedo, J. Antioxidant Enzyme Activities Following Acute or Chronic Methylphenidate Treatment in Young Rats. *Neurochem. Res.* **2008**, *33*, 1024–1027. [CrossRef] [PubMed]

136. Comim, C.M.; Gomes, K.M.; Reus, G.Z.; Petronilho, F.; Ferreira, G.K.; Streck, E.L.; Dal-Pizzol, F.; Quevedo, J. Methylphenidate Treatment Causes Oxidative Stress and Alters Energetic Metabolism in an Animal Model of Attention-Deficit Hyperactivity Disorder. *Acta Neuropsychiatr.* **2014**, *26*, 96–103. [CrossRef]

137. Motaghinejad, M.; Motevalian, M.; Shabab, B.; Fatima, S. Effects of Acute Doses of Methylphenidate on Inflammation and Oxidative Stress in Isolated Hippocampus and Cerebral Cortex of Adult Rats. *J. Neural Transm.* **2017**, *124*, 121–131. [CrossRef]

138. Smith, R.E.; Ozben, T.; Saso, L. Modulation of Oxidative Stress: Pharmaceutical and Pharmacological Aspects 2018. *Oxidative Med. Cell. Longev.* **2019**, *2019*, 6380473. [CrossRef]

139. Sun, Y.; Yang, T.; Leak, R.K.; Chen, J.; Zhang, F. Preventive and Protective Roles of Dietary Nrf2 Activators against Central Nervous System Diseases. *CNS Neurol. Disord. Drug Targets* **2017**, *16*, 326–338. [CrossRef]

140. Huang, C.; Wu, J.; Chen, D.; Jin, J.; Wu, Y.; Chen, Z. Effects of Sulforaphane in the Central Nervous System. *Eur. J. Pharmacol.* **2019**, *853*, 153–168. [CrossRef]

141. Shahripour, R.B.; Harrigan, M.R.; Alexandrov, A.V. N-Acetylcysteine (NAC) in Neurological Disorders: Mechanisms of Action and Therapeutic Opportunities. *Brain Behav.* **2014**, *4*, 108–122. [CrossRef] [PubMed]

142. Heshmati, J.; Morvaridzadeh, M.; Maroufizadeh, S.; Akbari, A.; Yavari, M.; Amirinejad, A.; Maleki-Hajiagha, A.; Sepidarkish, M.; Amirinejhad, A. Omega-3 Fatty Acids Supplementation and Oxidative Stress Parameters: A Systematic Review and Meta-Analysis of Clinical Trials. *Pharmacol. Res.* **2019**, *149*, 104462. [CrossRef] [PubMed]

Review

When Oxidative Stress Meets Epigenetics: Implications in Cancer Development

Álvaro García-Guede [1,2], Olga Vera [3,*] and Inmaculada Ibáñez-de-Caceres [1,2]

1   Epigenetics Laboratory, INGEMM, Hospital La PAZ. 28046 Madrid, Spain; alvarogr@ucm.es (Á.G.-G.); inma.ibanezca@salud.madrid.org (I.I.-d.-C.)
2   Experimental Therapies and Novel Biomarkers in Cancer, Instituto de Investigación Sanitaria del Hospital La Paz. IdiPAZ, 28046 Madrid, Spain
3   Department of Molecular Oncology, H. Lee Moffitt Cancer Center and Research Institute, Tampa, FL 33612, USA
*   Correspondence: olga.verapuente@moffitt.org

Received: 4 May 2020; Accepted: 29 May 2020; Published: 1 June 2020

Abstract: Cancer is one of the leading causes of death worldwide and it can affect any part of the organism. It arises as a consequence of the genetic and epigenetic changes that lead to the uncontrolled growth of the cells. The epigenetic machinery can regulate gene expression without altering the DNA sequence, and it comprises methylation of the DNA, histones modifications, and non-coding RNAs. Alterations of these gene-expression regulatory elements can be produced by an imbalance of the intracellular environment, such as the one derived by oxidative stress, to promote cancer development, progression, and resistance to chemotherapeutic treatments. Here we review the current literature on the effect of oxidative stress in the epigenetic machinery, especially over the largely unknown ncRNAs and its consequences toward cancer development and progression.

Keywords: epigenetics; cancer; oxidative stress; miR7/MAFG/Nrf2 axe; chemoresistance; cancer therapy

1. Introduction

1.1. Cancer

Cancer is a generic term defining a wide and heterogeneous group of diseases that can affect any part of the organism. It is considered a multiphasic disease primarily characterized by the appearance of abnormal cells in tissues, with uncontrolled growth beyond their limits and that can invade adjacent organs, disseminating to other parts of the body [1]. According to the World Health Organization, cancer is one of the leading causes of death being responsible for 9.6 million of deaths in 2018 (17% of the total death worldwide). As stated by the World Cancer Report in its edition of 2014, the five most frequently diagnosed tumors among men were lung, prostate, colorectal stomach and liver cancers, whereas among women were breast, colorectal, lung, cervix and stomach cancers [1,2].

Cancer development is a consequence of molecular alterations of genetic and/or epigenetic origin. These can be initiated by the accumulation of genetic changes in DNA that affect the DNA sequence, such as mutations and chromosomal rearrangements, or by modifications in DNA, histones and non-coding RNAs that do not change the original sequence (epigenetic modifications). All of these changes promote the clonal selection of those cells that show a more aggressive phenotype. In 2000, different diseases were first grouped together and collectively referred to as cancer based on their major molecular alterations: self-sufficiency in growth signals, insensitivity to antigrowth signals, evading apoptosis, limitless replicative potential, sustained angiogenesis and tissue invasion and metastasis [3]. Afterwards, in 2009, another five characteristics related to the metabolic, proteotoxic, mitotic and oxidative stress, and DNA damage were also considered when defining cancer [4]. The last revision to the molecular definition of a cancer cell was done in 2011 by Hanahan and Weinberg, who determined new features and defined the presence of a tumor microenvironment developed by the cells along the multiple steps of tumorigenesis [5]. Together, all these properties encompass the most up-to-date definition of the cancer cell.

1.2. Epigenetics

Epigenetics is the discipline that studies the inheritable changes of gene expression that are produced by chemical alterations of DNA, histones, and the involvement of non-coding RNAs, rather than changes in the original sequence of DNA. All these changes lead to the remodeling of chromatin to promote or impede gene expression. The silencing of gene expression at the chromatin level is necessary for the normal life of eukaryotic organisms, and it is particularly important in the regulation of biological processes such as the embryonic development, differentiation, or genomic imprinting [6].

There are three major mechanisms of epigenetic regulation described.

1.2.1. DNA Methylation

DNA methylation is the most studied and best known epigenetic mechanism. Generally, DNA methylation is a synonym of genic silencing, as it shapes an inaccessible state of the chromatin for the transcription process. This chemical modification consists of the addition of a methyl group (CH_3) in carbon 5 of a cytosine located, generally, in 5′-Cytosine-phosphate-Guanine-3′ (CpG) dinucleotides regions. CpG dinucleotide distribution is asymmetric along the genome and its accumulation, preferentially in promoter regions, is called CpG Island (CGI) [7,8]. In the human genome, there are approximately 30,000 unmethylated CGIs that warrant the potentially active configuration of constitutive genes. The methylation pattern of DNA is responsible for cell differentiation; thus, its dysregulation leads to a number of diseases, including cancer [9].

The process of DNA methylation is catalyzed by the DNA-Methyltransferases (DNMTs). There are four types of DNMTs, two involved in de novo methylation of DNA during development (DNMT3A and 3B), one responsible for maintaining the methylation patterns after DNA replication (DNMT1), and one without catalytic site that acts in conjunction with the de novo DNMTs to recruit chromatin remodeling complexes [8,10]. This process, which is essential in embryonic development, is related to the dosage compensation in mammals (X chromosome inactivation) [11–13] and genomic imprinting (selective silencing of either maternal or paternal genes) [14]. Although this silencing is mediated by DNA methylation, it is necessary for the involvement of the whole epigenetic machinery for the correct development of the process.

1.2.2. Histone Modifications

Another well-known epigenetic mechanism consists of the chemical modification of histones that are part of nucleosomes in chromatin. Similar to other proteins, histones can undergo posttranslational modifications such as acetylation, methylation, phosphorylation, ribosylation, ubiquitination, sumoylation, or glycosylation (reviewed in [15,16]), whose main function is the regulation of DNA accessibility. In general, phosphorylation [17–19] and ribosylation [20,21] promote euchromatin and gene transcription, whereas sumoylation triggers gene silencing [22,23] and ubiquitination can play a dual role [24–26]. In the present review, we will mainly address the acetylation and methylation of histones, which consist of the addition of NH_3 groups through the action of Histone Acetyl Transferases (HAT) and CH_3 groups by the Histone Methyltransferases (HMTs) in specific residues. Cooperation between DNA methylation and histones modifications induce the recruitment of other chromatin modifiers to promote an active or inactive conformation of chromatin [16] (Table 1).

Table 1. Relationship between the epigenetic modifications and the state of the chromatin.

Spot of Epigenetic Modification	Potentially Active Chromatin	Potentially Inactive Chromatin
DNA	DNA methylation outside CpG islands	DNA methylation in CpG islands of regulatory regions
Histones	Acetylated Unmethylated H3K4 Methylation	Deacetylated Methylated H3K4 unmethylation
Chromatin conformation	Open	Condensed Constitutive Heterochromatin

1.2.3. Non-Coding RNAs

Non-coding RNAs (ncRNAs) are a newly identified group of epigenetic regulators that can fine-tune gene expression without altering the DNA sequence. Prior to the discovery of their regulatory capacity, ncRNAs were considered junk sequences accumulated during evolution, since they occupy regions of the genome with no apparent coding function [27]. However, the Encyclopedia of DNA Elements (ENCODE) project contributed to the classification of non-coding RNAs with previously known functions (ribosomal or transcriptional RNA) [28] but also to the identification of novel ncRNAs [29,30]. Nowadays, transcriptional ncRNAs are classified into small ncRNAs and long ncRNAs (lncRNAs). Small ncRNAs can be divided into microRNAs (miRNAs), PIWI (P-element Induced Wimpy)-interfering RNAs (piRNAs), and small interfering RNAs (siRNAs).

2. Implication of Epigenetics in the Development of Cancer

A number of studies have shown that cancer cells experience global changes in chromatin that first affect the whole epigenome through a process of general hypomethylation leading to the genomic instability, and second affect the loss of function by the hypermethylation of specific tumor suppressor genes that regulate the signaling pathways involved in the cell differentiation process, such as *APC*, *GATA-4*, or *p16*, allowing the clonal growth and the abnormal survival of cells [31]. In hematological malignancies (leukemia, lymphomas, and myelomas) mutations in epigenetic modifiers such as *JAK2*, *DNMT3A*, *IDH2*, or *EZH2* are common hallmarks of these diseases [32–34]. This leads to alterations of the chromatin structure and the silencing of a number of genes such as *p15*, *RASSF1A* [33,34], *TET2*, *CYP1B1* [35], *PDZD2*, *CSDA* [36], *DAPK1*, *ZAP70* [37], *TERT*, or *TWIST2* [38] to promote the growth of cancer cells in these malignancies. The involvement of DNA methylation and histones modifications in cancer is a well-known event that has been long studied in the last decades [39–44]. However, the implication of non-coding RNAs and their regulation in cancer is still largely unknown, although in the past years this knowledge has increased considerably [45,46]. Therefore, in the present section, we will analyze the implication of non-coding RNAs in the development of solid tumors.

2.1. microRNAs

microRNAs are a group of non-coding small RNAs (measuring 19–23 nucleotides in length) that regulate gene expression through posttranscriptional regulation without altering their DNA sequence. Regulation occurs through miRNA binding to the 3′-untranslated region (3′UTR) of the messenger RNA (mRNA) of a target gene [47].

miRNAs are transcribed as primary microRNAs (pri-miRNAs) by RNA polymerase II from DNA. Afterwards, this molecule suffers two endonucleasic cuts mediated by the (1) Drosha and the Di George Syndrome critical region 8 gene (DGCR8) protein inside the nucleus [48], and (2) Dicer in the cytoplasm. The Drosha cut generates a hairpin of 60–100 nucleotides that is termed the miRNA precursor (pre-miRNA), which is exported to the cytoplasm by the nuclear membrane transporter 'Exportin-5'. Once in the cytoplasm, Dicer cuts this structure to generate a double-strand RNA molecule of 19–23 nucleotides. The RISC complex (RNA-Induced Silencing Complex) selects one of the strands, which provokes the degradation of the other one and searches for the homology region in the 3′UTR of a mRNA to block its transcription or favor its degradation [47].

miRNAs were first related to cancer in 2002; those that were downregulated were defined as tumor suppressor miRNAs, such as the miR-15a/16-1 cluster in chronic lymphocytic leukemia [49]. In addition, there is another type of microRNAs, such as the miR-17-92 cluster, whose induction increases cell proliferation, survival, and tumor angiogenesis. The gaining or loss of these miRNAs can increase or decrease the activity of several signaling pathways in cancer cells [50]. Moreover, miRNAs can be involved in epigenetic regulation through the activation or inactivation of DNA-methyltransferases. An example of this is the miR-29 family, which is a well-studied tumor-suppressive microRNA that targets the DNA-methyltransferases *DNMT1*, *DNMT3A* and *DNMT3B* [51] in a number of tumors such as ovarian [52], lung [53], liver [54], melanoma [55], and also hematological malignancies [56,57]. Some well-known tumor suppressors and oncogenes, such as c-MYC or p53, regulate the expression of miRNAs to promote cancer progression. For instance, the oncogene c-MYC activates miR-17, miR-19a, or miR-9 to promote cell cycle progression, inhibition of apoptosis, or metastasis (reviewed in [58]). The loss of p53 also promotes cell cycle progression, cell survival, or stemness phenotypes through the p53-activated miRNAs miR-34, miR-200, miR-15a/16-1 or miR-145 (reviewed in [59]). Although still not much is known about gene and miRNA epigenetic regulation in cancer and their implications for chemotherapy response, another regulatory mechanism of miRNA expression is the DNA methylation of CpGs close to the sequence of the miRNAs of intronic regions or even in regulatory promoter regions of the gene in which they are located [59–61].

2.2. piRNAs and siRNAs

PIWI-interfering RNAs and small interfering RNAs are groups of small non-coding RNAs of approximately 24–31 and 20–22 nucleotides long, respectively. The main difference between them resides in its processing: siRNAs mature from a double-strand RNA precursors such as miRNAs, whereas piRNAs are processed from single-strand RNA precursors [62]. piRNAs bind to PIWI proteins and siRNAs form the RISC (RNA interfering silent complex) complex. Both piRNA–PIWI [63] and RISC complexes [64,65] can transcriptionally repress gene expression by recruiting the chromatin silencing machinery (methyltransferases and deacetylases) or directly bind to the target mRNA by sequence complementary to induce post-transcriptional gene silencing.

piRNAs were initially discovered a short RNAs in Drosophila that inhibit the expression of the Stellate gene in a germ line [66]. However, follow-up studies showed that piRNA elements are conserved along evolution and also control biological processes in somatic cells (reviewed in [62]). Moreover, recent evidences suggest that piRNAs are related with human cancers. For example, piR-651 [67,68] and piR-55490 [69] promote tumor cell proliferation in lung cancer, piR-36712 [70] and piR-021285 [71] suppress cell proliferation and invasion in breast cancer, piR-1245 induces tumor growth and is a poor prognostic biomarker in colorectal cancer [72], and piR-52207 stimulates tumorigenesis in ovarian cancer [73].

siRNAs were discovered in the 1990s, after the injection of transgenes into Petunia plants [74] and Double-stranded (ds) RNA specific to a genomic region into *C. elegans* [75] that resulted in RNA interference (RNAi) associated gene expression changes. A few years later, a transient silencing with synthetic exogenous siRNA in mammalian cells was observed [76], confirming their existence and their gene-repressive function. Since then, the increasing knowledge about siRNAs has been directed more toward the development of the siRNA as therapeutic tools rather than as biomarkers, which have supposed a great advance for the understanding and therapeutics of cancer disease (reviewed in [77,78]).

2.3. Long Non-Coding RNAs

Long non-coding RNAs (lncRNAs) are RNA transcripts of more than 200 nucleotides in length that lack evident open reading frames [79]. lncRNAs are transcribed at lower levels than mRNAs, most of them are poorly conserved along evolution, and their expression seems to be more cell and tissue-specific than mRNAs. As RNA molecules, they show a dual activity that allows them to interact with proteins and other nucleic acids to form complex structures to enhance their regulatory role. Their implication in gene expression regulation is wider than the one exerted by microRNAs; therefore, their regulation is also stricter [80,81].

lncRNAs can be classified according to their chromosomal location and type of regulation regarding their associated coding genes. Those lncRNAs encoded within the sequence of a coding gene are classified as "overlapping lncRNAs", including sense, antisense, bidirectional, exonic, and intronic lncRNAs. Conversely, lncRNAs located between two genes are termed "long intergenic non-coding RNAs" (lincRNAs) [82–85]. The action of lncRNAs can act in cis when they regulate the expression of another gene encoded in the 1–300 kb upstream region [84,85]. Trans-acting lncRNAs can also regulate genes that are encoded anywhere in the genome. In addition, lncRNAs whose function is exclusively limited to the nucleus are guiders of chromatin modifying elements—such as DNMTs, Polycomb Repressor Complex (PRC), and/or HATs—to repress or activate the transcription. There is another group of lncRNAs that exert their function in the cytoplasm to regulate the expression of mRNAs and microRNAs in different ways [30,82,86].

The first lncRNAs identified are crucial for the correct embryonic development, since they regulate the chromosome dosage compensation (Xist) and the genomic imprinting and silencing of maternal or paternal genes (*H19*) [11,87,88], which exemplifies the important role of lncRNAs regulating the normal functioning of the cell and the organism. Thus, alterations in lncRNAs contribute to the development and progression of human diseases, including cancer. In cancer, the metastasis-associated lung adenocarcinoma transcript 1 (*MALAT1*), which regulates mRNA splicing, is the most studied cancer-associated lncRNA [89]. The expression of *MALAT1* is upregulated in Non Small Cell Lung Cancer (NSCLC) and ovarian cancer metastatic tumors [90,91], promotes aggressive phenotypes [91–93], and can be used as a prognostic biomarker in stage I NSCLC [90]. In addition, MALAT1 is reported to be overexpressed in uveal melanoma [94], melanoma [95], hematological malignancies [96–98], and other tumor types [99]. Moreover, recent research have identified a number of lncRNAs such as PVT1 (Plasmacytoma Variant Translocation 1), HOTAIR (HOX Transcript Antisense RNA), GAS5 (Growth Arrest Specific-5), SAMMSON (Survival Associated Mitochondrial Melanoma Specific Oncogenic Non-Coding RNA), CASC15 (Cancer Susceptibility 15), or MEG3 (Maternally Expressed Gene 3) that promote cancer development and progression and can be used as biomarkers of the disease (reviewed in [100–103]). Given the role of epigenetic mechanisms in cancer development and progression, studying how changes in the epigenetic machinery promote aggressive phenotypes would contribute to the development of new therapeutic approaches and better biomarkers in the clinic.

3. Oxidative Stress and Its Effect in the Epigenetic Machinery to Promote Cancer

Oxidative stress is defined as an imbalance between reactive oxygen species production (ROS) and the response of antioxidant enzymes. The major signaling pathway that regulates oxidative stress is NRF2 (Nuclear factor erythroid 2-related factor 2)–KEAP1 (Kelch-like ECH-associated protein 1) [104–107]. An intracellular ROS increase triggers the activation of NRF2 by KEAP1 inhibition. Nrf2 is translocated to nucleus and dimerizes with other proteins, such as small MAFs (musculoaponeurotic fibrosarcomas), to bind to the Antioxidant Response Element (ARE) sequences in the DNA. The Nrf2–sMAF (small MAF family) dimer works similar to a DNA transcription factor, recognizing AREs and activating several antioxidants genes [108–110]. Alteration of the NRF2–KEAP1 pathway is one of the most common and most studied events in cancer regarding pro-oncogenic disorders of oxidative stress. Furthermore, the increase of ROS triggers a multitude of response in different signaling pathways such as MAPK (Mitogen-activated protein kinase) [111], NF-κB (nuclear factor kappa-light-chain-enhancer of activated B cells) [112], STAT3 (Signal transducer and activator of transcription 3) [113,114], or PPARγ (Peroxisome proliferator-activated receptor gamma) [115] that promote antioxidant gene expression, proliferation, and survival in response to oxidative stress, which allow cancer cells to progress. While the activation of these pathways can control antioxidant gene expression independently, all these signaling pathways are also interconnected, highlighting the complex regulatory network that controls the response to oxidative stress.

Increased oxidative stress derived from the augmented metabolism of cancer cells is a common event in many tumor types to promote and maintain its tumorigenic potential. In fact, oxidative stress produces genomic instability and genomic damage that can lead to tumorigenesis [5,116,117]. It has been shown that the hypoxic state of tumor cells increases the oxidative stress situation, which leads to structural and epigenetic changes mediated by the Hypoxia-Inducible Factor (HIF)-1 [118,119]. In another study, continuous exposure to the oxidative stress of non-tumoral kidney cells results in malignant transformation [120]. Oxidative stress leads to an expression imbalance both at the level of histones (HDAC1, HMT1, and HAT1) and of epigenetic regulators (DNMT1, DNMT3a, and MBD4) in these cells. However, the acquired tumorigenic potential of these non-tumoral kidney cells decreased after treatment with the DNA demethylating agent 5-aza-2′-deoxycytidine [120], which supports the notion of an implication of the epigenetic machinery in tumor development. Moreover, several studies indicate a link between glutathione (GSH) metabolism and the control of epigenetics mechanisms at different levels. GSH is an important antioxidant enzyme that intervenes in several biological processes. Alterations in GSH synthesis or GSH depletion produce global DNA hypomethylation that could be due a decrease of S-adenosylmethionine (SAM) [121,122], which is a methyl group donor required for the action of DNMTs and HMTs [122,123]. Before DNMTs and HMTs obtain methyl groups from SAM, it suffers a catalytic transformation from methionine into SAM through the methionine adenosyltransferase (MAT) [122,123]. Both MAT and methionine synthase (MS) are very sensitive to oxidative stress and the balance of GSH synthesis, which explains a low activity of methyl transferase enzymes and a decrease of the genomic methylation level in redox imbalance situations (reviewed in [121,122]. All these changes in the oxidative state of the cell lead to the modulation of the epigenetic machinery at every level and alterations that promote tumor development.

3.1. Effect over DNA Methylation

Changes in DNA methylation derived from oxidative stress are mainly due to alterations of the activity and function of DNMTs (see above). Several studies suggest that the induction of oxidative stress mediated by hydrogen peroxide increases the activity of DNMT1 and its binding to the promoters of tumor suppressor genes as *RUNX3* [118]. In addition, hydroxyl radicals promote a global hypomethylation due to the interference in DNMTs–DNA binding capacity [124]. In addition, a high oxidative stress state induces changes in the catalytic cycle of iron and therefore to the inhibition of DNA demethylases of the TET family, thus increasing the levels of DNA methylation [119].

Moreover, high oxidative stress situations induced by ROS production increase 8-hydroxydeoxyguanosine (8-OHdG) levels in some cancer types. 8-OHdG triggers a conformational modification that changes the chromatin active state to chromatin repressive state. Therefore, 8-OHdG could promote tumorigenesis due to changes in the methylation patterns of tumor suppressor genes [125]. In addition, 8-OHdG blocks DNMTs–DNA binding, leading to a global hypomethylation of the genome [126,127].

One element that contributes enormously to the oxidative stress and thus in alterations of the epigenetic machinery is the tobacco smoke [128]. Apart from being a potent carcinogen, tobacco smoke induces high levels of ROS that result in the development of diseases such as the chronic obstructive pulmonary disease (COPD), which also leads to the alteration of the DNA methylation patterns [128,129].

3.2. Effect over Histone Modifications

It is also well known that oxidative stress induce changes in the acetylation and methylation of histones as it acts over the enzymes that maintain the chromatin state (see above). Oxidative stress induced by hydrogen peroxide can recruit histone modifiers complexes to promoters of active tumor suppressor genes to inhibit them [118,119]. Despite oxidative stress affecting the posttranslational modifications of histones that regulate the chromatin, it does not act in the same way due to the different sensibility to the oxidative stress of the HMTs, HDMs, and HATs [119].

In a similar manner to DNA methylation, one of the most remarkable changes in histone deacetylases (HDAC), which reduce their activity, is produced as consequence of cigarette smoke [130]. COPD patients show a decrease in HDAC2 activity that increases acetylation in histones H3 and H4 of the *NF-κB* promoter and thus to the dysregulation of proinflammatory genes [128].

In addition, there is evidence of a clear interaction between HIF1α and several HDACs and lysine acetyltransferases (KATs) in hypoxia context. The HIF1-directed transcriptional response appears to be responsible, in part, for the increased stabilization of HIF1α due to the action of HDACs and KATs [131,132].

3.3. Effect over Non-Coding RNAs

Similarly, transcriptional regulation mediated by non-coding RNAs is also altered in several ways by oxidative stress, as it has been shown for DNA and histones modifications.

Currently, there are a number of microRNAs whose alteration on their expression pattern is due to changes in the cellular oxidative stress [133,134]. For example, miR-200c is upregulated in epithelial cells as a result of increasing ROS and leads to an increase of apoptotic and senescent cells through the action of its target gene *ZEB1* [135]. Other miRNAs whose expression is induced by transcription factors that are sensitive to increased levels of ROS are miR-27a/b through c-MYC [136], miR-200 and miR-506 through p53 [137,138], and miR-206 through p38 (reviewed in [134]). In addition, the processing of miRNAs from their primary form is regulated by DGRC8–Drosha complexes. Recent reports demonstrate that increased oxidative situations decrease the processing capacity of DGRC8, which relies on Fe(III) for its action, and therefore the downregulation of the corresponding mature miRNA [139,140].

There are several miRNAs involved in the NRF2–ARE detoxification pathway, some due to the direct targeting of NRF2 or its natural inhibitor KEAP1, and others due to the indirect action over genes that regulate this signaling pathway (Table 2). For instance, miR-101 inhibits the expression of NRF2 in breast cancer cells to enhance their sensitivity to oxidative stress and suppress proliferation [141]. miR-432-3p binds directly to the KEAP1 coding region, downregulating it and upregulating the transcription of downstream genes of the NRF2–ARE pathway in esophageal squamous cell carcinoma [142]. miR-7 relieves the oxidative stress of neuroblastoma cells by targeting KEAP1, which promotes an increased expression of NRF2 and the transcription of antioxidant genes such as *HO-1* and *GLCGM* [143]. miR-200a binds to the KEAP1 3′UTR sequence leading to the degradation of its mRNA in breast cancer cell [144], hepatocellular carcinoma cells [145], and esophageal squamous cell carcinoma cells [146] (Figure 1). However, further research is needed to fully understand the complex regulatory network between miRNAs and the KEAP1/NRF2 axis.

Table 2. MicroRNAs involved in the antioxidant response in cancer disease.

Name	Cancer TYPE	Target	Effect over Antioxidant Response	Reference
MIR-101	Breast	*NRF2*	downregulated	[141]
MIR-28	Breast	*NRF2*	downregulated	[147]
MIR-153	Breast	*NRF2*	downregulated	[148]
MIR-432-3P	Esophageal Squamous	*KEAP1*	upregulated	[142]
MIR-200A	Breast, Hepatocelullar, Esophageal squamous	*KEAP1*	upregulated	[144–146]
MIR-23A	Leukemic	*KEAP1*	upregulated	[149]

Table 2. *Cont.*

Name	Cancer TYPE	Target	Effect over Antioxidant Response	Reference
MIR-7	Neuroblastoma	*KEAP1*	upregulated	[143]
MIR-7	Non small cell lung	*MAFG*	downregulated	[150]
MIR-148B	Endometrial	*ERMP1*	downregulated	[151]
MIR-500A-5P	Breast	*TXNRD1* and *NFE2L2*	downregulated	[152]
MIR-143	Colon	*SOD1*	downregulated	[153]
MIR-139-5P	Breast	*MAT2A*	downregulated	[154]
MIR-29B	Ovary	*SIRT1*	upregulated	[155]
MIR-31	Oral squamous	*SIRT3*	downregulated	[156]
MIR-33A	Glioma	*SIRT6*	downregulated	[157]
MIR-517A	Melanoma	JNK sig. path.	downregulated	[158]
MIR-144-3P	Lung	*NRF2*	downregulated	[159]
MIR-340	Hepatocelullar	*NRF2*	downregulated	[160]
MIR-141	Ovary	*KEAP1*	upregulated	[161]

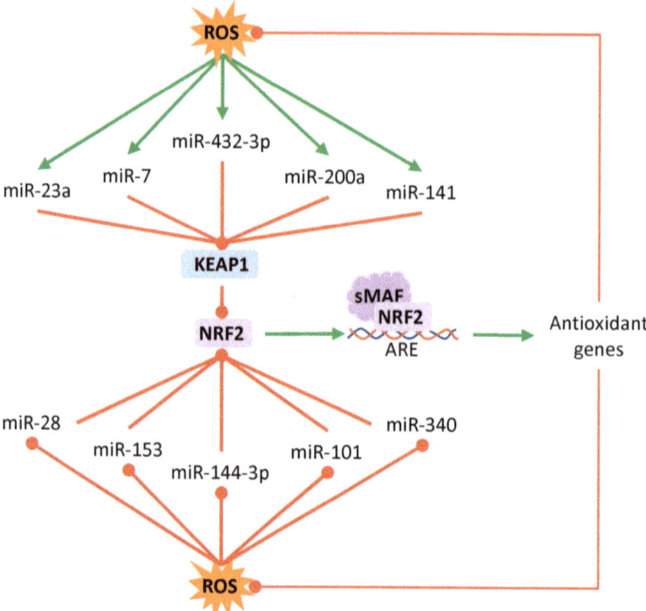

Figure 1. Regulation of NRF2/Keap1 by microRNAs. Oxidative stress can induce or repress some microRNAs that regulate the posttranscriptional expression of Keap1 or NRF2. Red lines with rounded end indicate inhibition. Green arrows indicate activation.

On the other hand, and despite that lncRNAs are a relative recent discovery, the alteration of these non-coding RNAs have also been linked to oxidative stress. The lncRNA "nuclear lung cancer associated transcript 1" (*NLUCAT1*) is upregulated in hypoxia in lung cancer cells through HIF2A, NF-κB, and NRF2 transcription factors [162]. This lncRNA promotes oncogenic abilities and cell survival in the presence of cisplatin treatment probably by upregulating the expression of antioxidant genes (*ALDH3A1*, *GPX2*, *GLRX*, and *PDK4*) through a mechanism still not understood [162]. In hepatocellular carcinoma cells, cell death produced by erastin, a strong inductor of ferroptosis, is mediated by *GABPB1-AS1* lncRNA [163]. This lncRNA blocks *GABPB1* mRNA recruitment to polysomes to decrease its protein expression and reduce the transcription of peroxidase proteins, which in turn increases ROS production and cell death [163]. The exposure of renal cell carcinoma cells to H2O2 to induce oxidative stress leads to a downregulation of the lncRNA 'secretory carrier membrane protein 1' (*SCAMP1*) and to apoptosis of the cells [164]. In this situation, *SCAMP1* acts as a competitive endogenous RNA (ceRNA) with *ZEB1* and *JUN* to sequester miR-429. Thus, the downregulation of *SCAMP1* increases the availability of miR-429 to target *ZEB1* and *JUN* and this way decreases cell viability [164]. This mechanism of "sponging" microRNAs through lncRNAs is a common feature to regulate gene expression by which RNA molecules sequester microRNAs to prevent them for targeting other RNAs. For instance, the expression of *HULC* and *H19* is triggered by oxidative stress to upregulate Interleukin-6 (*IL-6*) and the C-X-C chemokine receptor type 4 (*CXCR4*) sequestering Let-7a/b and miR-372/-373, respectively, to promote cholangiocarcinoma migration and invasion [165]. In multiple myeloma, *MALAT1* positively regulates NRF1 and NRF2 through the transcriptional inhibition of *Keap1* by EZH2 [96], which is a component of the PRC that induces H3K27me3 [166] through a mechanism still under study. In addition, *MALAT1* and EZH2 also interact to inhibit miR-29 in this disease [97], which is interesting due to the negative regulation of Keap1 through miR-29 [167] and its implication in the regulation of ROS [168], as it has been shown in other diseases. While all these reports clearly describe oxidative stress responses mediated by lncRNAs, as summarized in Table 3, a detailed regulation of these non-coding RNAs by reactive oxygen species and their role in cancer development and progression is still under study.

Table 3. Long non-coding RNAs (lncRNAs) regulated by the antioxidant response in cancer disease.

Name	Cancer TYPE	Intermediates/Effectors	Effect over Antioxidant Response	References
NLUCAT1	Lung cancer	HIF2A, NF-KB, NRF2	Upregulated	[162]
GABPB1-AS1	Hepatocellular carcinoma	GABPB1	Upregulated	[163]
SCAMP1	Renal cell carcinoma	miR-429/ZEB, JUN	Downregulated	[164]
HULC	Cholangiocarcinoma	Let-7a, Let-7b/IL-6	Upregulated	[165]
H19	Cholangiocarcinoma	miR-372, miR-373/CXCR4	Upregulated	[165]
UCA1	Non-small cell lung cancer	miR-495/NRF2	Upregulated	[169]
UCA1	Hepatocellular carcinoma	miR-184/OSGIN1	Upregulated	[170]

4. Chemotherapy-Induced Oxidative Stress: Epigenetics and Cisplatin Resistance

One of the main problems associated with cancer treatment is the development of resistance to different treatments. This resistant state in cancer cells can be explained by intrinsic and/or acquired mechanisms of desensitization to the drug's action. There are several ways proposed by which cancer cells develop drug resistance such as cellular plasticity, clonal selection due to pharmacological stress, signaling pathways plasticity, or the epigenetic mechanism, among others.

4.1. Cisplatin Resistance Mechanisms

Cisplatin (CDDP) is the first-line chemotherapeutic treatment in a multitude of tumors. Cisplatin is an alkylating agent that intercalates in DNA strands, forming adducts that cause genomic damage [171–174]. Moreover, one consequence of cisplatin administration is the production of high levels of ROS and nitrogen (RNS), which increase the oxidative stress in tumor cells that can promote alterations in the epigenetic machinery [175–177]. Although cisplatin is a potent inductor of cell apoptosis due to the increased intracellular oxidative imbalance, amongst other functions, the main limitation of its use is that the disease almost invariably progresses to a platinum-resistant state.

There are a number of events underlying the phenomenon of cisplatin resistance in cancer. One such event consists of alterations in DNA repair mediated by the activation of the epidermal growth factor receptor (EGFR) pathway resulting in cellular proliferation or the nuclear interaction of the receptor with proteins responsible for rejoining double-strand breaks [170]. Other proposed mechanisms include the reduced intracellular accumulation of cisplatin due to the overexpression of molecular ABC transporters that act by transporting the drug out of the cell, the activation of transcription factors that induce cell proliferation, the increase of antiapoptotic proteins, such as BCL-2, and alterations in the epigenetic regulatory machinery, mostly in DNA methyltransferases that modify the expression of a number of genes [172,178–180] (Figure 2). Despite the efforts to unravel the specific mechanisms for developing cisplatin resistance, it seems to be a multifactorial effect that includes several of the above-named mechanisms. Thus, understanding these molecular mechanisms of resistance development is a crucial step to improve the treatment of the disease.

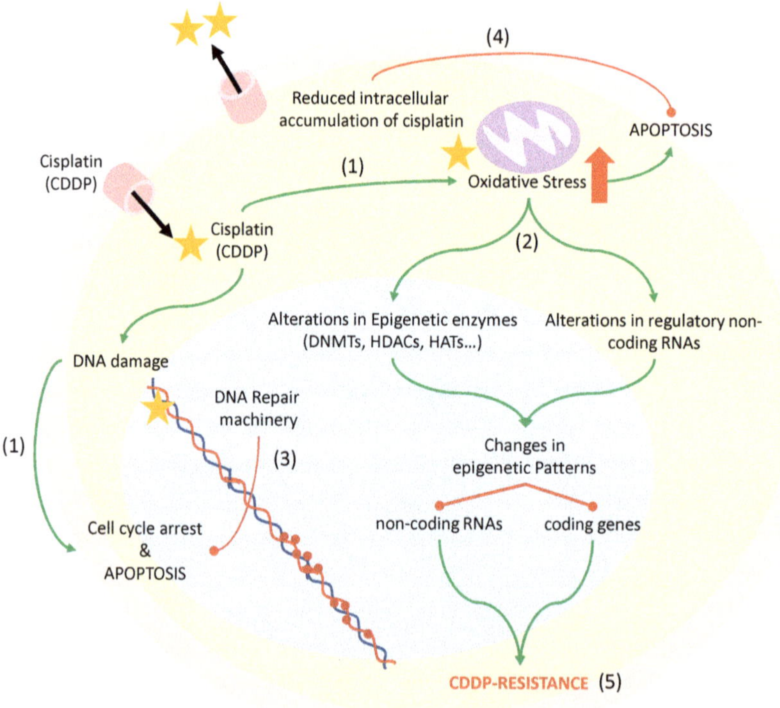

Figure 2. Effect of cisplatin inside the cell. Cisplatin is a platinum-derived drug that acts by generating aqueous species that bind to the N7 position in the guanine and causes inter and intra-strand cross-links in DNA that activate the mechanisms of cell cycle arrest or programmed cell death (1). In addition, cisplatin increases the oxidative stress of the cells, which leads to apoptosis (1), but also to alterations in the epigenetic enzymes and regulatory non-coding RNAs that change the epigenetic patterns (2). An increase of the DNA repair machinery (3), the reduced intracellular accumulation of cisplatin (4), or changes in the epigenetic machinery as a result of oxidative stress (5) lead to the development of cisplatin resistance and thus to the therapeutic limitation of its use. Red lines with rounded ends indicate inhibition. Green arrows indicate activation.

4.2. DNA Methylation and Histone Modifications

As we mentioned above, cisplatin-resistance development is related to the activation of DNMTs, whose activity can be modified due to the oxidative stress produced by cisplatin treatment. The process of DNA methylation in tumor cells leads to the silencing of specific genes that are crucial in normal conditions for the correct functioning of the cells. This silencing can be direct, if hypermethylation occurs in the promoter region of the gene that becomes inactive, or indirect in those cases where hypermethylation is on regulatory regions of coding and non-coding genes, silencing them and therefore increasing the expression of their target genes.

In fact, the transcriptional silencing of some genes, such as the arginine–succinate–synthetase, mediated by the hypermethylation of the CpG dinucleotides of its promoter, is a frequent epigenetic event associated to the development of cisplatin resistance in ovarian cancer [181]. Other studies describe the loss of expression of *IGFBP-3* (Insulin-like Growth Factor Binding Protein-3) in NSCLC as an effect of cisplatin administration. The silencing of this gene is produced by the hypermethylation of its promoter region in cisplatin-resistant cancer cells and leads to the development of cisplatin resistance through the activation of the Insulin-like growth factor receptor (IGF-IR)/AKT pathway [182,183]. There is increasing knowledge about the genes whose promoters are hypermethylated in cancer and that are related to cisplatin resistance as a consequence of the epigenetic silencing that they are suffering. However, the number of coding and non-coding genes identified up to date remains limited [184].

Alterations in the expression in histones deacetylases and demethylases can also contribute to the development of a cisplatin-resistant state in some tumor types. An example of this occurs in NSCLC in which the increased expression of these enzymes, the histone deacetylase-6 (HDAC6) specifically, decreases sensitivity to cisplatin through the reduction of apoptosis and prevention of DNA damage [185]. On the other side, oxidative stress induced by cisplatin administration leads to changes in histone demethylases, altering the methylation pattern of histones and being a silencing mechanism in cancer [183]. Moreover, cisplatin and other chemotherapeutic regimens also promote ROS induction and ubiquitination of the histone variant H2AX, which determine the sensitivity to the drug [186], reinforcing the importance on histones modifications in redox balance.

The interest of studying the link between epigenetic modifications of non-coding RNAs regulatory regions and the development of platinum-resistant phenotypes has increased in the recent past years. Therefore, in the next sections, we will discuss how alterations of non-coding RNAs promote resistance to cisplatin treatment.

4.3. microRNAs

One of the regulatory mechanisms of miRNAs expression is their silencing by the methylation of their regulatory regions, thereby increasing the expression of their target genes [45,187]. One of the first miRNAs identified to be regulated by this mechanism was miR-200c, whose hypermethylation of its regulatory region is responsible for the downregulation of this miRNA and therefore the subsequent development of resistance to chemotherapy in NSCLC cell lines [188]. The silencing of miR-493 by DNA hypermethylation is also involved in promoting cisplatin resistance in lung cancer cells by increasing the expression of its target gene *TCRP1* [189], which is a gene linked to cisplatin resistance in lung, ovarian, and tongue cancer cells [190,191].

There are also a number of miRNAs that regulate the expression of KEAP1/NRF2 to control cisplatin response. miR-144-3p is downregulated in lung cancer to increase *NRF2* expression and promote better cell survival in the presence of cisplatin treatment [159]. In cisplatin-resistant hepatocellular carcinoma cell lines, miR-340 is downregulated, which increases NRF2 activity, thus promoting resistance to the treatment [160]. In addition, miR-141, which targets *KEAP1*, is upregulated in ovarian cancer cell lines to promote cisplatin resistance [161]. Although the authors of this study conclude that NF-KB, and not the NRF2 pathway, is responsible for the observed phenotype, it might be worth studying the miR-141/KEAP1/NRF2 axis due to the known role of NRF2 in promoting cisplatin resistance in other tumor types.

miR-7/MAFG/ROS axis

Interestingly, some miRNAs can target the binding partners of NRF2 to affect cisplatin response. An example of this is the miR-7/*MAFG*/ROS axis that leads to cellular and biological responses that promote tumor development and progression in different tumor types. The main epigenetic mechanism involved in this aberrant alteration is the DNA methylation at the CpG Island surrounding the chromosomal location of miR-7 that has been associated with cancer progression [150]. In this study, the authors worked with cellular models of cisplatin resistance established after the chronic exposure to the cisplatin treatment of initially cisplatin-sensitive NSCLC and ovarian cancer cell lines [150,192]. These cisplatin-resistant cell lines share a common downregulation of miR-7 expression, compared to the parental sensitive cells, due to the hypermethylation of a CpG island located on its promoter [150]. Most importantly, lung adenocarcinoma patients harbor a hypermethylated miR-7 promoter when compared to healthy controls [193]. In addition, these patients show lower overall and progression-free survival rates, suggesting that the hypermethylation of miR-7 could be involved in the early establishment of the disease [193]. In fact, miR-7 hypermethylation constitutes a molecular event that accounts mainly at the expense of the emphysema patients, as analyzed from a huge COPD cohort. Patients with emphysema phenotype are considered the group of COPD patients that have a higher risk of developing lung cancer than other COPD phenotypes [194], indicating that the silencing of miR-7 through DNA hypermethylation might be one of the main determinants explaining the higher incidence of lung cancer in these patients [195]. Similarly, studies in ovarian cancer have also shown that the hypermethylation of miR-7 is associated with a worst response to platinum-derived chemotherapy, which is indicated by lower overall survival and progression-free survival rates when miR-7 promoter is methylated [150]. One of the effects of miR-7 silencing in cisplatin-resistant lung and ovarian cancer cell lines is the upregulation of its target gene MAFG (Musculoaponeurotic Fibrosarcoma Oncogene Family, protein G) [150,196] (Figure 3).

MAFG is a bZIP transcription factor that belongs to the small MAF family (sMAFs) of proteins. The sMAFs harbor a basic region motif for DNA binding and a leucine zipper motif for dimerization and are thought to have compensating roles. sMAFs can homodimerize with themselves or heterodimerize with other bZIP transcription factors such as the Cap'N'Collar (CNC) family or the Bach family to activate or repress gene expression in response to oxidative stress. Therefore, miR-7 silencing promotes low levels of ROS in cisplatin-resistant cancer cells due to the upregulation of *MAFG*. In fact, cisplatin-resistant NSCLC cell lines show lower ROS levels than the sensitive counterpart in response to cisplatin treatment (Figure 3). This effect can be overcome by sequestering MAFG through specific aptamers against the protein, as it results in a decrease of cell viability due to an increase of ROS production. In fact, knockout (KO) experiments in mice have shown that the small Mafs are essential for the activation of ARE-dependent genes. MafG KO mice showed mild thrombocytopenia and motor ataxia that improved with age, while MafK and MafF KO mice did not show an apparent phenotype [197], suggesting that the role of MafG is not well compensated by the other sMafs. In addition, the double KO MafG/MafF compromised the expression of *NQO1* and other oxidative stress-response genes [110]. Moreover, other authors have described the essential role of sMafs in regulating the expression of cytoprotective genes [110]. Importantly, ChIP-seq experiments to identify the binding sites of NRF2 and MAFG have demonstrated that, when heterodimerized, they can regulate genes involved in antioxidant and metabolic networks such as *NQO1*, *HMOX1*, *IDH1*, *PGD*, and *G6PD*, amongst others [109].

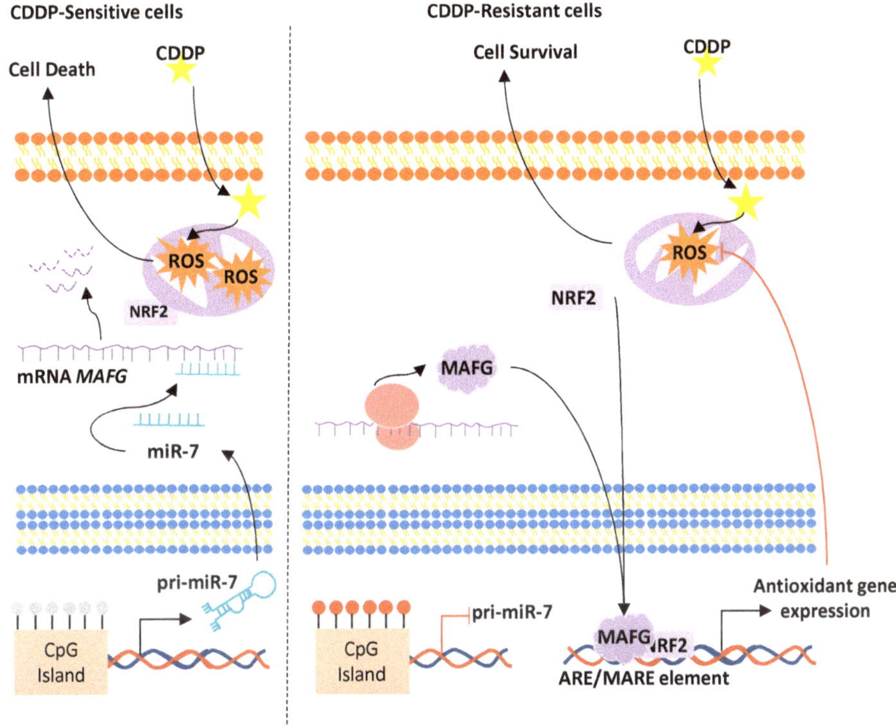

Figure 3. Proposed mechanism for the acquired resistance to cisplatin in NSCLC and ovarian cancer cells through miR-7 methylation and Musculoaponeurotic Fibrosarcoma Oncogene Family, protein G (MAFG) overexpression. Sensitive cells are represented on the left. The unmethylation of the miR-7 CpG Island allows its transcription, thus targeting the mRNA of MAFG, which promotes an increase of ROS and cell death. Cisplatin-resistant cells, right, harbor a methylated CpG island of miR-7, therefore allowing MAFG translation into protein, interaction with NRF2, and the transcription of antioxidant genes. CDDP, cisplatin; ROS, Reactive Oxygen Species.

Despite sMafs have been associated with cellular response, little is known about their implication in human pathologies. However, there are a number of reports that relate the overexpression of MAFG to cancer development and progression. Several studies demonstrate that MAFG promotes aggressive phenotypes in hepatic tumors [198–200], bladder cancer [201], and colon cancer [202,203].

In addition to its role in actively repressing or promoting gene expression, MAFG has recently been involved in the regulation of the methylator phenotype in colon and melanoma [204,205]. In both studies, hyperactivation of the MAPK pathway derived by oncogenic BRAF^{V600E} or EGFR^{G719S} prevents the ubiquitination of MAFG by phosphorylation of the S124 through ERK. This posttranslational modification leads to the formation of a protein complex orchestrated by MAFG stabilization and formed by BACH1, CDH8, and DNMT3B. The activation of these complexes leads to the increased methylation of CpG islands located in the tumor suppressive genes of both melanoma and colon cancer [204,205]. Therefore, the implications of MAFG in regulating gene expression are more complex than initially thought and worth studying in the future in order to identify potential biomarkers and therapeutic targets. Most importantly, since the KO mice of MAFG did not show a strong phenotype [197], targeting MAFG as a therapeutic approach, as it has been previously done to restore cisplatin sensitivity in lung cancer cells through DNA aptamers [196], provides promising results and potential therapeutic strategies in the future.

Apart from being stabilized by a posttranslational regulation, MAFG expression is also regulated at different levels. The ChIP-seq experiment mentioned above also confirmed that *MAFG* transcription can be regulated by NRF2 [109], which is an observation that was previously reported in mouse [206], suggesting a regulatory feedback between MAFG and NRF2 in response to oxidative stress in order to cope with high levels of ROS. However, MAFG can also be regulated at the post-transcriptional level by microRNAs. Although the coding sequence of MAFG is relatively small, the 3'UTR is over 4000 nucleotides long and over than 50 miRNAs are predicted to regulate MAFG. Some reported examples include miR-218 in smoking-induced disease processes in lung [207] and miR-7 in NSCLC and ovarian cancer cells [150].

Interestingly, as we mentioned above, miR-7 can also regulate the expression of genes that promote or reduce oxidative stress in the cell. One example is the regulation of *KEAP1* in neuroblastoma cells [143]. In these cells, miR-7 relieves oxidative stress by targeting *KEAP1*, which promotes an increased expression of NRF2 at the posttranslational level, which will promote the transcription of antioxidant genes such as *HO-1* and *GLCGM*. Besides, miR-7 can regulate the expression of proteins that promote oxidative stress situations. A-synuclein is a protein involved in synaptic activity through the regulation of vesicle docking, fusion, and neurotransmitter release [208–210]. One of the hallmarks of Parkinson's disease is the aggregation of a-syn in neurons to cause an intracellular toxic burden. This promotes mitochondrial dysfunction, the decreased activity of the superoxide dismutase 1 (*SOD1*), and low GSH, which in turn results in increased ROS production [211,212]. miR-7, whose expression is mostly present in neurons, can reduce a-syn expression by targeting the 3'UTR region of the mRNA, protecting the cells against the a-syn-mediated proteasome impairment and susceptibility to oxidative stress [209]. In addition, the role of miR-7 regulating a-syn has been studied in the ischemic brains of rodents. In this case, cerebral ischemia in rodents decreases miR-7 expression, which induces α-Syn expression and promotes cell death. Interestingly, the administration of an miR-7 mimic in pre-ischemic rats has a neuroprotective effect and decreased brain damage in post-ischemic brains, while these effects were lost in a-syn-/- mice [208]. Although these reports are not focused on cancer, due to the important role of miR-7 as a tumor suppressor (reviewed in [213]), it would be worth studying its role in regulating oxidative stress through these proteins in cancer.

4.4. Long Non-Coding RNA

Different CpG island methylation patterns have also been described between different types of lncRNAs depending on their chromosomal location in several tumor types as a result of chronic treatment with cisplatin [112]. However, the specific role of these silenced lncRNAs in cisplatin resistance remains unknown. Recent evidence suggests that lncRNAs are involved in chemoresistance to various anticancer therapies. One example is HOTTIP, an lncRNA regulating 5' HOXA gene transcription, which has been associated with cell proliferation, invasion, and chemoresistance in osteosarcoma, liver, and pancreatic cancers [113,114]. The lncRNAs UCA1 and ROR have been associated with the resistance of cancer cells to platinum-based treatments in bladder [115] and nasopharyngeal cancers [116], respectively. In addition, some miRNAs are regulatory intermediates between UCA1 and oxidative stress produced by cisplatin [117] or arsenic treatment [96] in which UCA1 acts as ceRNA with other microRNA target genes. In this context, UCA1 upregulation promotes a competing situation for miR-495 and miR-184 with NRF2 [117] and 'oxidative stress induced growth inhibitor 1' (OSGIN1) [96], respectively, to upregulate their expression and promote a better survival to drug treatments of NSCLC cells [117] and **hepatocellular carcinoma** (HCC) cells [96]. Although these reports show the implication of lncRNAs in promoting drug-resistant phenotypes, the specific mechanism by which lncRNAs regulate chemotherapy resistance is yet a novel field for exploration.

5. Conclusions and Future Directions

The relevance of epigenetic mechanisms in cancer has increased in the last years considerably. DNA methylation, histone modifications, and ncRNA are crucial elements that regulate both oncogenes and tumor suppressor genes. In this review, we bring out the importance of the relationship between these epigenetics mechanisms and oxidative stress in cancer disease. Indeed, when an increase of oxidative stress occurs, different pathways are activated—among them, the regulation of epigenetic mechanisms, in order to activate antioxidant response. This happens to maintain a high metabolism, which is a characteristic of cancer cells, and also to avoid apoptosis induced by oxidative stress and genomic damage, for example, in the development of therapy resistance.

Specifically, the way in which miRNAs and lncRNAs regulate key oxidative stress genes is a promising horizon that we can approach to lead anticancer therapies in the future. An example of this is the miR-7/MAFG axis regulation in NSCLC and ovarian cancer. miR-7 methylation is able to modulate the antioxidant response over ROS production by *MAFG* downregulation. Furthermore, MAFG seems to be leading a methylator program of specific genes in colorectal and melanoma cancer, so that its increase could affect the methylation profile of different tumor suppressor genes, promoting the cancer cell survival. Interestingly, ROS can both trigger *MAFG* transcription, to respond to oxidative stress, but it also can promote the MAPK pathway hyperactivation that leads to the stabilization of MAFG, thus suggesting a new molecular mechanism that changes the epigenetic patterns through ROS/MAFG.

Advances and the development of new technologies would allow in the future targeting these elements that are critical for the cell. For example, the identification of aptamers that block MAFG function has opened new research directions for re-sensitizing cisplatin-resistant cells in vitro. Although other aptamers have been used in preclinical models to target cellular membrane receptors, targeting MAFG would be challenging, as it mainly located in the nucleus. Therefore, the development of new delivery approaches will be needed to use MAFG aptamers as a functional therapy. For example, the use of nanoparticles would be a powerful approach to deliver these inhibiting molecules, as it has been shown for microRNAs.

Although the current review focused on ncRNAs and their effectors in response to oxidative stress in solid tumors, it is important to highlight the crucial role of the other well-known epigenetic modifications of DNA and histones, as well as alterations of the enzymes that affect these modifications, not only in solid tumors but also in hematological malignancies. There are a substantial number of epigenetic modifiers that are approved for their use in cancer treatment [214,215]. Inhibitors of DNMTs, HDACs, or other chromatin modifier complexes, such as EZH2 inhibitors, have increased the variety of therapeutic approaches for the treatment of cancer. In addition, in the past years, the use of miRNA mimics in clinical trials [216,217], such as miR-29 or mir-34, for the treatment of cancer and other diseases has opened new possibilities for therapeutic strategies. In line with this, another potential approach to re-sensitize cisplatin-resistant cells would be to overexpress miR-7 agonists through nanoparticles. Although these approaches have yet to be studied, their feasibility has proven to be effective with other microRNAs.

Conclusively, understanding the relationship between oxidative stress and epigenetics mechanisms is crucial to both understanding cancer development and gaining insight into the cancer disease, to finally allow new innovative approaches through the use of specific therapeutic strategies.

Author Contributions: I.I.-d.-C.: Conception design and draft, funding acquisition; Á.G.-G. and O.V.: wrote the manuscript; O.V., Á.G.-G. and I.I.-d.-C.: All authors reviewed and/or revised the manuscript. I.I.-d.-C. and O.V.: Approved the final version; O.V.: Agreed to be accountable for all aspects of the work in ensuring that questions related to the accuracy or integrity of any part of the work are appropriately investigated and resolved. All authors have read and agreed to the published version of the manuscript.

Funding: This research was funded by the Fondo de Investigación Sanitaria-Instituto de Salud Carlos III, grants numbers PI15/00186 and PI18/00050 and by MINECO, RTC-2016-5314-1 to I.I.C and the European Regional Development Fund/European Social Fund FIS [FEDER/FSE, Una Manera de Hacer Europa].

Conflicts of Interest: The authors declare no conflict of interest.

References

1. World Health Organ. 2018. Available online: http://www.who.int/cancer/en/ (accessed on 3 May 2020).
2. Bray, F.; Ferlay, J.; Soerjomataram, I.; Siegel, R.L.; Torre, L.A.; Jemal, A. Global cancer statistics 2018: GLOBOCAN estimates of incidence and mortality worldwide for 36 cancers in 185 countries. *CA Cancer J. Clin.* **2018**, *68*, 394–424. [CrossRef] [PubMed]
3. Hanahan, D.; Weinberg, R.A. The hallmarks of cancer. *Cell* **2000**, *100*, 57–70. [CrossRef]
4. Luo, J.; Solimini, N.L.; Elledge, S.J. Principles of cancer therapy: Oncogene and non-oncogene addiction. *Cell* **2009**, *136*, 823–837. [CrossRef] [PubMed]
5. Hanahan, D.; Weinberg, R.A. Hallmarks of cancer: The next generation. *Cell* **2011**, *144*, 646–674. [CrossRef] [PubMed]
6. Bishop, K.S.; Ferguson, L.R. The interaction between epigenetics, nutrition and the development of cancer. *Nutrients* **2015**, *7*, 922–947. [CrossRef]
7. Stirzaker, C.; Taberlay, P.C.; Statham, A.L.; Clark, S.J. Mining cancer methylomes: Prospects and challenges. *Trends Genet. TIG* **2014**, *30*, 75–84. [CrossRef]
8. Moore, L.D.; Le, T.; Fan, G. DNA methylation and its basic function. *Neuropsychopharmacol. Off. Publ. Am. Coll. Neuropsychopharmacol.* **2013**, *38*, 23–38. [CrossRef]
9. Zampieri, M.; Ciccarone, F.; Calabrese, R.; Franceschi, C.; Burkle, A.; Caiafa, P. Reconfiguration of DNA methylation in aging. *Mech. Ageing Dev.* **2015**, *151*, 60–70. [CrossRef]
10. Deplus, R.; Brenner, C.; Burgers, W.A.; Putmans, P.; Kouzarides, T.; de Launoit, Y.; Fuks, F. Dnmt3L is a transcriptional repressor that recruits histone deacetylase. *Nucleic Acids Res.* **2002**, *30*, 3831–3838. [CrossRef]
11. Brown, S.D. XIST and the mapping of the X chromosome inactivation centre. *Bioessays* **1991**, *13*, 607–612. [CrossRef]
12. Ng, K.; Pullirsch, D.; Leeb, M.; Wutz, A. Xist and the order of silencing. *EMBO Rep.* **2007**, *8*, 34–39. [CrossRef] [PubMed]
13. Boumil, R.M.; Ogawa, Y.; Sun, B.K.; Huynh, K.D.; Lee, J.T. Differential methylation of Xite and CTCF sites in Tsix mirrors the pattern of X-inactivation choice in mice. *Mol. Cell. Biol.* **2006**, *26*, 2109–2117. [CrossRef] [PubMed]
14. Reik, W.; Walter, J. Genomic imprinting: Parental influence on the genome. *Nat. Rev. Genet.* **2001**, *2*, 21–32. [CrossRef] [PubMed]
15. Portela, A.; Esteller, M. Epigenetic modifications and human disease. *Nat. Biotechnol.* **2010**, *28*, 1057–1068. [CrossRef] [PubMed]
16. Bannister, A.J.; Kouzarides, T. Regulation of chromatin by histone modifications. *Cell Res.* **2011**, *21*, 381–395. [CrossRef] [PubMed]
17. Yang, W.; Xia, Y.; Hawke, D.; Li, X.; Liang, J.; Xing, D.; Aldape, K.; Hunter, T.; Alfred Yung, W.K.; Lu, Z. PKM2 phosphorylates histone H3 and promotes gene transcription and tumorigenesis. *Cell* **2012**, *150*, 685–696. [CrossRef]
18. Lau, P.N.; Cheung, P. Histone code pathway involving H3 S28 phosphorylation and K27 acetylation activates transcription and antagonizes polycomb silencing. *Proc. Natl. Acad. Sci. USA* **2011**, *108*, 2801–2806. [CrossRef]
19. Dawson, M.A.; Bannister, A.J.; Gottgens, B.; Foster, S.D.; Bartke, T.; Green, A.R.; Kouzarides, T. JAK2 phosphorylates histone H3Y41 and excludes HP1alpha from chromatin. *Nature* **2009**, *461*, 819–822. [CrossRef]
20. Cohen-Armon, M.; Visochek, L.; Rozensal, D.; Kalal, A.; Geistrikh, I.; Klein, R.; Bendetz-Nezer, S.; Yao, Z.; Seger, R. DNA-independent PARP-1 activation by phosphorylated ERK2 increases Elk1 activity: A link to histone acetylation. *Mol. Cell* **2007**, *25*, 297–308. [CrossRef]
21. Moss, E.; Halkos, M.E. Cost effectiveness of robotic mitral valve surgery. *Ann. Cardiothorac. Surg.* **2017**, *6*, 33–37. [CrossRef]
22. Gill, G. Something about SUMO inhibits transcription. *Curr. Opin. Genet. Dev.* **2005**, *15*, 536–541. [CrossRef] [PubMed]
23. Yang, S.H.; Sharrocks, A.D. SUMO promotes HDAC-mediated transcriptional repression. *Mol. Cell* **2004**, *13*, 611–617. [CrossRef]
24. Meas, R.; Mao, P. Histone ubiquitylation and its roles in transcription and DNA damage response. *DNA Repair* **2015**, *36*, 36–42. [CrossRef]

25. Oya, E.; Nakagawa, R.; Yoshimura, Y.; Tanaka, M.; Nishibuchi, G.; Machida, S.; Shirai, A.; Ekwall, K.; Kurumizaka, H.; Tagami, H.; et al. H3K14 ubiquitylation promotes H3K9 methylation for heterochromatin assembly. *EMBO Rep.* **2019**, *20*, e48111. [CrossRef] [PubMed]

26. Kim, J.; Guermah, M.; McGinty, R.K.; Lee, J.S.; Tang, Z.; Milne, T.A.; Shilatifard, A.; Muir, T.W.; Roeder, R.G. RAD6-Mediated transcription-coupled H2B ubiquitylation directly stimulates H3K4 methylation in human cells. *Cell* **2009**, *137*, 459–471. [CrossRef]

27. Mattick, J.S.; Makunin, I.V. Non-coding RNA. *Hum. Mol. Genet.* **2006**, *15*, R17–R29. [CrossRef]

28. Gibb, E.A.; Brown, C.J.; Lam, W.L. The functional role of long non-coding RNA in human carcinomas. *Mol. Cancer* **2011**, *10*, 38. [CrossRef]

29. Djebali, S.; Davis, C.A.; Merkel, A.; Dobin, A.; Lassmann, T.; Mortazavi, A.; Tanzer, A.; Lagarde, J.; Lin, W.; Schlesinger, F.; et al. Landscape of transcription in human cells. *Nature* **2012**, *489*, 101–108. [CrossRef]

30. Ponting, C.P.; Oliver, P.L.; Reik, W. Evolution and functions of long noncoding RNAs. *Cell* **2009**, *136*, 629–641. [CrossRef]

31. Sanchez-Perez, I.; Murguia, J.R.; Perona, R. Cisplatin induces a persistent activation of JNK that is related to cell death. *Oncogene* **1998**, *16*, 533–540. [CrossRef]

32. McPherson, S.; McMullin, M.F.; Mills, K. Epigenetics in Myeloproliferative Neoplasms. *J. Cell. Mol. Med.* **2017**, *21*, 1660–1667. [CrossRef] [PubMed]

33. Machova Polakova, K.; Koblihova, J.; Stopka, T. Role of epigenetics in chronic myeloid leukemia. *Curr. Hematol. Malig. Rep.* **2013**, *8*, 28–36. [CrossRef] [PubMed]

34. Amodio, N.; D'Aquila, P.; Passarino, G.; Tassone, P.; Bellizzi, D. Epigenetic modifications in multiple myeloma: Recent advances on the role of DNA and histone methylation. *Expert Opin. Ther. Targets* **2017**, *21*, 91–101. [CrossRef] [PubMed]

35. Rahmani, M.; Talebi, M.; Hagh, M.F.; Feizi, A.A.H.; Solali, S. Aberrant DNA methylation of key genes and Acute Lymphoblastic Leukemia. *Biomed. Pharmacother. Biomed. Pharmacother.* **2018**, *97*, 1493–1500. [CrossRef]

36. Figueroa, M.E.; Lugthart, S.; Li, Y.; Erpelinck-Verschueren, C.; Deng, X.; Christos, P.J.; Schifano, E.; Booth, J.; van Putten, W.; Skrabanek, L.; et al. DNA methylation signatures identify biologically distinct subtypes in acute myeloid leukemia. *Cancer Cell* **2010**, *17*, 13–27. [CrossRef]

37. Guieze, R.; Wu, C.J. Genomic and epigenomic heterogeneity in chronic lymphocytic leukemia. *Blood* **2015**, *126*, 445–453. [CrossRef]

38. Mansouri, L.; Wierzbinska, J.A.; Plass, C.; Rosenquist, R. Epigenetic deregulation in chronic lymphocytic leukemia: Clinical and biological impact. *Semin. Cancer Biol.* **2018**, *51*, 1–11. [CrossRef]

39. Das, P.M.; Singal, R. DNA methylation and cancer. *J. Clin. Oncol.* **2004**, *22*, 4632–4642. [CrossRef]

40. Esteller, M. Cancer epigenomics: DNA methylomes and histone-modification maps. *Nat. Rev. Genet.* **2007**, *8*, 286–298. [CrossRef]

41. Koch, A.; Joosten, S.C.; Feng, Z.; de Ruijter, T.C.; Draht, M.X.; Melotte, V.; Smits, K.M.; Veeck, J.; Herman, J.G.; Van Neste, L.; et al. Analysis of DNA methylation in cancer: Location revisited. *Nat. Rev. Clin. Oncol.* **2018**, *15*, 459–466. [CrossRef]

42. Chi, P.; Allis, C.D.; Wang, G.G. Covalent histone modifications—miswritten, misinterpreted and mis-erased in human cancers. *Nat. Rev. Cancer* **2010**, *10*, 457–469. [CrossRef] [PubMed]

43. Morera, L.; Lubbert, M.; Jung, M. Targeting histone methyltransferases and demethylases in clinical trials for cancer therapy. *Clin. Epigenet.* **2016**, *8*, 57. [CrossRef] [PubMed]

44. Nacev, B.A.; Feng, L.; Bagert, J.D.; Lemiesz, A.E.; Gao, J.; Soshnev, A.A.; Kundra, R.; Schultz, N.; Muir, T.W.; Allis, C.D. The expanding landscape of 'oncohistone' mutations in human cancers. *Nature* **2019**, *567*, 473–478. [CrossRef]

45. Gulyaeva, L.F.; Kushlinskiy, N.E. Regulatory mechanisms of microRNA expression. *J. Transl. Med.* **2016**, *14*, 143. [CrossRef] [PubMed]

46. Ha, M.; Kim, V.N. Regulation of microRNA biogenesis. *Nat. Rev. Mol. Cell Biol.* **2014**, *15*, 509–524. [CrossRef]

47. Davis, B.N.; Hata, A. Regulation of MicroRNA Biogenesis: A miRiad of mechanisms. *Cell Commun. Signal. CCS* **2009**, *7*, 18. [CrossRef]

48. Landthaler, M.; Yalcin, A.; Tuschl, T. The human DiGeorge syndrome critical region gene 8 and Its D. melanogaster homolog are required for miRNA biogenesis. *Curr. Biol. CB* **2004**, *14*, 2162–2167. [CrossRef]

49. Calin, G.A.; Dumitru, C.D.; Shimizu, M.; Bichi, R.; Zupo, S.; Noch, E.; Aldler, H.; Rattan, S.; Keating, M.; Rai, K.; et al. Frequent deletions and down-regulation of micro- RNA genes miR15 and miR16 at 13q14 in chronic lymphocytic leukemia. *Proc. Natl. Acad. Sci. USA* **2002**, *99*, 15524–15529. [CrossRef]

50. Mendell, J.T.; Olson, E.N. MicroRNAs in stress signaling and human disease. *Cell* **2012**, *148*, 1172–1187. [CrossRef]

51. Kwon, J.J.; Factora, T.D.; Dey, S.; Kota, J. A Systematic Review of miR-29 in Cancer. *Mol. Ther. Oncolytics* **2019**, *12*, 173–194. [CrossRef]

52. Creighton, C.J.; Hernandez-Herrera, A.; Jacobsen, A.; Levine, D.A.; Mankoo, P.; Schultz, N.; Du, Y.; Zhang, Y.; Larsson, E.; Sheridan, R.; et al. Integrated analyses of microRNAs demonstrate their widespread influence on gene expression in high-grade serous ovarian carcinoma. *PLoS ONE* **2012**, *7*, e34546. [CrossRef]

53. Fabbri, M.; Garzon, R.; Cimmino, A.; Liu, Z.; Zanesi, N.; Callegari, E.; Liu, S.; Alder, H.; Costinean, S.; Fernandez-Cymering, C.; et al. MicroRNA-29 family reverts aberrant methylation in lung cancer by targeting DNA methyltransferases 3A and 3B. *Proc. Natl. Acad. Sci. USA* **2007**, *104*, 15805–15810. [CrossRef] [PubMed]

54. Parpart, S.; Roessler, S.; Dong, F.; Rao, V.; Takai, A.; Ji, J.; Qin, L.X.; Ye, Q.H.; Jia, H.L.; Tang, Z.Y.; et al. Modulation of miR-29 expression by alpha-fetoprotein is linked to the hepatocellular carcinoma epigenome. *Hepatology* **2014**, *60*, 872–883. [CrossRef]

55. Nguyen, T.; Kuo, C.; Nicholl, M.B.; Sim, M.S.; Turner, R.R.; Morton, D.L.; Hoon, D.S. Downregulation of microRNA-29c is associated with hypermethylation of tumor-related genes and disease outcome in cutaneous melanoma. *Epigenetics* **2011**, *6*, 388–394. [CrossRef] [PubMed]

56. Amodio, N.; Rossi, M.; Raimondi, L.; Pitari, M.R.; Botta, C.; Tagliaferri, P.; Tassone, P. miR-29s: A family of epi-miRNAs with therapeutic implications in hematologic malignancies. *Oncotarget* **2015**, *6*, 12837–12861. [CrossRef]

57. Mazzoccoli, L.; Robaina, M.C.; Apa, A.G.; Bonamino, M.; Pinto, L.W.; Queiroga, E.; Bacchi, C.E.; Klumb, C.E. MiR-29 silencing modulates the expression of target genes related to proliferation, apoptosis and methylation in Burkitt lymphoma cells. *J. Cancer Res. Clin. Oncol.* **2018**, *144*, 483–497. [CrossRef] [PubMed]

58. Bui, T.V.; Mendell, J.T. Myc: Maestro of MicroRNAs. *Genes Cancer* **2010**, *1*, 568–575. [CrossRef]

59. Hermeking, H. MicroRNAs in the p53 network: Micromanagement of tumour suppression. *Nat. Rev. Cancer* **2012**, *12*, 613–626. [CrossRef]

60. Furuta, M.; Kozaki, K.I.; Tanaka, S.; Arii, S.; Imoto, I.; Inazawa, J. miR-124 and miR-203 are epigenetically silenced tumor-suppressive microRNAs in hepatocellular carcinoma. *Carcinogenesis* **2010**, *31*, 766–776. [CrossRef]

61. Balaguer, F.; Link, A.; Lozano, J.J.; Cuatrecasas, M.; Nagasaka, T.; Boland, C.R.; Goel, A. Epigenetic silencing of miR-137 is an early event in colorectal carcinogenesis. *Cancer Res.* **2010**, *70*, 6609–6618. [CrossRef]

62. Iwasaki, Y.W.; Siomi, M.C.; Siomi, H. PIWI-Interacting RNA: Its Biogenesis and Functions. *Annu. Rev. Biochem.* **2015**, *84*, 405–433. [CrossRef] [PubMed]

63. Liu, Y.; Dou, M.; Song, X.; Dong, Y.; Liu, S.; Liu, H.; Tao, J.; Li, W.; Yin, X.; Xu, W. The emerging role of the piRNA/piwi complex in cancer. *Mol. Cancer* **2019**, *18*, 123. [CrossRef] [PubMed]

64. Hall, I.M.; Shankaranarayana, G.D.; Noma, K.; Ayoub, N.; Cohen, A.; Grewal, S.I. Establishment and maintenance of a heterochromatin domain. *Science* **2002**, *297*, 2232–2237. [CrossRef]

65. Martinez, J.; Patkaniowska, A.; Urlaub, H.; Luhrmann, R.; Tuschl, T. Single-stranded antisense siRNAs guide target RNA cleavage in RNAi. *Cell* **2002**, *110*, 563–574. [CrossRef]

66. Aravin, A.A.; Naumova, N.M.; Tulin, A.V.; Vagin, V.V.; Rozovsky, Y.M.; Gvozdev, V.A. Double-stranded RNA-mediated silencing of genomic tandem repeats and transposable elements in the D. melanogaster germline. *Curr. Biol. CB* **2001**, *11*, 1017–1027. [CrossRef]

67. Yao, J.; Wang, Y.W.; Fang, B.B.; Zhang, S.J.; Cheng, B.L. piR-651 and its function in 95-D lung cancer cells. *Biomed. Rep.* **2016**, *4*, 546–550. [CrossRef] [PubMed]

68. Li, D.; Luo, Y.; Gao, Y.; Yang, Y.; Wang, Y.; Xu, Y.; Tan, S.; Zhang, Y.; Duan, J.; Yang, Y. piR-651 promotes tumor formation in non-small cell lung carcinoma through the upregulation of cyclin D1 and CDK4. *Int. J. Mol. Med.* **2016**, *38*, 927–936. [CrossRef]

69. Peng, L.; Song, L.; Liu, C.; Lv, X.; Li, X.; Jie, J.; Zhao, D.; Li, D. piR-55490 inhibits the growth of lung carcinoma by suppressing mTOR signaling. *Tumour Biol. J. Int. Soc. Oncodev. Biol. Med.* **2016**, *37*, 2749–2756. [CrossRef]

70. Tan, L.; Mai, D.; Zhang, B.; Jiang, X.; Zhang, J.; Bai, R.; Ye, Y.; Li, M.; Pan, L.; Su, J.; et al. PIWI-interacting RNA-36712 restrains breast cancer progression and chemoresistance by interaction with SEPW1 pseudogene SEPW1P RNA. *Mol. Cancer* **2019**, *18*, 9. [CrossRef]

71. Fu, A.; Jacobs, D.I.; Hoffman, A.E.; Zheng, T.; Zhu, Y. PIWI-interacting RNA 021285 is involved in breast tumorigenesis possibly by remodeling the cancer epigenome. *Carcinogenesis* **2015**, *36*, 1094–1102. [CrossRef]

72. Weng, W.; Liu, N.; Toiyama, Y.; Kusunoki, M.; Nagasaka, T.; Fujiwara, T.; Wei, Q.; Qin, H.; Lin, H.; Ma, Y.; et al. Novel evidence for a PIWI-interacting RNA (piRNA) as an oncogenic mediator of disease progression, and a potential prognostic biomarker in colorectal cancer. *Mol. Cancer* **2018**, *17*, 16. [CrossRef]

73. Singh, G.; Roy, J.; Rout, P.; Mallick, B. Genome-wide profiling of the PIWI-interacting RNA-mRNA regulatory networks in epithelial ovarian cancers. *PLoS ONE* **2018**, *13*, e0190485. [CrossRef] [PubMed]

74. Jorgensen, R.A.; Cluster, P.D.; English, J.; Que, Q.; Napoli, C.A. Chalcone synthase cosuppression phenotypes in petunia flowers: Comparison of sense vs. antisense constructs and single-copy vs. complex T-DNA sequences. *Plant Mol. Biol.* **1996**, *31*, 957–973. [CrossRef]

75. Fire, A.; Xu, S.; Montgomery, M.K.; Kostas, S.A.; Driver, S.E.; Mello, C.C. Potent and specific genetic interference by double-stranded RNA in Caenorhabditis elegans. *Nature* **1998**, *391*, 806–811. [CrossRef] [PubMed]

76. Elbashir, S.M.; Harborth, J.; Lendeckel, W.; Yalcin, A.; Weber, K.; Tuschl, T. Duplexes of 21-nucleotide RNAs mediate RNA interference in cultured mammalian cells. *Nature* **2001**, *411*, 494–498. [CrossRef] [PubMed]

77. Singh, A.; Trivedi, P.; Jain, N.K. Advances in siRNA delivery in cancer therapy. *Artif. Cells Nanomed. Biotechnol.* **2018**, *46*, 274–283. [CrossRef]

78. Young, S.W.; Stenzel, M.; Yang, J.L. Nanoparticle-siRNA: A potential cancer therapy? *Crit. Rev. Oncol. Hematol.* **2016**, *98*, 159–169. [CrossRef]

79. Dey, B.K.; Mueller, A.C.; Dutta, A. Long non-coding RNAs as emerging regulators of differentiation, development, and disease. *Transcription* **2014**, *5*, e944014. [CrossRef]

80. Wilusz, J.E.; Sunwoo, H.; Spector, D.L. Long noncoding RNAs: Functional surprises from the RNA world. *Genes Dev.* **2009**, *23*, 1494–1504. [CrossRef]

81. Mercer, T.R.; Dinger, M.E.; Mattick, J.S. Long non-coding RNAs: Insights into functions. *Nat. Rev. Genet.* **2009**, *10*, 155–159. [CrossRef]

82. Wang, K.C.; Chang, H.Y. Molecular mechanisms of long noncoding RNAs. *Mol. Cell* **2011**, *43*, 904–914. [CrossRef] [PubMed]

83. Ma, L.; Bajic, V.B.; Zhang, Z. On the classification of long non-coding RNAs. *RNA Biol.* **2013**, *10*, 925–933. [CrossRef] [PubMed]

84. Guttman, M.; Rinn, J.L. Modular regulatory principles of large non-coding RNAs. *Nature* **2012**, *482*, 339–346. [CrossRef] [PubMed]

85. Chen, L.L. Linking Long Noncoding RNA Localization and Function. *Trends Biochem. Sci.* **2016**, *41*, 761–772. [CrossRef]

86. Fatica, A.; Bozzoni, I. Long non-coding RNAs: New players in cell differentiation and development. *Nat. Rev. Genet.* **2014**, *15*, 7–21. [CrossRef]

87. Bartolomei, M.S.; Zemel, S.; Tilghman, S.M. Parental imprinting of the mouse H19 gene. *Nature* **1991**, *351*, 153–155. [CrossRef]

88. Brannan, C.I.; Dees, E.C.; Ingram, R.S.; Tilghman, S.M. The product of the H19 gene may function as an RNA. *Mol. Cell. Biol.* **1990**, *10*, 28–36. [CrossRef]

89. Tripathi, V.; Ellis, J.D.; Shen, Z.; Song, D.Y.; Pan, Q.; Watt, A.T.; Freier, S.M.; Bennett, C.F.; Sharma, A.; Bubulya, P.A.; et al. The nuclear-retained noncoding RNA MALAT1 regulates alternative splicing by modulating SR splicing factor phosphorylation. *Mol. Cell* **2010**, *39*, 925–938. [CrossRef]

90. Ji, P.; Diederichs, S.; Wang, W.; Boing, S.; Metzger, R.; Schneider, P.M.; Tidow, N.; Brandt, B.; Buerger, H.; Bulk, E.; et al. MALAT-1, a novel noncoding RNA, and thymosin beta4 predict metastasis and survival in early-stage non-small cell lung cancer. *Oncogene* **2003**, *22*, 8031–8041. [CrossRef]

91. Zhou, Y.; Xu, X.; Lv, H.; Wen, Q.; Li, J.; Tan, L.; Sheng, X. The Long Noncoding RNA MALAT-1 Is Highly Expressed in Ovarian Cancer and Induces Cell Growth and Migration. *PLoS ONE* **2016**, *11*, e0155250. [CrossRef]

92. Formisano, S.; Brewer, H.B., Jr.; Osborne, J.C., Jr. Effect of pressure and ionic strength on the self-association of Apo-A-I from the human high density lipoprotein complex. *J. Biol. Chem.* **1978**, *253*, 354–359.

93. Hudson, A.M.; McMartin, C. Metabolism of tritiated 1-24 corticotrophin in peripheral tissues immediately after intravenous injection in the rat [proceedings]. *J. Endocrinol.* **1978**, *79*, 62P–63P. [PubMed]

94. Sun, L.; Sun, P.; Zhou, Q.Y.; Gao, X.; Han, Q. Long noncoding RNA MALAT1 promotes uveal melanoma cell growth and invasion by silencing of miR-140. *Am. J. Transl. Res.* **2016**, *8*, 3939–3946. [PubMed]

95. Li, F.; Li, X.; Qiao, L.; Liu, W.; Xu, C.; Wang, X. MALAT1 regulates miR-34a expression in melanoma cells. *Cell Death Dis.* **2019**, *10*, 389. [CrossRef] [PubMed]

96. Amodio, N.; Stamato, M.A.; Juli, G.; Morelli, E.; Fulciniti, M.; Manzoni, M.; Taiana, E.; Agnelli, L.; Cantafio, M.E.G.; Romeo, E.; et al. Drugging the lncRNA MALAT1 via LNA gapmeR ASO inhibits gene expression of proteasome subunits and triggers anti-multiple myeloma activity. *Leukemia* **2018**, *32*, 1948–1957. [CrossRef]

97. Stamato, M.A.; Juli, G.; Romeo, E.; Ronchetti, D.; Arbitrio, M.; Caracciolo, D.; Neri, A.; Tagliaferri, P.; Tassone, P.; Amodio, N. Inhibition of EZH2 triggers the tumor suppressive miR-29b network in multiple myeloma. *Oncotarget* **2017**, *8*, 106527–106537. [CrossRef]

98. Amodio, N.; Raimondi, L.; Juli, G.; Stamato, M.A.; Caracciolo, D.; Tagliaferri, P.; Tassone, P. MALAT1: A druggable long non-coding RNA for targeted anti-cancer approaches. *J. Hematol. Oncol.* **2018**, *11*, 63. [CrossRef]

99. Li, Z.X.; Zhu, Q.N.; Zhang, H.B.; Hu, Y.; Wang, G.; Zhu, Y.S. MALAT1: A potential biomarker in cancer. *Cancer Manag. Res.* **2018**, *10*, 6757–6768. [CrossRef]

100. Huarte, M. The emerging role of lncRNAs in cancer. *Nat. Med.* **2015**, *21*, 1253–1261. [CrossRef]

101. Bhan, A.; Soleimani, M.; Mandal, S.S. Long Noncoding RNA and Cancer: A New Paradigm. *Cancer Res.* **2017**, *77*, 3965–3981. [CrossRef]

102. Peng, W.X.; Koirala, P.; Mo, Y.Y. LncRNA-mediated regulation of cell signaling in cancer. *Oncogene* **2017**, *36*, 5661–5667. [CrossRef] [PubMed]

103. Schmitt, A.M.; Chang, H.Y. Long Noncoding RNAs in Cancer Pathways. *Cancer Cell* **2016**, *29*, 452–463. [CrossRef] [PubMed]

104. Betteridge, D.J. What is oxidative stress? *Metab. Clin. Exp.* **2000**, *49*, 3–8. [CrossRef]

105. Yamamoto, M.; Kensler, T.W.; Motohashi, H. The KEAP1-NRF2 System: A Thiol-Based Sensor-Effector Apparatus for Maintaining Redox Homeostasis. *Physiol. Rev.* **2018**, *98*, 1169–1203. [CrossRef]

106. DeNicola, G.M.; Karreth, F.A.; Humpton, T.J.; Gopinathan, A.; Wei, C.; Frese, K.; Mangal, D.; Yu, K.H.; Yeo, C.J.; Calhoun, E.S.; et al. Oncogene-induced Nrf2 transcription promotes ROS detoxification and tumorigenesis. *Nature* **2011**, *475*, 106–109. [CrossRef]

107. Chen, X.L.; Kunsch, C. Induction of cytoprotective genes through Nrf2/antioxidant response element pathway: A new therapeutic approach for the treatment of inflammatory diseases. *Curr. Pharm. Des.* **2004**, *10*, 879–891. [CrossRef]

108. Katsuoka, F.; Yamamoto, M. Small Maf proteins (MafF, MafG, MafK): History, structure and function. *Gene* **2016**, *586*, 197–205. [CrossRef]

109. Hirotsu, Y.; Katsuoka, F.; Funayama, R.; Nagashima, T.; Nishida, Y.; Nakayama, K.; Engel, J.D.; Yamamoto, M. Nrf2-MafG heterodimers contribute globally to antioxidant and metabolic networks. *Nucleic Acids Res.* **2012**, *40*, 10228–10239. [CrossRef]

110. Katsuoka, F.; Motohashi, H.; Ishii, T.; Aburatani, H.; Engel, J.D.; Yamamoto, M. Genetic evidence that small maf proteins are essential for the activation of antioxidant response element-dependent genes. *Mol. Cell. Biol.* **2005**, *25*, 8044–8051. [CrossRef]

111. Perez, S.; Rius-Perez, S.; Tormos, A.M.; Finamor, I.; Nebreda, A.R.; Talens-Visconti, R.; Sastre, J. Age-dependent regulation of antioxidant genes by p38alpha MAPK in the liver. *Redox Biol.* **2018**, *16*, 276–284. [CrossRef] [PubMed]

112. Reuter, S.; Gupta, S.C.; Chaturvedi, M.M.; Aggarwal, B.B. Oxidative stress, inflammation, and cancer: How are they linked? *Free Radic. Biol. Med.* **2010**, *49*, 1603–1616. [CrossRef]

113. Bourgeais, J.; Gouilleux-Gruart, V.; Gouilleux, F. Oxidative metabolism in cancer: A STAT affair? *JAK-STAT* **2013**, *2*, e25764. [CrossRef]

114. Jung, J.E.; Kim, G.S.; Narasimhan, P.; Song, Y.S.; Chan, P.H. Regulation of Mn-superoxide dismutase activity and neuroprotection by STAT3 in mice after cerebral ischemia. *J. Neurosci. Off. J. Soc. Neurosci.* **2009**, *29*, 7003–7014. [CrossRef]

115. Inoue, I.; Goto, S.; Matsunaga, T.; Nakajima, T.; Awata, T.; Hokari, S.; Komoda, T.; Katayama, S. The ligands/activators for peroxisome proliferator-activated receptor alpha (PPARalpha) and PPARgamma increase Cu2+,Zn2+-superoxide dismutase and decrease p22phox message expressions in primary endothelial cells. *Metab. Clin. Exp.* **2001**, *50*, 3–11. [CrossRef] [PubMed]

116. Schumacker, P.T. Reactive oxygen species in cancer cells: Live by the sword, die by the sword. *Cancer Cell* **2006**, *10*, 175–176. [CrossRef]

117. Aggarwal, V.; Tuli, H.S.; Varol, A.; Thakral, F.; Yerer, M.B.; Sak, K.; Varol, M.; Jain, A.; Khan, M.A.; Sethi, G. Role of Reactive Oxygen Species in Cancer Progression: Molecular Mechanisms and Recent Advancements. *Biomolecules* **2019**, *9*, 735. [CrossRef] [PubMed]

118. Nishida, N.; Kudo, M. Oxidative stress and epigenetic instability in human hepatocarcinogenesis. *Dig. Dis.* **2013**, *31*, 447–453. [CrossRef] [PubMed]

119. Kreuz, S.; Fischle, W. Oxidative stress signaling to chromatin in health and disease. *Epigenomics* **2016**, *8*, 843–862. [CrossRef] [PubMed]

120. Mahalingaiah, P.K.; Ponnusamy, L.; Singh, K.P. Oxidative stress-induced epigenetic changes associated with malignant transformation of human kidney epithelial cells. *Oncotarget* **2017**, *8*, 11127–11143. [CrossRef] [PubMed]

121. Cyr, A.R.; Domann, F.E. The redox basis of epigenetic modifications: From mechanisms to functional consequences. *Antioxid. Redox Signal.* **2011**, *15*, 551–589. [CrossRef]

122. Garcia-Gimenez, J.L.; Roma-Mateo, C.; Perez-Machado, G.; Peiro-Chova, L.; Pallardo, F.V. Role of glutathione in the regulation of epigenetic mechanisms in disease. *Free Radic. Biol. Med.* **2017**, *112*, 36–48. [CrossRef] [PubMed]

123. Lu, S.C. Regulation of glutathione synthesis. *Mol. Asp. Med.* **2009**, *30*, 42–59. [CrossRef] [PubMed]

124. Franco, R.; Schoneveld, O.; Georgakilas, A.G.; Panayiotidis, M.I. Oxidative stress, DNA methylation and carcinogenesis. *Cancer Lett.* **2008**, *266*, 6–11. [CrossRef] [PubMed]

125. Nishida, N.; Arizumi, T.; Takita, M.; Kitai, S.; Yada, N.; Hagiwara, S.; Inoue, T.; Minami, Y.; Ueshima, K.; Sakurai, T.; et al. Reactive oxygen species induce epigenetic instability through the formation of 8-hydroxydeoxyguanosine in human hepatocarcinogenesis. *Dig. Dis.* **2013**, *31*, 459–466. [CrossRef] [PubMed]

126. Udomsinprasert, W.; Kitkumthorn, N.; Mutirangura, A.; Chongsrisawat, V.; Poovorawan, Y.; Honsawek, S. Global methylation, oxidative stress, and relative telomere length in biliary atresia patients. *Sci. Rep.* **2016**, *6*, 26969. [CrossRef]

127. Ziech, D.; Franco, R.; Pappa, A.; Panayiotidis, M.I. Reactive oxygen species (ROS)–induced genetic and epigenetic alterations in human carcinogenesis. *Mutat. Res.* **2011**, *711*, 167–173. [CrossRef]

128. Arita, A.; Costa, M. Oxidative Stress and the Epigenome in Human Disease. *J. Genet. Genome Res.* **2014**, *1*, 5. [CrossRef]

129. Sundar, I.K.; Yao, H.; Rahman, I. Oxidative stress and chromatin remodeling in chronic obstructive pulmonary disease and smoking-related diseases. *Antioxid. Redox Signal.* **2013**, *18*, 1956–1971. [CrossRef]

130. Yao, H.; Rahman, I. Role of histone deacetylase 2 in epigenetics and cellular senescence: Implications in lung inflammaging and COPD. *Am. J. Physiol. Lung Cell. Mol. Physiol.* **2012**, *303*, L557–L566. [CrossRef]

131. Sabbagh, Z.; Vatanparast, H. Is calcium supplementation a risk factor for cardiovascular diseases in older women? *Nutr. Rev.* **2009**, *67*, 105–108. [CrossRef]

132. Lawless, M.W.; O'Byrne, K.J.; Gray, S.G. Oxidative stress induced lung cancer and COPD: Opportunities for epigenetic therapy. *J. Cell. Mol. Med.* **2009**, *13*, 2800–2821. [CrossRef] [PubMed]

133. Fuschi, P.; Maimone, B.; Gaetano, C.; Martelli, F. Noncoding RNAs in the Vascular System Response to Oxidative Stress. *Antioxid. Redox Signal.* **2017**. [CrossRef] [PubMed]

134. Lan, J.; Huang, Z.; Han, J.; Shao, J.; Huang, C. Redox regulation of microRNAs in cancer. *Cancer Lett.* **2018**, *418*, 250–259. [CrossRef] [PubMed]

135. Magenta, A.; Cencioni, C.; Fasanaro, P.; Zaccagnini, G.; Greco, S.; Sarra-Ferraris, G.; Antonini, A.; Martelli, F.; Capogrossi, M.C. miR-200c is upregulated by oxidative stress and induces endothelial cell apoptosis and senescence via ZEB1 inhibition. *Cell Death Differ.* **2011**, *18*, 1628–1639. [CrossRef] [PubMed]

136. Yang, H.; Li, T.W.; Zhou, Y.; Peng, H.; Liu, T.; Zandi, E.; Martinez-Chantar, M.L.; Mato, J.M.; Lu, S.C. Activation of a novel c-Myc-miR27-prohibitin 1 circuitry in cholestatic liver injury inhibits glutathione synthesis in mice. *Antioxid. Redox Signal.* **2015**, *22*, 259–274. [CrossRef] [PubMed]

137. Xiao, Y.; Yan, W.; Lu, L.; Wang, Y.; Lu, W.; Cao, Y.; Cai, W. p38/p53/miR-200a-3p feedback loop promotes oxidative stress-mediated liver cell death. *Cell Cycle* **2015**, *14*, 1548–1558. [CrossRef] [PubMed]

138. Yin, M.; Ren, X.; Zhang, X.; Luo, Y.; Wang, G.; Huang, K.; Feng, S.; Bao, X.; Huang, K.; He, X.; et al. Selective killing of lung cancer cells by miRNA-506 molecule through inhibiting NF-kappaB p65 to evoke reactive oxygen species generation and p53 activation. *Oncogene* **2015**, *34*, 691–703. [CrossRef] [PubMed]

139. Faller, M.; Matsunaga, M.; Yin, S.; Loo, J.A.; Guo, F. Heme is involved in microRNA processing. *Nat. Struct. Mol. Biol.* **2007**, *14*, 23–29. [CrossRef] [PubMed]

140. Barr, I.; Smith, A.T.; Chen, Y.; Senturia, R.; Burstyn, J.N.; Guo, F. Ferric, not ferrous, heme activates RNA-binding protein DGCR8 for primary microRNA processing. *Proc. Natl. Acad. Sci. USA* **2012**, *109*, 1919–1924. [CrossRef]

141. Yi, J.; Huang, W.Z.; Wen, Y.Q.; Yi, Y.C. Effect of miR-101 on proliferation and oxidative stress-induced apoptosis of breast cancer cells via Nrf2 signaling pathway. *Eur. Rev. Med. Pharmacol. Sci.* **2019**, *23*, 8931–8939. [CrossRef]

142. Akdemir, B.; Nakajima, Y.; Inazawa, J.; Inoue, J. miR-432 Induces NRF2 Stabilization by Directly Targeting KEAP1. *Mol. Cancer Res.* **2017**, *15*, 1570–1578. [CrossRef] [PubMed]

143. Kabaria, S.; Choi, D.C.; Chaudhuri, A.D.; Jain, M.R.; Li, H.; Junn, E. MicroRNA-7 activates Nrf2 pathway by targeting Keap1 expression. *Free Radic. Biol. Med.* **2015**, *89*, 548–556. [CrossRef]

144. Eades, G.; Yang, M.; Yao, Y.; Zhang, Y.; Zhou, Q. miR-200a regulates Nrf2 activation by targeting Keap1 mRNA in breast cancer cells. *J. Biol. Chem.* **2011**, *286*, 40725–40733. [CrossRef] [PubMed]

145. Petrelli, A.; Perra, A.; Cora, D.; Sulas, P.; Menegon, S.; Manca, C.; Migliore, C.; Kowalik, M.A.; Ledda-Columbano, G.M.; Giordano, S.; et al. MicroRNA/gene profiling unveils early molecular changes and nuclear factor erythroid related factor 2 (NRF2) activation in a rat model recapitulating human hepatocellular carcinoma (HCC). *Hepatology* **2014**, *59*, 228–241. [CrossRef] [PubMed]

146. Liu, M.; Hu, C.; Xu, Q.; Chen, L.; Ma, K.; Xu, N.; Zhu, H. Methylseleninic acid activates Keap1/Nrf2 pathway via up-regulating miR-200a in human oesophageal squamous cell carcinoma cells. *Biosci. Rep.* **2015**, *35*. [CrossRef]

147. Yang, M.; Yao, Y.; Eades, G.; Zhang, Y.; Zhou, Q. MiR-28 regulates Nrf2 expression through a Keap1-independent mechanism. *Breast Cancer Res. Treat.* **2011**, *129*, 983–991. [CrossRef]

148. Wang, B.; Teng, Y.; Liu, Q. MicroRNA-153 Regulates NRF2 Expression and is Associated with Breast Carcinogenesis. *Clin. Lab.* **2016**, *62*, 39–47. [CrossRef]

149. Khan, A.U.H.; Rathore, M.G.; Allende-Vega, N.; Vo, D.N.; Belkhala, S.; Orecchioni, S.; Talarico, G.; Bertolini, F.; Cartron, G.; Lecellier, C.H.; et al. Human Leukemic Cells performing Oxidative Phosphorylation (OXPHOS) Generate an Antioxidant Response Independently of Reactive Oxygen species (ROS) Production. *EBioMedicine* **2016**, *3*, 43–53. [CrossRef]

150. Vera, O.; Jimenez, J.; Pernia, O.; Rodriguez-Antolin, C.; Rodriguez, C.; Sanchez Cabo, F.; Soto, J.; Rosas, R.; Lopez-Magallon, S.; Esteban Rodriguez, I.; et al. DNA Methylation of miR-7 is a Mechanism Involved in Platinum Response through MAFG Overexpression in Cancer Cells. *Theranostics* **2017**, *7*, 4118–4134. [CrossRef]

151. Qu, J.; Zhang, L.; Li, L.; Su, Y. miR-148b Functions as a Tumor Suppressor by Targeting Endoplasmic Reticulum Metallo Protease 1 in Human Endometrial Cancer Cells. *Oncol. Res.* **2018**, *27*, 81–88. [CrossRef]

152. Degli Esposti, D.; Aushev, V.N.; Lee, E.; Cros, M.P.; Zhu, J.; Herceg, Z.; Chen, J.; Hernandez-Vargas, H. miR-500a-5p regulates oxidative stress response genes in breast cancer and predicts cancer survival. *Sci. Rep.* **2017**, *7*, 15966. [CrossRef]

153. Gomes, S.E.; Pereira, D.M.; Roma-Rodrigues, C.; Fernandes, A.R.; Borralho, P.M.; Rodrigues, C.M.P. Convergence of miR-143 overexpression, oxidative stress and cell death in HCT116 human colon cancer cells. *PLoS ONE* **2018**, *13*, e0191607. [CrossRef]

154. Pajic, M.; Froio, D.; Daly, S.; Doculara, L.; Millar, E.; Graham, P.H.; Drury, A.; Steinmann, A.; de Bock, C.E.; Boulghourjian, A.; et al. miR-139-5p Modulates Radiotherapy Resistance in Breast Cancer by Repressing Multiple Gene Networks of DNA Repair and ROS Defense. *Cancer Res.* **2018**, *78*, 501–515. [CrossRef] [PubMed]

155. Hou, M.; Zuo, X.; Li, C.; Zhang, Y.; Teng, Y. Mir-29b Regulates Oxidative Stress by Targeting SIRT1 in Ovarian Cancer Cells. *Cell. Physiol. Biochem. Int. J. Exp. Cell. Physiol. Biochem. Pharmacol.* **2017**, *43*, 1767–1776. [CrossRef] [PubMed]

156. Kao, Y.Y.; Chou, C.H.; Yeh, L.Y.; Chen, Y.F.; Chang, K.W.; Liu, C.J.; Fan Chiang, C.Y.; Lin, S.C. MicroRNA miR-31 targets SIRT3 to disrupt mitochondrial activity and increase oxidative stress in oral carcinoma. *Cancer Lett.* **2019**, *456*, 40–48. [CrossRef]

157. Chang, M.; Qiao, L.; Li, B.; Wang, J.; Zhang, G.; Shi, W.; Liu, Z.; Gu, N.; Di, Z.; Wang, X.; et al. Suppression of SIRT6 by miR-33a facilitates tumor growth of glioma through apoptosis and oxidative stress resistance. *Oncol. Rep.* **2017**, *38*, 1251–1258. [CrossRef]

158. Yang, C.; Yan, Z.; Hu, F.; Wei, W.; Sun, Z.; Xu, W. Silencing of microRNA-517a induces oxidative stress injury in melanoma cells via inactivation of the JNK signaling pathway by upregulating CDKN1C. *Cancer Cell Int.* **2020**, *20*, 32. [CrossRef] [PubMed]

159. Yin, Y.; Liu, H.; Xu, J.; Shi, D.; Zhai, L.; Liu, B.; Wang, L.; Liu, G.; Qin, J. miR1443p regulates the resistance of lung cancer to cisplatin by targeting Nrf2. *Oncol. Rep.* **2018**, *40*, 3479–3488. [CrossRef] [PubMed]

160. Shi, L.; Chen, Z.G.; Wu, L.L.; Zheng, J.J.; Yang, J.R.; Chen, X.F.; Chen, Z.Q.; Liu, C.L.; Chi, S.Y.; Zheng, J.Y.; et al. miR-340 reverses cisplatin resistance of hepatocellular carcinoma cell lines by targeting Nrf2-dependent antioxidant pathway. *Asian Pac. J. Cancer Prev.* **2014**, *15*, 10439–10444. [CrossRef]

161. Van Jaarsveld, M.T.; Helleman, J.; Boersma, A.W.; van Kuijk, P.F.; van Ijcken, W.F.; Despierre, E.; Vergote, I.; Mathijssen, R.H.; Berns, E.M.; Verweij, J.; et al. miR-141 regulates KEAP1 and modulates cisplatin sensitivity in ovarian cancer cells. *Oncogene* **2013**, *32*, 4284–4293. [CrossRef]

162. Moreno Leon, L.; Gautier, M.; Allan, R.; Ilie, M.; Nottet, N.; Pons, N.; Paquet, A.; Lebrigand, K.; Truchi, M.; Fassy, J.; et al. The nuclear hypoxia-regulated NLUCAT1 long non-coding RNA contributes to an aggressive phenotype in lung adenocarcinoma through regulation of oxidative stress. *Oncogene* **2019**, *38*, 7146–7165. [CrossRef] [PubMed]

163. Qi, W.; Li, Z.; Xia, L.; Dai, J.; Zhang, Q.; Wu, C.; Xu, S. LncRNA GABPB1-AS1 and GABPB1 regulate oxidative stress during erastin-induced ferroptosis in HepG2 hepatocellular carcinoma cells. *Sci. Rep.* **2019**, *9*, 16185. [CrossRef] [PubMed]

164. Shao, Q.; Wang, Q.; Wang, J. LncRNA SCAMP1 regulates ZEB1/JUN and autophagy to promote pediatric renal cell carcinoma under oxidative stress via miR-429. *Biomed. Pharmacother. Biomed. Pharmacother.* **2019**, *120*, 109460. [CrossRef] [PubMed]

165. Wang, W.T.; Ye, H.; Wei, P.P.; Han, B.W.; He, B.; Chen, Z.H.; Chen, Y.Q. LncRNAs H19 and HULC, activated by oxidative stress, promote cell migration and invasion in cholangiocarcinoma through a ceRNA manner. *J. Hematol. Oncol.* **2016**, *9*, 117. [CrossRef]

166. Kim, K.H.; Roberts, C.W. Targeting EZH2 in cancer. *Nat. Med.* **2016**, *22*, 128–134. [CrossRef]

167. Zhou, L.; Xu, D.Y.; Sha, W.G.; Shen, L.; Lu, G.Y.; Yin, X.; Wang, M.J. High glucose induces renal tubular epithelial injury via Sirt1/NF-kappaB/microR-29/Keap1 signal pathway. *J. Transl. Med.* **2015**, *13*, 352. [CrossRef]

168. Sun, L.; Zhang, J.; Li, Y. Chronic central miR-29b antagonism alleviates angiotensin II-induced hypertension and vascular endothelial dysfunction. *Life Sci.* **2019**, *235*, 116862. [CrossRef]

169. Li, C.; Fan, K.; Qu, Y.; Zhai, W.; Huang, A.; Sun, X.; Xing, S. Deregulation of UCA1 expression may be involved in the development of chemoresistance to cisplatin in the treatment of non-small-cell lung cancer via regulating the signaling pathway of microRNA-495/NRF2. *J. Cell. Physiol.* **2020**, *235*, 3721–3730. [CrossRef]

170. Gao, M.; Li, C.; Xu, M.; Liu, Y.; Liu, S. LncRNA UCA1 attenuates autophagy-dependent cell death through blocking autophagic flux under arsenic stress. *Toxicol. Lett.* **2018**, *284*, 195–204. [CrossRef]

171. Siddik, Z.H. Cisplatin: Mode of cytotoxic action and molecular basis of resistance. *Oncogene* **2003**, *22*, 7265–7279. [CrossRef]

172. Rabik, C.A.; Dolan, M.E. Molecular mechanisms of resistance and toxicity associated with platinating agents. *Cancer Treat. Rev.* **2007**, *33*, 9–23. [CrossRef] [PubMed]

173. Wang, D.; Lippard, S.J. Cellular processing of platinum anticancer drugs. *Nat. Rev. Drug Discov.* **2005**, *4*, 307–320. [CrossRef] [PubMed]

174. Dasari, S.; Tchounwou, P.B. Cisplatin in cancer therapy: Molecular mechanisms of action. *Eur. J. Pharmacol.* **2014**, *740*, 364–378. [CrossRef] [PubMed]

175. Marullo, R.; Werner, E.; Degtyareva, N.; Moore, B.; Altavilla, G.; Ramalingam, S.S.; Doetsch, P.W. Cisplatin induces a mitochondrial-ROS response that contributes to cytotoxicity depending on mitochondrial redox status and bioenergetic functions. *PLoS ONE* **2013**, *8*, e81162. [CrossRef] [PubMed]

176. Martins, N.M.; Santos, N.A.; Curti, C.; Bianchi, M.L.; Santos, A.C. Cisplatin induces mitochondrial oxidative stress with resultant energetic metabolism impairment, membrane rigidification and apoptosis in rat liver. *J. Appl. Toxicol. JAT* **2008**, *28*, 337–344. [CrossRef] [PubMed]
177. Schaaf, G.J.; Maas, R.F.; de Groene, E.M.; Fink-Gremmels, J. Management of oxidative stress by heme oxygenase-1 in cisplatin-induced toxicity in renal tubular cells. *Free Radic. Res.* **2002**, *36*, 835–843. [CrossRef]
178. Kartalou, M.; Essigmann, J.M. Mechanisms of resistance to cisplatin. *Mutat. Res.* **2001**, *478*, 23–43. [CrossRef]
179. Holzer, A.K.; Samimi, G.; Katano, K.; Naerdemann, W.; Lin, X.; Safaei, R.; Howell, S.B. The copper influx transporter human copper transport protein 1 regulates the uptake of cisplatin in human ovarian carcinoma cells. *Mol. Pharmacol.* **2004**, *66*, 817–823. [CrossRef]
180. Wu, C.; Wangpaichitr, M.; Feun, L.; Kuo, M.T.; Robles, C.; Lampidis, T.; Savaraj, N. Overcoming cisplatin resistance by mTOR inhibitor in lung cancer. *Mol. Cancer* **2005**, *4*, 25. [CrossRef]
181. Nicholson, L.J.; Smith, P.R.; Hiller, L.; Szlosarek, P.W.; Kimberley, C.; Sehouli, J.; Koensgen, D.; Mustea, A.; Schmid, P.; Crook, T. Epigenetic silencing of argininosuccinate synthetase confers resistance to platinum-induced cell death but collateral sensitivity to arginine auxotrophy in ovarian cancer. *Int. J. Cancer* **2009**, *125*, 1454–1463. [CrossRef]
182. Ibragimova, I.; Ibanez de Caceres, I.; Hoffman, A.M.; Potapova, A.; Dulaimi, E.; Al-Saleem, T.; Hudes, G.R.; Ochs, M.F.; Cairns, P. Global reactivation of epigenetically silenced genes in prostate cancer. *Cancer Prev. Res. (Phila.)* **2010**, *3*, 1084–1092. [CrossRef] [PubMed]
183. Cortes-Sempere, M.; de Miguel, M.P.; Pernia, O.; Rodriguez, C.; de Castro Carpeno, J.; Nistal, M.; Conde, E.; Lopez-Rios, F.; Belda-Iniesta, C.; Perona, R.; et al. IGFBP-3 methylation-derived deficiency mediates the resistance to cisplatin through the activation of the IGFIR/Akt pathway in non-small cell lung cancer. *Oncogene* **2013**, *32*, 1274–1283. [CrossRef] [PubMed]
184. Chang, X.; Monitto, C.L.; Demokan, S.; Kim, M.S.; Chang, S.S.; Zhong, X.; Califano, J.A.; Sidransky, D. Identification of hypermethylated genes associated with cisplatin resistance in human cancers. *Cancer Res.* **2010**, *70*, 2870–2879. [CrossRef] [PubMed]
185. Wang, L.; Xiang, S.; Williams, K.A.; Dong, H.; Bai, W.; Nicosia, S.V.; Khochbin, S.; Bepler, G.; Zhang, X. Depletion of HDAC6 enhances cisplatin-induced DNA damage and apoptosis in non-small cell lung cancer cells. *PLoS ONE* **2012**, *7*, e44265. [CrossRef]
186. Gruosso, T.; Mieulet, V.; Cardon, M.; Bourachot, B.; Kieffer, Y.; Devun, F.; Dubois, T.; Dutreix, M.; Vincent-Salomon, A.; Miller, K.M.; et al. Chronic oxidative stress promotes H2AX protein degradation and enhances chemosensitivity in breast cancer patients. *Embo Mol. Med.* **2016**, *8*, 527–549. [CrossRef]
187. Kim, N.H.; Choi, S.H.; Kim, C.H.; Lee, C.H.; Lee, T.R.; Lee, A.Y. Reduced MiR-675 in exosome in H19 RNA-related melanogenesis via MITF as a direct target. *J. Investig. Dermatol.* **2014**, *134*, 1075–1082. [CrossRef]
188. Ceppi, P.; Mudduluru, G.; Kumarswamy, R.; Rapa, I.; Scagliotti, G.V.; Papotti, M.; Allgayer, H. Loss of miR-200c expression induces an aggressive, invasive, and chemoresistant phenotype in non-small cell lung cancer. *Mol. Cancer Res.* **2010**, *8*, 1207–1216. [CrossRef]
189. Gu, Y.; Zhang, Z.; Yin, J.; Ye, J.; Song, Y.; Liu, H.; Xiong, Y.; Lu, M.; Zheng, G.; He, Z. Epigenetic silencing of miR-493 increases the resistance to cisplatin in lung cancer by targeting tongue cancer resistance-related protein 1(TCRP1). *J. Exp. Clin. Cancer Res.* **2017**, *36*, 114. [CrossRef]
190. Liu, X.; Feng, M.; Zheng, G.; Gu, Y.; Wang, C.; He, Z. TCRP1 expression is associated with platinum sensitivity in human lung and ovarian cancer cells. *Oncol. Lett.* **2017**, *13*, 1398–1405. [CrossRef]
191. Jia, X.; Zhang, Z.; Luo, K.; Zheng, G.; Lu, M.; Song, Y.; Liu, H.; Qiu, H.; He, Z. TCRP1 transcriptionally regulated by c-Myc confers cancer chemoresistance in tongue and lung cancer. *Sci. Rep.* **2017**, *7*, 3744. [CrossRef]
192. Ibanez de Caceres, I.; Cortes-Sempere, M.; Moratilla, C.; Machado-Pinilla, R.; Rodriguez-Fanjul, V.; Manguan-Garcia, C.; Cejas, P.; Lopez-Rios, F.; Paz-Ares, L.; de CastroCarpeno, J.; et al. IGFBP-3 hypermethylation-derived deficiency mediates cisplatin resistance in non-small-cell lung cancer. *Oncogene* **2010**, *29*, 1681–1690. [CrossRef] [PubMed]
193. Rodriguez-Antolin, C.; Felguera-Selas, L.; Pernia, O.; Vera, O.; Esteban, I.; Losantos Garcia, I.; de Castro, J.; Rosas-Alonso, R.; Ibanez de Caceres, I. miR-7 methylation as a biomarker to predict poor survival in early-stage non-small cell lung cancer patients. *Cell Biosci.* **2019**, *9*, 63. [CrossRef] [PubMed]

194. Young, R.P.; Hopkins, R.J.; Christmas, T.; Black, P.N.; Metcalf, P.; Gamble, G.D. COPD prevalence is increased in lung cancer, independent of age, sex and smoking history. *Eur. Respir. J.* **2009**, *34*, 380–386. [CrossRef] [PubMed]

195. Rosas-Alonso, R.; Galera, R.; Sanchez-Pascuala, J.J.; Casitas, R.; Burdiel, M.; Martinez-Ceron, E.; Vera, O.; Rodriguez-Antolin, C.; Pernia, O.; De Castro, J.; et al. Hypermethylation of Anti-oncogenic MicroRNA 7 is Increased in Emphysema Patients. *Arch. Bronconeumol.* **2019**. [CrossRef]

196. Vera-Puente, O.; Rodriguez-Antolin, C.; Salgado-Figueroa, A.; Michalska, P.; Pernia, O.; Reid, B.M.; Rosas, R.; Garcia-Guede, A.; SacristAn, S.; Jimenez, J.; et al. MAFG is a potential therapeutic target to restore chemosensitivity in cisplatin-resistant cancer cells by increasing reactive oxygen species. *Transl. Res.* **2018**, *200*, 1–17. [CrossRef]

197. Shavit, J.A.; Motohashi, H.; Onodera, K.; Akasaka, J.; Yamamoto, M.; Engel, J.D. Impaired megakaryopoiesis and behavioral defects in mafG-null mutant mice. *Genes Dev.* **1998**, *12*, 2164–2174. [CrossRef] [PubMed]

198. Yang, H.; Liu, T.; Wang, J.; Li, T.W.; Fan, W.; Peng, H.; Krishnan, A.; Gores, G.J.; Mato, J.M.; Lu, S.C. Deregulated methionine adenosyltransferase alpha1, c-Myc, and Maf proteins together promote cholangiocarcinoma growth in mice and humans(double dagger). *Hepatology* **2016**, *64*, 439–455. [CrossRef]

199. Fan, W.; Yang, H.; Liu, T.; Wang, J.; Li, T.W.; Mavila, N.; Tang, Y.; Yang, J.; Peng, H.; Tu, J.; et al. Prohibitin 1 suppresses liver cancer tumorigenesis in mice and human hepatocellular and cholangiocarcinoma cells. *Hepatology* **2017**, *65*, 1249–1266. [CrossRef]

200. Liu, T.; Yang, H.; Fan, W.; Tu, J.; Li, T.W.H.; Wang, J.; Shen, H.; Yang, J.; Xiong, T.; Steggerda, J.; et al. Mechanisms of MAFG Dysregulation in Cholestatic Liver Injury and Development of Liver Cancer. *Gastroenterology* **2018**, *155*, 557–571. [CrossRef]

201. Wang, J.P.; Leng, J.Y.; Zhang, R.K.; Zhang, L.; Zhang, B.; Jiang, W.Y.; Tong, L. Functional analysis of gene expression profiling-based prediction in bladder cancer. *Oncol. Lett.* **2018**, *15*, 8417–8423. [CrossRef]

202. Yang, H.; Li, T.W.; Peng, J.; Mato, J.M.; Lu, S.C. Insulin-like growth factor 1 activates methionine adenosyltransferase 2A transcription by multiple pathways in human colon cancer cells. *Biochem. J.* **2011**, *436*, 507–516. [CrossRef] [PubMed]

203. Torres, S.; Garcia-Palmero, I.; Marin-Vicente, C.; Bartolome, R.A.; Calvino, E.; Fernandez-Acenero, M.J.; Casal, J.I. Proteomic Characterization of Transcription and Splicing Factors Associated with a Metastatic Phenotype in Colorectal Cancer. *J. Proteome Res.* **2018**, *17*, 252–264. [CrossRef] [PubMed]

204. Fang, M.; Hutchinson, L.; Deng, A.; Green, M.R. Common BRAF(V600E)-directed pathway mediates widespread epigenetic silencing in colorectal cancer and melanoma. *Proc. Natl. Acad. Sci. USA* **2016**, *113*, 1250–1255. [CrossRef]

205. Fang, M.; Ou, J.; Hutchinson, L.; Green, M.R. The BRAF oncoprotein functions through the transcriptional repressor MAFG to mediate the CpG Island Methylator phenotype. *Mol. Cell* **2014**, *55*, 904–915. [CrossRef]

206. Katsuoka, F.; Motohashi, H.; Engel, J.D.; Yamamoto, M. Nrf2 transcriptionally activates the mafG gene through an antioxidant response element. *J. Biol. Chem.* **2005**, *280*, 4483–4490. [CrossRef] [PubMed]

207. Schembri, F.; Sridhar, S.; Perdomo, C.; Gustafson, A.M.; Zhang, X.; Ergun, A.; Lu, J.; Liu, G.; Bowers, J.; Vaziri, C.; et al. MicroRNAs as modulators of smoking-induced gene expression changes in human airway epithelium. *Proc. Natl. Acad. Sci. USA* **2009**, *106*, 2319–2324. [CrossRef]

208. Kim, T.; Mehta, S.L.; Morris-Blanco, K.C.; Chokkalla, A.K.; Chelluboina, B.; Lopez, M.; Sullivan, R.; Kim, H.T.; Cook, T.D.; Kim, J.Y.; et al. The microRNA miR-7a-5p ameliorates ischemic brain damage by repressing alpha-synuclein. *Sci. Signal.* **2018**, *11*. [CrossRef]

209. Junn, E.; Lee, K.W.; Jeong, B.S.; Chan, T.W.; Im, J.Y.; Mouradian, M.M. Repression of alpha-synuclein expression and toxicity by microRNA-7. *Proc. Natl. Acad. Sci. USA* **2009**, *106*, 13052–13057. [CrossRef]

210. Titze-de-Almeida, R.; Titze-de-Almeida, S.S. miR-7 Replacement Therapy in Parkinson's Disease. *Curr. Gene Ther.* **2018**, *18*, 143–153. [CrossRef]

211. Fields, C.R.; Bengoa-Vergniory, N.; Wade-Martins, R. Targeting Alpha-Synuclein as a Therapy for Parkinson's Disease. *Front. Mol. Neurosci.* **2019**, *12*, 299. [CrossRef]

212. Perfeito, R.; Ribeiro, M.; Rego, A.C. Alpha-synuclein-induced oxidative stress correlates with altered superoxide dismutase and glutathione synthesis in human neuroblastoma SH-SY5Y cells. *Arch. Toxicol.* **2017**, *91*, 1245–1259. [CrossRef] [PubMed]

213. Hansen, T.B.; Kjems, J.; Damgaard, C.K. Circular RNA and miR-7 in cancer. *Cancer Res.* **2013**, *73*, 5609–5612. [CrossRef] [PubMed]

Antioxidants **2020**, *9*, 468

214. Akar, R.O.; Selvi, S.; Ulukaya, E.; Aztopal, N. Key actors in cancer therapy: Epigenetic modifiers. *Turk. J. Biol. Turk. Biyol. Derg.* **2019**, *43*, 155–170. [CrossRef]

215. Cheng, Y.; He, C.; Wang, M.; Ma, X.; Mo, F.; Yang, S.; Han, J.; Wei, X. Targeting epigenetic regulators for cancer therapy: Mechanisms and advances in clinical trials. *Signal Transduct. Target. Ther.* **2019**, *4*, 62. [CrossRef] [PubMed]

216. Hanna, J.; Hossain, G.S.; Kocerha, J. The Potential for microRNA Therapeutics and Clinical Research. *Front. Genet.* **2019**, *10*, 478. [CrossRef]

217. Bonneau, E.; Neveu, B.; Kostantin, E.; Tsongalis, G.J.; De Guire, V. How close are miRNAs from clinical practice? A perspective on the diagnostic and therapeutic market. *Ejifcc* **2019**, *30*, 114–127.

 *antioxidants*

Article

Vitamin C Activates the Folate-Mediated One-Carbon Cycle in C2C12 Myoblasts

Armando Alcazar Magana [1,2], Ralph L. Reed [2,3], Rony Koluda [1], Cristobal L. Miranda [2,3], Claudia S. Maier [1,2] and Jan F. Stevens [2,3,*]

[1] Department of Chemistry, Oregon State University, 153 Gilbert Hall, Corvallis, OR 97331, USA; alcazara@oregonstate.edu (A.A.M.); koludar@oregonstate.edu (R.K.); claudia.maier@oregonstate.edu (C.S.M.)

[2] Linus Pauling Institute, Oregon State University, 2900 SW Campus way, Corvallis, OR 97331, USA; reedr@oregonstate.edu (R.L.R.); cristobal.miranda@oregonstate.edu (C.L.M.)

[3] Department of Pharmaceutical Sciences, Oregon State University, 1601 SW Jefferson Way, Corvallis, OR 97331, USA

* Correspondence: fred.stevens@oregonstate.edu; Tel.: +1-541-737-9534

Received: 30 January 2020; Accepted: 4 March 2020; Published: 5 March 2020

Abstract: Vitamin C (L-ascorbic acid, AA) is an essential cellular antioxidant and cofactor for several α-ketoglutarate-dependent dioxygenases. As an antioxidant, AA interacts with vitamin E to control oxidative stress. While several reports suggest an interaction of AA with folate (vitamin B9) in animals and humans, little is known about the nature of the interaction and the underlying molecular mechanisms at the cellular level. We used an untargeted metabolomics approach to study the impact of AA on the metabolome of C2C12 myoblast cells. Compared to untreated cells, treatment of C2C12 cells with AA at 100 µM resulted in enhanced concentrations of folic acid (2.5-fold) and 5-methyl-tetrahydrofolate (5-methyl-THF, 10-fold increase) whereas the relative concentrations of 10-formyl-tetrahydrofolate decreased by >90% upon AA pretreatment, indicative of increased utilization for the biosynthesis of active THF metabolites. The impact of AA on the folate-mediated one-carbon cycle further manifested itself as an increase in the levels of methionine, whose formation from homocysteine is 5-methyl-THF dependent, and an increase in thymidine, whose formation from deoxyuridine monophosphate (dUMP) is dependent on 5,10-methylene-THF. These findings shed new light on the interaction of AA with the folate-mediated one-carbon cycle and partially explain clinical findings that AA supplementation enhances erythrocyte folate status and that it may decrease serum levels of homocysteine, which is considered as a biomarker of cardiovascular disease risk.

Keywords: vitamin C; folic acid; one-carbon metabolism; C2C12 cells; metabolomics; mass spectrometry; ascorbic acid

---

1. Introduction

Vitamin C (ascorbic acid, AA) has important functions in animals and humans to maintain health and prevent disease [1]. AA acts as a cofactor for several α-ketoglutarate-dependent dioxygenases, such as prolyl hydroxylase involved in collagen synthesis [2]. Its capacity to donate electrons gives AA antioxidant properties [1] and ability to scavenge electrophiles in Michael additions [3,4]. The antioxidant properties of AA can manifest in many ways and affect a multitude of metabolic pathways. For example, AA supplementation in humans lowers the production and urinary excretion of lipid peroxidation-derived reactive aldehydes [5]. From metabolomics analyses of zebrafish fed a diet lacking sufficient AA, we established that AA deficiency, which causes oxidative stress, activates the purine nucleotide cycle to regenerate ATP [6]. Recently, we reported that AA prevents cellular nitrate tolerance in glyceryl trinitate-treated porcine renal epithelial cells, which is relevant to angina [7].

Using a mass spectrometry-based metabolomics approach, we found that AA protects xanthine oxidase from glyceryl trinitrate-induced inactivation, which is relevant because this enzyme plays a role in the conversion of glyceryl trinitrate into the vasoactive metabolite, nitric oxide [7]. These examples illustrate how a data-driven metabolomics approach can reveal novel functions of AA.

AA is well known to interact with vitamin E in cellular redox cycles to control oxidative stress [2] and prevent oxidative stress damage to proteins and DNA [8]. Interactions between AA and folic acid (vitamin B9) have also been reported [9,10], although the underlying mechanism remains poorly understood. Folic acid plays a key role as a one-carbon carrier molecule in methylation reactions—notably, the conversion of homocysteine (Hcy) into methionine and the conversion of deoxyuridine monophosphate (dUMP) into deoxythymidine monophosphate (dTMP) [11]. Given that elevated plasma Hcy is considered a risk factor for cardiovascular disease [12] and given the reports of an AA–folate interaction [9,10], AA supplementation has been explored as an attractive way to increase circulating levels of folic acid and to reduce Hcy levels, with mixed results [10,13], although the clinical significance of the interaction and the underlying molecular mechanism remain poorly understood.

Humans cannot synthesize AA and depend on dietary AA for their survival. Most non-primate animals produce AA in the liver [14]. Thus, human and most mammalian cell lines grown in culture do not synthesize AA. Unless cultured cells are supplemented with AA, they are less protected against oxidative stress than their in vivo counterparts. To elucidate the biochemical pathways modulated by AA supplementation, we examined the changes in the metabolome of mouse C2C12 cells in response to AA supplementation. We selected these cells because they are metabolically active and susceptible to oxidative stress [15]. Using an unbiased, untargeted mass spectrometry-based metabolomics approach, we measured major shifts in pool sizes of tetrahydrofolate (THF) metabolites and of substrates of methylation reactions, which prompted us to focus our investigation on the effects of AA supplementation on the folate-mediated one-carbon cycle and associated pathways. Because metabolomics measures relative levels of individual metabolites, the technique provides clues as to which pathways and which enzymatic steps in those pathways are modulated by AA treatment. Here we report that AA facilitates the reductive steps in the conversion of 10-formyl-THF into 5-methyl-THF, which provides new insights in the interaction between AA and folic acid and how the interaction promotes methylation reactions to provide substrates for amino acid and DNA synthesis.

2. Materials and Methods

2.1. Reagents

LC–MS-grade methanol and water were purchased from EMD Millipore (Burlington, MA, USA). Formic acid ACS reagent was from Fisher Chemicals (Suwanee, GA, USA). ACS reagents folic acid, folinic acid (5-formyltetrahydrofolic acid), 5-methyltetrahydrofolic acid, methionine-(methyl-d_3) (used as internal standard) and butylated hydroxytoluene were purchased from Sigma Aldrich (St. Louis, MO, USA). Sodium ascorbate was from Merck. $^{13}C_6$-ascorbic acid was from Omicron Biochemicals, Inc. (South Bend, IN, USA). DMEM was from Life Technologies (Grand Island, NY, USA). Penicillin, streptomycin and fetal bovine serum were from Invitrogen (Carlsbad, CA, United States).

2.2. Cell Culture and Treatment

C2C12 cells were obtained from ATCC (Manassas, VA, USA) and were first propagated in 75 cm^2 flasks using a culture medium consisting of DMEM, 10% fetal bovine serum, 100 units/mL penicillin, and 100 µg/mL streptomycin. The cells were harvested from the flask using trypsin and seeded in 10 cm culture dishes containing 10 mL of the supplemented DMEM medium, which contains 14 amino acids, including methionine, serine, and glycine, and nine vitamins including folic acid but no vitamin C. Immediately after seeding, the cells were treated with 100 µM sodium ascorbate or 100 µM $^{13}C_6$-AA by adding 10 µL of the respective 100 mM stock solution. The rationale behind using the two isotopologues of AA was that this approach offered us a tool to assign mass spectral features to AA or its metabolites in the metabolomics dataset. At this concentration, AA supplementation did not change the pH of the medium. Control cells were grown in the absence of AA. All three groups (biological triplicates) were maintained under sterile conditions with 5% CO_2 at 37 °C. After 48 h of incubation, cells were scraped, counted using a Hemacytometer Counting Chamber Bright-line Neubauer Cell counter (Hausser Scientific, Horsham, PA, USA) according to the manufacturer's instructions, and spun down at 4 °C at 500× *g* for 5 min. The pellets (equivalent to 3.5×10^6 cells per dish) were washed three times with cold PBS and transferred to 1.7 mL Eppendorf tubes. The pellets were frozen and stored at –80 °C until sample preparation for LC–MS/MS analysis.

2.3. LC–MS/MS Analysis

Cells were resuspended in 300 µL of ice-cold 50:50 (*v:v*) methanol:ethanol containing 0.02% *w/v* of butylated hydroxytoluene (to prevent post-harvest oxidation) and disrupted by sonication for 3×15 s in a Fisher Scientific sonifier (1/2 inch disruptor horn, 4 kHz) on ice with 30 s between steps. Samples were spun for 10 min at 14,000× *g* at 4 °C and the supernatant was transferred to mass spectrometry vials (Microsolv, Leland, NC, USA). LC–MS/MS was carried out with a method previously described [6,7,15] with some differences. Briefly, data-dependent acquisition (DDA) in both the positive and the negative ion mode was conducted using a Shimadzu Nexera UPLC system connected to an AB SCIEX TripleTOF® 5600 mass spectrometer (AB SCIEX, Concord, Canada). Samples were randomized before injections. A quality control (QC) sample, obtained by mixing equal aliquots of all samples, was analyzed every five LC runs. The injection volume was 5 µL. Chromatographic separation was performed using an Inertsil Phenyl-3 column (4.6 × 150 mm, 100 Å, 5 µm; GL Sciences, Rolling Hills Estates, CA, USA). A gradient with two mobile phases (A, water with 0.1% *v/v* formic acid and B, methanol with 0.1% *v/v* formic acid) was used. The elution gradient was as follows: 0 min, 5% B; 1 min, 5% B; 11 min, 30% B; 20 min, 100% B; 25 min, 100% B; 30 min, 5% B; and 35 min, 5% B. Period cycle time was 950 ms; accumulation time 100 ms; *m/z* scan range 100–1200; and collision energy 35 V with collision energy spread (CES) of 15 V. Mass calibration was automatically performed after every fifth LC run.

2.4. Metabolomics Data Processing

Raw data processing was performed using Progenesis QI™ software with Metlin™ plugin V1.0.6499.51447 (NonLinear Dynamics, United Kingdom) for peak picking, retention time correction, peak alignment, data normalization and metabolite annotations. Annotation confidence was achieved in accordance with reporting criteria for chemical analysis suggested by the Metabolomics Standards Initiative (MSI) [16,17]. Thus, level 1 (L1) annotations were made based on accurate mass (error < 10 ppm), fragment ion spectral pattern (library score > 50), isotope pattern (library score > 80), and retention time (error < 5%) comparison of analytes with those of authentic standards in our in-house library (> 700) analyzed under identical conditions.

For metabolites not present in our in-house library, level 2 (L2, tentative) annotations were made according to the MSI guidelines [16,17] and based on accurate mass, fragment ion spectral pattern, and isotope pattern comparison with online data available in METLIN (which has MS/MS experimental data) and the Human Metabolome Database (which contains in silico generated MS/MS data). Ions generated from QC samples were retained for annotation and included in the dataset if the coefficient of variation (CV) of their abundance did not exceed 25% (QC distribution is shown in supplementary Figure S1). When metabolites were detected in both ion modes, the one with the highest signal-to-noise ratio for a particular metabolite was kept. Relative quantities of metabolites were determined by calculating their corresponding peak areas.

2.5. Statistical Analysis

The area under the curve of molecular ions was selected for relative quantitation of annotated metabolites. Only annotated metabolites were used for statistical analysis. Figures (principal component analysis, heatmaps, dendrograms and biplots) were generated with MetaboAnalyst 4.0 (Montreal, Quebec, Canada) [18] and PowerPoint 2016 (Microsoft, Redmond, WA, USA). Plots were generated with GraphPad Prism 8.3.0 (La Jolla, CA, United States). Metabolite changes were tested by a one-way analysis of variance followed by Tukey's HSD (honestly significant difference) post-hoc analysis and Holm FDR (false discovery rate) correction, with a p-value of < 0.05 indicating significance. Unless otherwise stated, data are presented as the mean ± SD of three independent experiments ($n = 3$/group). In the tables, the two vitamin C treatment groups were collapsed into a single AA group ($n = 6$). To compensate for unequal variance or non-normal distribution, data were logarithmically transformed and Pareto scaled (mean centered and divided by the square root of the standard deviation). No outliers were excluded from the statistical analyses.

3. Results

Metabolomics analysis was performed on mouse skeletal muscle-derived C2C12 myoblasts that were AA deficient or AA supplemented. We selected C2C12 cells because we have experience with metabolome measurements of this cell line, they are metabolically active, and because they are susceptible to oxidative stress [15]. We exposed the cells to two isotopologue forms of AA, sodium ascorbate and $^{13}C_6$-ascorbic acid, because the approach allowed us to distinguish between exposure compounds (AA and its metabolites or degradation products) and the effects of the exposure on the metabolome. We did not find evidence that exogenous AA supplementation interfered with annotation of endogenous metabolites. We also did not detect AA in the non-supplemented cells, which is expected because mice produce AA in the liver and not in muscle tissue (Figure S2).

The identity of 102 metabolites was assigned using our in-house library (L1 metabolites, supplementary Table S1), while an additional 102 metabolites were assigned by comparison with spectral data available from METLIN and the HMDB (L2 metabolites, supplementary Table S1). The resulting 204 metabolites were further investigated to determine how and to what extent AA exposure altered the cellular metabolome. Figure 1 shows an analysis of similarities and dissimilarities among the treatments. Principal component analysis (PCA) grouped together supplemented cells and separated them from AA-deficient cells (Figure 1a), which demonstrates that the cells responded similarly to exposure to either sodium ascorbate or $^{13}C_6$-ascorbic acid.

A PCA biplot, a representation that shows an overlay of the score and loading plots, visualizes the changing metabolome in response to AA supplementation (Figure 1b). It revealed that the following metabolites were most prominently associated with the group separation of the AA-deficient (controls) from the AA-supplemented cells in the scatter plot of the PCA (Figure 1a): folic acid, 10-formyl-tetrahydrofolate (10-formyl-THF), 5-methyl-THF, homocysteic acid (oxidized form of homocysteine, Hcy), and uridine-5′-diphosphate (UDP) (Figure 1b). A heatmap of the top 40 most differentiating metabolites also shows that the AA-supplemented cells responded differently from the control group (Figure 1c). Using the variation in all 204 annotated metabolites, a similar dendrogram representation emerged (Figure 1d), in which the non-supplemented cells clustered together as did the AA-supplemented cells without distinction between sodium ascorbate- and $^{13}C_6$-ascorbic acid-treated cells.

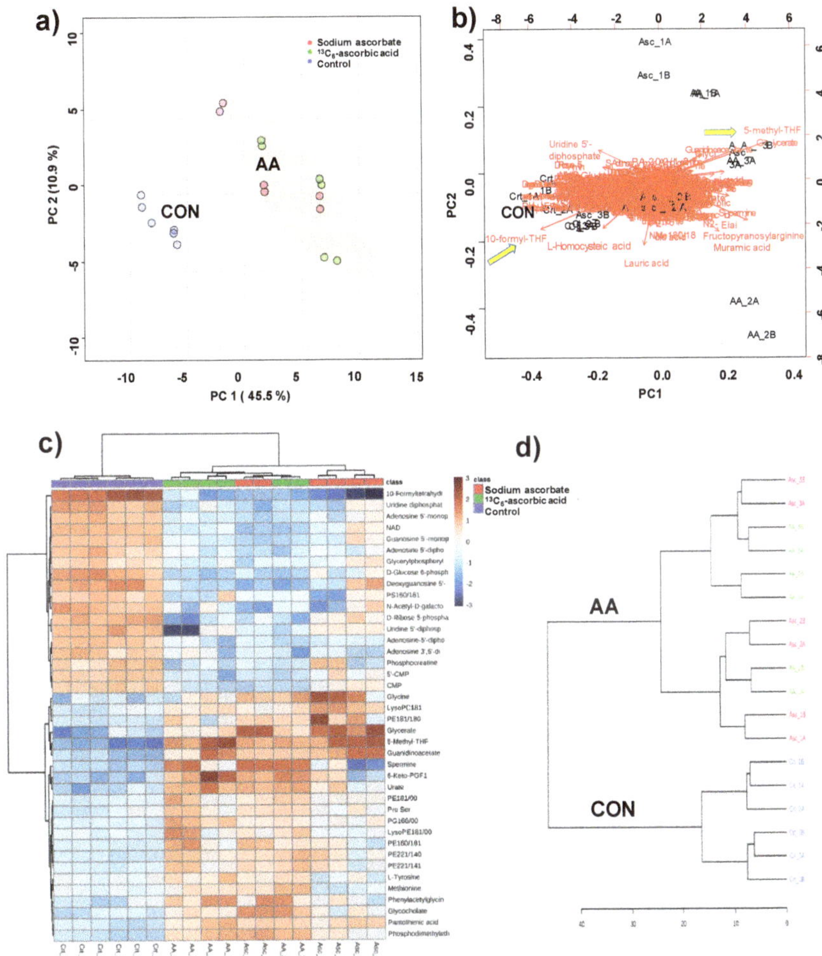

Figure 1. Analysis of treated C2C12 cells with ascorbic acid and with ascorbate 100 μM. (**a**) Principal component analysis (PCA) scores plot, where each dot represents a sample analysis (thus, 6 dots per experiment resulting from biological triplicates and technical duplicates); (**b**) PCA biplot visualizes which metabolites contribute most to the separation of the experimental groups; (**c**) heatmap visualizing the top 40 most differentiating metabolites. Color coding indicates greater deviation from the mean of all samples for a particular metabolite; (**d**) dendrogram indicating the degree of similarity among samples, constructed on the basis of all 204 metabolites. The analysis was performed using MetaboAnalyst V4.0.

Of the three folic acid metabolites we detected, the formyl derivative of THF was not identical with an authentic standard of 5-formyl-THF with regard to retention time and MS/MS fragmentation pattern (Figure 2). Comparison of the spectra of both formyl derivatives revealed that 5-formyl-THF produced a low-mass fragment with *m/z* 120.0437 which was absent in the spectrum of the regioisomer shown in Figure 2a. Instead, the corresponding fragment of the regioisomer appeared at *m/z* 132.0427. We identified the *m/z* 120.0437 fragment of 5-formyl-THF as $^+H_2N=(C_6H_4)=C=O$ and the *m/z* 132.0427 fragment ion of the regioisomer as $CH_2=N-(C_6H_4)-C\equiv O^+$, the difference corresponding to an imino substituent on the nitrogen atom. As this nitrogen atom corresponds to atom position 10 in the precursor ion and considering that the imino group is formed upon loss of a H_2O molecule from the formyl substituent, we can identify the regioisomer as 10-formyl-THF.

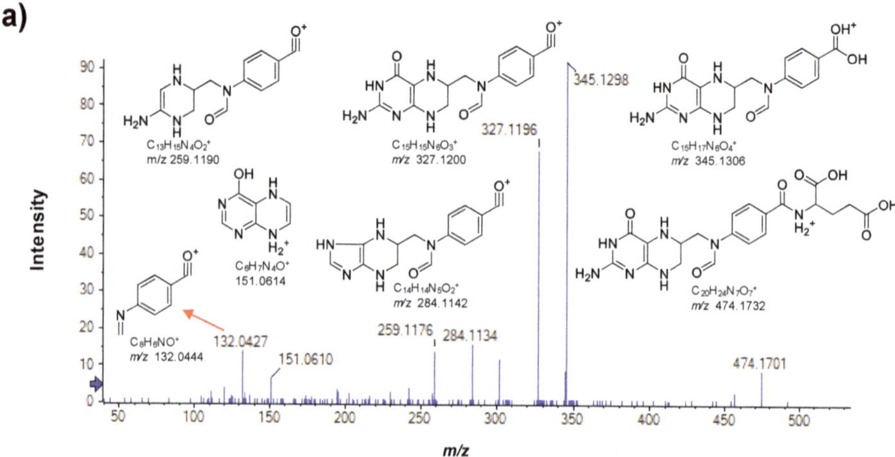

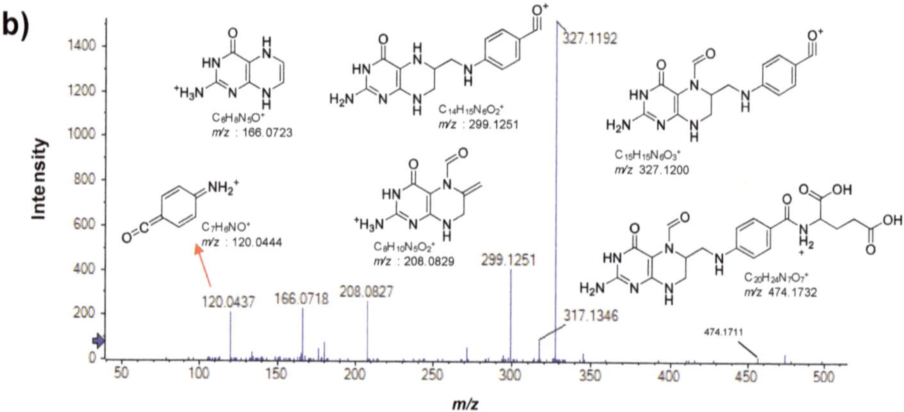

Figure 2. Positive ion mode MS/MS spectra of the two detected formyl-tetrahydrofolate (THF) isomers and the proposed structures of their fragment ions (CE 35 V; CES of 15 V). (**a**) 10-formyl-tetrahydrofolate (10-formyl-THF) and (**b**) 5-formyl-tetrahydrofolate (5-formyl-THF, authentic standard). Peak labels denote accurate masses and fragment ion *m/z* values denote exact masses.

One-way ANOVA of the top 50 differentiating metabolites (Table 1) revealed that the relative cellular levels of folic acid, 5-methyl-THF, 10-formyl-THF, methionine sulfoxide, UDP, glycine, and thymidine changed significantly in response to AA-treatment ($n = 3$/group, $p < 0.05$, Figure 2 and supplementary Table S1). Although homocysteine was below the limit of integration, its oxidized form (homocysteic acid) was quantifiable but the change did not reach statistical significance (FC 0.69, $p = 0.08$; AA/Control, supplementary Table S1). These metabolites are either part of the folate-mediated one-carbon cycle or associated pathways (Figure 3). Their change in relative cellular levels in response to AA-treatment, shown in bar graphs, indicates that AA facilitates the reductive conversion of 10-formyl-THF into 5-methyl-THF.

Table 1. Top 50 annotated compounds (one-way ANOVA followed by Tukey's HSD (honestly significant difference) post-hoc analysis and Holm FDR (false discovery rate) correction; FC > 1.2, $p \leq 0.005$) sorted by the significance of p-value for the comparison AA/Control. The complete list for the 204 annotated compounds in alphabetical order is presented in Table S2.

Compound	m/z	Retention Time (min)	Formula	Error [1]	FC [2]	p-Value
5-methyl-THF	460.1938	14.28	$C_{20}H_{25}N_7O_6$	0.26	9.97	8.07×10^{-9}
10-formyl-THF	474.1731	16.96	$C_{20}H_{23}N_7O_7$	0.05	0.08	6.31×10^{-7}
Pantothenic acid	218.1035	10.51	$C_9H_{17}NO_5$	1.08	1.60	8.35×10^{-7}
Guanidinoacetate	118.0603	4.74	$C_3H_7N_3O_2$	4.32	2.67	4.17×10^{-6}
Methionine sulfoxide	166.0526	4.99	$C_5H_{11}NO_3S$	4.08	0.71	4.90×10^{-5}
Phosphodimethylethanolamine	192.0320	5.10	$C_4H_{12}NO_4P$	0.92	1.95	7.95×10^{-5}
PE(22:1/14:0)	746.5671	26.99	$C_{41}H_{80}NO_8P$	1.19	1.57	1.91×10^{-4}
Urate	169.0351	5.97	$C_5H_4N_4O_3$	0.63	2.69	2.63×10^{-4}
LysoPC(18:1)	522.3545	25.88	$C_{26}H_{52}NO_7P$	0.78	1.47	4.31×10^{-4}
Phenylacetylglycine	192.0665	17.88	$C_{10}H_{11}NO_3$	0.59	1.66	7.95×10^{-4}
PE(18:1/18:0)	744.5530	26.96	$C_{41}H_{80}NO_8P$	0.11	1.77	8.59×10^{-4}
Adenosine-5'-diphosphoglucose	428.0365	4.87	$C_{10}H_{15}N_5O_{10}P_2$	0.44	0.60	9.18×10^{-4}
6-keto-PGF1	393.2249	22.40	$C_{20}H_{34}O_6$	0.34	3.37	9.93×10^{-4}
PE(22:1/14:1)	744.5529	26.99	$C_{41}H_{78}NO_8P$	0.33	1.43	1.22×10^{-3}
D-glucose 6-phosphate	259.0225	4.25	$C_6H_{13}O_9P$	0.16	0.38	1.31×10^{-3}
Adenosine 5'-monophosphate	346.0558	5.41	$C_{10}H_{14}N_5O_7P$	0.15	0.57	1.61×10^{-3}
Glycerate	105.0188	5.03	$C_3H_6O_4$	2.27	4.38	1.65×10^{-3}
Uridine diphosphate (UDP)	405.0094	4.07	$C_9H_{14}N_2O_{12}P_2$	1.36	0.42	1.73×10^{-3}
Thymidine	241.0831	11.03	$C_{10}H_{14}N_2O_5$	3.33	1.90	1.81×10^{-3}
Glycocholate	464.3009	23.90	$C_{26}H_{43}NO_6$	6.89	1.81	1.85×10^{-3}
LysoPG(16:0)	507.2694	24.33	$C_{22}H_{45}O_9P$	0.42	1.38	1.96×10^{-3}
LysoPE(18:1)	478.2934	24.90	$C_{23}H_{46}NO_7P$	2.54	1.39	1.97×10^{-3}
Glycerol 2-phosphate	171.0060	4.44	$C_3H_9O_6P$	0.25	1.28	2.08×10^{-3}
Hippurate	180.0650	17.08	$C_9H_9NO_3$	1.02	1.46	2.13×10^{-3}
Adenosine 5'-diphosphate	426.0221	4.85	$C_{10}H_{15}N_5O_{10}P_2$	0.21	0.60	2.38×10^{-3}
LysoPC(16:1)	538.3144	25.28	$C_{24}H_{48}NO_7P$	1.77	1.35	2.46×10^{-3}
Glycine	76.0385	4.79	$C_2H_5NO_2$	1.84	2.31	2.86×10^{-3}
5'-CMP	324.0591	4.77	$C_9H_{14}N_3O_8P$	0.03	0.65	2.97×10^{-3}
Adenosine 3',5'-diphosphate	450.0186	4.97	$C_{10}H_{15}N_5O_{10}P_2$	0.20	0.58	4.39×10^{-3}
NAD	662.1024	6.27	$C_{21}H_{27}N_7O_{14}P_2$	3.00	0.50	4.51×10^{-3}
Guanosine 5'-monophosphate	362.0507	5.26	$C_{10}H_{14}N_5O_8P$	2.98	0.60	4.63×10^{-3}
Aminoadipic acid	162.0755	4.92	$C_6H_{11}NO_4$	4.20	0.78	4.70×10^{-3}
N-acetyl-D-galactosamine	220.0826	5.16	$C_8H_{15}NO_6$	1.29	0.46	4.80×10^{-3}
Deoxyguanosine 5'-monophosphate	346.0556	6.99	$C_{10}H_{14}N_5O_7P$	0.55	0.36	5.54×10^{-3}
PE(16:0/18:1)	716.5231	26.78	$C_{39}H_{76}NO_8P$	0.62	1.54	6.36×10^{-3}
PS(16:0/16:1)	732.4822	26.53	$C_{38}H_{72}NO_{10}P$	0.07	0.67	7.14×10^{-3}
PE(18:0/18:1)	766.5385	27.08	$C_{41}H_{80}NO_8P$	1.01	1.38	8.03×10^{-3}
Succinate	117.0188	6.72	$C_4H_6O_4$	2.85	0.64	8.12×10^{-3}
CMP	322.0447	4.74	$C_9H_{14}N_3O_8P$	0.49	0.66	8.79×10^{-3}
Glycerylphosphorylethanolamine	214.0486	4.49	$C_5H_{14}NO_6P$	11.19	0.75	1.19×10^{-2}
Phosphocreatine	212.0427	4.97	$C_4H_{10}N_3O_5P$	3.82	0.69	1.19×10^{-2}
PE-NMe(22:5/18:1)	850.5601	29.08	$C_{46}H_{80}NO_8P$	0.10	1.50	1.29×10^{-2}
D-ribose 5-phosphate	229.0119	4.46	$C_5H_{11}O_8P$	0.08	0.44	1.29×10^{-2}
Indolelactic acid	204.0666	18.62	$C_{11}H_{11}NO_3$	0.13	1.33	1.31×10^{-2}
Pyridoxine	170.0806	6.74	$C_8H_{11}NO_3$	0.66	1.29	1.38×10^{-2}
Cytidine diphosphate choline	489.1149	5.33	$C_{14}H_{26}N_4O_{11}P_2$	0.64	0.68	1.40×10^{-2}
N-acetyl-D-glucosamine	222.0968	5.17	$C_8H_{15}NO_6$	0.50	0.80	1.42×10^{-2}
Cytidine 2',3'-cyclic monophosphate	306.0486	5.17	$C_9H_{12}N_3O_7P$	0.14	1.37	1.44×10^{-2}
Folic acid	442.1468	17.85	$C_{19}H_{19}N_7O_6$	0.34	2.45	1.45×10^{-2}
PE(15:0/24:1)	832.6058	28.87	$C_{44}H_{86}NO_8P$	0.73	1.34	1.52×10^{-2}

Abbreviations: PE—phosphatidylethanolamine, PG—phosphatidylglycerol, and PS—phosphatidylserine. [1] Error in ppm. [2] Fold change AA/control.

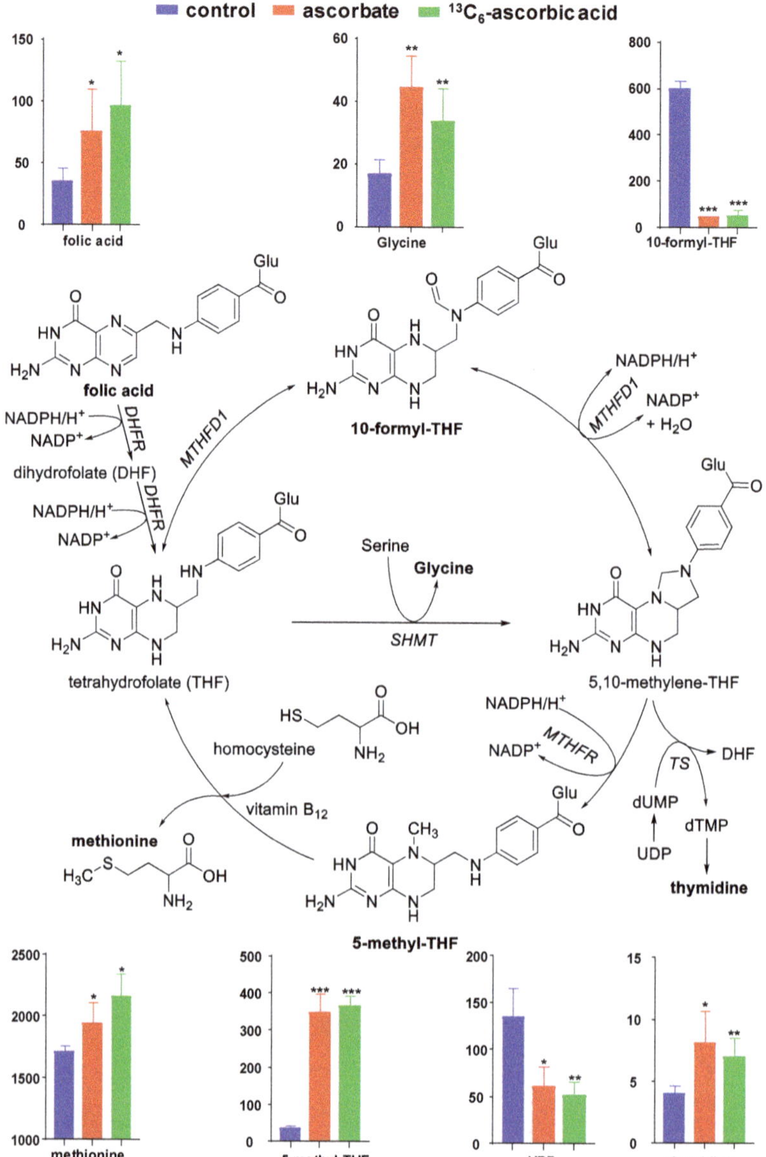

Figure 3. Simplified one-carbon cycle and associated pathways. Relative cellular levels of indicated metabolites are presented in the bar graphs (y-axes indicate peak areas expressed as 10^3 counts per second). Enzyme abbreviations: DHFR, dihydrofolate reductase; Glu, glutamic acid; MTHFD1, 5,10-methylene-THF dehydrogenase 1; MTHFR, 5,10-methylene-THF reductase; SHMT, serine hydroxymethyl transferase; TS, thymidylate synthase. Asterisks (*) indicate statistical significance between the control and treatment (*: $p \leq 0.05$, **: $p \leq 0.01$, ***: $p \leq 0.001$).

4. Discussion

The most significant changes in our untargeted metabolomics data as the result of exposure to AA were related to the folate-dependent one-carbon cycle and associated pathways that branch off this cycle (Figure 3). The observed AA-related decrease in 10-formyl-THF and concomitant increase in the cellular levels of 5-methyl-THF indicates that AA increases the activity of the folate-dependent one-carbon cycle. Stokes made a similar observation in a scurvy patient whose major urinary folate was 10-formyl-THF before AA treatment and 5-methyl-THF after AA therapy, concluding that AA "keeps the folate metabolism pool in action" [19]. Although we did not detect 5,10-methylene-THF, the intermediate in the conversion of 10-formyl-THF into 5-methyl-THF, we did observe an AA-related increase in the 5,10-methylene-THF-dependent formation of thymidine (a dTMP metabolite) and decrease in UDP. In DNA synthesis, UDP forms dUMP, which is the substrate of thymidylate synthase to form dTMP. THF can be converted directly into 5-methyl-THF by the enzyme serine hydroxymethyl transferase (SHMT, Figure 3), which consumes serine and produces glycine [20]. Although we observed a significant increase in the levels of glycine following AA treatment with no change in serine levels, we cannot draw conclusions regarding this pathway because both amino acids are components of the DMEM growth medium and because both amino acids are non-essential and can be formed and consumed in a multitude of other pathways. Regarding 5-methyl-THF as a co-substrate for the methionine synthase-mediated conversion of homocysteine into methionine, we observed an increase in the levels of methionine (Figure 3). This increase in methionine reflects its regeneration from homocysteine because the total amount of methionine, an essential amino acid, was provided by the growth medium. Taken together, the AA-induced changes in the metabolome of C2C12 cells appear to be a reflection of AA facilitating the folate-dependent one-carbon cycle and associated biochemical pathways (Figure 3). As the steps involved in the reductive conversion of 10-formyl-THF into 5-methyl-THF via 5,10-methylene-THF, are catalyzed by the NAD(P)H-dependent enzymes MTHFD and MTHFR, respectively, it is conceivable that AA plays a role in sparing NAD(P)H for reductive folate cycling. In scorbutic guinea pigs, for instance, cells can utilize a bioavailable form of glutathione (GSH) to delay the fall in tissue AA concentrations [21]. The authors conclude that AA and GSH "can spare each other" [21]. Because both AA and GSH require reducing equivalents from NADPH for the enzymatic recovery of their reduced forms in redox reactions [8], AA supplementation effectively spares NADPH by partially covering for GSH. The available NADPH can be used for a multitude of dehydrogenases involved in the detoxification ROS as well as their products [8]. We have previously observed that THP1 monocytes grown in the absence of AA showed greater cellular protein damage than AA-adequate cells due to AA's ability to facilitate the detoxification of the lipid peroxidation product, 4-hydroxy-2(*E*)-nonenal [22], by enhancing the activities of the NADPH-dependent enzyme aldo-keto reductase and the NADH-dependent enzyme alcohol dehydrogenase [23]. AA can also directly scavenge the lipid peroxidation-derived acrolein by forming a Michael adduct [3] which would spare NADPH for other biochemical pathways, including the reductive part of the folate cycle. Conversely, Fan et al. [24] have shown that in proliferating HEK293T cells, grown in the absence of AA, the folate cycle turns counterclockwise to generate 10-formyl-THF and NADPH. This finding is in agreement with our results that in AA-deficient cells the levels of 10-formyl-THF are higher than in AA-adequate cells. We attribute the lack of change in cellular GSH and NADPH levels following AA supplementation to compensatory effects.

The distribution of folate metabolite pools in our study is in agreement with the principal folate metabolites found in foods, i.e., 10-formyl-THF and 5-formyl-THF [20]. However, food folates and intracellular folates of mammalian cells are primarily present in the form of polyglutamic acid derivatives that contain 3–9 Glu units conjugated with the carboxyl group of the *p*-aminobenzoic acid moiety of folate. In vivo, these polyglutamate units are hydrolyzed to mono-Glu forms which are recognized by specific folate receptors for cellular uptake [20]. We did not detect oligo- or poly-Glu forms of the folate metabolites in the lysates of C2C12 cells, which indicates that the form provided by the growth medium, i.e., the mono-Glu form of oxidized folic acid, is not converted into poly-Glu forms in C2C12 cells, which requires pteroylpolyglutamate synthetase (ligase) enzymes. By contrast, human fibroblasts and Chinese hamster ovary cells grown in culture accumulate polyglutamate folates [25,26]. The higher levels of oxidized monoglutamate folic acid in the AA-supplemented cells cannot be explained by higher ligase activity in the AA-deficient cells and probably reflects higher degradation of folates into pteridine and *p*-aminobenzoic acid products in the control cells.

10-Formyl-THF is a co-substrate for the de novo biosynthesis of purines as it delivers two carbon atoms of the purine ring structure [11]. Although we observed lower levels of 10-formyl-THF in the AA-treated cells consistent with its utilization for purine biosynthesis, the expected increase in levels of the purine, guanine, did not reach significance (1.3-fold, $p = 0.055$). There are multiple explanations for the metabolic changes because purine biosynthesis involves other substrates, including glutamine, glycine, and aspartate, and because there are multiple ways to interconvert purine forms such as their (deoxy)ribonucleoside and (deoxy)ribonucleotides [11]. We detected only one *N*-formyl derivative of THF which we identified as 10-formyl-THF by careful analysis of its MS/MS fragmentation pattern which unequivocally placed the formyl substituent on the nitrogen atom of the *p*-aminobenzoic acid moiety of THF, i.e., position 10 in THF. We excluded the possibility that the *N*-formyl form was 5-formyl-THF based on chromatographic and mass spectral comparison with an authentic sample of 5-formyl-THF. The detection of only one *N*-formyl regioisomer is somewhat unexpected because both forms can be reversibly interconverted by isomerization [11] and thus both would be detected assuming similar detection limits.

Despite the increase in the relative cellular levels of methionine following AA exposure, we observed an AA-related decrease in methionine sulfoxide. Methionine is highly sensitive to oxidation by reactive oxygen species (ROS) and thus the levels of methionine sulfoxide can be considered a cellular biomarker of oxidative stress. The lower levels of methionine sulfoxide in the AA-treated cells compared to the control cells reflects the cellular antioxidant activity of AA (Figure S3). This effect can be explained by AA's ability to scavenge ROS, thus preventing oxidation of methionine, or by AA facilitating the enzymatic recovery of methionine from methionine sulfoxide, which is catalyzed by methionine sulfoxide reductase (Msr) in a thioredoxin (Trx)- and NADPH-dependent manner [8,27,28]. As thioredoxin is integral to preserving an adequate redox status of cells as well, for instance by recovering reduced glutathione (GSH) from oxidized glutathione (GSSG) [29], AA indirectly assists Trx in its antioxidant activity by scavenging ROS thereby sparing NADPH needed for the regeneration of Trx from its oxidized form.

Elevated plasma levels of homocysteine (Hcy) are associated with cardiovascular disease while a causal relationship has not been established and the underlying mechanism(s) remain(s) poorly understood. Several clinical studies have tested the hypothesis that lowering Hcy by nutritional or pharmacological means reduces cardiovascular risk [30]. One such approach is folate therapy (with or without co-supplementation with vitamins B6 and B12) aimed at facilitating the conversion of Hcy into methionine by increasing the activity of the 5-methyl-THF-dependent enzyme, methionine synthase. A meta-analysis of several such trials revealed an average 25% reduction in plasma Hcy levels but the folate supplementation had no significant effect on major vascular events [30]. On the other hand, the China Stroke Primary Prevention Trial with 20,702 individuals reported a significant beneficial effect of folate supplementation in primary stroke prevention [31]. Unsatisfactory evidence-based support for the folate/Hcy hypothesis has led to the notion that perhaps Hcy is not the root cause of cardiovascular disease but that it creates or reflects a prooxidant environment characterized by decreased bioavailability of nitric oxide which negatively affects vascular function. Others have suggested that prolonged exposure to high plasma Hcy levels induces proatherogenic epigenetic alterations that persist long after subsidence of plasma Hcy [32]. Therefore, low-cost therapies that lower Hcy levels may be attractive for patients with hyperhomocysteinemia.

In an observational study with 20 men and 40 women older than 50 years, serum Hcy correlated negatively with circulating levels of ascorbate ($r = -0.30$, $p < 0.05$) and folate ($r = -0.31$, $p < 0.05$) [33]. AA supplementation (500 mg/day) resulted in a significant 40% increase in erythrocyte folate in a study population comprising 65 male and 35 female Italian smokers without lowering Hcy levels [10]. High-dose AA supplementation (4.5 g/day) in 44 patients with coronary heart disease and without clinical ascorbate deficiency did not change plasma Hcy levels despite a marked increase in plasma ascorbate [34]. Our present cell culture study suggests that co-supplementation with folic acid and AA may activate the folate-mediated one-carbon cycle thereby promoting the conversion of Hcy into methionine while reducing cellular oxidative stress. ClinicalTrials.gov lists some 20 studies aimed at lowering Hcy with folic acid, combinations of B vitamins, prescription drugs, or with vitamin-prescription drug combinations, but none of the studies include(d) co-supplementation with AA. [35]. Acknowledging the limitation that the experiments were performed with one non-human-derived cell line, we propose a mechanism-based rationale for co-supplementation with folic acid and AA in humans, especially in elderly patients diagnosed with hyperhomocysteinemia and who are at risk for developing inadequate vitamin C status [2].

5. Conclusions

Our experiments with C2C12 myoblasts grown in the presence and absence of AA suggest that AA activates the folate-mediated one-carbon cycle thereby promoting the 5,10-methylene-THF-dependent formation of thymidine and facilitating the 5-methyl-THF-dependent conversion of Hcy into methionine. From a mechanistic perspective, AA appears to spare NAD(P)H and so increase availability of reducing equivalents for the reductive steps from 10-formyl-THF to 5-methyl-THF via 5,10-methylene-THF. The findings are relevant to nutritional or pharmacological treatment of hyperhomocysteinemia because the activity of the folate cycle appears to depend on AA concentration. Furthermore, our mechanistic findings provide a rationale for exploring the benefits of low-cost co-supplementation with AA in folate therapy of hyperhomocysteinemia.

Supplementary Materials: The following are available online at http://www.mdpi.com/2076-3921/9/3/217/s1, **Figure S1:** Principal component analysis of the metabolomes of sham- and AA-treated C2C12 cells. Each set of dots represents a sample analyzed by LC–MS/MS. QC samples were analyzed every fifth LC run. Metabolites with CV > 25% were excluded from the analysis. **Figure S2:** Extracted ion chromatograms showing the total ion currents of the $[M - H]^-$ ion of AA. A representative sample from each experimental group is displayed. The m/z 175.028 $[M - H]^-$ ion of AA was absent in the control sample (blue) and present only in the sodium ascorbate treatment group (pink). The m/z 181.045 $[M - H]^-$ ion corresponding to $^{13}C_6$-ascorbic acid was only observed in samples treated with this AA isotopologue. MS/MS spectra recorded for the two AA isotopologues are displayed in the panels on the right. **Figure S3. (a)** Relative levels of methionine and methionine sulfoxide (Met-SO), and **(b)** Relative levels of Met-SO scaled to 100%. Panel **(a)** shows that the AA-related increase in methionine does not account for the decrease in Met-SO in the AA-treated cells. Asterisks (*) indicate statistical significance between the control and treatment (*: $p \leq 0.05$, ***: $p \leq 0.001$). **Table S1:** Compounds with CV < 25% on QC samples injected every five LC runs. Level 1 annotations were obtained from an in-house compound library consisting of >700 authentic IROA standards (IROA Technology, Bolton, MA). Level 2 annotations were obtained by Progenesis QI^{TM} software with Metlin^{TM} plugin V1.0.6499.51447 (NonLinear Dynamics, United Kingdom) using experimental MS/MS fragmentation, and Human Metabolome Database (HMDB) using in silico MS/MS fragmentation. Adduct declustering was implemented by using the Progenesis deconvolution algorithm. **Table. S2.** One-way analysis of variance followed by Tukey's HSD post-hoc analysis and Holm FDR correction. Annotated compounds sorted by alphabetical order. Fold change and p-value are presented for the comparison of AA/Control.

Author Contributions: Conceptualization, A.A.M. and J.F.S.; methodology, A.A.M., C.L.M., C.S.M., and J.F.S.; formal analysis, A.A.M. and J.F.S.; investigation, A.A.M. and R.K.; resources, J.F.S. and C.S.M.; data curation, A.A.M.; writing—original draft preparation, A.A.M., R.L.R., and J.F.S.; writing—review and editing, all authors.; visualization, A.A.M.; supervision, J.F.S. and C.S.M.; project administration, J.F.S; funding acquisition, J.F.S., C.S.M., and C.L.M. All authors have read and agreed to the published version of the manuscript.

Funding: This research was funded by The National Institutes of Health, grant number S10RR027878, and the OSU Foundation Buhler-Wang Research Fund.

Acknowledgments: The authors thank Jeffrey Morré (OSU Mass Spectrometry Center) for technical assistance.

Conflicts of Interest: The authors declare no conflict of interest.

References

1. Higdon, J.V.; Frei, B. *Vitamin C: An Introduction, in The Antioxidant Vitamins C and E*; Packer, L., Traber, M.G., Kraemer, K., Frei, B., Eds.; AOAC Press: Champaign, IL, USA, 2002; pp. 11–16.
2. Traber, M.G.; Stevens, J.F. Vitamins C and E: Beneficial effects from a mechanistic perspective. *Free Radic Biol. Med.* **2011**, *51*, 1000–1013. [CrossRef]
3. Kesinger, N.G.; Langsdorf, B.L.; Yokochi, A.F.; Miranda, C.L.; Stevens, J.F. Formation of a vitamin C conjugate of acrolein and its paraoxonase-mediated conversion into 5,6,7,8-tetrahydroxy-4-oxooctanal. *Chem. Res. Toxicol.* **2010**, *23*, 836–844. [CrossRef]
4. Kesinger, N.G.; Stevens, J.F. Covalent interaction of ascorbic acid with natural products. *Phytochemistry* **2009**, *70*, 1930–1939. [CrossRef] [PubMed]
5. Kuiper, H.C.; Bruno, R.S.; Traber, M.G.; Stevens, J.F. Vitamin C supplementation lowers urinary levels of 4-hydroperoxy-2-nonenal metabolites in humans. *Free Radic Biol. Med.* **2011**, *50*, 848–853. [CrossRef] [PubMed]
6. Kirkwood, J.S.; Lebold, K.M.; Miranda, C.L.; Wright, C.L.; Miller, G.W.; Tanguay, R.L.; Barton, C.L.; Traber, M.G.; Stevens, J.F. Vitamin C deficiency activates the purine nucleotide cycle in zebrafish. *J. Biol. Chem.* **2012**, *287*, 3833–3841. [CrossRef] [PubMed]
7. Axton, E.R.; Cristobal, E.; Choi, J.; Miranda, C.L.; Stevens, J.F. Metabolomics-Driven Elucidation of Cellular Nitrate Tolerance Reveals Ascorbic Acid Prevents Nitroglycerin-Induced Inactivation of Xanthine Oxidase. *Front Pharmacol.* **2018**, *9*, 1085. [CrossRef] [PubMed]
8. Halliwell, B.; Gutteridge, C.J.M. *Free Radicals in Biology and Medicine*, 3rd ed.; Oxford University Press: Oxford, UK, 1999; p. 936.
9. Lucock, M.; Yates, Z.; Boyd, L.; Naylor, C.; Choi, J.H.; Ng, X.; Skinner, V.; Wai, R.; Kho, J.; Tang, S.; et al. Vitamin C-related nutrient-nutrient and nutrient-gene interactions that modify folate status. *Eur. J. Nutr.* **2013**, *52*, 569–582. [CrossRef]
10. Cafolla, A.; Dragoni, F.; Girelli, G.; Tosti, M.E.; Costante, A.; De Luca, A.M.; Funaro, D.; Scott, C.S. Effect of folic acid and vitamin C supplementation on folate status and homocysteine level: A randomised controlled trial in Italian smoker-blood donors. *Atherosclerosis* **2002**, *163*, 105–111. [CrossRef]
11. Berg, J.M.; Tymoczko, J.L.; Stryer, L. *Biochemistry*, 5th ed.; W.H. Freeman and Company: New York, NY, USA, 2001.
12. Verhoef, P.; Stampfer, M.J. Prospective studies of homocysteine and cardiovascular disease. *Nutr. Rev.* **1995**, *53*, 283–288. [CrossRef]
13. Mix, J.A. Do megadoses of vitamin C compromise folic acid's role in the metabolism of plasma homocysteine? *Nutr. Res.* **1999**, *19*, 161–165. [CrossRef]
14. Drouin, G.; Godin, J.R.; Page, B. The genetics of vitamin C loss in vertebrates. *Curr. Genom.* **2011**, *12*, 371–378. [CrossRef] [PubMed]
15. Kirkwood, J.S.; Legette, L.L.; Miranda, C.L.; Jiang, Y.; Stevens, J.F. A metabolomics-driven elucidation of the anti-obesity mechanisms of xanthohumol. *J. Biol. Chem.* **2013**, *288*, 19000–19013. [CrossRef] [PubMed]
16. Sumner, L.W.; Amberg, A.; Barrett, D.; Beale, M.H.; Beger, R.; Daykin, C.A.; Fan, T.W.; Fiehn, O.; Goodacre, R.; Griffin, J.L.; et al. Proposed minimum reporting standards for chemical analysis Chemical Analysis Working Group (CAWG) Metabolomics Standards Initiative (MSI). *Metabolomics* **2007**, *3*, 211–221. [CrossRef] [PubMed]
17. Viant, M.R.; Kurland, I.J.; Jones, M.R.; Dunn, W.B. How close are we to complete annotation of metabolomes? *Curr. Opin. Chem. Biol.* **2017**, *36*, 64–69. [CrossRef]
18. Chong, J.; Wishart, D.S.; Xia, J. Using MetaboAnalyst 4.0 for Comprehensive and Integrative Metabolomics Data Analysis. *Curr. Protoc. Bioinform.* **2019**, *68*, e86. [CrossRef]
19. Stokes, P.; Melikian, V.; Leeming, R.L.; Portman-Graham, H.; Blair, J.A.; Cooke, W.T. Folate metabolism in scurvy. *Am. J. Clin. Nutr.* **1975**, *28*, 126–129. [CrossRef]
20. Gropper, S.; Smith, J.; Groff, J. Folic acid. In *Advanced Nutrition and Human Metabolism*; Thomson Wadsworth: Belmont, CA, USA, 2005; pp. 301–309.
21. Mårtensson, J.; Han, J.; Griffith, O.W.; Meister, A. Glutathione ester delays the onset of scurvy in ascorbate-deficient guinea pigs. *Proc. Natl. Acad. Sci. USA* **1993**, *90*, 317–321. [CrossRef]

22. Chavez, J.; Chung, W.G.; Miranda, C.L.; Singhal, M.; Stevens, J.F.; Maier, C.S. Site-specific protein adducts of 4-hydroxy-2(E)-nonenal in human THP-1 monocytic cells: Protein carbonylation is diminished by ascorbic acid. *Chem. Res. Toxicol.* **2010**, *23*, 37–47. [CrossRef]

23. Miranda, C.L.; Reed, R.L.; Kuiper, H.C.; Alber, S.; Stevens, J.F. Ascorbic acid promotes detoxification and elimination of 4-hydroxy-2(E)-nonenal in human monocytic THP-1 cells. *Chem. Res. Toxicol.* **2009**, *22*, 863–874. [CrossRef]

24. Fan, J.; Ye, J.; Kamphorst, J.J.; Shlomi, T.; Thompson, C.B.; Rabinowitz, J.D. Quantitative flux analysis reveals folate-dependent NADPH production. *Nature* **2014**, *510*, 298–302. [CrossRef]

25. Hilton, J.G.; Cooper, B.A.; Rosenblatt, D.S. Folate polyglutamate synthesis and turnover in cultured human fibroblasts. *J. Biol. Chem.* **1979**, *254*, 8398–8403. [PubMed]

26. Cichowicz, D.J.; Foo, S.K.; Shane, B. Folylpoly-gamma-glutamate synthesis by bacteria and mammalian cells. *Mol. Cell Biochem.* **1981**, *39*, 209–228. [CrossRef] [PubMed]

27. Lee, B.C.; Gladyshev, V.N. The biological significance of methionine sulfoxide stereochemistry. *Free Radic Biol. Med.* **2011**, *50*, 221–227. [CrossRef] [PubMed]

28. Moskovitz, J. Methionine sulfoxide reductases: Ubiquitous enzymes involved in antioxidant defense, protein regulation, and prevention of aging-associated diseases. *Biochim. Biophys. Acta* **2005**, *1703*, 213–219. [CrossRef] [PubMed]

29. Lu, J.; Holmgren, A. The thioredoxin antioxidant system. *Free Radic Biol. Med.* **2014**, *66*, 75–87. [CrossRef]

30. Santilli, F.; Davi, G.; Patrono, C. Homocysteine, methylenetetrahydrofolate reductase, folate status and atherothrombosis: A mechanistic and clinical perspective. *Vascul. Pharmacol.* **2016**, *78*, 1–9. [CrossRef]

31. Huo, Y.; Li, J.; Qin, X.; Huang, Y.; Wang, X.; Gottesman, R.F.; Tang, G.; Wang, B.; Chen, D.; He, M.; et al. Efficacy of folic acid therapy in primary prevention of stroke among adults with hypertension in China: The CSPPT randomized clinical trial. *JAMA* **2015**, *313*, 1325–1335. [CrossRef]

32. Krishna, S.M.; Dear, A.; Craig, J.M.; Norman, P.E.; Golledge, J. The potential role of homocysteine mediated DNA methylation and associated epigenetic changes in abdominal aortic aneurysm formation. *Atherosclerosis* **2013**, *228*, 295–305. [CrossRef]

33. Cascalheira, J.F.; Parreira, M.C.; Viegas, A.N.; Faria, M.C.; Domingues, F.C. Serum homocysteine: Relationship with circulating levels of cortisol and ascorbate. *Ann. Nutr. Metab.* **2008**, *53*, 67–74. [CrossRef]

34. Bostom, A.G.; Yanek, L.; Hume, A.L.; Eaton, C.B.; McQuade, W.; Nadeau, M.; Perrone, G.; Jacques, P.F.; Selhub, J. High dose ascorbate supplementation fails to affect plasma homocyst(e)ine levels in patients with coronary heart disease. *Atherosclerosis* **1994**, *111*, 267–270. [CrossRef]

35. Savini, I.; Catani, M.V.; Duranti, G.; Ceci, R.; Sabatini, S.; Avigliano, L. Vitamin C homeostasis in skeletal muscle cells. *Free Radic Biol. Med.* **2005**, *38*, 898–907. [CrossRef] [PubMed]

 *antioxidants*

Review

Potential Protective Role Exerted by Secoiridoids from *Olea europaea* L. in Cancer, Cardiovascular, Neurodegenerative, Aging-Related, and Immunoinflammatory Diseases

María Luisa Castejón, Tatiana Montoya, Catalina Alarcón-de-la-Lastra and Marina Sánchez-Hidalgo *

Department of Pharmacology, School of Pharmacy, University of Seville, 41012 Sevilla, Spain;
mcastejon1@us.es (M.L.C.); tmontoya@us.es (T.M.); calarcon@us.es (C.A.-d.-l.-L.)
* Correspondence: hidalgosanz@us.es; Tel.: +34954557469; Fax: +34954556074

Received: 30 December 2019; Accepted: 7 February 2020; Published: 10 February 2020

Abstract: Iridoids, which have beneficial health properties, include a wide group of cyclopentane [*c*] pyran monoterpenoids present in plants and insects. The cleavage of the cyclopentane ring leads to secoiridoids. Mainly, secoiridoids have shown a variety of pharmacological effects including anti-diabetic, antioxidant, anti-inflammatory, immunosuppressive, neuroprotective, anti-cancer, and anti-obesity, which increase the interest of studying these types of bioactive compounds in depth. Secoiridoids are thoroughly distributed in several families of plants such as Oleaceae, Valerianaceae, Gentianaceae and Pedialaceae, among others. Specifically, *Olea europaea* L. (Oleaceae) is rich in oleuropein (OL), dimethyl-OL, and ligstroside secoiridoids, and their hydrolysis derivatives are mostly OL-aglycone, oleocanthal (OLE), oleacein (OLA), elenolate, oleoside-11-methyl ester, elenoic acid, hydroxytyrosol (HTy), and tyrosol (Ty). These compounds have proved their efficacy in the management of diabetes, cardiovascular and neurodegenerative disorders, cancer, and viral and microbial infections. Particularly, the antioxidant, anti-inflammatory, and immunomodulatory properties of secoiridoids from the olive tree (*Olea europaea* L. (Oleaceae)) have been suggested as a potential application in a large number of inflammatory and reactive oxygen species (ROS)-mediated diseases. Thus, the purpose of this review is to summarize recent advances in the protective role of secoiridoids derived from the olive tree (preclinical studies and clinical trials) in diseases with an important pathogenic contribution of oxidative and peroxidative stress and damage, focusing on their plausible mechanisms of the action involved.

Keywords: immunomodulation; inflammation; olive tree; oxidative stress; secoirioids

---

1. Introduction

Iridoids, which have beneficial health properties, include a wide group of cyclopentane [*c*] pyran monoterpenoids present in plants and insects. The name iridoid derived from *Iridomyrmex*, a genus of fornices from which iridomirmecin and iridodial compounds were isolated. These products have been considered as defensive compounds. In fact, the biosynthesis of these derivatives of monoterpenes takes place in the different organisms by similar pathways; defense is its main role, and in the case of insects, they are used as sex pheromones [1].

Iridoids were first isolated in the latter part of the 19th century, but Halpern and Schmid proposed the basic skeleton of the iridoids in their investigation of the structure of plumieride in 1958 [2]. Particularly, they are secondary metabolites of terrestrial and marine flora and fauna, being found in a large number of plants families, usually as glycosides. For this reason, some of them are chemotaxonomically useful as markers of genus in various plant families. Besides, they exhibit a wide range of bioactivities including anti-inflammatory, antibacterial, anti-carcinogenic, and antiviral activities [3]. In fact, they are used as bitter tonics, sedatives, antipyretics, cough drugs, remedies for skin disorders, and as hypotensive agents. In addition, they are useful as an antidote in mushroom intoxications by *Amanita* type.

The cleavage of the cyclopentane ring of iridoids leads to secoiridoids. Mainly, secoiridoids have shown a large variety of pharmacological properties including anti-diabetic, anti-inflammatory, immunosuppressive, neuroprotective, anti-cancer, and anti-obesity. This fact encouraged us to study the bioactivities of these phytochemicals in depth and update the latest preclinical and clinical data on its bioactivity and potential therapeutic uses.

1.1. Structure and Classification

Several classifications have been developed over the years, given the variety and complexity of iridoids and secoiridoids [4,5].

From 1980 to date, bibliographic data has used the classification proposed by El-Naggar and Beal [2], who categorize these compounds according to the number of carbons included in their structure:

- Group 1: C_8 iridoids (di-nor-iridoids)
- Group 2: C_9 iridoids (nor-iridoids)
- Group 3: C_{10} iridoids, which occur mainly as glycosides
- Group 4: Aglycones and some iridoids included in the other three groups lacking a sugar residue in their structure
- Group 5: Iridoids derivatives. This group comprises compounds derived from the opening of the pyran ring
- Group 6: Included bis-iridoids as a result of condensation of two monomers, (a) directly as in iridolinalin A, or (b) through a sugar residue as in globuloside A.

At the same time, there are other classifications of secoiridoids according to the presence of these compounds in certain families, including the Oleaceae family. In fact, a total of 232 secoiridoids (aglycones, glycosides, derivatives, and dimers) have been isolated from nine genus of the family Oleaceae. These genera include Fontanesia, Fraxinus, Jasminum, Ligustrum, Olea, Osmanthus, Phillyrea, Picconia, and Syringa, and these secoiridoids were classified into other five groups [6]:

- Simple secoiridoids: Generally, for the simple secoiridoids, positions C_7 and C_{11} have either a free carboxylic acid group or a methyl ethyl ester derivative of the acid. The configurations of the positions C_1 and C_5 are *S*.
- Conjugated secoiridoids: This group of compounds is the most numerous secoiridoids isolated from the Oleaceae family. The name of the class derives from the type of compound that is linked or conjugated to the secoiridoid nucleus. Based on this, this class is further categorized into seven subgroups: aromatic-conjugated, sugar-conjugated, terpene-conjugated, cyclopentane-conjugated, coumarin-conjugated, lignans-conjugated, and other secoiridoids. Normally, the conjugations occur in C_7 due to this position, which is is usually oxidized to a carboxylic acid and esterified with diverse groups.
- 10-Oxyderivative of oleoside secoiridoids: This group contains the oleoside nucleus with distinct structural differences. The C_8 and C_9 positions exist as double bonds, with a hydroxy group at the C_8 position or an ester formed by an oxygen atom with different groups. A total of 40 10-Oxyderivative of oleoside secoiridoids have been isolated from the Oleaceae family.
- Z-Secoiridoids: This class of secoiridoids is characterized by the presence of double-bond geometry at the C_8 in Z-configuration; however, only five compounds have been isolated from the Oleaceae family.

- Secologanosides and oxidized secologanoside secoiridoids: Compounds of this class are based on the secologanoside nucleus. They are differentiated by the positions on the C–C double bond between C_8 and C_{10} and C_{10} oxidation level.

1.2. Main Naturally Occurring Iridoids and Secoiridoids Present in Olea europaea L

Iridoids and secoriridoids are thoroughly distributed in the plants of class Magnoliopside, concretely belonging to the following families: *Scropulariaceae, Verbenaceae, Lamiaceae, Apocynaceae, Loganiaceae, Bignoniaceae, Plantaginaceae, Rubiaceae, Pedaliaceae, Cornaceae, Acantheceae, Loasaceae, Lentibulariaceae, Gentianaceae, Oleaceae, Nyctanthaceae, Caprifoliaceae, Dispsacaceae*, and *Valerianaceae*.

For instance, *Valeriana officinalis* L. (Valerianaceae), *Harpagophytum procumbens* L. (Pedaliaceae), *Genciana lutea* L. (Gentianaceae), *Fraxinus excelsior* L. (Oleaceae) and *Olea europaea* L. (Oleaceae) are the most representative medicinal plants commonly used in medicine, due to their iridod/secoiridoid content [3]. Particularly, *Olea europaea* L. (Oleaceae) is a small evergreen tree with firm branches and a grayish bark. The leaves are lanceolate, opposite, short-petioled, mucronate, green above and hoary on the underside. On the other hand, the flowers are small, short, erect racemes, axillary, very much shorter than the leaves, and the fruit is a small smooth, purple, or green drup, with a nauseous, bitter flesh, enclosing a sharp-pointed stone [7].

Olea europaea L. preparations have been traditional used in folk medicine in the European Mediterranean area, Arabia peninsula, India and other tropical and subtropical regions, as a diuretic, emollient, hypotensive, and for urinary and bladder infections [8]. Most of the plant parts of *Olea europaea* L. are used in the traditional system of medicine around the world. Oil is taken with lemon juice to treat gall stones [9]. Leaves are taken orally for stomach and intestinal diseases and used as mouth cleanser [10], and the decoction of dried leaves is taken orally for diabetes [11]. An extract of the fresh leaves is taken orally to treat hypertension and to induce diuresis [12], whereas an infusion of the fresh leaves is taken orally as an alternative treatment for inflammatory diseases [13]. Similarly, essential oil extracted from the fruit is also used to treat rheumatism, promote blood circulation [14], and as a laxative [15].

The main biophenol secoiridoids found in the olive tree include: oleuropein (OL), dimethyl-OL, ligstroside, and their hydrolysis derivatives such as OL-aglycone, oleocanthal (OLE), oleacein (OLA), elenolate, oleoside-11-methyl ester, elenoic acid, hydroxytyrosol (HTy), and tyrosol (Ty) (Figure 1).

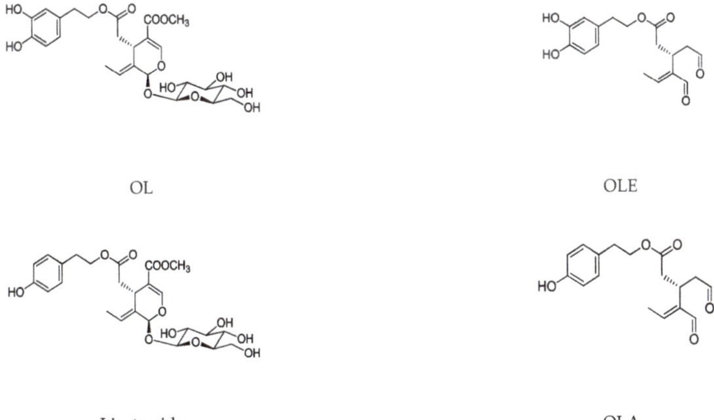

Figure 1. Chemical structures of secoiridoids most abundant in olive trees. OL: oleuropein, OLA: oleacein, OLE: oleocanthal.

1.3. Biosynthesis and Biotransformation of Secoiridoids in Olive Tree

The amount and distribution of secoiridoids present in olive tissues depend on various environmental factors such as the ripening cycle, geographical origin, and cultivation practices, among others. Besides, the content of phenolic glycosides as patterns and the activity of endogenous enzymes can play a role in the quantitative composition of secoiridoids in the olive tree [16].

The fact that secoiridoids are mostly present in early stages is due to the enzymatic and chemistry reactions that take place in the maturation time. In this sense, three different states in fruit maturation have been described: growth phase, green maturation phase, and black maturation phase, which is characterized by the presence of anthocyans.

OL is mainly abundant in early stages, although its levels decrease during the maturation process. In fact, OL decreases quickly in black crops and is not present in some varieties of Oleaceae.

The main precursor of OL and ligstroside is oleoside 11-methyl ester (elenolic acid glucoside). Firstly, geraniol synthase (GES) catalyzes the transformation of genaryl diphosphate to geraniol, which is converted to 10-hydroxygeraniol by the geraniol 10-hydroxylase enzyme. The iridioids in Oleaceae must be formed from this point with 10-hydroxygeraniol as the starting compound via irididal and iridotrial up to deoxyloganic acid, which is the precursor of loganin and loganic acid, as well as secologanin and secologanic acid [17]. From this point, up to five routes have been proposed to explain the origin of all iridoids found in this family. However, it is known that most of the secoiridoids present in *Olea europaea L.* are derived from deoxyloganic acid as a common intermediate [17,18]. Following this line, nicotinamide adenine dinucleotide deshydrogenase (NADH) acts on 10-hydroxygeraniol to form deoxyloganic acid aglucone. The transfer of glucosyl groups to deoxyloganic acid aglucone (precursor of monoterpene indolic alcaloids and OL) is catalyzed by glucosyltransferase (GT). Deoxyloganic acid experiments a 7-α-hydroxylation of the cyclopentane ring and forms 7-epiloganic acid, which quickly goes to 7-ketologanic acid through hydroxyl group oxidation. Loganic acid methyltransferase catalyzes 7-ketologanin syntheses. In this point, secologanin synthase (SLS) oxides a ketonic group to form 11-methyl oleoside, which is immediately glucosylated by GT. Finally, 7-β-1-D-glucopyranosyl-11-methyl oleoside is esterified with Ty to produce ligstroside, and then OL is formed [17,19] (Figure 2).

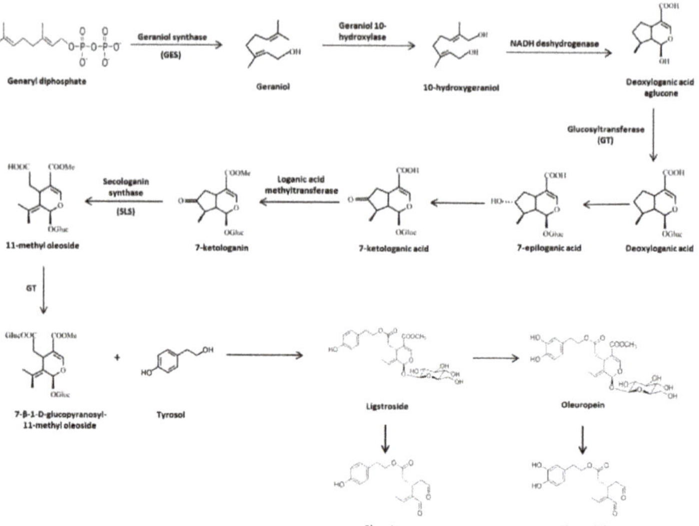

Figure 2. Biosynthesis and biotransformation of secoiridoids in *Olea europaea* L.

Secoiridoids are distributed throughout the tissues of the olive tree, but their nature and concentration change among different parts of the plant. Thus, biosynthetic or mechanicals transformation during production are decisive to quantify alterations of the bioactive small molecules [18]. Particularly, OL is the major secoiridoid constituent of unripe drupes (peel, pulp, and seed). The amount of OL decreases along fruit maturation, as commented above, whereas its aglycon form increases its levels. OL-aglycone is formed by the cleavage of the glycosidic bond mediated by β-glucosidase activity. Ligstroside has been described as a common phenolic component in different olive tissues (leaf, fruit pulp, and stone) and olive oil, but it has been rarely found in olive seeds [20].

In the course of maturation, OL and ligstroside are considered pattern components. Both of them are present in the olive fruit, but they are almost non-existent in olive oil (85–95% reduction) [21,22]. β-glucosidase acts by decreasing OL and ligstroside levels, aglycon forms from OL, and ligstroside can be detected as isomers due to the keto-enolic tautomeric equilibrium of the elenolic acid moiety [23,24].

Other dialdehydics structurally related to these secoiridoid precursors are OLA and OLE. Different authors have reported that both OLA and OLE levels increase during ripening due to OL and ligstroside degradation, respectively [25]. Thus, they concluded that OL and ligstroside are natural precursors of OLA and OLE as breakdown products resulting from enzymatic activity during the extraction and maturation processes [26–29].

OL and ligstroside have been detected in olive leaves. In turn, OLA and OLE levels are augmented in mature fruits, such as OL and ligstroside aglycons. Moreover, OLA and OL-aglycone are more plentiful in olive oil [16].

1.4. Functional and Physiological Chemistry of the Main Secoiridoids of Potential Medical Interest

ROS have a remarkable role in the development of oxidative stress and also in the pathology of numerous diseases. For example, oxidative stress is one of the major cellular features in the onset of many pathological conditions such as Alzheimer's and Parkinson disease, renal injury, diabetes, cardiovascular diseases, cancer and aging; it occurs when excessive ROS accumulation produced during the normal cell metabolic processes is unbalanced by the antioxidant defence system [30], and it may induce the oxidative modification of cellular macromolecules including lipids, proteins, and nucleic acids [31].

Previous epidemiological studies show that the Mediterranean diet is associated with a low incidence of cardiovascular disease or cancer. This may reflect the nutritional effects of the bioactive compounds contained in its major source of fatty acids, i.e., extra-virgin olive oil (EVOO), which is rich in phenolic compounds [32,33]. The phenolic compounds contained in EVOO include HTy, Ty, and their secoiridoids precursors such as OL, OLE, or OLA, among others.

A large number of studies highlighted the antioxidant properties of these compounds including their abilities to promote the activity of ROS-detoxifying enzymes, such as superoxide dismutase (SOD), catalase (CAT), glutathione reductase (GSR) and glutathione S-transferase (GST), to compete with coenzyme Q as an electron carrier in the mitochondrial electron transport chain, which is a site of ROS generation, to act as free-radical-scavenging antioxidants and to inhibit lipid peroxidation [31,34]. In particular, it was shown that dietary OL and OLE treatments were associated with reductions in the production of superoxide anion radical in human monocytes from healthy donors [35]. Similarly, OL administration significantly increased the SOD and glutathione peroxidase (GPx) activity levels in the obstructed kidneys from rats with unilateral ureteral obstruction induced-kidney injury [36]. These results are in accordance with other reports showing that OL was able to increase the amount of enzymes such as GPx and SOD in gentamicin-induced renal toxicity and cisplatin-induced renal injury models [37,38]. The phenolic compounds present in *Olea europaea* L. activated enzymatic and non-enzymatic antioxidant defense mechanisms, particularly preventing cell membrane damage by high-dose UV-B rays [39].

Vitamin E (α-tocopherol) is an essential micronutrient in the diet of all mammals, and it is a potent antioxidant in biological systems. This compound is the main chain-breaking antioxidant that prevents the propagation of free radicals reactions, and consequently prevents the tissue damage. In this sense, it has been reported that supplementing the diet with OL for 21 days maintained higher levels of α-tocopherol in liver of female Wistar rats [40].

The most important antioxidant activity of the olive tree phenolic compounds is related to the free-scavenging ability because they inhibit the propagation chain during the oxidation process through the donation of radical hydrogen to alkylproxyl radicals and the formation of stable derivatives during this reaction. These compounds also act as metal chelators, preventing the generation of high concentrations of hydroxyl radicals [41]. This capacity may be reduced by the presence of the $-COOOH_3$ fragment in several secoirioids structures, because it seems to cause a decrease in the antioxidant activity which is related to the inability of this group to act as an H-donor [42]. In fact, OL-aglycone presents better radical-scavenging capacity than single hydroxyl substitutions, such as Ty, and also protects low-density lipoprotein (LDL) from oxidation [43].

The relationship between oxidative stress and inflammation has been established by many authors. The pathogenic role of mixed advanced glycoxidation products (AGE) and advanced lipid peroxidation products (ALE) generated in the course of oxidative stress and their adducts with cell biomolecules, such as proteins and nucleic acids, in a number of chronic inflammatory and autoimmune diseases is well documented [44].

On the other hand, in the past years, the leaves of *Olea europaea* L. have been considered as an important source of antioxidant compounds. In this case, OL is the most predominant and active phenolic compound (60–90 mg/g dried olive leaves), and it is usually considered as an antioxidant reference in comparison with other secoirioids [45,46]. OL contains active components in its molecule with conferred a potential antioxidant activity. It has been suggested that these properties are related to the H-atom donation from the phenolic groups present in OL. For example, OL administration showed a protective effect against ROS production in endothelial cells since OL showed good cytocompatibility and antioxidant activity, which revealed effectiveness in controlling the oxidative stress upon exposure to H_2O_2 [47]. In addition, OL presented slightly weaker radical scavenging activity than HTy by 2,7-dichlorodihydrofluorescein diacetate (DCFH-DA) and ABTS methods [48] and the ability to inhibit LDL oxidation by scavenging free radicals [49].

Although the studies about the antioxidant activity of OLE are limited, it has been demonstrated that OLE could inhibit nicotinamide adenine dinucleotide phosphate oxidase (NOX) in isolated human monocytes, and also OLE was able to reduce intracellular ROS levels in SH-SY5Y cells. Recently, Montoya et al. have also showed that OLE produced a potent reduction of intracellular ROS and nitrites production in LPS-induced murine peritoneal macrophages [50].

OLA is one of the major phenolic compounds present in the olive tree, but its antioxidant profile is more unknown in comparison with other secoirioids, such as OL. However, OLA has demonstrated to be a potent scavenger of HOCl and myeloperoxidase release and exerts a stronger inhibitory capacity of neutrophil's oxidative in comparison to OL [32].

1.5. Pharmacokinetics of Secoiridoids from the Olive Tree

Numerous investigations have revelead the beneficial effects of olive leaves and olive oil for the treatment of many diseases. The possibility that their constituents may achieve any biological effects depends on the chance of reaching molecular targets in a specific tissue or organ at a sufficient dose, which is dependent on their metabolism and bioavailability [51].

The mechanism of intestinal absorption of olive oil phenols is unclear, as it does not appear to be strictly dependent on the polarity of these compounds. For example, OL, which is apolar, is able to diffuse through the lipid bilayer of the epithelial cell membrane. However, also, the more polar HTy appears to be absorbed via a passive bidirectional diffusion mechanism [52,53].

To date, reports regarding the bioavailability of the phenolic compounds present in olive oil are extended, whereas bioavailability studies concerning secoiridoids derivatives such as OL, OLE, and OLA have been scarcely studied. Thus, data on the bioavailability of these secoiridoids compounds in human and even animals would be of great interest in order to establish their potential health benefits.

Generally, secoiridoids from *Olea Europaea L.* are mainly present in glycosylated forms (OL and ligstroside). For this reason, after oral administration, enzyme from saliva starts a hydrolysis process continued in the stomach by digestive enzymes and β-glycosidases. The unmodified forms could be absorbed to the small intestine or colon, where they are hydrolyzed, finally [54]. The result of this hydrolytic process is the formation of derived aglycon forms. At the same time, the aglycons that were formed could be absorbed in the small intestine or colon. The chemical structure and vehicle of administration are decisive for the rate and extension of gastrointestinal absorption. Structural modifications occur via conjugation in small intestinal epithelial cells or in liver, after transport through the portal system. Metabolites reach the general blood circle, from which they are excreted in the urine [55].

Related to OL, some authors suggest that it could be absorbed in the small intestine or colon. In this sense, Kendall et al. reported that OL diffuses in the stomach and remained stable and intact at the gastric level during digestion being absorbed in the small intestine in healthy young adults [56]. Similarly, several studies have established that OL is stable at the gastric level during digestion, since the bioavailability of its main metabolite HTy is higher [57]. On the contrary, Corona et al. determined that OL was absorbed in the colon and degraded by gut microbiota in rat intestinal segment that produces HTy, which expresses biological activity [58]. Further studies developed in rats confirmed that OL and HTy were present in plasma, feces, and urine as such, and also conjugated as glucuronide after oral OL administration [59–61]. Besides, there are several studies that focus on the study of the colonic pathway of phenolic compounds present in the olive tree. For example, Mosele et al. have demonstrated that OL hydrolysis formed OL-aglycone, elenoic acid, HTy and HTy-Ac in vitro, which is probably a consequence of the hydrolytic transformation making all of them less resistant to the gastric acidic hydrolysis in comparison to OL and more sensitive to temperature, pH, and enzyme activity. However, OL administration detected other related metabolites such as homovanillic acid in rats [62]. These divergences could be explained taking into account the differences found between both murine and human models.

Surprisingly, it has been postulated that plasma peak at 1 hour (h) after OL administration for humans or 2 h for a murine model, although OL could be detected after 10 minutes (min) [56,63]. In addition, glucuronides and OL sulfate derivatives could be detected in plasma (at 23 min) or in the urine (at 8 h) after ingestion. Particularly, De Bock and colleagues described a heterogeneous effect in OL leaf extract bioavailability and metabolism that was dependent on a number of factors, including preparation (capsule/liquid) and gender [63]. In fact, they found that compared to capsules, OL leaf extract in a liquid formulation led to a greater OL peak levels and area under the curve (AUC) in plasma and described that males may be more efficient at conjugating OL.

The production of metabolites during the metabolism process of these compounds could be interesting given that these new compounds may also exert beneficial effects on the organism. The metabolism of secoiridoids can be carried out by phase I (hydrogeneration, hydroxylation, hydratation, etc) or phase II reactions (glucuronidation, methylation, sulfation, etc.). A perfused rat intestinal model determined that the major small intestine metabolites from OL-aglycone were glucoronide conjugates, so they are not absorbed in parental form [64]. OL-aglycone and ligstroside aglycone present a 55%–66% rate of absorption in humans. The aglycon forms were excreted in urine as HTy or Ty [65]. Particularly, during gastric digestion, OL, OL-aglycone, and other phenolic compounds are transformed into HTy, which is throughout transformed into its phase II metabolites (glucuronide and sulfate conjugates) and into HTy-Ac by effect of the acetyl-CoA enzyme [66]. OL and ligstroside formed sulfate and glucuronide conjugates during absoption after undergoing extensive first-pass intestinal/hepatic metabolism, whereas concentrations of their free forms are not detected in the body fluids. These compounds were absorbed quickly after oral administration and then were metabolized and excreted in the urine mainly as glucuronide [67]. The catabolism of OL could produce phenylacetic and phenylpropionic catabolites which then could be absorbed and subsequently transforme into phase II metabolites. Thus, compounds such as ligstroside or OL could be absorbed as such or reabsorbed in the form of their respective glucuronide conjugates [58,68].

The bioavilability of OLE is still very limited. Ligstroside aglycone and OLE are hydrolyzed in the acidic gastric environment in the stomach, leaving free Ty after 30 min. OLE is absorbed in the small intestine mediated by passive diffusion through the membrane, which is favorable given an adequate coefficient of partition (log P = 1.02) and stability in acid gastric [69]. In fact, Romero et al. verified that OLE was stable in gastric acid conditions at 37 °C for 4 h [70]. Although it has been postulated that OLE is mainly metabolized by Phase I (hydrogenation, hydroxylation, and hydration) [71,72], some hydrogenated metabolites pass to Phase II of metabolism as glucoronidated forms [54], whereas methylated or sulfate forms of OLE have not been detected in any study in humans to date [71,72]. The report published by García-Villalba et al. revealed that OLE and several secoiridoids metabolites were excreted in human urine between 2 and 6 h after olive oil ingestion [72].

On the other hand, OLA may be absorbed in the small intestine by passive diffusion through the membrane due to its favorable partition coafficient. OLA was found to be stable at gastric acid remaing unalterated after 4 h of incubation [73]. After the consumption of olive oil, De las hazas et al. [66] described that OLA was hydrolyzed into HTy and elenoic acid in the digestive system and further metabolized by the enzymatic systems.

In conclusion, there are a lot of plants that have been used as medicines since time immemorial; specificallys, *Olea europaea* L. is a species rich in compounds that have proven their efficacy in the management of several complex diseases including cardiovascular disorders, diabetes, and viral and microbial infections, but additional works are still really necessary to explore the evidences for other traditional uses of this plant. Particularly, the antioxidant, anti-inflammatory, and immunomodulatory properties of secoiridoids from the olive tree (leaves and fruits) have been suggested as a potential application in several oxidative stress-mediated diseases including cancer, cardiovascular disorders, neurodegeneration, the aging process and immunoinflammatory diseases. Thus, the purpose of this review is summarize recent advances in the protective role of these secoiridoids derived from olive tree (preclinical and clinical studies) in these pathologies derived from oxidative stress and focusing on their plausible mechanisms of action involved.

2. Protective Role of the Olive Tree Secoiridois in Diseases with an Important Pathogenic Contribution of Oxidative and Peroxidative Damage

2.1. Olive Tree Secoiridoids and Cancer

Cancer is a complex chronic degenerative disease characterized by a multistep process in which normal cells turn into malignant cells, acquiring several properties such as abnormal proliferation and reduced apoptosis.

The main factors that cause the majority of cancer cases are tobacco and dietary habits. In fact, there is an estimation that around 30% of all cancers may be avoidable by changing food intake [74,75]. Therefore, the identification and characterization of foods and their components, which could prevent the incidence and development of cancer, is an important objective for modern nutritional research [74,76]. In this sense, it is important to take into account that populations who are living near to the Mediterranean area have a lower incidence of cancer compared to other regions. This fact is probably due to the consumption of the diet known as the Mediterranean diet [77]. Besides, it is well-known that the pathophysiology of common diseases states such as cancer, cardiovascular disease, arthritis, and neurodegenerative diseases, among others, are associated with chronic inflammation [78].

There are a large numbers of studies that support the chemopreventive role of natural compounds derived from EVOO and the olive tree such as OL, OLE, OLA, or ligstroside against different cancers and inflammation process. Particularly, the role of secoiridoids derived from *Olea europaea* L. has been investigated in different types of cancerous processes (Tables 1 and 2).

Table 1. Most recent in vitro studies that corroborate the important role of secoiridoids from the olive tree in the control and progression of different types of cancer.

Phenolic Compound	Cell Line	Concentration	Effects	Reference
	NL-Fib: normal human skin fibroblasts; LN-18: poorly differentiated glioblastoma; TF-1a: erythroleukemia; 786O: renal cell adenocarcinoma; T-47D: infiltrating ductal carcinoma of breast-pleural effusion; MCF-7: human breast cancer; RPMI-7951: malignant melanoma skin-lymphoide metastasis; LoVo: colorectal adenocarcinoma-supraclavicular region metastasis	0.005%, 0.01% and 0.025% of OL in fibroblast tissue culture medium	OL inhibited cell growth, motility and invasiveness	[79]
	HT29 and SW260 human colon adenocarcinoma cell line	[0–100 μM]	OL might induce anti-proliferative and pro-apoptotic effects	[80]
	HT29	200, 400, and 800 μM	OL limited the growth and induced apoptosis via the p53 pathway	[81]
	MDA-MB-231 human breast cancer cell line	200 μg/mL	OL produced the up-regulating of TIMPs gene expression and the down-regulation MMPs overexpression gene	[82]
OL	MDA-MB-231; MCF-10A and MCF-7 human breast cancer cell lines	[0–300 μM]	OL exhibited specific cytoxicity against breast cancer cells, which is probably mediated through the induction of apoptosis via mitochondrial pathway	[83]
	SKBR3 breast cancer cell line	100 μM	OL worked as G-protein-coupled receptor (GPER) inverse agonists in estrogen receptor (ER)-negative and GPER-positive SKBR3	[84]
	MCF-7	100 and 200 μM	OL induced apoptosis in breast tumor cells via p53-dependent pathway	[85]
	MDA-MB-231 and MCF-7	[0–100 μM]	OL inhibited the viability of breast cancer cells and induced apoptosis via modulating NF-κB activation cascade	[86]
	MCF-7	[0–1200 μg/mL]	OL suppressed cells migration through suppression of epithelial-mesenchymal transition and could reduce DOX-induced side effects by reducing its effective dose	[87]
	MCF-7	[0–100 μM]	OL decreased the expression of both HDAC2 and HDAC3, induced apoptosis, and retarded cell migration and cell invasion in a dose-dependent manner	[88]
	MCF-7	200, 400, 600, and 1000 μM	OL inhibited the proliferation and invasion of cells by inducing apoptosis	[89]
	MDA-MB-231	[0–100 μM]	OL reduced cell viability in a dose-dependent manner; suppressed HGF or 3-MA, and induced cell migration and invasion	[90]

Table 1. *Cont.*

Phenolic Compound	Cell Line	Concentration	Effects	Reference
	MCF-7	[0–250 μM]	OL inhibited protein tyrosine phosphatase 1B (PTB1B)	[91]
	HepG2 and Huh7 human HCC cell lines	[0–100 μM]	OL induced apoptosis in HCC cells via the suppression of PI3K/Akt	[92]
	HepG2	100, 200 and 300 μM	OL could control the influencing of pro-nerve growth factor (NGF) and NGF balance via affecting MMP-7 activity without affecting the gene expression of NGF in HCC.	[93]
	LNCaP human prostate cancer androgen-responsive and DU145 androgen non-responsive cell lines	100 and 500 μM	OL reduced cell viability and induced thiol group modification	[94]
	TCP-1 and BCPAP thyroid tumor cell line	10, 50, and 100 μM	OL was able to inhibit in vitro thyroid cancer cell proliferation acting on the growth-promoting signal pathway	[95]
	HeLa human cervical carcinoma cell line	150 and 200 μM	OL-induced apoptosis was activated by the JNK/SPAK signal pathway	[96]
	SH-SY5Y human neuroblastoma cell line	350 μM	OL caused cell cycle arrest by down-regulating CyclinD1, CyclinD2, CyclingD3, CDK4, and CDK6 and up-regulating p53 and CDKN2A, CDKN2B, CDKN1A gene expressions. OL also induced apoptosis	[97]
	U251 and A172 human glioma cancer cell lines	0, 200, and 400 μM	OL inhibited cell viability and reduced the expression levels of MMP-2 and MMP-9. In addition, a specific PI3K inhibitor enhanced the pro-apoptotic and anti-invasive effects induced by OL	[98]
	HNE1 and HONE1 human nasopharyngeal carcinoma (NPC) cell lines	0 and 200 μM	OL treatments reduced the activity of the HIF-1α-miR-519d-PDRG1 pathway, which is essential to the radio-sensitizing effect of OL	[99]
	A549 human non-small cell lung cancer (NSCLC)	[0–200 μM]	OL caused a decrease in mithocondrial membrane potential, increase in Bax/Bcl2 ratio, release of mithocondrial cytochrome C, and activation of caspase 9 and caspase 3	[100]
	H1299 lung cancer cell line	[0–200 μM]	OL-induced apoptosis via the mitochondrial apoptotic cascade was activated by the p38 MAPK signaling pathway in H1299 cells	[101]
	A549 and BEAS-2B human noncancerous cell line	50 and 150 μM	OL induced apoptosis in A549 cells	[102]
	MIA PaCa-2, BxPC-3, and CFPAC-1 pancreatic cancer and HPDE non-tumorigenic pancreas cell lines	200 μM	OL arrested cell cycle, increased the Bax/Bcl-2 ratio, increased the activation of caspase 3/7, and induced apoptosis in MIA-PaCa-2	[103]
	A375 human melanoma cell line	[250–500 μM]	OL was able to stimulate apoptosis (500 μM), while at a dose of 250 μM it affected cell proliferation and induced the down-regulation of the pAkt/pS6 pathway	[104]
	OE-19 human esophagical cancer (EC) cell line	200 μM	OL inhibited the growth of EC cells as well as inhibiting HIF-1α and up-regulating BTG anti-proliferation F factor 3 (BTG3) expressions	[105]

Table 1. *Cont.*

Phenolic Compound	Cell Line	Concentration	Effects	Reference
	143B human osteosarcoma (OS) cell line	100 μM	OL showed alone and in combination with 2-methoxyestradiol a potent anti-cancer potential in highly metastatic OS cell	[106]
	AGS Human gastric adenocarcinoma cell line	[0–1000 μg/mL]	Magnetic nano-OL could trigger apoptosis in the AGS cell line	[107]
OL-aglycone	SH-SY5Y and RIN-5F insulinoma cell lines	100 μM	OL-aglycone triggered autophagy in cultured cells through the Ca^{2+}-CAMKKβ–AMPK axis.	[108]
	HT29 and HCT-116 human colon adenocarcinoma cell line	1, 2, 5, and 10 μg/mL	OLE produced an inhibition of AP1 activity and cyclooxygenase 2 (COX2) expression in HT29 cells	[109]
	MDA-MB-231, MCF-7, and PC3 prostate cancer cell lines	[0–20 μM]	OLE inhibited the proliferation, migration, and invasion of the epithelial human breast and prostate cancer cell lines and demonstrated anti-angiogenic activity	[110]
	BT-474, MDA-MB-231, and MCF-7	[0–60 μM]	OLE reduced the c-Met kinase activity, cell growth, migration, and invasion of breast cancer cells and induced G1 cell cycle arrest and apoptosis, as well as, inhibited c-Met-dependent signaling	[111]
	MDA-MB-231	[0–10 μM]	OLE showed strong anti-proliferative and down-regulated the expression of phosphorylated mTOR	[112]
	BT-474	[0–100 μg/mL]	OLE reduced breast cancer progression and locoregional recurrence models	[113]
OLE	MDA-MB-231	5 mg/mL	OLE was able to control breast cancer progression	[114]
	BT-474 and MDA-MB-231	[0–200 μM]	OLE with the dual HER2/EGFR inhibitor, LP, induced synergistic tumor growth inhibition	[115]
	MCF-10A, MDA-MB-231, and MCF-7	1, 10, and 20 μM	OLE could be responsible for the selective activation of TRCP6-dependent Ca^{2+} influx and TRCP6 down-regulation at low μM concentrations	[116]
	Huh-7, HepG2, and HCCLM3 HCC cancer cell lines	[0–80 μM]	OLE inhibited proliferation and cell cycle progression and also inhibited HCC cell migration and invasion	[117]
	Huh-7, HepG2, and HCCLM3	5 and 10 μM	OLE reduced cell proliferation and increased cell death	[118]
	U937 hystocytic lymphoma cancer cell line	30 μM	OLE significantly inhibited the expression of Hsp90, a chaperone with a key role in cancer and neurodegeneration	[119]
	A375: A2058; HUVEC and HaCat cancer cell lines	[0–60 μM]	OLE suppressed STAT3 phosphorylation, decreased STAT3 nuclear localization, and inhibited STAT3 transcriptional activity	[120]
	Inmortalized human keratinocytes stimulated with epidermal growth factor	[0–100 μM]	OLE promoted the inhibition of ERK and Akt phosphorylation and the suppression of B-raf expression	[121]

Table 1. *Cont.*

Phenolic Compound	Cell Line	Concentration	Effects	Reference
OLA	Inmortalized human keratinocytes stimulated with epidermal growth factor	[0–100 μM]	OLA promoted the inhibition of Erk and Akt phosphorylation and the suppression of B-raf expression	[121]
	HL60 human promyelocytic leukemia cell line	[0–10 μM]	OLA reduced the DNA damage at concentrations as low as 1 μM when co-incubated in the medium with H_2O_2	[43]
	NCI-H929; RPMI-8226; U266; MM1S and IIN3 human MM cancer cell lines	2.5, 5 and 10 μM	OLA elicited significant antitumor activity by promoting cell cycle arrest and apoptosis either with a simple agent or in combination with Carfilzomib	[122]

Table 2. Most recent in vivo studies that corroborate the important role of secoiridoids from the olive tree in the control and progression of different types of cancer.

Phenolic Compound	Animal Model	Doses	Effects	Reference
OL	Swiss albino with soft tissue sarcomas	1% OLE in drinking water	OL inhibited cell growth, motility, and invasiveness	[79]
	Male hairless mice (5 weeks old) were UVB irratied (36–180 mJ/cm^2)	10 and 25 mg/Kg/day	OL increased the skin thickness and reductions in skin elasticity, skin carcinogenesis, and tumor growth	[123]
	DSS-induced CRC in C57BL/6 mice	50 and 100 mg/Kg	OL prevented the development of colonic neoplasia in by ameliorating colon inflammatory processes and limiting the activation of the main transcription factors involved	[124]
	Male Sprague–Dawley rats that received an injection of cisplatin (7 mg/Kg)	50, 100, and 200 mg/Kg/day	OL enhanced antioxidant activity and prevented oxidative stress, which it turn reduced 8-hydroxy-2′deoxy-guanosine (8-OH-dG) levels in lymphocytes of cisplatin-treated animals	[125]
	HNE1 and HONE1 injected into 6–8-week-old BalB/c mice	[0–200 μM]	OL was a radiation-sensitizing agent of NPC cells in an in vivo model	[99]
	Four-week-old C57BL/6N mice with HFD with or without OL and which were injected with B16F10 melanoma cells	0.02% and 0.04% enriched-diets	OL suppressed HFD-induced solid tumor growth and reduced HFD-induced expression of angiogenesis, lymphangiogenesis, and hypoxia markers	[126]
	Male Sprague–Dawley rats that received an injection of cisplatin (7 mg/Kg)	50, 100, and 200 mg/Kg/day	OL significantly decreased the formations of DNA damage and the level of malondialdehyde (MDA), and it increased the levels of total antioxidant status in pancreas tissue samples	[127]
	BalB/c OlaHsd-foxn1 injected with MDA-MB-231	50 mg/Kg	The combined treatment with OL and DOX downregulated the antiapoptosis and proliferation protein, nuclear transcription factor-kappa B (NF-κB), and its main oncogenic target Cyclin D1. It also inhibited the expression of Bcl-2	[128]
	Severe combined immunodeficiency mice (6 weeks-old) that received a subcutaneous injection of OE-19 cancer cells	200 μM	OL inhibited the growth of xenograft EC tumor as well as inhibited HIF-1α and upregulated B-cell translocation gene 3 (BTG3) expressions	[105]

Table 2. *Cont.*

Phenolic Compound	Animal Model	Doses	Effects	Reference
OL-aglycone	Transgenic hemizygous CRND8 mice harboring a double-mutant gene of APP695 and wild-type control lettermates with 4 and 10 months of age	100 μM	In OL-fed animals, there was a reduction of phospho-mTOR immunoreactivity and phosphorylated mTOR substrate p70 S6K levels	[108]
OLE	Swiss albino mice (6 weeks old)	10 mg/Kg/day	OLE reduced breast cancer progression and locoregional recurrence models	[113]
	Female athymic nude mice (Foxn1^{nu}/Foxn1^{+}) (4-5 weeks-old) inyected with BT-474 and MDa-MB-231	10 mg/Kg/day	OLE inhibited locoregional recurrence in luminal HER^{2+}/ER^{+} BT-474 tumors	[129]
	Orthotopic tumor model of HCC in BalB/c mice	0, 5 and 10 mg/Kg/day	OLE suppressed tumor growth and impeded HCC metastasis in an in vivo lung metastasis model. OLE inhibited STAT3 activation and increased the activity of protein tyrosine hosphatase	[117]

OL is the most abundant of the phenolic compounds in olives. It could scavenge reactive oxygen and nitrogen species as well as promote nitric oxide (NO) production in macrophages. In fact, it has been postulated that OL may be the major factor responsible for the beneficial effects of the Mediterranean diet against tumor growth [79].

There are several studies describing the potential role of OL in breast cancer. For example, researchers have established the beneficial effects of OL in MCF-7, MCF-10A, and MDA-MB-231 human breast cancer cells where OL was able to decrease the expression of histone deacetylase II (HDAC2), HDAC3, and HDAC4, induce apoptosis, and retard cell migration and invasion in a dose-dependent manner [88,89]. Besides, OL produced tissue inhibitors of metalloproteinases (TIMPs) overexpression and metalloproteinases (MMPs) genes down-regulation, which could help in the prevention of breast cancer metastasis [82]. Similarly, OL treatment reduced MDA-MB-231 cell viability in a dose-dependent manner and significantly suppressed hepatocyte growth factor (HGF) and 3-methyladenine (3-MA) inducing cell migration and invasion [90].

In human HT29 colon adenocarcinoma cell line, OL limited the growth and induced apoptosis via p53 pathway activation, adapting the hypoxia-inducible factor 1-α (HIF-1α) response to hypoxia [81]. In addition, OL induced anti-proliferative and pro-apoptotic effects in a range of doses from 0 to 100 μM in HT29 cells [80].

On the other hand, OL induced apoptosis in hepatocellular carcinoma (HCC) via the suppression of the phosphatidylinositol 3-kinase and protein kinase B (PI3K/Akt) pathway [130]. In fact, in combination with other compounds, such as cisplatin, OL could lead to more effective chemotherapeutic combination against HCC [130]. Similarly, in HeLa human cervical carcinoma cells, OL-induced apoptosis was activated by the c-Jun N-terminal kinase (JNK)/Ste20-like proline alanine rich kinase (SPAK) signal pathways [96], and both OL and its peracetylated derivative were able to inhibit thyroid cancer cell proliferation acting on the growth-promoting signal pathway in the TCP-1 and BCPAP thyroid tumor cell lines [95]. Likewise, OL reduced cell viability in human prostate cancer androgen-responsive cells [94], induced autophagy and caused cell cycle arrest in SH-SY5Y human neuroblastoma cells [97,108], and reduced cell viability in U251 and A172 human glioma cancer cells [98].

Similarly, there are several in vivo studies that have showed the beneficial effects of OL in different cancer models. For example, Giner et al. have demonstrated that OL prevented the development of colonic neoplasia in dextran sulfate sodium (DSS)-induced colorectal cancer (CRC) in mice by ameliorating colon inflammatory processes [124]. Besides, there was a relation between the consumption of a high-fat diet (HFD) and the development of solid tumors, which was satisfactorily suppressed with OL-enriched diets reducing the HFD-induced expression of angiogenesis, lymphangiogenesis, and hypoxia markers [126]. Elamin et al. have shown that OL and doxorubicin (DOX) combined treatment down-regulated the antiapoptosis and proliferation protein in a murine model of breast cancer [83].

OLE is a bioactive micronutrient in the Mediterranean diet that may be associated with positive findings in some epidemiological studies that suggest a fewer incidences of breast, colon cancer, and other malignancies compared to western and other populations. There are several studies to enhance the beneficial role of OLE in this type of disease (Tables 1 and 2). In fact, authors believe that consuming more EVOO with high OLE content is a prudent dietary approach to prevent cancer with the caveat that dietary oils convey calories and consequently other caloric sources will have to yield to avoid obesity [131]. OLE is a unique tyrosine-protein kinase Met (c-MET) inhibitor for the control of breast cancer progression and loco regional recurrence [113,114]. The combination of OLE with lapatinib (LP) treatment would allow the use of reduced dose of targeted therapies such as LP, which would reduce future resistance emergence and drug toxicity while maintaining maximal therapeutic activity [115]. Besides, OLE has been able to reduce c-MET kinase activity, cell growth, and the migration and invasion of breast cancer cells; induce G1 cell cycle arrest and apoptosis; as well as inhibit c-MET-dependent signaling in cultured breast cancer cells and tumorigenicity in in vivo murine models [111].

OLE showed the capacity to inhibit breast cancer locoregional recurrence in luminal HER^{2+}/ER^+ BT-474 tumors. The prevention of tumor recurrence was associated with the down-regulation of MET and HER^2 receptors and suppression of receptor activation [129]. OLE showed a strong anti-proliferative role against several breast cancer cell lines; for example, OLE was able to down-regulate the expression of phosphorylated mammalian target of rapamycin (mTOR) in metastatic MDA-BD-231 breast cancer cell lines [112] and inhibit the proliferation, migration, and invasion of the epithelial human breast cancer and prostate cancer cell lines (MCF7; MDA-BD-231; and PC3). In addition, LeGendre et al. demonstrated that OLE selectively and rapidly induces cell death in cancer cells without being cytotoxic to noncancerous cells. OLE induced cell death by entering the lysosome and inhibiting acid sphingomyelinase (ASM) activity, which induced lyposomal membrane permeabilization (LMP). Even the consumption of 10 mg/Kg oral daily OLE water emulsion treatment significantly suppressed the MDA-BD-231 tumor growth by 90% [132].

Multiple myeloma (MM) is another disease that would be approached with OLE treatment. MM is a plasma cell malignancy that causes devastating bone destruction by activating osteoclasts in the bone marrow milieu. OLE produced the inhibition of macrophage inflammatory 1 alpha (MIP-α) expression and secretion in MM cells proliferation by inducing the activation of apoptosis mechanisms and by down-regulating extracellular signal-regulated kinase (ERK)$_{1/2}$ and protein kinase B (Akt) signal transduction pathways [133]. Besides, OLE was able to inhibit hepatocellular cancer tumor growth and metastasis by preventing signal transducer and activator of transcription (STAT)3 in in vitro and in vivo models. In fact, OLE inhibited proliferation and cell cycle progression in different HCC cell lines, and it also inhibited HCC cells migration and invasion in in vitro models [117].

OLA, also known as 3,4-(dihydrophenyl) ethanol (3,4-DHPEA-EDA) has antioxidant, anti-inflammatory, and anti-microbial activities well documented in previous studies, but its effects on tumor biology are still poorly defined [122]. Particularly, published studies reporting the beneficial properties of OLA in the development of several cancers have been summarized in Tables 1 and 2. In this sense, OLA was able to reduce the DNA damage in HL60 promyelocytic leukemia cells when co-incubated with H_2O_2 in the medium [43] and also presented similar effects to OLE in the reduction of viability and migration of non-melanoma skin cancer cells and in the inhibition of proliferation of ERK and Akt phosphorylation and particularly through the reduction of B-Raf expression [121]. Likewise, OLA reduced the viability of MM primary samples and cell lines even in the presence of bone marrow stromal cells (BMSCs) [84].

2.2. Olive Tree Secoiridoids and Cardiovascular Diseases

The Global Burden of Disease indicates that cardiovascular diseases are still the main cause of global death, representing about 31% of total deaths in the world in 2015 [134]. These pathologies affect heart and vessels, as coronary heart disease, cerebrovascular disease, peripheral arterial disease, and pulmonary embolism, among others [135]. There is evidence that suggests a possible link between inflammation, endothelial dysfunction, and cardiovascular diseases are increased by oxidative stress. Oxidative stress plays a critical role in the development and progression of atherosclerosis and their complications including characteristics affections such as the regulation of vascular tone, vascular smooth muscle growth, monocyte adhesion, platelet function, and fibrinolytic activity, among others [136]. In terms of risk factors, the three world leading factors for cardiovascular diseases are (i) high systolic blood pressure (SBP), (ii) smoking, and (iii) high body mass index (BMI). Proper nutrition habits and healthy lifestyle play a major preventive role [134].

It has been widely reported that secoiridoids play a beneficial role against cardiovascular diseases based on their antioxidant and anti-inflammatory activities. Interesting studies performed with animal and cell models suggest that secoiridoids intake may be beneficial for the prevention and adjuvant treatment of such diseases (Tables 3 and 4). Catalán et al. confirmed changes at proteomic level in cardiovascular tissues (aorta and heart tissues), down-regulating proteins related to the proliferation and migration of endothelial cells and occlusion of blood vessels in the aorta, and proteins related to heart failure in heart tissue in Wistar rats fed a secoiridoids-enriched diet [137].

Table 3. Beneficial effects of secoiridoids from the olive tree in the control and progression of different types of cardiovascular diseases: in vitro studies.

Phenolic Compound	Cell Line	Concentration	Effects	Reference
OL	Healthy human LDL	10 mM	OL inhibited LDL levels, lipid peroxides, malondial dehydelysine, 4-hydroxynonenal lysine adducts expression	[138]
	LPS-stimulates mouse macrophages		OL reduced superoxide anion generation, neutrophils respiratory burst, and hypochlorous acid	[139]
	Endothelial progenitors cells (CD31+ and VEGFR-2$^+$)	[1–10 μM]	OL reduced senescent cells and reactive oxygen species (ROS) formation; restoration of migration, adhesion, tube formation, and the up-regulation of Nrf-2 and HO-1 expressions.	[140]
OL-aglycone	Human umbilical vascular endotelial cells	5 and 25 mM	OL-aglycone reduced cell surface expressions and mRNA levels of ICAM-1 and VCAM-1	[141]
	Mouse atrial myocites HL-1	60 mM	OL-aglycone inhibited tranthyretin toxicity	[142]
OLA	Human neutrophils and monocytes	[1–10 μM]	OLA proved to be stronger in the reduction of formyl-met-leu-phenylalanine and phorbol-myristate-acetate-induced oxidative bursts in neutrophils and myeloperoxidase release	[32]
	Human neutrophils	50 and 100 mM	OLA reduced elastase release, IL-8, MMP-9, and NEP activity	[143]
	Human macrophages	10–20 mM	OLA increased IL-10, HO-1, and CD163 expression	[144]

Table 4. Beneficial effects of secoiridoids from the olive tree in the control and progression of different types of cardiovascular diseases: in vivo and clinical studies.

In Vivo Studies				
Phenolic compound	Animal Model	Doses	Effects	Reference
OL	Ischemia–reperfusion in insolated rat hearts	20 µg/g	OL reduced the creatine kinase, glutathione release, membrane lipid peroxidation	[145]
	Ischemic-treated hypercholesterolemic rabbits	10 or 20 mg/Kg/day	OL reduced the infarct size, total cholesterol, triglyceride concentration, and lipid peroxidation	[146]
	Doxorubicin-induced acute cardiotoxicity rats	100 or 200 mg/Kg	OL modulated the CPK, lactate deshydrogenase, aspartate and alanine aminotransferase, and lipid peroxidation	[147]
	Doxorubicin-induced acute cardiotoxicity in rats	100 or 200 mg/Kg	OL reduced the acetate and succinate levels. Restore metabolic changes	[148]
	Doxorubicin-induced chronic cardiomyopathy in rats	1000 or 2000 mg/Kg	OL controlled cardiac histopathology, nitro-oxidative stress, IL-6, myocardial metabolomics	[149]
	Rabbit model of atherosclerosis	100 mg/Kg	OL decreased lipids, cholesterol, LDL levels, TNF-α, NF-kB, ICAM-1, and VCAM-1 expressions	[150]
	Obesity-induced cardiac metabolic changes	0.023 mg/Kg/day	OL increased oxygen consumption, fat oxidation, and myocardial β-hydroxyacyl coenzyme A dehydrogenase activity and the up-regulation of antioxidant enzyme expression	[151]
	Renovascular hypertension and diabetes 2 rats	20, 40, or 60 mg/Kg/day	OL reduced blood pressure, blood glucose, serum total cholesterol, LDL, and triglycerides levels. Raised HDL levels.	[152]
	Diabetic hypertensive rats	20, 40, or 60 mg/Kg/day	OL lowered blood pressure, MDA, creatine kinase, and the induction of HDL levels	[153]
	Diabetic hypertensive rats	20, 40, or 60 mg/Kg/day	OL decreased blood pressure, glucose, and serum MDA levels. OL increased of HDL and erythrocyte SOD	[154]
	Spontaneous hypertensive rats	10 mg/Kg	OL reduced the oxidative stress, carotid and renal hemodynamics, blood pressure, and heart rate	[155]
	Rats fed with high-cholesterol diet	3 mg/Kg	OL modulated total cholesterol, triglycerides, LDL and HDL levels, and liver antioxidant enzymes	[156]
OL-aglycone	Neonatal rats ventricular myocytes with MAO-A enzyme overexpressed	100 µM	OL-aglycone decreased oxidative stress, autophagic flux blockade and cell necrosis	[157]
	Mature and progenitor endotelial cells	10 µM	OL-aglycone down-regulated NF-kB, IL-8, vascular endothelial growth factor (VEGF), MMP-2, and MMP-9	[158]
	Rats fed with high-cholesterol diet	3 mg/Kg	OL-aglycone modulated total cholesterol, triglycerides, LDL and HDL levels, and liver antioxidant enzymes	[156]
Clinical Trials				
Phenolic Compound	Experimental System	Concentration	Effects	Reference
OL	232 hypertensive patients	500 mg twice daily	OL lowered systolic and diastolic blood pressure, triglycerides, and LDL levels	[159]

One of the best-documented cardiovascular protector secoiridoids is OL. OL inhibited in a dose-dependent manner the copper sulfate-induced oxidation of LDL, reducing the formation of lipid peroxides and malondhial dehydelysine and 4-hydroxynonenol-lysine adducts [138]. These data indicated the protection of the apoprotein layer. Additionally, OL was able to scavenge superoxide anions generated by either polymorphonuclear cells or by the xanthine/xanthine oxidase system [139]. In this line, pretreatment with 20 μg/g of OL before ischemia–reperfusion in isolated rats hearts resulted in a significant reduction in creatine kinase and glutathione release, supporting experimental evidences of a direct cardioprotective effect of OL [145]. More recently, Parzonko and colleagues described that OL-treated endothelial progenitors cells (type CD31+ and VEGFR-2$^+$) showed a decrease in the percentage of senescent cells and ROS formation and restored migration, adhesion, and tube formation. This effect was related to nuclear factor E2-related factor 2 (Nrf2) and heme oxigenase-1 (HO-1) expressions [140] (Table 3).

Relating to animal models, Andreadou and co-workers have developed several in vivo experimental models with OL treatment, defining the potential effect of this secoiridoid as a cardioprotector. Ischemia-treated rabbits fed with 10 or 20 mg/Kg/day OL-supplemented diets showed a reduction in infarct size, total cholesterol, and triglyceride concentrations [146]. Similarly, OL administrated via intravenous decreased some markers of cardiovascular disease in DOX-induced acute cardiotoxic rats such as creatine phosphokinase (CPK), lactate deshydrogenase, aspartate and alanine aminotransferase, and lipid peroxidation in myocardial tissue [147]. Completing this study, these authors revealed that OL down-regulated acetate and succinate levels and restored metabolic changes to normal levels in myocardial tissue [148]. Finally, Andreadou et al. concluded that OL administration could also prevent cardiomyopathy [149] (Table 4).

Concerning studies of atherosclerosis, OL could decreased serum lipids and tumor necrosis factor alpha (TNF-α) levels, which was accompanied by a down-regulation of monocyte chemostactic protein-1 and vascular cell adhesion molecule [150]. Ebaid et al. studied the effects of OL intake in obesity-induced cardiac metabolic changes. They found that an OL diet showed an increase of oxygen consumption, fat oxidation, and myocardial β-hydroxyacyl coenzyme A dehydrogenase activity and a reduction in the levels of lipid hydroperoxide and up-regulation of antioxidant enzyme confirmed that OL improved myocardial oxidative stress in standard-fed conditions [151].

With regard to OL antihypertensive effects, there are several studies performing different animal models. In this line, diabetic and hypetensive rats receiving 20, 40, or 60 mg/Kg/day of OL presented significantly reduced blood pressure, blood glucose, serum total cholesterol, LDL, triglyceride, MDA, coronary effluent creatine kinase, and coronary resistance. The animals also had high-density lipoprotein (HDL), erythrocyte SOD, left ventricular develop pressure, rate of rise, and rate of decrease of ventricular pressure [152–154]. Antihypertensive activity has also been supported by Ivanov et al., who reported significant changes in carotid and renal hemodynamics, reducing cardiovascular risk and improving vascular resistance in spontaneous hypertensive rat oxidative stress [155].

Regarding lipid regulation, the administration of OL and its aglycone form significantly down-regulated the serum levels of total cholesterol, triglycerides, and LDL and up-regulated HDL and liver antioxidant enzymes in Wistar rats fed a cholesterol-rich diet. These results demonstrated that secoiridoids administration could control the lipid peroxidation process, enhancing antioxidant enzyme activity [156].

The clinical trials of the effects of secoiridoids on cardiovascular diseases are scant. In this regard, 232 hypertension patients were involved in a clinical study subjected to a 500-mg oral dose of an OL-enriched extract administration twice daily for 8 weeks. The patients presented a significant reduction of systolic and diastolic blood pressure as well as the levels of triglycerides and LDL [159,160] (Table 4). Stock and colleagues measured cholesterol efflux capacity from free cholesterol-enriched macrophages to apolipoprotein B-depleted serum as the cholesterol acceptor in patients with coronary artery disease. OL showed a positive behavior against LDL oxidation, promoting cholesterol efflux and suggesting preventive effects against coronary artery diseases and enhanced atheroprotective actions [161].

According to OL aglycone, only few studies have been carried out. Dell'agli et al. described that OL aglycone exerted a modulation in early atherogenesis, reducing cell surface expressions of intracellular and vascular cell adhesion molecules (ICAM-1 and VCAM-1) in human umbilical vascular endothelial cells [141]. Similar to OL, the aglycone form was studied in rats fed with a cholesterol-rich diet by Jemai el at. The results suggested that the hypocholesterolemic effect of OL-aglycone might be due to its abilities to lower serum total cholesterol, triglycerides, and LDL cholesterol levels, slowing the lipid peroxidation process and enhancing SOD and CAT antioxidant enzyme activities, exhibiting a cardioprotective role against lipid oxidation and cholesterol efflux [156].

More recently, Leri et al. reported that OL-aglycone was able to reduce transthyretin toxicity in mouse atrial myocytes, so it could be used as treatment for severe cardiac symptoms [142]. In addition, Miceli and coworkers explored the effects of OL-aglyone in myocytes with an overexpression of monoamine oxidase-A (MAO-A), which is an enzyme that causes oxidative stress, autophagy flux blockade, and cell necrosis as a model of cardiac stress characterized by autophagy dysfunction. They observed that OL-aglycone counteracted the cytotoxic MAO-A effects [157]. Margheri et al. also reported the effects of OL aglycone on capillary morphogenesis induced by MRC5 fibroblast "senescense associated secretory phenotype" and progenitor endothelial cells, establishing that this secoiridoid could modulate angiogenesis indirectly on senescent fibroblasts [158] (Table 4).

To date, OLE cardioprotective activity has been slightly investigated. Even so, some authors defend its cardioprotective property based on its capacity to inhibit COX-1 and COX-2 expression. It is well-known that thrombotic and cardiovascular disorders are linked to an imbalance in prostanoid homeostasis, particularly prostaglandin or thromboxane production, which are involved in vasodilatation or vasoconstriction, respectively, and platelet aggregation [162]. OLE has exerted strong inhibitory effects on COX-1 and COX-2 in several studies [50,163,164]; nevertheless, future studies are needed to confirm the property of OLE in cardiovascular disorders (Table 3).

The cardiovascular protection effects of OLA were tested in vitro in human neutrophils and monocytes. This compound was able to scavenge O_2^-, H_2O_2, and NO levels, among other parameters, which are implicated in tissue injury and chronic diseases, as atherosclerosis [32,165]. In a similar work, Czerwinska et al. studied the capacity of OLA on neutral endopeptidase (NEP) activity and other functions of human neutrophils, such as elastase, MMP-9 and interleukin (IL)-8 production, which was markedly increased in patients with myocardial infarction [32]. The authors concluded that OLA could play a role in the cardiovascular protective effects described by olive oil by inhibiting NEP activity, adhesion molecules expression, and elastase release. Likewise, Filipek and colleagues showed that OLA increased CD163 expression in human macrophages, supporting the significant role in attenuation of plaque destabilization induced by hemorrhages [144]. Later, these authors reported the beneficial effects of OLA in attenuating the destabilization of carotid plaque in 20 patients with hypertension. This work revealed the ability of OLA to modulate IL-10, HO-1, MMP-9, and high mobility group protein-1, which is a specific biomarker of cell lethality [166]. Thus, this compound could be potentially useful in the reduction of ischemic stroke risk. Concluding, the preventive and curative role of OLA, in terms of cardiovascular injury, could be attributed to its ability to regulate LDL oxidation and MPO activity, to reduce the expression of adhesion molecules, as angiotensin II production, and to confer protection to erythrocytes from oxidative hemolysis [33,73].

2.3. Olive Tree Secoiridoids and Neurodegeneration

Neurodegeneration is a process that leads to a progressive loss of structure or function of neurons, irreversible neuronal damage, death, and a common final pathway present in aging and neurodegenerative diseases. In addition, oxidative stress induced by impaired mitochondrial functions has been also reported [167]. Particularly, superoxide anion formation and the production of hydrogen peroxide are triggered by the induction of NADPH oxidase (NOX) subunit. This condition together with a high NO level, produced by the induction of inducible nitric oxide synthase (iNOS) results in the formation of peroxynitrite and nitrative stress [69]. Examples of neurodegenerative diseases include Alzheimer's disease (AD), Parkinson's disease (PD), Huntington's disease, amyotrophic lateral sclerosis, frontotemporal dementia, and the spinocerebellar ataxias [168]. These diseases represent a primary health problem, especially in the aging population. Powerful experimental model organisms such as the mouse, fruit fly, nematode worm, and even baker's yeast have been used for many years to explore neurodegenerative diseases and have provided key insights into these brain disorders.

Several epidemiological and observational studies support the belief that traditional alimentary regimens such as the Mediterranean diet where olive oil is the primary source of added fat is associated with improved aging and a reduced incidence of age-related diseases, including cardiovascular diseases, cancer, and cognitive decline [169]. Particularly, olive leaves and EVOO contain many functional phenolics that have been demonstrated to be able to reduce risk and offer protection against several aging and lifestyle-related diseases, including neurodegeneration, in both animal and human's models. In fact, EVOO consumption has well-documented antioxidant, anti-inflammatory, anti-proliferative, anti-carcinogenic, and antibacterial effects [170]. Among the 200 different chemical compounds detected in olive oil, quantitatively, the class of secoiridoids is the most abundant. A number of different studies investigated the effects of secoiridoids from olive trees in both in vitro and in vivo models of AD and PD (Tables 5 and 6).

Table 5. In vitro studies that corroborate the effects of secoiridoids from the olive tree in different types of neurodegeneration processes.

Phenolic Compound	Cell Line	Concentration	Effects	Reference
OL	6-OHDA-induced toxicity in rat adrenal pheochromocytoma (PC12) cells	20 and 25 μg/mL	OL decreased cell damage and reduce biochemical markers of PC12 cell death	[171]
	PC12 cells exposed to the potent parkinsonian toxin 6-OHDA	10^{-12} M	OL showed neuroprotective effects in an in vitro model of PD when administered preventively as a pretreatment. OL significantly decreased neuronal death. OL could also reduce the mitochondrial production of ROS resulting from blocking SOD activity	[172]
OL-aglycone	SH-SY5Y	[0–25 μM]	OL-aglycone prevented the growth of toxic Aβ1-42 oligomers and blocked their successive growth into mature fibrils following its interaction with the peptide N-terminus	[173]
	Exposure of SH-SY5Y cells with Aβ42	[10–1000 μM]	OL were able to attenuate cell death caused by Aβ42, copper-Aβ42, and [laevodihydroxyphenylalanine (l-DOPA)] l-DOPA-Aβ42-induced toxicity after 24 h	[174]
	NBM of adult male Wistar rats	450 μM	An apparent reduction in the amount of soluble A11-positive oligomers was detected in the NBM injected with Aβ42 aggregated with OL as compared with the NBM injected with Aβ42 alone	[175]
OLE	Mouse brain endothelial cells (bEnd3)	25 and 50 μM	Treatment of bEnd3 cells with OLE resulted in significant increase in P-gp and LRP1 levels	[173]

Table 6. In vivo studies that corroborate the effects of secoiridoids from the olive tree in different types of neurodegeneration processes.

Phenolic Compound.	Animal Model	Doses	Effects	Reference
OL-aglycone	Double transgenic TgCRND8 mice, a model of amyloid-ß deposition	8 weeks dietary supplementation of OL-aglycone (50 mg/Kg of diet)	Dietary supplementation of OL-aglycone strongly improved the cognitive performance of young/middle-aged TgCRND8 mice, with respect to age-matched littermates with unsupplemented diet	[176]
	Transgenic mice (APPswe/PS1dE9)	50 mg/Kg of OL-aglycone containing olive leaf extracts (OLE) from 7 to 23 weeks of age.	Treatment mice (OL-aglycone) were showed significantly reduced amyloid plaque deposition ($p < 0.001$) in cortex and hippocampus in comparison	[174]
	Transgenic CL2006 and CL4176 strains of *C. elegans*	50 and 100 μM	OL-aglycone-fed CL2006 worms displayed reduced Aβ plaque deposition, less abundant toxic Aβ oligomers, remarkably decreased paralysis, and increased lifespan	[177]
	Systemic amyloidosis murine model	15 μM	OL-aglycone hindered amyloid aggregation of Aβ(1-42) and its cytotoxicity and eliminated the appearance of early toxic oligomers, favoring the formation of stable harmless protofibrils, which were structurally different from the typical Aβ(1-42) fibrils	[178]
	TgCRND8 mice	50 mg/Kg of diet during 8 weeks	OL-aglycone was active against glutaminylcyclase-catalyzed pE3-Aß generation, reducing enzyme expression and interfering both with Aß42 and pE3-Aß aggregation	[175]
	TgCRND8 (Tg) mice AD	Diet supplementation with OL-aglycone at 12.5 or 0.5 mg kg-1of diet	An OL-aglycone supplementation diet and the mix of polyphenols were found to improve significantly cognitive functions ($p < 0.0001$). Aß42 and pE-3Aß plaque area and number were significantly reduced in the cortex	[179]
OLE	5xFAD mouse model of AD	EVOO rich with OLE	EVOO-rich OLE consumption in combination with donepezil significantly reduced Aβ load and related pathological changes	[180]
	TgSwDI mice	Daily i.p. with 5 mg/Kg OLE at 4 age of months and continued for 4 weeks.	OLE significantly decreased amyloid load in the hippocampal parenchyma and microvessels, which was associated with enhanced cerebral clearance of Aβ across the BBB	[181]
	C57BL/6 wild-type male mice	10 mg/Kg of OLE twice daily from 7 to 8 weeks of age andcontinued for 2 weeks (i.p.)	OLE enhanced clearance of Aβ from the brain. A significant increase in the expression of P-gp and LRP1 was also observed in the brain microvessels	[182]

AD is characterized by the increased accumulation of intracellular neurofibrillary tangles (NFTs) of hyperphosphorylated tau protein and of extracellular Aβ protein deposits (Aβ plaques) derived from amyloid precursor protein (APP) cleavage by γ-secretase and β-secretase. Dietary supplementation of OL (50 mg/Kg of diet) strongly improved the cognitive performance of young/middle-aged/aged TgCRND8 mice, and it also reduced ß-amyloid levels and plaque deposits. Moreover, OL-aglycone-fed mice brain displayed an astonishingly intense autophagy reaction [175,176,179]. Similar results were described in transgenic mice (APPswe/PS1dE9), where OL treatment showed significantly reduced amyloid plaque deposition in the cortex and hippocampus as compared to control mice [174]. Moreover,

OL hindered the amyloid aggregation of Aβ (1-42) and its cytotoxicity and eliminated the appearance of early toxic oligomers, favoring the formation of stable harmless protofibrils, which were structurally different from the typical Aβ (1-42) fibrils [178]. Using transgenic CL2006 and CL4176 strains of *C. elegans* strains expressing Aβ42, as a simplified invertebrate model of AD, Diomede et al. evidenced that 50–100 μM OL-fed CL2006 worms displayed reduced Aβ plaque deposition, less abundant toxic Aβ oligomers, remarkably decreased paralysis, and increased lifespan with respect to untreated animals. A protective effect was also observed in CL4176 worms but only when OL was administered before the induction of the Aβ transgene expression [177] (Table 6).

In vitro studies have revealed that OL prevented the growth of toxic Aβ1-42 oligomers and blocked their successive growth into mature fibrils following its interaction with the peptide N-terminus and attenuated SH-SY5Y cell death caused by Aβ42, copper-Aβ42, and laevodihydroxyphenylalanine (l-DOPA)-Aβ42-induced toxicity after 24 h treatment, and a marked attenuated Aβ-induced astrocytes and microglia reaction was also found in the nucleus basalis magnocellularis (NBM) from adult male Wistar rats injected with Aβ42 aggregated with OL [171,174,177] (Table 5).

The potential protective effect of OLE in AD has been also investigated in TgSwDI mice. Mice treated for 4 weeks with OLE significantly decreased amyloid load in the hippocampal parenchyma and microvessels. This reduction was associated with enhanced cerebral clearance of Aβ across the blood–brain barrier (BBB), which was accompanied by an increase of P-glycoprotein (P-gp) and low density lipoprotein receptor-related protein 1 (LRP1) expressions, and activated the ApoE-dependent amyloid clearance pathway in the mice brains. The anti-inflammatory effect of OLE in the brains of these mice was also obvious where it was able to reduce astrocytes activation and IL-1β levels [181]. Similarly, 10 mg/kg of OLE administered twice daily from 7 to 8 weeks of age and continued for 2 weeks (i.p.) enhanced the clearance of Aβ from C57BL/6 wild-type male mice brain and significantly increased the expression of P-gp and LRP1 [182]. In mouse brain endothelial cells (bEnd3), 25 and 50 μM OLE treatment resulted in a significant increase in P-gp and LRP1 levels [182]. Moreover, in a 5xFAD mouse model of AD, OLE-rich EVOO consumption, in combination with donepezil, significantly reduced Aβ load and related pathological changes, up-regulated synaptic proteins, enhanced BBB tightness, and reduced neuroinflammation associated with Aβ pathology [180] (Tables 5 and 6).

PD is characterized by a progressive loss of dopaminergic neurons in the midbrain region known as *substantia nigra pars compacta* and by the presence of cytoplasmic protein aggregates called the Lewy body as well as Lewy neurites in remaining neurons.

Previous studies showed that OL inhibited αSN amyloidogenesis by directing αSN monomers into small αSN oligomers with lower toxicity, thereby suppressing the subsequent fibril growth phase [183]. The neuroprotective effect of OL has been explored in PC12 cells exposed to the potent parkinsonian toxin 6-hydroxydopamine (6-OHDA). OL treatment significantly decreased neuronal death and reduced the mitochondrial production of ROS resulting from blocking superoxide dismutase activity. Moreover, the quantification of autophagy and acidic vesicles in the cytoplasm alongside the expression of specific autophagy markers uncovered a regulatory role for OL against autophagy flux impairment induced by bafilomycin A1 [171,172].

2.4. Olive Tree Secoiridoids and Ageing

Aging is a natural biological process that involves the gradual decline of physiological function and the eventual failure of organism homeostasis followed by death. Aging is the process of accumulation of damages to cells, tissues, and organs of an individual that is universal and unique, thereby reducing the overall health of the organism. It is evident that aging can induce stress inside the system in the form of ROS or other stressors, reduce overall health, and induce age-associated neurological diseases [184]. The maintenance of homeostasis between the formation and elimination of damaged proteins is a key process in the development and growth of organisms [185]. To date, very few studies concerning the anti-aging effects of secoiridoids have been performed. However, secoiridoids suggest a potential age-related damage regulation based on their antioxidant, anti-inflammatory, and neuroprotector effects (Tables 7 and 8).

Several in vitro studies have supported the potential of OL on the proteasome, which regulates the balance of cellular viability and is crucial in stress, aging, or senescent conditions [185]. The treatment of cell lysates from human embryonic fibroblast IMR90 enhanced three major proteasome catalytic activities: the chymo-trypsin-like (ch-L), the peptidylglutamyl-peptide hydrolase (PGPH) activity, and the trypsin-like (T-L). This activity was supported by Katsiki et al. OL-treated cells retained proteasome function during replicative senescence, and human embryonic fibroblast cultures exhibited a delay appearance of senescence morphology [186]. Santiago-Mora et al. reported the effects of OL on osteoblastogenesis and adipogenesis in mesenchymal stem cells from human bone marrow. OL stimulated osteoclastogenesis rising cellular matrix mineralization and inhibited bone desorption [187] (Table 7).

In terms of epigenetic, it has been postulated that oxidative damage to mitochondrial DNA (mtDNA) is one of several signs of age-related physiological consequences [188]. Fabiani and colleagues reported that OL and OL-algycone form counteracted DNA alterations in HL60 cells and peripheral blood mononuclear cell (PMBC) H_2O_2-induced DNA damage [43]. Moreover, OL counteracted bone loss and reduced α-1-acid glycoprotein plasma concentrations in senile osteoporosis rats [189]. Nikou et al. studied the effects of OLE and OLA in *Drosophila* flies, reporting that dietary administration of both of them was able to increase the T-L proteasome activity and 20S and 19S proteosomal subunits expression, leading to a significant reduction of ROS levels. Subsequently, it was reported that OLA up-regulated the gene expression of the proteasome, antioxidant response, and molecular chaperones in human skin fibroblasts [190] (Table 8). These data confirmed a potential anti-aging of both secoiridoids and suggested that these compounds could be critical nutraceuticals as preventive and therapeutic treatment of different age diseases. Nevertheless, clinical studies that confirm these suggestions need to be developed in the future.

Table 7. Potential role of secoiridoids obtained from the olive tree in anti-aging: in vitro studies.

Phenolic Compound	Cell Line	Concentration	Effects	Reference
	Human embryonic fibroblast (IMR90)	[0.1–50 mM]	OL enhanced ch-L, PGPH, and PGPH proteasome activity	[185]
	Human embryonic fibroblast		OL retained proteasome function during replicative senescence and delayed in the appearance of senescence morphology	[186]
OL	Mesenchymal stem cells from human bone marrow	[1–100 μM]	OL enhanced osteogenic gene expression markers and osteoblast phenotypic characteristic	[187]
	Human promyelocitic leukemia cells (HL60)	10 μM	OL restored DNA damage	[43]
OL-aglycone	Human promyelocitic leukemia cells (HL60)	10 μM	OL-aglycone restored DNA damage	[43]
OLE	Normal human skin fibroblasts	50, 100, 150, and 200 μM	OLE up-regulated genes' expression of proteasome, antioxidant responses, and molecular chaperones genes	[190]
OLA	Normal human skin fibroblasts	50, 100, 150, and 200 μM	OLA up-regulated genes' expression of proteasome, antioxidant responses, and molecular chaperones genes	[190]

Table 8. Potential role of secoiridoids obtained from the olive tree in anti-aging processes: in vivo studies.

Phenolic Compound	Animal Model	Doses	Effects	Reference
OL	Senile osteoporosis rats model	15 mg/kg	OL counteracted bone loss and reduce α-1-acid glycoprotein plasma concentration	[189]
OLE	*Drosophila* in vivo model	Dietary supplementation of OLE: 400 nM, 200 nM, and 100nM.	OLE increased the ch-L proteasome activity and the expression of 20S and 19S proteasomal subunits and decreased of ROS levels in somatic tissues of *Drosophila* flies	[190]
OLA	*Drosophila* in vivo model	Dietary supplementation of OLA: 400 nM, 200 nM, and 100 nM	OLA increased the ch-L proteasome activity and the expression of 20S and 19S proteasomal subunits and decreased of ROS levels in somatic tissues of *Drosophila* flies.	[190]

2.5. Secoiridoids Olive Tree in Autoimmune Diseases

Autoimmune diseases are heterogeneous groups of diseases whose condition is that your immune system mistakenly attacks your body. In the normal state, the immune system is able to differentiate between foreign cells, such as viruses and bacteria, and its own cells. However, in an autoimmune disease, the immune system recognizes our own cells as foreign cells and it releases proteins called autoantibodies that attack healthy cells. These types of diseases could appear at any stage of life with higher or lower severity. For example, there are types of autoimmune disease that target only one organ, such as type 1 diabetes mellitus (T1DM), which damages the pancreas, whereas there are other diseases, such as systemic lupus erythematosus (SLE), which affect the whole body.

The National Institutes of Health (NIH) estimates that up to 23.5 million Americans suffer from autoimmune disease and that the prevalence rose in 2019. For this reason, autoimmune diseases are recognized as a major health problem. Nowadays, researchers have identified 80–100 different autoimmune diseases and suspect at least 40 more diseases of having an autoimmune basis. These diseases are chronic, can be life-threatening, and are responsible for death in female children and women in all age groups up to 64 years old [191]. The most characteristic examples of these kinds of autoimmune diseases are rheumatoid arthritis (RA), SLE, Crohn's disease, ulcerative colitis (UC), and T1DM.

RA can be defined as a chronic inflammatory disease with a systemic autoimmune component, and it is mainly characterized by aggressive synovial hyperplasia, synovitis, the progressive destruction of cartilage, and bone erosion with the painful swelling of small joints, fatigue, prolonged stiffness and fever caused by immune responses, and specific innate inflammatory processes [192]. Worldwide, the annual incidence of RA is approximately three cases per 10,000 people, and the prevalence rate is approximately 1% increasing with age and peaking between the ages of 35 and 50 years old. RA affects all populations, although it is much more prevalent in some groups (e.g., 5–6% in some Native American groups) and much less prevalent in others (e.g., black persons from the Caribbean region) [193].

SLE can be defined as a chronic inflammatory and autoimmune disease that can affect multiple organ systems, including skin, joints, kidneys, and the brain, among others [194]. SLE is characterized by a deposition of immune complexes, which are formed in large amounts as antinuclear antibodies bind to the abundant nuclear material in blood and tissues, along with disturbances in both innate and adaptive immunity and T-cell signaling. In addition, SLE is characterized by its clinical and pathogenic complexity, difficult diagnosis, and the high number of complications that can affect the patient's quality of life [195]. There are worldwide differences in the incidence and prevalence of SLE that vary with sex, age, ethnicity, and time. The highest estimates of incidence and prevalence of SLE were in North America: 23.2/10000 persons/years and 24/10000, people respectively. The lowest incidences of SLE were reported in Africa and the Ukraine (0.3/10000 persons/years), and the lowest prevalence was observed in Northern Australia (0 cases in sample of 847 people). Women were more frequently affected than men for every age and ethnic group. Incidence peaked in middle adulthood and occurred later for men. People of black ethnicity had the highest incidence and prevalence of SLE, whereas those with white ethnicity had the lowest incidence and prevalence. This appeared to be an increasing trend of SLE prevalence with time [196].

Crohn's disease and UC are important chronic inflammatory disorders of the gastrointestinal system that contribute to the inflammatory bowel conditions, such as diarrhea with or without blood, abdominal pain, fever, weight loss, inflammation, and ulcers. These diseases have uncertain etiology but can be associated with multifactorial conditions in terms of immunity, genetics, and non-immune conditions such as environmental factors [197]. The total number of new cases of Crohn's disease diagnosed each year (incidence) was 10.7 per 100,000 people or approximately 33,000 new cases per year. The total number of new cases of UC diagnosed each year was 12.2 per 100,000 people or approximately 38,000 new cases per year [198].

T1DM is considered an autoimmune disease that results from the destruction of pancreatic β-cells and is mediated by the immune system. This is caused by an autoimmune reaction where the body's defense system attacks the cells that produce insulin. As a result, the body produces very little or no insulin. Multiple genetic and environmental factors found in variable combinations in individual patients are involved in the development of T1DM. Genetic risk is defined by the presence of particular allele combinations, which in the major susceptibility locus (the HLA region) affect T-cell recognition and tolerance to foreign and autologous molecules. T1DM can affect people at any age, but it usually develops in children or young adults. Around 10% of all people with diabetes have T1DM [199].

Due to the high prevalence and rising incidence of this kind of disease, nowadays, there is a requirement to investigate to develop palliative remedies or treatments that help us improve the symptoms and management of these types of diseases to improve the quality of life of patients. A new source of news alternative for autoimmune diseases is based on the use of natural compounds obtained from natural resources, such as *Olea europaea* L., which is traditionally used as diuretic, hypotensive, emollient, laxative, febrifuge, skin cleanser, and it is also used for the treatment of urinary infections, gallstones, bronchial asthma, and diarrhea. The published studies related to the effects of secoiridoids from the olive tree in these autoimmune diseases are summarized in Tables 9 and 10.

Table 9. Effective mechanisms and concentrations of bioactive secoiridoids from the olive tree in in vitro models of immunoinflammatory diseases.

Phenolic Compound	Cell Line	Concentration	Effects	Reference
OL	LPS-stimulated murine peritoneal macrophages.	25 and 50 μM	OL reduced pro-inflammatory cytokines levels and interferon (IFN)-γ, as well as iNOS and COX-2 overexpressions	[200]
	Human synovial fibroblasts cell line (SW982)	50 and 100 μM	OL pre-treatment down-regulated mitogen active protein kinase (MAPK)s and NF-κB and induction of Nrf2-linked HO-1 signaling pathways	[201]
OLE	LPS-stimulated murine peritoneal macrophages.	[25–100 μM]	OLE showed a potent reduction of ROS, nitrites, and pro-inflammatory cytokines levels. OLE inhibited canonical and noncanonical inflammasome signaling pathways	[50]
	J774 LPS-stimulated macrophages	50 μM	OLE inhibited LPS-induced NO production without affecting cell viability	[117]

Table 10. Effective mechanisms and concentrations of bioactive secoiridoids from the olive tree in in vivo models and clinical trials of immunoinflammatory diseases.

In Vivo Studies				
Phenolic compound	Animal Model	Doses	Effects	Reference
OL	Diabetes type I model induced by subcutaneous alloxan monohydrate injection in Sprague-Dawley male rats	Oral gavage 15 mg/Kg/day of OL	OL significantly decreased leucocyte infiltration and glomerulosclerosis. OL decreased the levels of urea, nitrite, and creatinine and decreased MPO activity	[202]
	Experimental autoimmune myocarditis (EAM) model induced by porcine cardiac myosin in Lewis rats	Oral gavage 20 mg/Kg/day of OL	OL improved cardiac functions and attenuate inflammatory cell infiltration and cytokine expression levels	[203]
	Chronic colitis model induced by DSS (1% in first and second cycles and 2% in third and fourth cycle) in female C57BL/6 mice (6–8 weeks at age weighting 18–20 g)	Diet supplemented with 0.25% OL	OL exhibited a decrease of inflammatory symptoms and decreased inflammatory cell recruitment	[204]
	Acute colitis model induced by DSS (5%) for 7 days in BALB/c mice (6–8 weeks at age weigthing 18–20 g)	Diet supplemented with 1% OL	Oral administration of OL attenuated the extent and severity of acute colitis and reduced production of inflammatory mediators	[205]
OL-aglycone	Collagen type II-induced arthritis (CIA) in three-week-old male DBA-J/1	Oral gavage 40 mg/Kg/day of OL	OL prevented joints inflammation and reduced inflammatory mediators overexpression and cytokines levels	[206]
Clinical Trials				
Phenolic Compound	Cells	Concentration	Effects	Reference
OL	14 outpatients biopsies with Ulcerative Colitis	3 µM OL from olive leaves from *Olea europaea* L.	OL reduced the expression of COX-2 and IL-17 in samples treated with OL. In addition, OL ameliorated inflammatory tissular damage	[207]

3. Conclusions

There is evidence indicating that secoiridoids from the olive tree has a large potential as a therapy for a wide variety of ROS-related diseases. Interesting studies performed with animal and cell models suggest that secoiridods intake may be beneficial for the prevention and adjuvant treatment of such diseases. Particularly, dietary supplementation of OL, OL-aglycone, or OLE strongly improved the cognitive performance as well as reduced β-amyloid levels and plaque deposits favoring the formation of stable harmless protofibrils in transgenic mice models of AD. Likewise, using transgenic strains of *C. elegans*, OL-fed CL2006 worms displayed reduced Aβ plaque deposition, less abundant toxic Aβ oligomers, remarkably decreased paralysis, and increased lifespan. Besides, OL prevented the growth of toxic Aβ1-42 oligomers and cell death in SH-SY5Y cells, increased P-gp and LRP1 levels in mouse brain endothelial cells, and a marked attenuated Aβ-induced astrocytes and microglia reaction was also described in the NBM of adult male Wistar rats injected with Aβ42 aggregated with OL. The neuroprotective effect of OL secoiridoid has been slightly explored in PD. OL treatment has been demonstrated to inhibit αSN amyloidogenesis, suppressing the subsequent fibril growth phase, decreasing neuronal death, and reducing the mitochondrial production of ROS resulting from blocking superoxide dismutase activity in PC12 cells exposed to 6-OHDA.

The best-documented cardiovascular protector secoiridoid is OL. The literature reviewed here validates that the treatment of cells and animal models with OL could be beneficial in combating oxidative stress and thereby protect individuals from cardiovascular diseases. The beneficial effects of secoiridoids have been attributed to their antioxidant capacity and their ability to modulate cellular antioxidant defense mechanisms. In addition, OL has been shown to modulate a variety of targets, which include iNOS and NO, TNFα, IL-8 and MMP-2 and MMP-9 in addition to VCAM-1 and ICAM-1, and modulating signaling pathways by altering MAPK, NFκB, and Nrf2/HO-1, among others. Finally, secoiridoid supplemented diets exerted a reduction in infarct size, total cholesterol, and triglyceride concentrations. On the contrary, the cardiovascular protection effects of OLE and OLA as well as clinical trials of the effects of these secoiridoids on cardiovascular diseases are very scant, and future studies are needed to confirm them in cardiovascular disorders.

OL has also shown considerable anti-cancer effects against many types of cancer, including breast cancer, colorectal cancer, prostate cancer, pancreatic cancer, cervical carcinoma, and thyroid cancer both in vitro and in vivo. OL is believed to exert its anticancer activity via multiple mechanisms, interfering with different cellular pathways and inducing/inhibiting the production of various types of cytokines, enzymes, or growth factors such as MAPKs, NF-κB, Akt, COX-2, and STAT3. On the contrary, the studies reporting the anti-cancer effects of OLE and OLA are very circumscribed to breast cancer, hepatocellular carcinoma, and hematologic neoplasias.

On the other hand, the remarkable anti-inflammatory and immunomodulatory effects of these bioactive compounds have been reported in several in vitro and experimental models of RA, SLE, inflammatory bowel disease (IBD), T1DM, and multiple sclerosis (MS). Particularly, OL, OLA, and OLE may exert a remarkable inmmunodulatory and anti-inflammatory effects reducing the induced inflammatory response in murine macrophages and human fibroblasts. This was accompanied by amelioration of the production of essential pro-inflammatory cytokines involved in the regulation of the immune system response through the prevention of MAPKs and NF-κB pathways activation. Moreover, OL and OLA secoiridoids were effective in preventing the induced immuno-inflammatory response in animal experimental models of UC, MS, T1DM, and SLE. Finally, very few studies about secoiridoids' anti-aging effects have been performed to date. Nevertheless, secoiridoids suggest posing a potential age-related damage regulation based on their antioxidant, anti-inflammatory, and neuroprotector effects.

In conclusion, the published data revealed, in general, consistent and very satisfactory results; however, the knowledge is very limited, especially in clinical trials. In this sense, further efforts are needed to mechanistically clarify the underlying biochemical and biological activities and pharmacokinetics/pharmacodynamics of secoiridoids from the olive tree in additional preclinical and clinical studies of ROS-related diseases.

Author Contributions: M.S.-H. and C.A.-d.-l.-L. conceived the idea. M.L.C., T.M., M.S.-H., and C.A.-d.-l.-L. performed the review of papers. M.S.-H., L.C., and T.M. wrote the first draft of the manuscript. All authors made meaningful contributions to the final manuscript and approved the manuscript for final submission. All authors have read and agreed to the published version of the manuscript.

Funding: The authors gratefully acknowledge the financial support by a research grant (Reference: AG2017-89342-P) from the Ministerio de Economía y Competitividad of Spain.

Conflicts of Interest: The authors declare no conflict of interest.

Abbreviations

3:4-DHPEA-EDA	3,4-(dihidrophenyl)etanol
3-MA	3-methyladenine
6-OHDA	6-hydroxydopamine
8-OH-dG	8-hydroxy-2′deoxy-guanosine
AD	Alzheimer's disases
AGE	advanced glycoxidation products
Akt	Protein kinase B
ALE	Advanced lipid peroxidation products
APP	Amyloid precursor protein
ASM	acid sphingomyelinase
AUC	area under the curve

BBB	Blood-brain barrier
BMI	Body mass index
BMSC	Bone marrow stromal cells
BTG3	B-cell translocation gene 3
ch-L	Chymo-trypsin-like
CIA	Collagen-induced arthritis
c-MET	Tyrosine-protein kinase Met
COX	Cyclooxigenase
CPK	Creatine phosphokinase
CRC	Colorrectal cancer
DSS	Dextran sulfate sodium
EAM	Experimental autoimmune myocarditis
EC	Esophagical cancer
ER	Estrogen receptor
ERK	Extracellular signal-regulated kinases
EVOO	Extra virgin olive oil
GES	Geraniol synthase
GPER	G-protein coupled receptor
GPx	Glutathione peroxidase
GT	Glucosyltransferase
h	hours
HCC	Hepatocellular carcinoma
HDAC	Histone deacetylase
HDL	High-density lipoprotein
HFD	High fat diet
HGF	Hepatocyte growth factor
HIF-1α	Hypoxia-inducible factor 1α
HO-1	Heme oxigenase-1
HTy	Hydroxytyrosol
IBD	Inflammatory bowel disease
ICAM-1	Intracellular adhesión molecule-1
IL	Interleukin
iNOS	Inducible nitric oxide synthase
i.p.	intraperitoneally
JNK	c-Jun N-terminal kinase
LDL	Low-density lioprotein
LMP	Lyposomal membrane permeabilization
LP	Lapatinib
LRP1	Lipoprotein receptor-related protein 1
MAO-A	Monoamine oxidase-A
MDA	Malodialdehyde
min	minutes
MIP-α	Macrophage inflammatory 1α
MM	Multiple mieloma
MMPs	Metalloproteinases
MPO	Myeloperoxidase
MS	Multiple sclerosis
mtDNA	mitochondrial DNA
mTOR	mammalian target of rapamycin
NADH	Nicotinamide adenine dinucleotide reduced form
NBM	Nucleus basalis magnocellularis
NEP	Neutral endopeptidase
NFTs	Neurofibrillary tangles

NF-κB	Nuclear transcription factor-kappa B
NGB	Pro-nerve growth factor
NIH	National institute of Health
NO	Nitric oxide
NOX	NADPH oxidase
NPC	Nasopharingeal carcinoma
Nrf2	Nuclear factor E2-related factor 2
NSCLC	Non-small cell lung cáncer
OL	Oleuropein
OLA	Olacein
OLE	Oleocanthal
OS	Osteosarcoma
PD	Parkinson's disease
P-gp	P-glycoprotein
PGPH	Peptidylgutamyl-peptide hydrolase
PI3K	Phosphatidylinositol 3-kinase
PBMC	Peripheral blood mnonuclear cells
PTB1B	Protein tyrosine phosphatase 1B
RA	Rheumatoid arthritis
ROS	Reactive oxygen species
SBP	Systolic blood pressure
SLS	Secologanin synthase
SPAK	Ste20-like proline alanine rich kinase
SLE	Systemic lupus erythematosus
STAT	Signal transducer and activator of transcription
T1DM	Type 1 Diabetes mellitus
TIMPs	Tissue inhibitors of metalloproteinases
T-L	Trypsin-like
Ty	Tyrosol
UC	Ulcerative colitis
VCAM-1	Vascular cell adhesión molecule-1
VEGF	Vascular endotelial growth factor

References

1. Boros, C.A.; Stermitz, F.R. Iridoids. updated part. *J. Nat. Prod.* **1990**, *53*, 1055–1147. [CrossRef]
2. El-Naggar, L.J.; Beal, J.L. Iridoids: A Review. *J. Nat. Prod.* **1980**, *43*, 649–707. [CrossRef] [PubMed]
3. Dinda, B.; Debnath, S.; Harigaya, Y. Naturally Occurring Secoiridoids and Bioactivity of Naturally Occurring Iridoids and Secoiridoids: A Review, Part 2. *Chem. Pharm. Bull. (Tokyo)* **2007**, *55*, 689–728. [CrossRef] [PubMed]
4. Hegnauer, R.; Kooiman, P. The Taxonomic Significance of Iridoids of Tubiflorae Sensu Wettstein (Author's Transl). *Planta Med.* **1978**, *33*, 1–33. [CrossRef] [PubMed]
5. Sticher, O.V.; Junod-Busch, U. The Iridoid Glucoside and Its Isolation. *Pharm. Acta Helvetiae* **1975**, *50*, 127–144.
6. Huang, Y.-L.; Oppong, M.B.; Guo, Y.; Wang, L.-Z.; Fang, S.-M.; Deng, Y.-R.; Gao, X.-M. The Oleaceae Family: A Source of Secoiridoids with Multiple Biological Activities. *Fitoter* **2019**, *136*, 104155. [CrossRef]
7. Ghanbari, R.; Anwar, F.; Alkharfy, K.M.; Gilani, A.-H.; Saari, N. Valuable Nutrients and Functional Bioactives in Different Parts of Olive (Olea europaea L.)—A Review. *Int. J. Mol. Sci.* **2012**, *13*, 3291–3340. [CrossRef]
8. Somova, L.; Shode, F.; Ramnanan, P.; Nadar, A. Antihypertensive, Antiatherosclerotic and Antioxidant Activity of Triterpenoids Isolated from Olea Europaea, Subspecies Africana Leaves. *J. Ethnopharmacol.* **2003**, *84*, 299–305. [CrossRef]
9. Sheth, A.; Mitaliya, K.D.; Joshi, S. *The herbs of Ayurveda*, 1st ed.; Ashok Sheth: Gujarat, India, 2005.
10. Bellakhdar, J.; Claisse, R.; Fleurentin, J.; Younos, C. Repertory of Standard Herbal Drugs in the Moroccan Pharmacopoea. *J. Ethnopharmacol.* **1991**, *35*, 123–143. [CrossRef]

11. Alarcon-Aguilara, F.J.; Roman-Ramos, R.; Perez-Gutierrez, S.; Aguilar-Contreras, A.; Contreras-Weber, C.C.; Flores-Saenz, J.L. Study of the Anti-Hyperglycemic Effect of Plants Used as Antidiabetics. *J. Ethnopharmacol.* **1998**, *61*, 101–110. [CrossRef]

12. Ribeiro, R.D.A.; De Melo, M.R.; De Barros, F.; Gomes, C.; Trolin, G. Acute Antihypertensive Effect in Conscious Rats Produced by Some Medicinal Plants Used in the State of São Paulo. *J. Ethnopharmacol.* **1986**, *15*, 261–269. [CrossRef]

13. Pieroni, A.; Heimler, D.; Pieters, L.; Van Poel, B.; Vlietinck, A.J. In Vitro Anti-Complementary Activity of Flavonoids from Olive (*Olea europaea* L.) Leaves. *Die Pharm.* **1996**, *51*, 765–768.

14. De Feo, V.; Aquino, R.; Menghini, A.; Ramundo, E.; Senatore, F. Traditional Phytotherapy in the Peninsula Sorrentina, Campania, Southern Italy. *J. Ethnopharmacol.* **1992**, *36*, 113–125. [CrossRef]

15. Ghazanfar, S.A.; Al-Al-Sabahi, A.M. Medicinal Plants of Northern and Central Oman (Arabia). *Econ. Bot.* **1993**, *47*, 89–98. [CrossRef]

16. Termentzi, A.; Halabalaki, M.; Skaltsounis, A.L. From Drupes to Olive Oil: An Exploration of Olive Key Metabolites. In *Olive and Olive Oil Bioactive Constituents*; Boskou, D., Ed.; Elsevier: Amsterdam, Netherlands, 2015; pp. 147–177.

17. Perez-Martinez, P.; López-Miranda, J.; Blanco-Colio, L.; Bellido, C.; Jimenez, Y.; Moreno, J.A.; Delgado-Lista, J.; Egido, J.; Pérez-Jiménez, F. The Chronic Intake of a Mediterranean Diet Enriched in Virgin Olive Oil, Decreases Nuclear Transcription Factor κB Activation in Peripheral Blood Mononuclear Cells from Healthy Men. *Atherosclerosis* **2007**, *194*, e141–e146. [CrossRef]

18. Jensen, S. Chemotaxonomy of the Oleaceae: Iridoids as Taxonomic Markers. *Phytochemistry* **2002**, *60*, 213–231. [CrossRef]

19. Obied, H.K.; Prenzler, P.D.; Ryan, D.; Servili, M.; Taticchi, A.; Esposto, S.; Robards, K. Biosynthesis and Biotransformations of Phenol-Conjugated Oleosidic Secoiridoids from *Olea europaea* L. *Nat. Prod. Rep.* **2008**, *25*, 1167. [CrossRef]

20. Maestri, D.; Barrionuevo, D.; Bodoira, R.; Zafra, A.; Jiménez-López, J.; Alché, J.D.D. Nutritional Profile and Nutraceutical Components of Olive (*Olea europaea* L.) Seeds. *J. Food Sci. Technol.* **2019**, *56*, 4359–4370. [CrossRef]

21. Kanakis, P.; Termentzi, A.; Michel, T.; Gikas, E.; Halabalaki, M.; Skaltsounis, A.-L. From Olive Drupes to Olive Oil: An HPLC-Orbitrap-based Qualitative and Quantitative Exploration of Olive Key Metabolites. *Planta Med.* **2013**, *79*, 1576–1587. [CrossRef] [PubMed]

22. Gómez-Rico, A.; Inarejos-García, A.M.; Salvador, M.D.; Fregapane, G. Effect of Malaxation Conditions on Phenol and Volatile Profiles in Olive Paste and the Corresponding Virgin Olive Oils (*Olea europaea* L. Cv. Cornicabra). *J. Agric. Food Chem.* **2009**, *57*, 3587–3595. [CrossRef]

23. Caruso, D.; Colombo, R.; Patelli, R.; Giavarini, F.; Galli, G. Rapid Evaluation of Phenolic Component Profile and Analysis of Oleuropein Aglycon in Olive Oil by Atmospheric Pressure Chemical Ionization-Mass Spectrometry (APCI-MS). *J. Agric. Food Chem.* **2000**, *48*, 1182–1185. [CrossRef] [PubMed]

24. Obied, H.K.; Bedgood, D.R.; Prenzler, P.D.; Robards, K.D.R.B., Jr. Chemical Screening of Olive Biophenol Extracts by Hyphenated Liquid Chromatography. *Anal. Chim. Acta* **2007**, *603*, 176–189. [CrossRef] [PubMed]

25. Gutierrez-Rosales, F.; Romero, M.P.; Casanovas, M.; Motilva, M.J.; Mínguez-Mosquera, M.I. Metabolites Involved in Oleuropein Accumulation and Degradation in Fruits of Olea europaea L.: Hojiblanca and Arbequina Varieties. *J. Agric. Food Chem.* **2010**, *58*, 12924–12933. [CrossRef] [PubMed]

26. Ryan, D.; Antolovich, M.; Herlt, T.; Prenzler, P.D.; Lavee, S.; Robards, K. Identification of Phenolic Compounds in Tissues of the Novel Olive Cultivar Hardy's Mammoth. *J. Agric. Food Chem.* **2002**, *50*, 6716–6724. [CrossRef]

27. Ryan, D.; Prenzler, P.D.; Lavee, S.; Antolovich, M. and Robards, K. Quantitative Changes in Phenolic Content during Physiological Development of the Olive (Olea europaea) Cultivar Hardy's Mammoth. *J. Agric. Food Chem.* **2003**, *51*, 2532–2538. [CrossRef]

28. Savarese, M.; Demarco, E.; Sacchi, R. Characterization of Phenolic Extracts from Olives (*Olea europaea* cv. Pisciottana) by Electrospray Ionization Mass Spectrometry. *Food Chem.* **2007**, *105*, 761–770. [CrossRef]

29. Servili, M.; Baldioli, M.; Selvaggini, R.; Miniati, E.; Macchioni, A.; Montedoro, G. High-Performance Liquid Chromatography Evaluation of Phenols in Olive Fruit, Virgin Olive Oil, Vegetation Waters, and Pomace and 1D- and 2D-Nuclear Magnetic Resonance Characterization. *J. Am. Oil Chem. Soc.* **1999**, *76*, 873–882. [CrossRef]

30. Presti, G.; Guarrasi, V.; Gulotta, E.; Provenzano, F.; Provenzano, A.; Giuliano, S.; Monfreda, M.; Mangione, M.; Passantino, R.; Biagio, P.S.; et al. Bioactive Compounds from Extra Virgin Olive Oils: Correlation between Phenolic Content and Oxidative Stress Cell Protection. *Biophys. Chem.* **2017**, *230*, 109–116. [CrossRef]

31. Cárdeno, A.; Sánchez-Hidalgo, M.; Alarcón-De-La-Lastra, C. An Up-Date of Olive Oil Phenols in Inflammation and Cancer: Molecular Mechanisms and Clinical Implications. *Curr. Med. Chem.* **2013**, *20*, 4758–4776. [CrossRef]

32. Czerwińska, M.; Kiss, A.K.; Naruszewicz, M. A Comparison of Antioxidant Activities of Oleuropein and Its Dialdehydic Derivative from Olive Oil, Oleacein. *Food Chem.* **2012**, *131*, 940–947. [CrossRef]

33. Paiva-Martins, F.; Fernandes, J.; Santos, V.; Silva, L.; Borges, F.; Rocha, S.; Belo, L.; Santos-Silva, A. Powerful Protective Role of 3,4-dihydroxyphenylethanol-elenolic acid Dialdehyde against Erythrocyte Oxidative-Induced Hemolysis. *J. Agric. Food Chem.* **2010**, *58*, 135–140. [CrossRef] [PubMed]

34. De La Lastra, C.A.; Villegas, I. Resveratrol as an Antioxidant and Pro-Oxidant Agent: Mechanisms and Clinical Implications. *Biochem. Soc. Trans.* **2007**, *35*, 1156–1160. [CrossRef] [PubMed]

35. Rosignoli, P.; Fuccelli, R.; Fabiani, R.; Servili, M.; Morozzi, G. Effect of Olive Oil Phenols on the Production of Inflammatory Mediators in Freshly Isolated Human Monocytes. *J. Nutr. Biochem.* **2013**, *24*, 1513–1519. [CrossRef] [PubMed]

36. Kaeidi, A.; Sahamsizadeh, A.; Allahtavakoli, M.; Fatemi, I.; Rahmani, M.; Hakimizadeh, E.; Hassanshahi, J. The Effect of Oleuropein on Unilateral Ureteral Obstruction Induced-Kidney Injury in Rats: The Role of Oxidative Stress, Inflammation and Apoptosis. *Mol. Biol. Rep.* **2020**, *47*, 1371–1379. [CrossRef] [PubMed]

37. Tavafi, M.; Ahmadvand, H.; Toolabi, P. Inhibitory Effect of Olive Leaf Extract on Gentamicin-Induced Nephrotoxicity in Rats—Yafteh. *Iran. J. Kidney Dis.* **2012**, *6*, 25–32.

38. Beiranvand, A.; Rasoulian, B.; Alizerai, M.; Hashemi, P.; Pilevarian, A.A.; Ezatpor, B.; Tavafi, M.; Chash, S. Pretreatment with Olive Leaf Extract Partially Attenuates Cisplatin-Induced Nephrotoxicity in Rats—Yafteh. Available online: http://yafte.lums.ac.ir/browse.php?a_id=190&sid=1&slc_lang=en (accessed on 15 January 2020).

39. Dias, M.C.; Pinto, D.C.G.A.; Freitas, H.; Santos, C.; Silva, A.M.S. The Antioxidant System in Olea Europaea to Enhanced UV-B Radiation also depends on Flavonoids and Secoiridoids. *Phytochemistry* **2020**, *170*, 112199. [CrossRef]

40. Bars-Cortina, D.; Hazas, M.-C.L.D.L.; Benavent-Vallés, A.; Motilva, M.-J. Impact of Dietary Supplementation with Olive and Thyme Phenols on alpha-tocopherol Concentration in the Muscle and Liver of Adult Wistar Rats. *Food Funct.* **2018**, *9*, 1433–1443. [CrossRef]

41. Gouvinhas, I.; Machado, N.; Sobreira, C.; Domínguez-Perles, R.; Gomes, S.; Rosa, E.; Barros, A.; Gouvinhas, I.; Machado, N.; Sobreira, C.; et al. Critical Review on the Significance of Olive Phytochemicals in Plant Physiology and Human Health. *Molecules* **2017**, *22*, 1986. [CrossRef]

42. Visioli, F.; Bellomo, G.; Galli, C. Free Radical-Scavenging Properties of Olive Oil Polyphenols. *Biochem. Biophys. Res. Commun.* **1998**, *247*, 60–64. [CrossRef]

43. Fabiani, R.; Rosignoli, P.; De Bartolomeo, A.; Fuccelli, R.; Servili, M.; Montedoro, G.F.; Morozzi, G. Oxidative DNA Damage Is Prevented by Extracts of Olive Oil, Hydroxytyrosol, and Other Olive Phenolic Compounds in Human Blood Mononuclear Cells and HL60 Cells. *J. Nutr.* **2008**, *138*, 1411–1416. [CrossRef] [PubMed]

44. Hussain, T.; Tan, B.; Yin, Y.; Blachier, F.; Tossou, M.C.B.; Rahu, N. Oxidative Stress and Inflammation: What Polyphenols Can Do for Us? *Oxid. Med. Cell. Longev.* **2016**, *2016*. [CrossRef] [PubMed]

45. Benavente-García, O.; Castillo, J.; Lorente, J.; Ortuño, A.; Del Rio, J. Antioxidant Activity of Phenolics Extracted from *Olea europaea* L. leaves. *Food Chem.* **2000**, *68*, 457–462. [CrossRef]

46. Sarica, S.; Toptas, S. Effects of Dietary Oleuropein Supplementation on Growth Performance, Serum Lipid Concentrations and Lipid Oxidation of Japanese Quails. *J. Anim. Physiol. Anim. Nutr.* **2014**, *98*, 1176–1186. [CrossRef] [PubMed]

47. De La Ossa, J.G.; Felice, F.; Azimi, B.; Salsano, J.E.; Digiacomo, M.; Macchia, M.; Danti, S.; Di Stefano, R. Waste Autochthonous Tuscan Olive Leaves (*Olea europaea* var. Olivastra seggianese) as Antioxidant Source for Biomedicine. *Int. J. Mol. Sci.* **2019**, *20*, 5918. [CrossRef] [PubMed]

48. Tripoli, E.; Giammanco, M.; Tabacchi, G.; Di Majo, D.; Giammanco, S.; La Guardia, M. The Phenolic Compounds of Olive Oil: Structure, Biological Activity and Beneficial Effects on Human Health. *Nutr. Res. Rev.* **2005**, *18*, 98–112. [CrossRef] [PubMed]

49. Visioli, F.; Galli, C. Oleuropein Protects Low Density Lipoprotein from Oxidation. *Life Sci.* **1994**, *55*, 1965–1971. [CrossRef]

50. Montoya, T.; Castejón, M.L.; Sánchez-Hidalgo, M.; González-Benjumea, A.; Fernández-Bolaños, J.G.; De-La-Lastra, C.A. Oleocanthal Modulates LPS-Induced Murine Peritoneal Macrophages Activation via Regulation of Inflammasome, Nrf-2/HO-1, and MAPKs Signaling Pathways. *J. Agric. Food Chem.* **2019**, *67*, 5552–5559. [CrossRef]

51. Soler-Rivas, C.; Marín, F.R.; Santoyo, S.; García-Risco, M.R.; Señoráns, F.J.; Reglero, G. Testing and Enhancing the in vitro Bioaccessibility and Bioavailability of Rosmarinus Officinalis Extracts with a High Level of Antioxidant Abietanes. *J. Agric. Food Chem.* **2010**, *58*, 1144–1152. [CrossRef]

52. Vissers, M.N.; Zock, P.L.; Roodenburg, A.J.C.; Leenen, R.; Katan, M.B. Olive Oil Phenols Are Absorbed in Humans. *J. Nutr.* **2002**, *132*, 409–417. [CrossRef]

53. Imran, M.; Nadeem, M.; Gilani, S.A.; Khan, S.; Sajid, M.W.; Amir, R.M. Antitumor Perspectives of Oleuropein and Its Metabolite Hydroxytyrosol: Recent Updates. *J. Food Sci.* **2018**, *83*, 1781–1791. [CrossRef]

54. Francisco, V.; Ruiz-Fernandez, C.; Lahera, V.; Lago, F.; Pino, J.; Skaltsounis, L.; González-Gay, M.A.; Mobasheri, A.; Gomez, R.; Scotece, M.; et al. Natural Molecules for Healthy Lifestyles: Oleocanthal from Extra Virgin Olive Oil. *J. Agric. Food Chem.* **2019**, *67*, 3845–3853. [CrossRef] [PubMed]

55. D'Archivio, M.; Filesi, C.; Varì, R.; Scazzocchio, B.; Masella, R. Bioavailability of the Polyphenols: Status and Controversies. *Int. J. Mol. Sci.* **2010**, *11*, 1321–1342. [CrossRef]

56. Kendall, M.; Batterham, M.; Callahan, D.L.; Jardine, D.; Prenzler, P.D.; Robards, K.; Ryan, D.; Kendall, M.; Batterham, M.; Callahan, D.L.; et al. Title: Randomized Controlled Study of the Urinary Excretion of Biophenols Following Acute and Chronic Intake of Olive Leaf Supplements. *Food Chem.* **2012**, *130*, 651–659. [CrossRef]

57. Karkovíc, A.K.; Markovíc, M.; Toríc, J.T.; Barbaríc, M.B.; Jakobušíc, C.; Brala, J. Molecules Hydroxytyrosol, Tyrosol and Derivatives and Their Potential Effects on Human Health. *Molecules* **2019**, *24*, 2001. [CrossRef]

58. Corona, G.; Tzounis, X.; Dessì, M.A.; Deiana, M.; Debnam, E.S.; Visioli, F.; Spencer, J.P.E. The Fate of Olive Oil Polyphenols in the Gastrointestinal Tract: Implications of Gastric and Colonic Microflora-Dependent Biotransformation. *Free. Radic. Res.* **2006**, *40*, 647–658. [CrossRef] [PubMed]

59. Del Boccio, P.; Di Deo, A.; De Curtis, A.; Celli, N.; Iacoviello, L.; Rotilio, D. Liquid Chromatography-Tandem Mass Spectrometry Analysis of Oleuropein and Its Metabolite Hydroxytyrosol in rat Plasma and Urine after Oral Administration. *J. Chromatogr. B* **2003**, *785*, 47–56. [CrossRef]

60. Bazoti, F.N.; Gikas, E.; Tsarbopoulos, A. Simultaneous Quantification of Oleuropein and its Metabolites in Rat Plasma by Liquid Chromatography Electrospray Ionization Tandem Mass Spectrometry. *Biomed. Chromatogr.* **2010**, *24*, 506–515. [CrossRef]

61. Lin, P.; Qian, W.; Wang, X.; Cao, L.; Li, S.; Qian, T. The Biotransformation of Oleuropein in Rats. *Biomed. Chromatogr.* **2013**, *27*, 1162–1167. [CrossRef] [PubMed]

62. Mosele, J.I.; Macià, A.; Motilva, M.-J. Molecules Metabolic and Microbial Modulation of the Large Intestine Ecosystem by Non-Absorbed Diet Phenolic Compounds: A Review. *Molecules* **2015**, *20*, 17429–17468. [CrossRef]

63. De Bock, M.; Thorstensen, E.B.; Derraik, J.G.B.; Henderson, H.V.; Hofman, P.L.; Cutfield, W.S. Human Absorption and Metabolism of Oleuropein and Hydroxytyrosol Ingested as Olive (*Olea europaea* L.) Leaf Extract. *Mol. Nutr. Food Res.* **2013**, *57*, 2079–2085. [CrossRef]

64. Pinto, J.; Paiva-Martins, F.; Corona, G.; Debnam, E.S.; Oruna-Concha, M.J.; Vauzour, D.; Gordon, M.H.; Spencer, J.P.E. Absorption and Metabolism of Olive Oil Secoiridoids in the Small Intestine. *Br. J. Nutr.* **2011**, *105*, 1607–1618. [CrossRef]

65. Cicerale, S.; Conlan, X.A.; Sinclair, A.J.; Keast, R.S.J. Chemistry and Health of Olive Oil Phenolics. *Crit. Rev. Food Sci. Nutr* **2009**, *49*, 218–236. [CrossRef] [PubMed]

66. Hazas, M.-C.L.D.L.; Piñol, C.; Macià, A.; Romero, M.-P.; Pedret, A.; Solà, R.; Rubio, L.; Motilva, M.-J. Differential Absorption and Metabolism of Hydroxytyrosol and its Precursors Oleuropein and Secoiridoids. *J. Funct. Foods* **2016**, *22*, 52–63. [CrossRef]

67. Tan, H.-W.; Tuck, K.L.; Stupans, I.; Hayball, P.J. Simultaneous Determination of Oleuropein and Hydroxytyrosol in Rat Plasma Using Liquid Chromatography with Fluorescence Detection. *J. Chromatogr. B* **2003**, *785*, 187–191. [CrossRef]

68. Hazas, M.-C.L.D.L.; Piñol, C.; Macià, A.; Motilva, M.-J. Hydroxytyrosol and the Colonic Metabolites Derived from Virgin Olive Oil Intake Induce Cell Cycle Arrest and Apoptosis in Colon Cancer Cells. *J. Agric. Food Chem.* **2017**, *65*, 6467–6476. [CrossRef] [PubMed]

69. Angeloni, C.; Malaguti, M.; Barbalace, M.C.; Hrelia, S. Bioactivity of Olive Oil Phenols in Neuroprotection. *Int. J. Mol. Sci.* **2017**, *18*, 2230. [CrossRef] [PubMed]

70. Romero, C.; Medina, E.; Vargas, J.; Brenes, M.; De Castro, A. In Vitro Activity of Olive Oil Polyphenols againstHelicobacter pylori. *J. Agric. Food Chem.* **2007**, *55*, 680–686. [CrossRef]

71. Silva, S.; Garcia-Aloy, M.; Figueira, M.E.; Combet, E.; Mullen, W.; Bronze, M.R. High Resolution Mass Spectrometric Analysis of Secoiridoids and Metabolites as Biomarkers of Acute Olive Oil Intake—An Approach to Study Interindividual Variability in Humans. *Mol. Nutr. Food Res.* **2018**, *62*, 1700065. [CrossRef]

72. García-Villalba, R.; Carrasco-Pancorbo, A.; Oliveras-Ferraros, C.; Vázquez-Martín, A.; Menendez, J.A.; Carretero, A.S.; Fernández-Gutiérrez, A. Characterization and Quantification of Phenolic Compounds of Extra-Virgin Olive Oils with Anticancer Properties by a Rapid and Resolutive LC-ESI-TOF MS Method. *J. Pharm. Biomed. Anal.* **2010**, *51*, 416–429. [CrossRef]

73. Naruszewicz, M.; Czerwinska, M.; Kiss, A. Oleacein. Translation from Mediterranean Diet to Potential Antiatherosclerotic Drug. *Curr. Pharm. Des.* **2014**, *21*, 1205–1212. [CrossRef]

74. Willett, W.C.; Sacks, F.; Trichopoulou, A.; Drescher, G.; Ferro-Luzzi, A.; Helsing, E.; Trichopoulos, D. Mediterranean Diet Pyramid: A Cultural Model for Healthy Eating. *Am. J. Clin. Nutr.* **1995**, *61*, 61. [CrossRef] [PubMed]

75. Dydjow-Bendek, D.; Zagoźdźon, P. Total Dietary Fats, Fatty Acids, and Omega-3/Omega-6 Ratio as Risk Factors of Breast Cancer in the Polish Population—A Case-Control Study. *In Vivo* **2020**, *34*, 423–431. [CrossRef] [PubMed]

76. Wang, J.; Wang, C. Dietary Fat Intake and Risk of Bladder Cancer: Evidence from a Meta-Analysis of Observational Studies. *Cell. Mol. Boil.* **2019**, *65*, 5–9. [CrossRef]

77. Grosso, G.; Buscemi, S.; Galvano, F.; Mistretta, A.; Marventano, S.; La Vela, V.; Drago, F.; Gangi, S.; Basile, F.; Biondi, A. Mediterranean Diet and Cancer: Epidemiological Evidence and Mechanism of Selected Aspects. *BMC Surg.* **2013**, *13*, S14. [CrossRef]

78. Solinas, G.; Germano, G.; Mantovani, A.; Allavena, P. Tumor-Associated Macrophages (TAM) as Major Players of the Cancer-Related Inflammation. *J. Leukoc. Boil.* **2009**, *86*, 1065–1073. [CrossRef]

79. Hamdi, H.K.; Castellon, R. Oleuropein, A Non-Toxic Olive Iridoid, Is an Anti-Tumor Agent and Cytoskeleton Disruptor. *Biochem. Biophys. Res. Commun.* **2005**, *334*, 769–778. [CrossRef]

80. Notarnicola, M.; Pisanti, S.; Tutino, V.; Bocale, D.; Rotelli, M.T.; Gentile, A.; Memeo, V.; Bifulco, M.; Perri, E.; Caruso, M.G. Effects of Olive Oil Polyphenols on Fatty Acid Synthase Gene Expression and Activity in Human Colorectal Cancer Cells. *Genes Nutr.* **2011**, *6*, 63–69. [CrossRef]

81. Cárdeno, A.; Sánchez-Hidalgo, M.; Rosillo, M.A.; De La Lastra, C.A. Oleuropein, a Secoiridoid Derived from Olive Tree, Inhibits the Proliferation of Human Colorectal Cancer Cell Through Downregulation of HIF-1α. *Nutr. Cancer* **2013**, *65*, 147–156. [CrossRef]

82. Hassan, Z.K.; Elamin, M.H.; Daghestani, M.H.; Omer, S.A.; Al-Olayan, E.M.; Elobeid, M.A.; Virk, P.; Mohammed, O.B. Oleuropein Induces Anti-Metastatic Effects in Breast Cancer. *Asian Pac. J. Cancer Prev.* **2012**, *13*, 4555–4559. [CrossRef]

83. Elamin, M.H.; Daghestani, M.H.; Omer, S.A.; A Elobeid, M.; Virk, P.; Al-Olayan, E.M.; Hassan, Z.K.; Mohammed, O.B.; Aboussekhra, A. Olive Oil Oleuropein Has Anti-Breast Cancer Properties with Higher Efficiency on ER-Negative Cells. *Food Chem. Toxicol.* **2013**, *53*, 310–316. [CrossRef]

84. Chimento, A.; Casaburi, I.; Rosano, C.; Avena, P.; De Luca, A.; Campana, C.; Martire, E.; Santolla, M.F.; Maggiolini, M.; Pezzi, V.; et al. Oleuropein and Hydroxytyrosol Activate GPER/ GPR30-Dependent Pathways Leading to Apoptosis of ER-Negative SKBR3 Breast Cancer Cells. *Mol. Nutr. Food Res.* **2014**, *58*, 478–489. [CrossRef] [PubMed]

85. Hassan, Z.K.; Elamin, M.H.; Omer, S.A.; Daghestani, M.H.; Al-Olayan, E.S.; Elobeid, M.A.; Virk, P. Oleuropein Induces Apoptosis via the p53 Pathway in Breast Cancer Cells. *Asian Pac. J. Cancer Prev.* **2014**, *14*, 6739–6742. [CrossRef] [PubMed]

86. Liu, L.; Ahn, K.S.; Shanmugam, M.K.; Wang, H.; Shen, H.; Arfuso, F.; Chinnathambi, A.; Alharbi, S.A.; Chang, Y.; Sethi, G.; et al. Oleuropein Induces Apoptosis via Abrogating NF-κB Activation Cascade in Estrogen Receptor–Negative Breast Cancer Cells. *J. Cell. Biochem.* **2018**, *120*, 4504–4513. [CrossRef] [PubMed]

87. Choupani, J.; Alivand, M.R.M.; Derakhshan, S.; Zaeifizadeh, M.S.; Khaniani, M. Oleuropein Inhibits Migration Ability Through Suppression of Epithelial-Mesenchymal Transition and Synergistically Enhances Doxorubicin-Mediated Apoptosis in MCF-7 Cells. *J. Cell. Physiol.* **2019**, *234*, 9093–9104. [CrossRef]

88. Bayat, S.; Derakhshan, S.M.; Derakhshan, N.M.; Khaniani, M.S.; Alivand, M.R. Downregulation ofHDAC2andHDAC3via oleuropein as a Potent Prevention and Therapeutic Agent in MCF-7 Breast Cancer Cells. *J. Cell. Biochem.* **2019**, *120*, 9172–9180. [CrossRef]

89. Mansouri, N.; Alivand, M.R.; Bayat, S.; Khaniani, M.S.; Derakhshan, S.M. The Hopeful Anticancer Role of Oleuropein in Breast Cancer through Histone Deacetylase Modulation. *J. Cell. Biochem.* **2019**, *120*, 17042–17049. [CrossRef]

90. Lu, H.-Y.; Zhu, J.-S.; Zhang, Z.; Shen, W.-J.; Jiang, S.; Long, Y.-F.; Wu, B.; Ding, T.; Huan, F.; Wang, S.-L. Hydroxytyrosol and Oleuropein Inhibit Migration and Invasion of MDA-MB-231 Triple-Negative Breast Cancer Cell via Induction of Autophagy. *Anticancer Agents Med. Chem.* **2019**, *19*, 1983–1990. [CrossRef]

91. Przychodzen, P.; Kuban-Jankowska, A.; Wyszkowska, R.; Barone, G.; Bosco, G.L.; Celso, F.L.; Kamm, A.; Daca, A.; Kostrzewa, T.; Gorska-Ponikowska, M. PTP1B Phosphatase as a Novel Target of Oleuropein Activity in MCF-7 Breast Cancer Model. *Toxicol. Vitr.* **2019**, *61*, 104624. [CrossRef]

92. Yan, C.-M.; Chai, E.-Q.; Cai, H.-Y.; Miao, G.-Y.; Ma, W. Oleuropein Induces Apoptosis via Activation of Caspases and Suppression of Phosphatidylinositol 3-kinase/protein kinase B Pathway in HepG2 Human Hepatoma Cell Line. *Mol. Med. Rep.* **2015**, *11*, 4617–4624. [CrossRef]

93. Sherif, I.O.; Al-Gayyar, M.M. Oleuropein Potentiates Anti-Tumor Activity of Cisplatin Against HepG2 through Affecting proNGF/NGF Balance. *Life Sci.* **2018**, *198*, 87–93. [CrossRef]

94. Vanella, L.; Acquaviva, R.; Di Giacomo, C.; Sorrenti, V.; Galvano, F.; Santangelo, R.; Cardile, V.; Gangia, S.; D'Orazio, N.; Abraham, N.G. Antiproliferative Effect of Oleuropein in Prostate Cell Lines. *Int. J. Oncol.* **2012**, *41*, 31–38. [CrossRef] [PubMed]

95. Bulotta, S.; Corradino, R.; Celano, M.; Maiuolo, J.; D'Agostino, M.; Oliverio, M.; Procopio, A.; Filetti, S.; Russo, D. Antioxidant and Antigrowth Action of Peracetylated Oleuropein in Thyroid Cancer Cells. *J. Mol. Endocrinol.* **2013**, *51*, 181–189. [CrossRef] [PubMed]

96. Yao, J.; Wu, J.; Yang, X.; Yang, J.; Zhang, Y.; Du, L. Oleuropein Induced Apoptosis in HeLa Cells via a Mitochondrial Apoptotic Cascade Associated with Activation of the c-Jun NH2-Terminal Kinase. *J. Pharmacol. Sci.* **2014**, *125*, 300–311. [CrossRef] [PubMed]

97. Seçme, M.; Eroğlu, C.; Dodurga, Y.; Bağcı, G. Investigation of Anticancer Mechanism of Oleuropein via Cell Cycle and Apoptotic Pathways in SH-SY5Y Neuroblastoma Cells. *Gene* **2016**, *585*, 93–99. [CrossRef] [PubMed]

98. Liu, M.; Wang, J.; Huang, B.; Chen, A.; Li, X. Oleuropein Inhibits the Proliferation and Invasion of Glioma Cells via Suppression of the AKT Signaling Pathway. *Oncol. Rep.* **2016**, *36*, 2009–2016. [CrossRef] [PubMed]

99. Xu, T.; Xiao, D. Oleuropein Enhances Radiation Sensitivity of Nasopharyngeal Carcinoma by Downregulating PDRG1 through HIF1α-repressed microRNA-519d. *J. Exp. Clin. Cancer Res.* **2017**, *36*, 3. [CrossRef]

100. Cao, S.; Zhu, X.; Du, L. P38 MAP Kinase is Involved in Oleuropein-Induced Apoptosis in A549 Cells by a Mitochondrial Apoptotic cascade. *Biomed. Pharmacother.* **2017**, *95*, 1425–1435. [CrossRef]

101. Wang, W.; Wu, J.; Zhang, Q.; Li, X.; Zhu, X.; Wang, Q.; Cao, S.; Du, L. Mitochondria-Mediated Apoptosis Was Induced by Oleuropein in H1299 Cells Involving Activation of p38 MAP Kinase. *J. Cell. Biochem.* **2019**, *120*, 5480–5494. [CrossRef]

102. Antognelli, C.; Frosini, R.; Santolla, M.F.; Peirce, M.J.; Talesa, V.N. Oleuropein-Induced Apoptosis Is Mediated by Mitochondrial Glyoxalase 2 in NSCLC A549 Cells: A Mechanistic Inside and a Possible Novel Nonenzymatic Role for an Ancient Enzyme. *Oxidative Med. Cell. Longev.* **2019**, *2019*, 8576961. [CrossRef]

103. Goldsmith, C.D.; Bond, D.R.; Jankowski, H.; Weidenhofer, J.; Stathopoulos, C.E.; Roach, P.D.; Scarlett, C.J. The Olive Biophenols Oleuropein and Hydroxytyrosol Selectively Reduce Proliferation, Influence the Cell Cycle, and Induce Apoptosis in Pancreatic Cancer Cells. *Int. J. Mol. Sci.* **2018**, *19*, 1937. [CrossRef]

104. Ruzzolini, J.; Peppicelli, S.; Andreucci, E.; Bianchini, F.; Scardigli, A.; Romani, A.; La Marca, G.; Nediani, C.; Calorini, L. Oleuropein, the Main Polyphenol of Olea europaea Leaf Extract, Has an Anti-Cancer Effect on Human BRAF Melanoma Cells and Potentiates the Cytotoxicity of Current Chemotherapies. *Nutrients* **2018**, *10*, 1950. [CrossRef] [PubMed]

105. Zhang, F.; Zhang, M. Oleuropein Inhibits Esophageal Cancer through Hypoxic Suppression of BTG3 mRNA. *Food Funct.* **2019**, *10*, 978–985. [CrossRef] [PubMed]

106. Przychodzen, P.; Wyszkowska, R.; Gorzynik-Debicka, M.; Kostrzewa, T.; Kuban-Jankowska, A.; Gorska-Ponikowska, M. Anticancer Potential of Oleuropein, the Polyphenol of Olive Oil, With 2-Methoxyestradiol, Separately or in Combination, in Human Osteosarcoma Cells. *Anticancer. Res.* **2019**, *39*, 1243–1251. [CrossRef] [PubMed]

107. Barzegar, F.; Zaefizadeh, M.; Yari, R.; Salehzadeh, A. Synthesis of Nano-Paramagnetic Oleuropein to Induce KRAS Over-Expression: A New Mechanism to Inhibit AGS Cancer Cells. *Medicina (Kaunas)* **2019**, *55*, E388. [CrossRef]

108. Rigacci, S.; Miceli, C.; Nediani, C.; Berti, A.; Cascella, R.; Pantano, D.; Nardiello, P.; Luccarini, I.; Casamenti, F.; Stefani, M. Oleuropein Aglycone Induces Autophagy via the AMPK/mTOR Signalling Pathway: A Mechanistic Insight. *Oncotarget* **2015**, *6*, 35344–35357. [CrossRef]

109. Khanal, P.; Oh, W.-K.; Yun, H.J.; Namgoong, G.M.; Ahn, S.-G.; Kwon, S.-M.; Choi, H.-K.; Choi, H.S. p-HPEA-EDA, a Phenolic Compound of Virgin Olive Oil, Activates AMP-Activated Protein Kinase to Inhibit Carcinogenesis. *Carcinogenesis* **2011**, *32*, 545–553. [CrossRef]

110. Elnagar, A.; Sylvester, P.; El Sayed, K. (−)-Oleocanthal as a c-Met Inhibitor for the Control of Metastatic Breast and Prostate Cancers. *Planta Medica* **2011**, *77*, 1013–1019. [CrossRef]

111. Akl, M.R.; Ayoub, N.M.; Mohyeldin, M.M.; Busnena, B.A.; Foudah, A.I.; Liu, Y.-Y.; Sayed, K.A.E. Olive Phenolics as c-Met Inhibitors: (-)-Oleocanthal Attenuates Cell Proliferation, Invasiveness, and Tumor Growth in Breast Cancer Models. *PLOS ONE* **2014**, *9*, e97622. [CrossRef]

112. Khanfar, M.A.; Bardaweel, S.K.; Akl, M.R.; Sayed, K.A. El Olive Oil-derived Oleocanthal as Potent Inhibitor of Mammalian Target of Rapamycin: Biological Evaluation and Molecular Modeling Studies. *Phyther. Res.* **2015**, *29*, 1776–1782. [CrossRef]

113. Tajmim, A.; Siddique, A.B.; Sayed, E.; El Sayed, K. Optimization of Taste-Masked (−)-Oleocanthal Effervescent Formulation with Potent Breast Cancer Progression and Recurrence Suppressive Activities. *Pharmaceutics* **2019**, *11*, E515. [CrossRef]

114. Qusa, M.H.; Siddique, A.B.; Nazzal, S.; El Sayed, K.A. Novel Olive Oil Phenolic (−)-Oleocanthal (+)-Xylitol-Based Solid Dispersion Formulations with Potent Oral Anti-Breast Cancer Activities. *Int. J. Pharm.* **2019**, *569*, 118596. [CrossRef] [PubMed]

115. Siddique, A.B.; Ebrahim, H.Y.; Akl, M.R.; Ayoub, N.M.; Goda, A.A.; Mohyeldin, M.M.; Nagumalli, S.K.; Hananeh, W.M.; Liu, Y.-Y.; Meyer, S.A.; et al. (−)-Oleocanthal Combined with Lapatinib Treatment Synergized against HER-2 Positive Breast Cancer in Vitro and in Vivo. *Nutrients* **2019**, *11*, 412. [CrossRef] [PubMed]

116. Diez-Bello, R.; Jardin, I.; Lopez, J.J.; El Haouari, M.; Ortega-Vidal, J.; Altarejos, J.; Salido, G.M.; Salido, S.; Rosado, J.A. (−)-Oleocanthal Inhibits Proliferation and Migration by Modulating Ca2+ Entry through TRPC6 in Breast Cancer Cells. *Biochim. Biophys. Acta - Mol. Cell Res.* **2019**, *1866*, 474–485. [CrossRef]

117. Pei, T.; Meng, Q.; Han, J.; Sun, H.; Li, L.; Song, R.; Sun, B.; Pan, S.; Liang, D.; Liu, L. (−)-Oleocanthal Inhibits Growth and Metastasis by Blocking Activation of STAT3 in Human Hepatocellular Carcinoma. *Oncotarget* **2016**, *7*, 43475–43491. [CrossRef] [PubMed]

118. De Stefanis, D.; Scimè, S.; Accomazzo, S.; Catti, A.; Occhipinti, A.; Bertea, C.M.; Costelli, P.; Stefanis, D. Anti-Proliferative Effects of an Extra-Virgin Olive Oil Extract Enriched in Ligstroside Aglycone and Oleocanthal on Human Liver Cancer Cell Lines. *Cancers* **2019**, *11*, 1640. [CrossRef] [PubMed]

119. Margarucci, L.; Cassiano, C.; Mozzicafreddo, M.; Angeletti, M.; Riccio, R.; Monti, M.C.; Tosco, A.; Casapullo, A. Chemical Proteomics-Driven Discovery of Oleocanthal as an Hsp90 Inhibitor. *Chem. Commun.* **2013**, *49*, 5844. [CrossRef]

120. Gu, Y.; Wang, J.; Peng, L. (−)-Oleocanthal Exerts Anti-Melanoma Activities and Inhibits STAT3 Signaling Pathway. *Oncol. Rep.* **2017**, *37*, 483–491. [CrossRef]

121. Polini, B.; Digiacomo, M.; Carpi, S.; Bertini, S.; Gado, F.; Saccomanni, G.; Macchia, M.; Nieri, P.; Manera, C.; Fogli, S. Oleocanthal and Oleacein Contribute to the in Vitro Therapeutic Potential of in Oil-Derived Extracts in Non-Melanoma Skin Cancer. *Toxicol. Vitr.* **2018**, *52*, 243–250. [CrossRef]

122. Juli, G.; Oliverio, M.; Bellizzi, D.; Cantafio, M.E.G.; Grillone, K.; Passarino, G.; Colica, C.; Nardi, M.; Rossi, M.; Procopio, A.; et al. Anti-tumor Activity and Epigenetic Impact of the Polyphenol Oleacein in Multiple Myeloma. *Cancers* **2019**, *11*, 990. [CrossRef]

123. Kimura, Y.; Sumiyoshi, M. Olive Leaf Extract and Its Main Component Oleuropein Prevent Chronic Ultraviolet B Radiation-Induced Skin Damage and Carcinogenesis in Hairless Mice. *J. Nutr.* **2009**, *139*, 2079–2086. [CrossRef]

124. Giner, E.; Recio, M.C.; Ríos, J.L.; Cerdá-Nicolás, J.M.; Giner, R.M. Chemopreventive Effect of Oleuropein in Colitis-Associated Colorectal Cancer in c57bl/6 Mice. *Mol. Nutr. Food Res.* **2016**, *60*, 242–255. [CrossRef] [PubMed]

125. Geyikoğlu, F.; Çolak, S.; Türkez, H.; Bakır, M.; Koç, K.; Hosseinigouzdagani, M.K.; Çeriğ, S.; Sönmez, M. Oleuropein Ameliorates Cisplatin-induced Hematological Damages Via Restraining Oxidative Stress and DNA Injury. *Indian J. Hematol. Blood Transfus.* **2017**, *33*, 348–354. [CrossRef] [PubMed]

126. Song, H.; Lim, Y.; Jung, J.I.; Cho, H.J.; Park, S.Y.; Kwon, G.T.; Kang, Y.-H.; Lee, K.W.; Choi, M.-S.; Park, J.H.Y. Dietary Oleuropein Inhibits Tumor Angiogenesis and Lymphangiogenesis in the B16F10 Melanoma Allograft Model: A Mechanism for the Suppression of High-Fat Diet-Induced Solid Tumor Growth and Lymph Node Metastasis. *Oncotarget* **2017**, *8*, 32027–32042. [CrossRef] [PubMed]

127. Koc, K.; Bakir, M.; Geyikoglu, F.; Cerig, S. Therapeutic Effects of Oleuropein on Cisplatin-Induced Pancreas Injury in Rats. *J. Cancer Res. Ther.* **2018**, *14*, 671. [CrossRef]

128. Elamin, M.H.; Elmahi, A.B.; Daghestani, M.H.; Al-Olayan, E.M.; Al-Ajmi, R.A.; Alkhuriji, A.F.; Hamed, S.S.; Elkhadragy, M.F. Synergistic Anti-Breast-Cancer Effects of Combined Treatment with Oleuropein and Doxorubicin in Vivo. *Altern. Ther. Health Med.* **2019**, *25*, 17–24.

129. Siddique, A.B.; Ayoub, N.M.; Tajmim, A.; Meyer, S.A.; Hill, R.A.; El Sayed, K.A. (−)-Oleocanthal Prevents Breast Cancer Locoregional Recurrence After Primary Tumor Surgical Excision and Neoadjuvant Targeted Therapy in Orthotopic Nude Mouse Models. *Cancers* **2019**, *11*, 637. [CrossRef]

130. Sherif, I.O. The Effect of Natural Antioxidants in Cyclophosphamide-Induced Hepatotoxicity: Role of Nrf2/HO-1 Pathway. *Int. Immunopharmacol.* **2018**, *61*, 29–36. [CrossRef]

131. Goren, L.; Zhang, G.; Kaushik, S.; Breslin, P.A.S.; Du, Y.-C.N.; Foster, D.A. (-)-Oleocanthal and (-)-Oleocanthal-Rich Olive Oils Induce Lysosomal Membrane Permeabilization in Cancer Cells. *PLOS ONE* **2019**, *14*, e0216024. [CrossRef]

132. Siddique, A.B.; Ebrahim, H.; Mohyeldin, M.; Qusa, M.; Batarseh, Y.; Fayyad, A.; Tajmim, A.; Nazzal, S.; Kaddoumi, A.; El Sayed, K. Novel Liquid-Liquid Extraction and Self-Emulsion Methods for Simplified Isolation of Extra-Virgin Olive Oil Phenolics with Emphasis on (−)-Oleocanthal and Its Oral Anti-Breast Cancer Activity. *PLOS ONE* **2019**, *14*, e0214798. [CrossRef]

133. Scotece, M.; Gómez, R.; Conde, J.; Lopez, V.; Gomez-Reino, J.J.; Lago, F.; Smith, A.B.; Gualillo, O. Further Evidence for the Anti-Inflammatory Activity of Oleocanthal: Inhibition of MIP-1α and IL-6 in J774 Macrophages and in ATDC5 Chondrocytes. *Life Sci.* **2012**, *91*, 1229–1235. [CrossRef]

134. Wang, H.; Naghavi, M.; Allen, C.; Barber, R.M.; A Bhutta, Z.; Carter, A.; Casey, D.C.; Charlson, F.J.; Chen, A.Z.; Coates, M.M.; et al. Global, Regional, and National Life Expectancy, All-Cause Mortality, and Cause-Specific Mortality for 249 Causes of Death, 1980–2015: A Systematic analysis for the Global Burden of Disease Study 2015. *Lancet* **2016**, *388*, 1459–1544. [CrossRef]

135. Reboredo-Rodríguez, P.; González-Barreiro, C.; Cancho-Grande, B.; Forbes-Hernandez, T.Y.; Gasparrini, M.; Afrin, S.; Cianciosi, D.; Carrasco-Pancorbo, A.; Simal-Gándara, J.; Giampieri, F.; et al. Characterization of Phenolic Extracts from Brava Extra Virgin Olive Oils and Their Cytotoxic Effects on MCF-7 Breast Cancer Cells. *Food Chem. Toxicol.* **2018**, *119*, 73–85. [CrossRef] [PubMed]

136. Crespo, M.C.; Tomé-Carneiro, J.; Dávalos, A.; Visioli, F. Pharma-Nutritional Properties of Olive Oil Phenols. Transfer of New Findings to Human Nutrition. *Foods* **2018**, *7*, 90. [CrossRef] [PubMed]

137. Catalán, Ú.; Rubio, L.; Hazas, M.-C.L.D.L.; Herrero, P.; Nadal, P.; Canela, N.; Pedret, A.; Motilva, M.-J.; Solà, R. Hydroxytyrosol and Its Complex Forms (Secoiridoids) Modulate Aorta and Heart Proteome in Healthy Rats: Potential Cardio-Protective Effects. *Mol. Nutr. Food Res.* **2016**, *60*, 2114–2129. [CrossRef] [PubMed]

138. Visioli, F.; Bellomo, G.; Montedoro, G.; Galli, C. Low Density Lipoprotein Oxidation Is Inhibited in Vitro by Olive Oil Constituents. *Atherosclerosis* **1995**, *117*, 25–32. [CrossRef]

139. Visioli, F.; Bellosta, S.; Galli, C. Oleuropein, the Bitter Principle of Olives, Enhances Nitric Oxide Production by Mouse Macrophages. *Life Sci.* **1998**, *62*, 541–546. [CrossRef]

140. Parzonko, A.; Czerwińska, M.E.; Kiss, A.K.; Naruszewicz, M. Oleuropein and Oleacein may Restore Biological Functions of Endothelial Progenitor Cells Impaired by Angiotensin II via Activation of Nrf2/heme Oxygenase-1 Pathway. *Phytomedicine* **2013**, *20*, 1088–1094. [CrossRef]

141. Dell'Agli, M.; Fagnani, R.; Mitro, N.; Scurati, S.; Masciadri, M.; Mussoni, L.; Galli, G.V.; Bosisio, E.; Crestani, M.; De Fabiani, E.; et al. Minor Components of Olive Oil Modulate Proatherogenic Adhesion Molecules Involved in Endothelial Activation. *J. Agric. Food Chem.* **2006**, *54*, 3259–3264. [CrossRef]

142. Leri, M.; Nosi, D.; Natalello, A.; Porcari, R.; Ramazzotti, M.; Chiti, F.; Bellotti, V.; Doglia, S.M.; Stefani, M.; Bucciantini, M. The Polyphenol Oleuropein Aglycone Hinders the Growth of Toxic Transthyretin Amyloid Assemblies. *J. Nutr. Biochem.* **2016**, *30*, 153–166. [CrossRef]

143. Czerwińska, M.E.; Kiss, A.K.; Naruszewicz, M. Inhibition of Human Neutrophils NEP Activity, CD11b/CD18 Expression and Elastase Release by 3,4-dihydroxyphenylethanol-elenolic Acid Dialdehyde, Oleacein. *Food Chem.* **2014**, *153*, 1–8. [CrossRef]

144. Filipek, A.; Czerwińska, M.E.; Kiss, A.K.; Wrzosek, M.; Naruszewicz, M. Oleacein Enhances Anti-Inflammatory Activity of Human Macrophages by Increasing CD163 Receptor Expression. *Phytomedicine* **2015**, *22*, 1255–1261. [CrossRef] [PubMed]

145. Manna, C.; Migliardi, V.; Golino, P.; Scognamiglio, A.; Galletti, P.; Chiariello, M.; Zappia, V. Oleuropein Prevents Oxidative Myocardial Injury Induced by Ischemia and Reperfusion. *J. Nutr. Biochem.* **2004**, *15*, 461–466. [CrossRef] [PubMed]

146. Andreadou, I.; Iliodromitis, E.K.; Mikros, E.; Constantinou, M.; Agalias, A.; Magiatis, P.; Skaltsounis, A.L.; Kamber, E.; Tsantili-Kakoulidou, A.; Kremastinos, D.T. The Olive Constituent Oleuropein Exhibits Anti-Ischemic, Antioxidative, and Hypolipidemic Effects in Anesthetized Rabbits. *J. Nutr.* **2006**, *136*, 2213–2219. [CrossRef] [PubMed]

147. Andreadou, I.; Sigala, F.; Iliodromitis, E.K.; Papaefthimiou, M.; Sigalas, C.; Aligiannis, N.; Savvari, P.; Gorgoulis, V.; Papalabros, E.; Kremastinos, D.T. Acute Doxorubicin Cardiotoxicity Is Successfully Treated with the Phytochemical Oleuropein through Suppression of Oxidative and Nitrosative Stress. *J. Mol. Cell. Cardiol.* **2007**, *42*, 549–558. [CrossRef]

148. Andreadou, I.; Papaefthimiou, M.; Zira, A.; Constantinou, M.; Sigala, F.; Skaltsounis, A.-L.; Tsantili-Kakoulidou, A.; Iliodromitis, E.K.; Kremastinos, D.T.; Mikros, E. Metabonomic Identification of Novel Biomarkers in Doxorubicin Cardiotoxicity and Protective Effect of the Natural Antioxidant Oleuropein. *NMR Biomed.* **2009**, *22*, 585–592. [CrossRef]

149. Andreadou, I.; Mikros, E.; Ioannidis, K.; Sigala, F.; Naka, K.; Kostidis, S.; Farmakis, D.; Tenta, R.; Kavantzas, N.; Bibli, S.-I.; et al. Oleuropein Prevents Doxorubicin-Induced Cardiomyopathy Interfering with Signaling Molecules and Cardiomyocyte Metabolism. *J. Mol. Cell. Cardiol.* **2014**, *69*, 4–16. [CrossRef]

150. Wang, L.; Geng, C.; Jiang, L.; Gong, D.; Liu, D.; Yoshimura, H.; Zhong, L. The Anti-Atherosclerotic Effect of Olive Leaf Extract Is Related to Suppressed Inflammatory Response in Rabbits with Experimental Atherosclerosis. *Eur. J. Nutr.* **2008**, *47*, 235–243. [CrossRef]

151. Ebaid, G.M.; Seiva, F.R.; Rocha, K.K.; A Souza, G.; Novelli, E.L. Effects of Olive Oil and Its Minor Phenolic Constituents on Obesity-Induced Cardiac Metabolic Changes. *Nutr. J.* **2010**, *9*, 46. [CrossRef]

152. Khalili, A.; Nekooeian, A.A.; Khosravi, M.B. Oleuropein Improves Glucose Tolerance and Lipid Profile in Rats with Simultaneous Renovascular Hypertension and Type 2 Diabetes. *J. Asian Nat. Prod. Res.* **2017**, *19*, 1011–1021. [CrossRef]

153. Nekooeian, A.A.; Khalili, A.; Khosravi, M.B. Oleuropein Offers Cardioprotection in Rats with Simultaneous Type 2 Diabetes and Renal Hypertension. *Indian J. Pharmacol.* **2014**, *46*, 398–403. [CrossRef]

154. Nekooeian, A.A.; Khalili, A.; Khosravi, M.B. Effects of Oleuropein in Rats with Simultaneous Type 2 Diabetes and Renal Hypertension: A Study of Antihypertensive Mechanisms. *J. Asian Nat. Prod. Res.* **2014**, *16*, 953–962. [CrossRef] [PubMed]

155. Ivanov, M.; Vajic, U.-J.; Mihailovic-Stanojevic, N.; Miloradovic, Z.; Jovovic, D.; Grujic-Milanovic, J.; Karanovic, D.; Dekanski, D. Highly Potent Antioxidant Olea europaea L. Leaf Extract Affects Carotid and Renal Haemodynamics in Experimental Hypertension: The Role of Oleuropein. *Excli J.* **2018**, *17*, 29–44. [PubMed]

156. Jemai, H.; Bouaziz, M.; Fki, I.; El Feki, A.; Sayadi, S. Hypolipidimic and Antioxidant Activities of Oleuropein and Its Hydrolysis Derivative-Rich Extracts from Chemlali Olive Leaves. *Chem. Interactions* **2008**, *176*, 88–98. [CrossRef] [PubMed]

157. Miceli, C.; Santin, Y.; Manzella, N.; Coppini, R.; Berti, A.; Stefani, M.; Parini, A.; Mialet-Perez, J.; Nediani, C. Oleuropein Aglycone Protects against MAO-A-Induced Autophagy Impairment and Cardiomyocyte Death through Activation of TFEB. *Oxidative Med. Cell. Longev.* **2018**, *2018*, 8067592. [CrossRef]

158. Margheri, F.; Menicacci, B.; Laurenzana, A.; Del Rosso, M.; Fibbi, G.; Cipolleschi, M.G.; Ruzzolini, J.; Nediani, C.; Mocali, A.; Giovannelli, L. Oleuropein Aglycone Attenuates the Pro-Angiogenic Phenotype of Senescent Fibroblasts: A Functional Study in Endothelial Cells. *J. Funct. Foods* **2019**, *53*, 219–226. [CrossRef]

159. Covas, M.-I.; Nyyssönen, K.; E Poulsen, H.; Kaikkonen, J.; Zunft, H.-J.F.; Kiesewetter, H.; Gaddi, A.; De La Torre, R.; Mursu, J.; Bäumler, H.; et al. The Effect of Polyphenols in Olive Oil on Heart Disease Risk Factors: A Randomized Trial. *Ann. Intern. Med.* **2006**, *145*, 333–341. [CrossRef]

160. Susalit, E.; Agus, N.; Effendi, I.; Tjandrawinata, R.R.; Nofiarny, D.; Perrinjaquet-Moccetti, T.; Verbruggen, M. Olive (Olea europaea) Leaf Extract Effective in Patients with Stage-1 Hypertension: Comparison with Captopril. *Phytomedicine* **2011**, *18*, 251–258. [CrossRef]

161. Stock, J. HDL Function and Cardiovascular Risk: Debate Continues . . . . *Atherosclerosis* **2013**, *229*, 38–41. [CrossRef]

162. Casapullo, A.; Del Gaudio, F.; Capolupo, A.; Cassiano, C.; Chiara Monti, M. Multi-Target Profile of Oleocanthal, An Extra-Virgin Olive Oil Component. *Curr. Bioact. Compd.* **2016**, *12*, 3–9. [CrossRef]

163. Beauchamp, G.K.; Keast, R.S.J.; Morel, D.; Lin, J.; Pika, J.; Han, Q.; Lee, C.-H.; Smith, A.B.; Breslin, P.A.S. Ibuprofen-Like activIty in Extra-Virgin Olive Oil. *Nature* **2005**, *437*, 45–46. [CrossRef]

164. Scotece, M.; Conde, J.; Abella, V.; López, V.; Francisco, V.; Ruiz, C.; Campos, V.; Lago, F.; Gomez, R.; Pino, J.; et al. Oleocanthal Inhibits Catabolic and Inflammatory Mediators in LPS-Activated Human Primary Osteoarthritis (OA) Chondrocytes Through MAPKs/NF-κB Pathways. *Cell. Physiol. Biochem.* **2018**, *49*, 2414–2426. [CrossRef] [PubMed]

165. Vinten-Johansen, J. Involvement of Neutrophils in the Pathogenesis of Lethal Myocardial Reperfusion Injury. *Cardiovasc. Res.* **2004**, *61*, 481–497. [CrossRef] [PubMed]

166. Filipek, A.; Czerwińska, M.E.; Kiss, A.K.; Polański, J.A.; Naruszewicz, M. Oleacein may Inhibit Destabilization of Carotid Plaques from Hypertensive Patients: Impact on high Mobility Group Protein-1. *Phytomedicine* **2017**, *32*, 68–73. [CrossRef] [PubMed]

167. Manoharan, S.; Guillemin, G.J.; Abiramasundari, R.S.; Essa, M.M.; Akbar, M. The Role of Reactive Oxygen Species in the Pathogenesis of Alzheimer's Disease, Parkinson's Disease, and Huntington's Disease: A Mini Review. *Oxidative Med. Cell. Longev.* **2016**, *2016*, 1–15. [CrossRef] [PubMed]

168. Wyss-Coray, T. Ageing, neurodegeneration and brain rejuvenation. *Nature* **2016**, *539*, 180–186. [CrossRef] [PubMed]

169. Stefani, M.; Rigacci, S. Beneficial Properties of Natural Phenols: Highlight on Protection against Pathological Conditions Associated with Amyloid Aggregation. *BioFactors* **2014**, *40*, 482–493. [CrossRef] [PubMed]

170. Capurso, A.; Crepaldi, G.; Capurso, C. Extra-Virgin Olive Oil, the Mediterranean Diet, and Neurodegenerative Diseases. In *Benefits of the Mediterranean Diet in the Elderly Patient*; Springer: Cham, Switzerland, 2018; pp. 81–95.

171. Pasban-Aliabadi, H.; Esmaeili-Mahani, S.; Sheibani, V.; Abbasnejad, M.; Mehdizadeh, A.; Yaghoobi, M.M. Inhibition of 6-Hydroxydopamine-Induced PC12 Cell Apoptosis by Olive (Olea europaea L.) Leaf Extract Is Performed by Its Main Component Oleuropein. *Rejuvenation Res.* **2013**, *16*, 134–142. [CrossRef]

172. Achour, I.; Arel-Dubeau, A.-M.; Renaud, J.; Legrand, M.; Attard, E.; Germain, M.; Martinoli, M.-G. Oleuropein Prevents Neuronal Death, Mitigates Mitochondrial Superoxide Production and Modulates Autophagy in a Dopaminergic Cellular Model. *Int. J. Mol. Sci.* **2016**, *17*, 1293. [CrossRef]

173. Leri, M.; Natalello, A.; Bruzzone, E.; Stefani, M.; Bucciantini, M. Oleuropein aglycone and hydroxytyrosol interfere differently with toxic Aβ1-42 aggregation. *Food Chem. Toxicol.* **2019**, *129*, 1–12. [CrossRef]

174. Omar, S.H.; Scott, C.J.; Hamlin, A.S.; Obied, H.K. Olive Biophenols Reduces Alzheimer's Pathology in SH-SY5Y Cells and APPswe Mice. *Int. J. Mol. Sci.* **2019**, *20*, E125. [CrossRef] [PubMed]

175. Luccarini, I.; Dami, T.E.; Grossi, C.; Rigacci, S.; Stefani, M.; Casamenti, F. Oleuropein Aglycone Counteracts Aβ42 Toxicity in the Rat Brain. *Neurosci. Lett.* **2014**, *558*, 67–72. [CrossRef] [PubMed]

176. Grossi, C.; Rigacci, S.; Ambrosini, S.; Dami, T.E.; Luccarini, I.; Traini, C.; Failli, P.; Berti, A.; Casamenti, F.; Stefani, M. The Polyphenol Oleuropein Aglycone Protects TgCRND8 Mice against Aß Plaque Pathology. *PLOS ONE* **2013**, *8*, e71702. [CrossRef] [PubMed]

177. Diomede, L.; Rigacci, S.; Romeo, M.; Stefani, M.; Salmona, M. Oleuropein Aglycone Protects Transgenic C. elegans Strains Expressing Aβ42 by Reducing Plaque Load and Motor Deficit. *PLOS ONE* **2013**, *8*, e58893. [CrossRef] [PubMed]

178. Rigacci, S.; Guidotti, V.; Bucciantini, M.; Nichino, D.; Relini, A.; Berti, A.; Stefani, M. Aβ(1-42) Aggregates into Non-Toxic Amyloid Assemblies in the Presence of the Natural Polyphenol Oleuropein Aglycon. *Curr. Alzheimer Res.* **2011**, *8*, 841–852. [CrossRef] [PubMed]

179. Pantano, D.; Luccarini, I.; Nardiello, P.; Servili, M.; Stefani, M.; Casamenti, F. Oleuropein Aglycone and Polyphenols from Olive Mill Waste Water Ameliorate Cognitive Deficits and Neuropathology. *Br. J. Clin. Pharmacol.* **2017**, *83*, 54–62. [CrossRef]

180. Batarseh, Y.S.; Kaddoumi, A. Oleocanthal-Rich Extra-Virgin Olive Oil Enhances Donepezil Effect by Reducing Amyloid-β Load and Related Toxicity in a Mouse Model of Alzheimer's Disease. *J. Nutr. Biochem.* **2018**, *55*, 113–123. [CrossRef]

181. Qosa, H.; Batarseh, Y.S.; Mohyeldin, M.M.; El Sayed, K.A.; Keller, J.N.; Kaddoumi, A. Oleocanthal Enhances Amyloid-β Clearance from the Brains of TgSwDI Mice and in Vitro across a Human Blood-Brain Barrier Model. *ACS Chem. Neurosci.* **2015**, *6*, 1849–1859. [CrossRef]

182. Abuznait, A.H.; Qosa, H.; Busnena, B.A.; El Sayed, K.A.; Kaddoumi, A. Olive-Oil-Derived Oleocanthal Enhances β-Amyloid Clearance as a Potential Neuroprotective Mechanism against Alzheimer's Disease: In Vitro and in Vivo Studies. *ACS Chem. Neurosci.* **2013**, *4*, 973–982. [CrossRef]

183. Mohammad-Beigi, H.; Aliakbari, F.; Sahin, C.; Lomax, C.; Tawfike, A.; Schafer, N.P.; Amiri-Nowdijeh, A.; Eskandari, H.; Møller, I.M.; Hosseini-Mazinani, M.; et al. Oleuropein Derivatives from Olive Fruit Extracts Reduce α-Synuclein Fibrillation and Oligomer Toxicity. *J. Boil. Chem.* **2019**, *294*, 4215–4232. [CrossRef]

184. Prasanth, M.I.; Brimson, J.M.; Chuchawankul, S.; Sukprasansap, M.; Tencomnao, T. Antiaging, Stress Resistance, and Neuroprotective Efficacies of Cleistocalyx nervosum var. paniala Fruit Extracts Using Caenorhabditis elegans Model. *Oxidative Med. Cell. Longev.* **2019**, *2019*, 1–14. [CrossRef] [PubMed]

185. Chondrogianni, N.; Chinou, I.; Gonos, E.S. Anti-aging Properties of the Olive Constituent Oleuropein in Human Cells. In *Olives and Olive Oil in Health and Disease Prevention*; Elsevier: Amsterdam, The Netherlands, 2010; pp. 1335–1343.

186. Katsiki, M.; Chondrogianni, N.; Chinou, I.; Rivett, A.J.; Gonos, E.S. The Olive Constituent Oleuropein Exhibits Proteasome Stimulatory Properties in Vitro and Confers Life Span Extension of Human Embryonic Fibroblasts. *Rejuvenation Res.* **2007**, *10*, 157–172. [CrossRef] [PubMed]

187. Santiago-Mora, R.; Casado-Díaz, A.; De Castro, M.D.; Quesada-Gómez, J.M. Oleuropein Enhances Osteoblastogenesis and Inhibits Adipogenesis: The Effect on Differentiation in Stem Cells Derived from Bone Marrow. *Osteoporos. Int.* **2011**, *22*, 675–684. [CrossRef] [PubMed]

188. Remmen, H. Oxidative Damage to Mitochondria and Aging. *Exp. Gerontol.* **2001**, *36*, 957–968. [CrossRef]

189. Puel, C.; Quintin, A.; Agalias, A.; Mathey, J.; Obled, C.; Mazur, A.; Davicco, M.J.; Lebecque, P.; Skaltsounis, A.L.; Coxam, V. Olive Oil and Its Main Phenolic Micronutrient (Oleuropein) Prevent Inflammation-Induced Bone Loss in the Ovariectomised Rat. *Br. J. Nutr.* **2004**, *92*, 119–127. [CrossRef] [PubMed]

190. Nikou, T.; Liaki, V.; Stathopoulos, P.; Sklirou, A.D.; Tsakiri, E.N.; Jakschitz, T.; Bonn, G.; Trougakos, I.P.; Halabalaki, M.; Skaltsounis, L.A. Comparison Survey of EVOO Polyphenols and Exploration of Healthy Aging-Promoting Properties of Oleocanthal and Oleacein. *Food Chem. Toxicol.* **2019**, *125*, 403–412. [CrossRef]

191. Lerner, A.; Jeremias, P.; Matthias, T. The World Incidence and Prevalence of Autoimmune Diseases Is Increasing. *Int. J. Celiac Dis.* **2016**, *3*, 151–155. [CrossRef]

192. Rosillo, M.A.; Sánchez-Hidalgo, M.; Sánchez-Fidalgo, S.; Aparicio-Soto, M.; Villegas, I.; Alarcón-de-la-Lastra, C. Dietary Extra-Virgin Olive Oil Prevents Inflammatory Response and Cartilage Matrix Degradation in Murine Collagen-Induced Arthritis. *Eur. J. Nutr.* **2016**, *55*, 315–325. [CrossRef]

193. Smolen, J.S.; Aletaha, D.; McInnes, I.B. Rheumatoid Arthritis. *Lancet* **2016**, *388*, 2023–2038. [CrossRef]

194. Noble, P.W.; Bernatsky, S.; Clarke, A.E.; Isenberg, D.A.; Ramsey-Goldman, R.; Hansen, J.E. DNA-Damaging Autoantibodies and Cancer: The Lupus Butterfly Theory. *Nat. Rev. Rheumatol.* **2016**, *12*, 429–434. [CrossRef]

195. Aparicio-Soto, M.; Sánchez-Hidalgo, M.; Cárdeno, A.; González-Benjumea, A.; Fernández-Bolaños, J.G.; Alarcón-De-La-Lastra, C. Dietary Hydroxytyrosol and Hydroxytyrosyl Acetate Supplementation Prevent Pristane-Induced Systemic Lupus Erythematous in Mice. *J. Funct. Foods* **2017**, *29*, 84–92. [CrossRef]

196. Rees, F.; Doherty, M.; Grainge, M.J.; Lanyon, P.; Zhang, W. The Worldwide Incidence and Prevalence of Systemic Lupus erythematosus: A Systematic Review of Epidemiological Studies. *Rheumatology (Oxford)* **2017**, *56*, 1945–1961. [CrossRef] [PubMed]

197. De Oliveira, G.A.L.; De La Lastra, C.A.; Rosillo, M.Á.; Martinez, M.L.C.; Sánchez-Hidalgo, M.; Medeiros, J.V.R.; Villegas, I. Preventive Effect of Bergenin against the Development of TNBS-Induced Acute Colitis in Rats Is Associated with Inflammatory Mediators Inhibition and NLRP3/ASC Inflammasome Signaling Pathways. *Chem. Interactions* **2019**, *297*, 25–33. [CrossRef] [PubMed]

198. Shivashankar, R.; Tremaine, W.J.; Harmsen, W.S.; Loftus, E.V. Incidence and Prevalence of Crohn's Disease and Ulcerative Colitis in Olmsted County, Minnesota from 1970 through 2010. *Clin. Gastroenterol. Hepatol.* **2017**, *15*, 857–863. [CrossRef]

199. Ilonen, J.; Lempainen, J.; Veijola, R. The Heterogeneous Pathogenesis of Type 1 Diabetes Mellitus. *Nat. Rev. Endocrinol.* **2019**, *15*, 635–650. [CrossRef]

200. Castejon, M.L.; Sánchez-Hidalgo, M.; Aparicio-Soto, M.; González-Benjumea, A.; Fernández-Bolaños, J.G.; Alarcón-De-La-Lastra, C. Olive Secoiridoid Oleuropein and Its Semisynthetic Acetyl-Derivatives Reduce LPS-Induced Inflammatory Response in Murine Peritoneal Macrophages via JAK-STAT and MAPKs Signaling Pathways. *J. Funct. Foods* **2019**, *58*, 95–104. [CrossRef]

201. Castejón, M.L.; Rosillo, M.Á.; Montoya, T.; González-Benjumea, A.; Fernández-Bolaños, J.G.; Alarcón-De-La-Lastra, C. Correction: Oleuropein Down-Regulated IL-1β-Induced Inflammation and Oxidative Stress in Human Synovial Fibroblast Cell Line SW982. *Food Funct.* **2017**, *8*, 2341.

202. Ahmadvand, H.; Shahsavari, G.; Tavafi, M.; Bagheri, S.; Moradkhani, M.R.; Kkorramabadi, R.M.; Khosravi, P.; Jafari, M.; Zahabi, K.; Eftekhar, R.; et al. Protective Effects of Oleuropein against Renal Injury Oxidative Damage in Alloxan-Induced Diabetic Rats; a Histological and Biochemical study. *J. Nephropathol.* **2017**, *6*, 204–209. [CrossRef]

203. Zhang, J.-Y.; Yang, Z.; Fang, K.; Shi, Z.-L.; Ren, D.-H.; Sun, J. Oleuropein Prevents the Development of Experimental Autoimmune Myocarditis in Rats. *Int. Immunopharmacol.* **2017**, *48*, 187–195. [CrossRef]

204. Giner, E.; Recio, M.-C.; Ríos, J.-L.; Giner, R.-M. Oleuropein Protects Against Dextran Sodium Sulfate-Induced Chronic Colitis in Mice. *J. Nat. Prod.* **2013**, *76*, 1113–1120. [CrossRef]

205. Giner, E.; Andújar, I.; Recio, M.C.; Ríos, J.L.; Cerdá-Nicolás, J.M.; Giner, R.M. Oleuropein Ameliorates Acute Colitis in Mice. *J. Agric. Food Chem.* **2011**, *59*, 12882–12892. [CrossRef]

206. Impellizzeri, D.; Esposito, E.; Mazzon, E.; Paterniti, I.; Di Paola, R.; Morittu, V.M.; Procopio, A.; Britti, D.; Cuzzocrea, S. Oleuropein Aglycone, an Olive Oil Compound, Ameliorates Development of Arthritis Caused by Injection of Collagen Type II in Mice. *J. Pharmacol. Exp. Ther.* **2011**, *339*, 859–869. [CrossRef]

207. LaRussa, T.; Oliverio, M.; Suraci, E.; Greco, M.; Placida, R.; Gervasi, S.; Marasco, R.; Imeneo, M.; Paolino, D.; Tucci, L.; et al. Oleuropein Decreases Cyclooxygenase-2 and Interleukin-17 Expression and Attenuates Inflammatory Damage in Colonic Samples from Ulcerative Colitis Patients. *Nutrients* **2017**, *9*, 391. [CrossRef]

 *antioxidants*   

Review

Glucose as a Major Antioxidant: When, What for and Why It Fails?

Andriy Cherkas [1,*,†], **Serhii Holota** [2,3], **Tamaz Mdzinarashvili** [4], **Rosita Gabbianelli** [5] and **Neven Zarkovic** [6]

1   Department of Internal Medicine # 1, Lviv National Medical University, 79010 Lviv, Ukraine
2   Department of Pharmaceutical, Organic and Bioorganic Chemistry, Lviv National Medical University, 79010 Lviv, Ukraine; golota_serg@yahoo.com
3   Department of Organic Chemistry and Pharmacy, Lesya Ukrainka Eastern European National University, 43025 Lutsk, Ukraine
4   Institute of Medical and Applied Biophysics, I. Javakhishvili Tbilisi State University, 0128 Tbilisi, Georgia; tamaz.mdzinarashvili@tsu.ge
5   Unit of Molecular Biology, School of Pharmacy, University of Camerino, 62032 Camerino, Italy; rosita.gabbianelli@unicam.it
6   Laboratory for Oxidative Stress (LabOS), Institute "Rudjer Boskovic", HR-10000 Zagreb, Croatia; zarkovic@irb.hr
*   Correspondence: cherkasandriy@yahoo.com
†   Present address: Team Early Projects Type 1 Diabetes, Therapeutic Area Diabetes and Cardiovascular Medicine, Research & Development, Sanofi-Aventis Deutschland GmbH., Industriepark Höchst-H831, Frankfurt am Main 65926, Germany; andriy.cherkas@sanofi.com

Received: 10 December 2019; Accepted: 3 February 2020; Published: 5 February 2020

Abstract: A human organism depends on stable glucose blood levels in order to maintain its metabolic needs. Glucose is considered to be the most important energy source, and glycolysis is postulated as a backbone pathway. However, when the glucose supply is limited, ketone bodies and amino acids can be used to produce enough ATP. In contrast, for the functioning of the pentose phosphate pathway (PPP) glucose is essential and cannot be substituted by other metabolites. The PPP generates and maintains the levels of nicotinamide adenine dinucleotide phosphate (NADPH) needed for the reduction in oxidized glutathione and protein thiols, the synthesis of lipids and DNA as well as for xenobiotic detoxification, regulatory redox signaling and counteracting infections. The flux of glucose into a PPP—particularly under extreme oxidative and toxic challenges—is critical for survival, whereas the glycolytic pathway is primarily activated when glucose is abundant, and there is lack of $NADP^+$ that is required for the activation of glucose-6 phosphate dehydrogenase. An important role of glycogen stores in resistance to oxidative challenges is discussed. Current evidences explain the disruptive metabolic effects and detrimental health consequences of chronic nutritional carbohydrate overload, and provide new insights into the positive metabolic effects of intermittent fasting, caloric restriction, exercise, and ketogenic diet through modulation of redox homeostasis.

Keywords: glucose; pentose phosphate pathway; NADPH; redox balance; glycogen; glycolysis; stress resistance; insulin resistance

---

1. Introduction

The glucose level in blood is one of the most important homeostatic parameters and is strictly regulated [1]. The complex interplay of signals from central and autonomic branches of the nervous system, and the impact of multiple hormones and cytokines, all support coordinated glucose flows within the body according to the actual needs and availability, in order to maintain its concentration in a narrow range [2]. Since severe alterations of glucose metabolism take place in many diseases—including diabetes, that affects hundreds of millions of patients worldwide—there is a wealth of information about health effects and biochemical changes due to high (over 10.0 mM) or low (under 3.5 mM) glucose levels. The pathways of glucose metabolism and its regulation such as glycolysis/glycogenolysis, pentose phosphate pathway (PPP), gluconeogenesis, polyol pathway, insulin signaling pathway and many others are very well studied and their physiology and pathophysiology are firmly established. It is well documented that hyperglycemia is associated with oxidative stress and that the severity of diabetes correlates with the levels of accumulation of lipid peroxidation products, oxidatively modified proteins and advanced glycation products; therefore, glucose itself is recognized by many prooxidant factor. Short-time higher than physiological glucose levels (more than 10.0 mM) cause certain degree of damage due to increased rate of non-enzymatic glycation of proteins but are usually not life-threatening if blood glucose does not exceed 20.0 mM and is associated with diabetic ketoacidosis due to insulin insufficiency. In contrast, low blood concentrations (2.5 mM and lower) can cause severe brain damage and potentially death within the periods of time as short as 5–6 h [3]. Brain and particularly neurons are the most sensitive to glucose deprivation, while other tissues and cells show a wide divergence in resistance to hypoglycemia [1] that is very much dependent on their function, peculiarities blood flow and capability to store glucose in the form of glycogen.

In the present review we would like to focus on the other aspects of glucose metabolism that are not sufficiently addressed in the literature, namely physiological aspects of involvement of glucose and its stores in the form of glycogen in regulation/maintenance of redox balance in cells and tissues. The role of glucose as a fundamental source of reducing equivalents to antioxidant intracellular machinery very often underestimated or ignored due to its reputation of primary source of energy. Understanding the involvement of glucose in redox processes will enable not only better explanation of its metabolic role, but also will open new possibilities to address poorly understood nature of insulin resistance, and metabolic changes in diabetes overall. Our interpretation is also consistent with the mounting epidemiological evidence and can explain deleterious health effects of excessive dietary consumption of carbohydrates and sedentary lifestyle.

2. Basic Overview of Glucose Metabolism and Its Role in Maintenance of Redox Balance

It is well-known that most of the glucose in human metabolism is utilized intracellularly in glycolytic pathway with further degradation of products in the tricarboxylic acid (TCA) cycle in order to produce NADH and ATP [4]. Glycolysis is effectively activated by insulin in conditions of glucose abundance, and a number of intermediates are also used for synthesis of needed amino and fatty acids as well as other important metabolites [2]. However, in conditions of limited glucose supply and/or excessive metabolic needs there are numerous alternative ways to generate enough NADH and ATP, for example by the oxidation of fatty and amino acids, and the utilization of ketone bodies. Flexibility and interchangeability of cellular energy supply provides sustainable and at the same time variable flow of metabolites that is capable to accumulate them when the nutrients are in abundance and consume them in a most effective way when there is their deficit. In periods of starvation or glucose deficit, the activation of catabolic programs is capable to maintain energy production in most of the organs [5]. Since neurons do not accumulate glycogen and total accumulation of glycogen in central nervous system being extraordinarily low is limited to astrocytes, neurons rely on glucose supply from the bloodstream [6]. Interestingly, in periods of starvation the brain can effectively use ketone bodies as a primary fuel accounting for more than 75% of its energetic needs, pointing out the possibility that glucose may be used for other purposes in this case [7]. This is further confirmed by the observation that glycolysis in neurons is actively downregulated by proteasomal degradation of 6-phosphofructo-2-kinase/fructose-2,6-bis- phosphatase-3, preventing the utilization of glucose for bioenergetics purposes. This mechanism, as suggested by authors, spares glucose in neurons for maintaining antioxidant status, especially in conditions of limited glucose supply [8].

The importance of other major pathway of glucose metabolism, which to certain degree is an alternative or parallel to upper glycolysis, a pentose phosphate pathway (PPP) is also long known. It is believed that its major function is the generation of reducing equivalents in the form of NADPH needed for de novo lipogenesis, synthesis of DNA and aromatic amino acid [9]. Indeed, proliferating cells use most of the NADPH for DNA and fatty acid synthesis [10]. The other major functions of NADPH are the reduction in oxidized thiols and glutathione, generation of superoxide anion and hydrogen peroxide during respiratory burst to fight infections and to provide redox signals to regulate cell functions. In addition, it is also needed for detoxification of xenobiotics [11]. A growing number of publications point out rerouting of glucose into a PPP as a major protective mechanism employed to counteract acute and severe oxidative stress [9,12,13]. According to the calculations, the full oxidation of one molecule of glucose in the PPP yields 12 molecules of NADPH reduced from $NADP^+$ [14]. This aspect highlights the extraordinary efficiency and prompt responsiveness of this mechanism in balancing redox homeostasis in conditions of acute oxidative challenge. Indeed, activation of redox-sensitive transcription factors such as Nrf2 or FOXOs, in response to oxidative stress will result in induction of antioxidant enzymes within hours [15], while rerouting of glucose into the PPP to generate reducing power for antioxidant enzymes takes place almost immediately [14]. With the use of ^{13}C flux analysis in neurons, it was recently shown that glucose metabolism through the PPP may be much more significant than was previously estimated [16]. Moreover, the authors demonstrated that about 73% of produced labeled pyruvate was exported from neurons as lactate [16]. This may indicate that neurons remove glucose that cannot be fully utilized in TCA away from the cells. In case of increased functional activity, oxidative stress or glucose deficit during starvation, most of the glucose flux may be redirected into the PPP.

The PPP is a major source of NADPH; however, it is not the only one. Substantial amounts of NADPH are generated in folate-dependent NADPH-producing pathway [10] as well as by cytosolic isocitrate dehydrogenase and malic enzyme [11]. However, these sources are often coupled with synthetic pathways; for example, isocitrate is in abundance when glycolysis is activated and contributes to fatty acid synthesis, therefore, it is difficult to expect their substantial contribution to the regeneration of NADPH in cases of oxidative stress. To a certain degree, the metabolism of amino acids can compensate for functional lack of glucose, and contribute to the maintenance of NADPH, but this seems to be the mechanism with limited power under extreme exposures. Noteworthily, recently it was shown that malic enzyme and 6-phosphogluconate dehydrogenase (6PGD) form a hetero-oligomer to promote the activity of 6PGD, independently on activity of malic enzyme [17]. It is likely that the other structural and functional interactions may exist in the cells in order to couple synergistic metabolic processes in response to oxidative stress. The activity of alternative pathways provides robustness of NADPH supply and, to some extent, compensates for the deficit of PPP flux in patients with glucose-6 phosphate dehydrogenase (G6PD) deficiency—one of the most common genetic diseases in humans [18]. Patients with G6PD deficiency generally have no symptoms and their lifespan is not affected by disease, but it was shown that in addition to increased hemolysis they are less resistant to some poisonings [18] and have higher risk of diabetes and metabolic syndrome [19]. Glucose-6 phosphate (G6P) is an exclusive substrate for G6PD, a rate limiting enzyme of the PPP that can be supplied from extracellular space in the form of glucose, then phosphorylated by hexokinase in human organisms. Alternatively, glucose-1 phosphate released from glycogen—if the latter is available in the cell—is converted by phosphoglucomutase to G6P. Some tissues—namely in the liver, kidneys or intestine [7], and to some extent the glial cells—can generate glucose via gluconeogenesis. Furthermore, tumor cells may reverse glycolysis in order to maintain their biosynthesis in glucose-free conditions [20]. Noteworthily, the expression of most of the enzymes in the PPP is controlled by Nrf2, a redox sensitive transcription factor involved in the upregulation of antioxidant and detoxifying genes, the degradation of damaged proteins and metabolic reprogramming during stress [21], pointing out the tight conjugation of redox balance maintenance and glucose metabolism. In other words, on the cellular and organism levels, responses to local or systemic oxidative stress are associated with increased glucose release/production by the liver and subsequently hyperglycemia, which may be physiological adaptive response in healthy subjects and may also take place as a chronic metabolic deterioration in diabetic patients.

3. Oxidative PPP Is Thermodynamically More Favorable Compared to Upper Glycolysis under Conditions of Limited Glucose Supply

In physiological conditions (without metabolic/oxidative stress) the ratio of reduced and oxidized forms of this coenzyme NADPH/NADP$^+$ is very high (in the range of approximately 100/1, but is highly tissue-dependent) and the lack of free NADP$^+$ prevents G6P from entering the PPP [14]. However, as soon as NADPH is oxidized (e.g., in conditions of oxidative stress) the increased availability of NADP$^+$ immediately redirects metabolic flow to the PPP and suppresses further steps of glycolysis and the downstream utilization of glucose metabolites in TCA [8,13]. In other words, cells prioritize the metabolism of glucose through the PPP over standard reactions of upper glycolysis in order to maintain a sufficient NADPH/NADP$^+$ ratio needed for the counteraction of acute oxidative challenge (prompt enzymatic reduction in glutathione and other oxidized thiols), biosynthesis and/or generation of superoxide during immune responses or as physiological redox signaling (Figure 1). Glucose availability for the PPP—either from extracellular space, or from intracellular glycogen stores—is essential for acute antioxidant responses ensuring the survival of cells and organism(s) in extreme conditions. Stable and robust liver glucose output in cases of severe stress (including oxidative stress), infection, starvation and extreme exercise is protected by the development of insulin resistance in order to provide sufficient glucose flow to balance redox homeostasis. In absolute quantities, especially at rest, glucose flow into the PPP may be low compared to standard upper glycolysis, but under conditions when the NADPH/NADP$^+$ ratio drops, amounts of glucose entering the PPP will correspond to the degree of NADPH depletion.

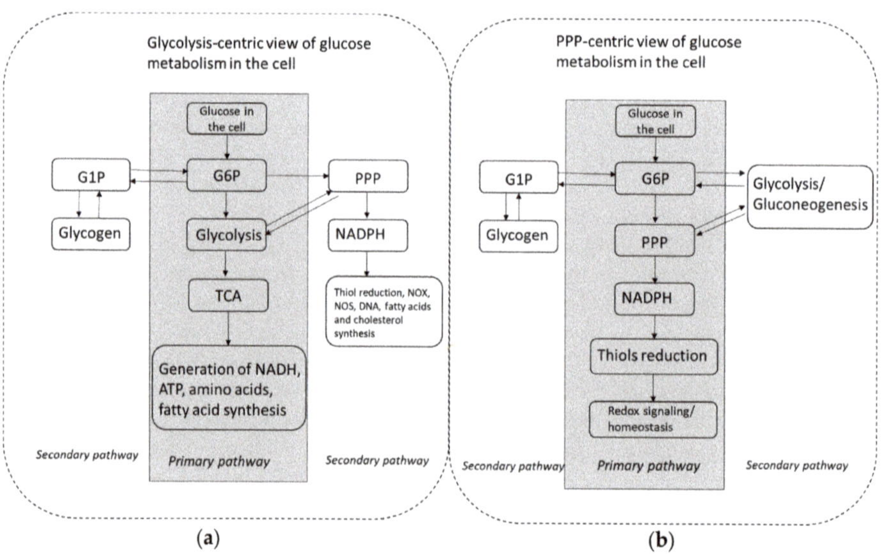

Figure 1. Schematic presentation of the conventional (**a**) and pentose phosphate pathway-centric (**b**) views of glucose metabolism. Abbreviations: G1P—glucose 1 phosphate, G6P—glucose 6 phosphate, PPP—pentose phosphate pathway, TCA—tricarboxylic acid cycle, NOX—NADPH oxidase, NOS—nitric oxide synthase.

A key factor leading G6P into the PPP is the presence of NADP$^+$. The first reactions of the PPP as well as other reactions producing NADPH are energetically very favorable and are basically irreversible [22]. In contrast, most of the reactions of glycolysis are reversible. The activation of glucose utilization through glycolysis in physiological conditions takes place when the NADPH/NADP$^+$ ratio is high (usually 50–100/1), glucose is relatively abundant and the insulin signaling is not compromised.

Glycolysis serves as a source of pyruvate for TCA cycle and a number of anabolic intermediates in conditions of high carbohydrate availability, and is supposed to be fully activated only occasionally under physiological conditions, often switching to the oxidation of fatty acids when glucose availability is limited. This shift takes place in concert with activation of transcription factors involved in antioxidant defense (for example Nrf2 and isoforms of FOXO transcription factor) as a part of systemic antioxidant response aimed to balance redox homeostasis, where gluconeogenesis and activation of the PPP are fundamental parts of it (Figure 2). A sedentary lifestyle plus an abundance of carbohydrates in food lead to an imbalance of nutritional consumption and actual metabolic needs; therefore, redox dysregulation contributes to metabolic syndrome and diabetes type 2, and is an important factor that increases the incidence of cancer that is discussed in more detail in the next section.

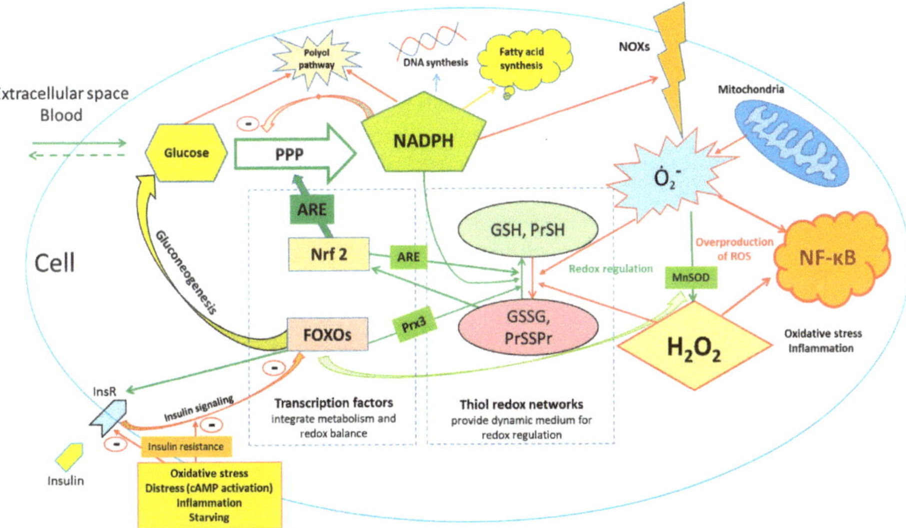

Figure 2. Glucose availability is a major factor in the maintenance of redox homeostasis through the reduction in oxidized NADP$^+$, which is used for the subsequent reduction in oxidized glutathione and thiols. At the same time NADPH is used for synthesis of DNA and fatty acid synthesis and is needed for activities of NADPH oxidases, NO-synthase and other processes. The scheme is simplified and regulatory networks functioning in living systems are much more complex and include other mechanisms and feedback loops. For example, it was recently shown that the deletion of Nrf2 in mice can be—to a large extent—compensated by other adaptive mechanisms in conditions of caloric restriction [23]. This suggests that robust regulatory network beyond Nrf2 and FOXO transcription factors exists in order to maintain redox balance. Abbreviations: PPP—pentose phosphate pathway, NOX—NADPH oxidase, NOS—nitric oxide synthase, GSH—glutathione, GSSG—glutathione disulfide, Pr—protein, PrSSPr—disulfide bonds between/within the proteins and other molecules, ARE—antioxidant response element, InsR—insulin receptor, cAMP—cyclic adenosine monophosphate, FOXO—forheadkbox O transcription factors, Prx3—peroxiredoxin 3, Nrf2—Nuclear factor (erythroid-derived 2)-like 2 transcription factor, MnSOD—manganese superoxide dismutase, NF-kB—nuclear factor kappa-light-chain-enhancer of activated B cells.

4. The Epidemiological Evidence of Glucose Overload in Human Population: Current vs. Historical Nutritional/Behavioral Patterns and a Growing Potential of Pharmacological Interventions

As currently observed, massive nutritional carbohydrate overload associated with dramatic decreases in physical activity has caused an epidemic of non-communicable diseases such as obesity, metabolic syndrome, type 2 diabetes, atherosclerosis, hypertension and cancer [24]. A large-scale epidemiological cohort study pointed out that high carbohydrate consumption is a major factor for all-cause mortality, while different types of fat were not associated with increased mortality. Thus, authors question the current dietary guidelines suggesting an urgent need for their reconsideration [25]. On the other hand, the results of a large prospective cohort study indicate that health outcomes depend rather on quality of food rather than just limiting carbohydrate or saturated fat in the diet. In addition, "unhealthy" low carbohydrate and/or fat diet may increase all-cause mortality in studied population [26]. Together with other evidences, including numerous animal studies, a serious demand for strategies on how to counteract carbohydrate overload is indicated. Different approaches, including pharmacological interventions [27] as well as public health and food regulation policies, are increasingly discussed in the literature [28]. The careful evaluation of multiplicity of factors must be in place in order to determine the optimal nutritional pattern for health preservation and the prevention of age-related diseases.

In fact, an abundance of carbohydrates in food during human evolution was rather rare and more of a short-term and often seasonal luxury [29]. Human neonates grow very rapidly and their need for carbohydrate supply is probably higher than in adults than in other species due to the relatively large brain and rapid development of the nervous system; however, human breast milk "only" contains about 6.7 g per 100 mL of lactose accounting for up to 40% of calories—still the highest compared to other mammals [30,31]. It is not likely that adults need more carbohydrates than babies as a percent of calories intake in usual conditions without regular vigorous physical activity. The excessive consumption of carbohydrates and low physical activity are major contributors to increasing rates of metabolic syndrome and obesity in children and adolescents [32].

At high growth rate, the intensive physical activity as well as strong immunity needed for resistance to infections and occasional poisonings requires the maintenance of sufficient levels of glucose in blood, despite prolonged periods of carbohydrate deficit. Dietary glucose—as well as other carbohydrates—were precious food components with limited, often seasonal availability promoting survival under extreme conditions [29]. Therefore, an evolutionary sweet taste developed to detect sources of digestible carbohydrates [33]. Only the development of agriculture about 10,000 years ago enabled higher consumption of grain, increasing the share of carbohydrate in the diet. Even though carbohydrates became more available, any possible overload would rather not take place considering intensive physical activity of most of the people at that time. Thus, the gradual increase in basic food availability and elimination of physical work created a massive nutritional carbohydrate overload at the organism's level causing respective health consequences [25]. Consistent with this, a metabolic core model was recently used to evaluate how increased glycolytic utilization of glucose together with glutamine-dependent lactate production promotes cancer growth [34]. Hyperglycemia may directly contribute to increased risk of cancer as it was recently shown by Wu et al. [35]. Overall, there is growing evidence—both mechanistic and epidemiological—that confirms previous predictions of interrelationships between risk of cardiovascular/metabolic diseases and cancer risks [36–38].

Consistent with this is recent data generated on a *C. elegans* model regarding the integration of stress-induced responses of nervous system with metabolic adaptations [39]. It was shown that the flight response mediated by tyramine (worm analog of catecholamines) in the end leads to the stimulation of insulin-IGF-1 signaling and has the opposite effect to longevity-promoting stress responses to heat, starvation, glucose restriction or exercise [40–42]. If hypothetically translated to modern humans, it can provide accurate mechanistical explanations as to why emotional stress accompanied with sedentary behaviors may have detrimental health consequences associated with both excessive insulin and adrenaline signaling causing atherosclerosis [24], contributing among the other factors to insulin resistance and accelerated aging [43], but can be reversed by exercise or fasting [44]. In this sense psycho-emotional stress prepares the organism to the impact of "expected" extreme factor in near future, and in cases of "false alarms", the excess of glucose needs to be utilized by the activation of insulin signaling contributing to pathologic continuum that leads to atherosclerosis [24].

In regard to glucose balance/overload and its crucial role in metabolic diseases, new data obtained in clinical trials is very important, where patients were exposed to sodium glucose co-transporter 2 (SGLT-2) inhibitors. Designed initially to improve glycemia in patients with type 2 diabetes that are not optimally controlled by metformin monotherapy—canagliflozin, dapagliflozin, empagliflozin and others are continuously surprising clinicians and researchers by new positive effects far beyond glycemia, including—but not limited to—the improvement of insulin resistance, reducing body weight [45], preventing acute cardiovascular events [46], exerting a normalizing effect on blood pressure, kidney function [47] and reversing manifestations of heart failure [46]. A plethora of positive effects convincingly confirmed in strictly controlled clinical trials according to the highest standards of evidence-based medicine by simply getting rid of approximately 50–70 g of (excessive) glucose per day makes a significant difference for patients, and may potentially find its place among preventive interventions. It is worth pointing that some contribution to the effect may potentially come from sodium reabsorption inhibition, however, it is clear that the role of glucose excretion is prominent.

Thus, from the evolutionary point of view, the human organism was rather not used to the consumption of large amounts of glucose and has had to optimize its metabolism in order to be able to produce its sufficient amounts in accordance with metabolic needs. The importance of the PPP evolved in order to provide resistance to oxidative challenges that are crucial for survival in acute extreme conditions in multicellular organisms. The upregulation of PPP under oxidative stress is tightly coupled with enhanced glucose output from glycogen stores and/or stimulation of gluconeogenesis. Glucose 6-phosphate is a specific substrate for the PPP that makes glucose so important and strictly regulated in maintaining redox homeostasis in human organisms [1].

5. Glycogen Protects against (Not Only) Oxidative Stress

According to our hypothesis, availability of intracellular glycogen is supposed to be protective against oxidative stress, and vice-versa; its absence exposes cells to higher risk. Indeed, neurons, which are unable to accumulate glycogen appear to be among the most sensitive cells to oxidative stress, and they apply sophisticated mechanisms to direct the flow of glucose into the PPP in order to protect themselves [8]. Severe hypoglycemia may result in seizures, loss of consciousness, coma and—if glucose is not administered/ingested for longer periods (more than 5–6 h)—death [48]. Very similar clinical manifestations have been observed in cases of hyperbaric oxygen exposure (so called oxygen poisoning) that also causes severe redox imbalance in brain [49].

A recent *C. elegans* study demonstrated the crucial role of glycogen stores in resistance to acute oxidative stress [50]. Moreover, the excessive accumulation of glycogen from a high-glucose diet and with impaired glycogen degradation resulted in decreased lifespan of the worms [50]. Insecticide poisonings causing oxidative stress in the fruit-eating bat *Artibeus lituratus* causes glycogen stores depletion [51]. It was recently shown that hawkmoths, who have one of the highest metabolic rates among known animals, use nectar sugar directed through the PPP to counteract oxidative damage resulting from flight (extremely intensive exercise) [52]. In humans, the inability to deplete muscular glycogen in patients with glycogen phosphorylase deficiency (McArdle disease) is associated with severe exercise-induced oxidative stress and a risk of rhabdomyolysis [53]. This points out the possibility that the function of glycogen in muscles is not only as an energy store during periods of intensive contraction, but also for counteracting oxidative challenges associated with exercise.

It was noted that main life- and health-span promoting interventions such as caloric restriction, intermittent fasting and exercise have in common that the depletion of glycogen stores [44], thus reducing the protective capacity of glycogen and exposing the cells to moderate hormetic oxidative stress. Glycogen stores are not simply an intracellular source of glucose, they also have an important signaling function [54] and are protective against a number of stressful situations, namely hyper/hypo osmotic stress [55], anoxia/hypoxia [56]. In addition, growing evidence indicates that a metabolic switch from utilization of glucose, which is abundant in western diets, to ketone bodies use derived from fatty acids is an evolutionarily conserved trigger-point responsible for health effects from intermittent fasting, caloric restriction and exercise [57]. Furthermore, a complex interplay of hormones including insulin, glucagon, leptin, adiponectin and others regulate metabolic adjustments in conditions of food abundance and deficit to provide needed glucose levels and energy in the organism [58–60].

6. Epigenetics and Posttranslational Protein Modification Modulate Oxidative Stress Responses

Epigenetics regulates gene expression, modifying DNA methylation and chromatin structure. This regulatory mechanism works differently in each tissue to guarantee specific genetic responses to environmental factors (i.e., nutrition, chemicals, stress), without any changes in the sequence of nucleotides [61,62]. Epigenetics plays a key role starting from early life, where it is the master director of cell differentiation, X-inactivation and the programming of adult health [63]; epigenetic changes can be transferred to the progenies and, sometimes, they can be reverted [64].

DNA methylation consists of the methylation of Cytosine at CpG islands in the promoter region of genes which has been associated with gene silencing, while different responses (activation or inhibition of gene expression) derives from the methylation of CpG islands located in the regulatory regions of genes. Histone modifications are changes that are more complex, because functional groups (i.e., acetyl, methyl, P, etc.) deriving from oxidation of nutrients, can be added to histones' amino acid residues, thus remodeling chromatin. The final result of histone modification is chromatin remodeling at specific genes, leading to increased/decreased gene expression associated with healthy or unhealthy regulatory responses [63].

Oxidative stress related with metabolic responses linked to hyperglycemia can enhance DNA methylation interfering with S-adenosyl-L-methionine (SAM), the key methyl donor for the DNA methyltransferases (DNMTs) which catalyze CpG methylation [65]. In particular, the deprotonation by superoxide anion of cytosine C5 at CpG islands can support the formation of DNA-SAM complex leading to the final cytosine methylation; furthermore, DNA methylation has been associated with the increase in DNMT1 and DNMT3B expression due to reactive oxygen species (ROS) [65,66]. Glucose can mediate epigenetic modification not only through ROS, but also because the high level of glucose can interfere with DNA demethylation via TET2 and AMPK [67].

However, oxidation at the level of guanine leading to 8-hydroxydeoxyguanosine (8-OHdG) in CpG islands, can also decrease cytosine methylation and reduce the binding of transcription factors to the promoter region. The oxidation of 5-methylcytosine (5mC) due to Ten-Eleven Translocation (TET) proteins leads to 5-hydroxymethylcytosine (5hmC) formation, which is deaminated to 5-hydroxymethyluracil and then replaced with unmethylated cytosine [68]. Oxidative stress can also inhibit the NAD^+ dependent deacetylase SIRT1 that controls inflammatory responses, lipid storage, telomerase activity, mitochondrial respiration and ROS production [69,70]. In this context, a high-fat/glucose diet that decreases NAD^+ content can negatively regulate Sirtuin activity.

The regulation of responses to oxidative stress is complex and includes many mechanisms [71] such as oxidative modifications of macromolecules by ROS [72], signaling through lipid peroxidation and their products [73] and involvement of different transcription factors (i.e., mentioned above FOXOs and Nrf2) [74,75]. Considering ubiquitous expression of these transcription factors as well as their crucial cellular functions it is very difficult to modulate them by pharmacological interventions [72,76]. Similarly, lipid peroxidation products play important physiological functions, for example in gastrointestinal tract [77]. Considering the significance of the regulatory functions of 4-hydroxynonenal and other lipid peroxidation products they also start attracting interest as a target for pharmacological interventions in major stress-associated disorders [78].

7. The Evidence from Glucose-6 Phosphate Dehydrogenase Deficiency

G6PD deficiency is one of the most common genetic human diseases and affects more than 400 million people worldwide [18], therefore, much can be learned from the published evidence. Since G6PD is a gateway to such an important metabolic pathway as the PPP, dramatic consequences to the patients could be expected. However, this is very often not the case. As was already mentioned in the introductory section, these patients have few or no symptoms, with generally positive prognoses and their expected lifespan is no different than that of the general population [18]. There are two basic explanations for this evidence: first, most of the patients have a moderate degree of G6PD deficiency and the PPP is still functioning at some level and, also, alternative pathways generate sufficient amounts of NADPH; second, humans with modern lifestyles are exposed to relatively low intensity stressors and there is simply no need for acute responses to stress. In contrast, severe G6PD deficiency does indeed have a detrimental effect on the immune system and causes higher susceptibility to infections [79]. In addition, G6PDH-deficient athletes and patients with this genetic defect may have severe hemolytic crises after physical exertion [80]; however, the severity of susceptibility of individual subjects may vary widely [81]. Complete G6PD knockout in mammals is incompatible with life, but in embryonic mice stem cells it led to the severe susceptibility of cells to oxidative stress induced by hydrogen peroxide or diamide and reduced cloning efficiency. However, the later was restored when the oxygen concentration was reduced [82]. Therefore, in conditions of substantial oxidative challenge, the proper function of the PPP and generation of NADPH are essential for survival [83]. Conversely, G6PD overexpression may be expected to increase resistance to oxidative stress. Indeed, recent reports indicate that G6PD overexpression extends the lifespan of *Drosophila melanogaster* [84], which is consistent with some of the results obtained using a G6PD overexpressing mouse model, where it leads to the extension of the health-span of mice and increased resistance to oxidative damage [85].

8. Redox Dependence of Pancreatic Regulation of Blood Glucose Levels

A growing body of evidence indicates that the release of insulin from pancreatic β-cells depends on the function of the PPP. It was shown that insulin levels in G6PD deficient patients are lower compared to unaffected controls, and these patients have significantly reduced insulin responses to the elevation of blood glucose [86]. More recently, with the use of metabolomics approach it was shown that insulin release is controlled by the direct implication of the PPP [87]. According to a recent review, among the most important amplifiers/regulators of insulin secretion by β-cells are high levels of NADPH and glutathione [88]. So, insulin release is taking place in conditions of "metabolic welfare" and oxidative stress may reduce the ability of β-cells to release insulin [89]. α-cells also have their intrinsic mechanisms of glucose sensing relying on intracellular redox balance, but they are activated by pro-oxidant situations [90]. Interestingly, under physiological conditions β-cells do not accumulate glycogen but are able to do so under prolonged hyperglycemia and may prolong insulin secretion even after normalization of glucose concentration. In contrast, α-cells do not accumulate glycogen, so when the concentration of blood glucose drops, they can be quickly activated without delay [91], which is extremely important in case of emergencies such as hypoglycemia.

One may argue that there are many other mechanisms for the regulation of insulin and glucagon that can either enhance or inhibit respective secretion, including paracrine δ-cells secreting somatostatin [92,93], effects of glucagon-like peptide-1 (GLP-1) [94], glucose-dependent insulinotropic peptide (GIP) [95], leptin/adiponectin axis [96], and autonomic nervous system [97,98]. However, the effects of all these regulators are integrated at the level of α- and β-cells and their metabolism resulting shifts of redox potential [87,88,90].

9. Inflammation, Insulin Resistance and Redox Homeostasis

NADPH produced by the PPP or by other pathways is also used by NADPH oxidases and nitric oxide synthase to produce superoxide anion. This is important for proper functioning of the immune system and for redox regulation of multiple processes in the tissues including endothelial function [99]. It was shown, that pro-inflammatory interleukin 1β enhances glucose uptake under hyperglycemic conditions in cultured human aortic smooth muscle cells. It also activates PPP and promotes the production of superoxide by NADPH oxidase contributing to vascular damage [100] pointing out a particularly dangerous combination of inflammation and hyperglycemia. Since immune cells require glucose for their function, they send regulatory signals, for example TNF-α [101] or microRNAs [102] to the liver to enhance hepatic glucose output and thus, chronic inflammatory conditions may cause insulin resistance [103]. The opposite effects are mediated by anti-inflammatory interleukin-10 [104] and the spleen plays a particularly important role in these regulatory interactions [105]. The autonomic nervous system may also be involved in regulation of interactions of local inflammatory conditions and oxidative stress [106]. Autonomic output is actively involved in the regulation of glucose homeostasis and can adjust the rates of glucose production and utilization independently of hormonal influences [107]. The healing of oxidative stress-associated conditions, therefore, may improve autonomic balance [108]. Chronic carbohydrate overload and reduced physical activity cause obesity and metabolic syndrome, and for these conditions, insulin resistance is very typical [109]. Thus, there exist complex multilevel regulatory interactions to provide a sufficient flow of glucose to tissues in order to maintain a redox balance. Prompt adjustments of redox homeostasis are critical for the immune defense where the PPP plays a major role. Insulin resistance developing in this case seems to be adaptive and to some extent a protective mechanism [110], but, if dysregulated, it leads to detrimental health consequences.

10. Glucose—"Oxidant" or "Antioxidant" after All? *Sola Dosis Facit Venenum*

There is a certain degree of confusion in the literature concerning the role of glucose in the maintenance of redox homeostasis. It was long known, that diabetes and hyperglycemia obviously cause redox dysregulation, oxidative stress and accelerated ageing [111]. On the other hand, glycogen clearly protects against the oxidative stress and at the same time decreases lifespan and health span [50,112]. Starvation and stress-induced gluconeogenesis clearly support survival and improve health and lifespans [112,113], but are associated with the increased generation of ROS [40]. Moreover, the complete withdrawal of glucose hence leads to short-term fall in ATP production, and surprisingly causes a rise in ATP and increased mitochondrial content shortly after that is associated with the activation of the protein deacetylase SIRT1. At the same time increased ROS production or possibly insufficient ROS utilization is documented [114]. This suggests that there are a variety of effective metabolic adaptations to compensate lack of glucose for ATP production, but a subsequent deficit of reducing power to counteract to increase in ROS production is much more difficult to compensate. Exercise, intermittent fasting/caloric restriction all lead to functional glucose/glycogen depletion through activation of autophagy and support healthy aging that requires certain degree of oxidative stress that leads to metabolic shift to catabolism and activate endogenous protective mechanisms that include enhanced protein quality control, stimulation of endogenous antioxidant defense including gluconeogenesis [57,111,115].

The other interesting aspect of glucose involvement in redox homeostasis is that supraphysiological concentrations of glucose in the cells may actually lead to the increased production of hydrogen peroxide by mitochondria through the inhibition of mitochondria-bound hexokinase [116]. Similarly, chronically high levels of NADPH may contribute to the enhanced generation of superoxide and/or hydrogen peroxide by NADPH-oxidases (NOX) as well as enhance the reduction in glucose in polyol pathway (Figure 2) [99]. Involvement of NOX in regulation and maintenance of redox homeostasis is very complex and depends on its isoforms that are expressed differently in tissues. For example, cardiomyocytes express NOX2 and NOX4 isoforms and use the generation of hydrogen peroxide for the regulation of cellular metabolism and contractile function (NOX4 activation acts similarly as beta-blockers decreasing inotropy) [117]. NOX isoforms are increasingly studied as potential therapeutic targets in order to modulate redox balance in the cells during cardiovascular and metabolic diseases [118].

Diabetes, both type 1 and 2 are well documented diseases associated with oxidative stress. The fundamental feature of diabetes is chronic hyperglycemia due to excessive glucose production and/or impaired its utilization by the tissues. There are several mechanisms contributing to oxidative stress in conditions of hyperglycemia that include but are not limited to non-enzymatic glycation and formation of advanced glycation products, the activation of polyol pathway that results in a depletion in NADPH (decreased rate of enzymatic reduction in oxidized glutathione and thiol groups of proteins), an increase in NADH and a depletion in NAD$^+$ (increased superoxide anion production in mitochondria), the inhibition of histone deacylation and excessive histone acetylation due to the accumulation of acetyl-CoA, as well as the generation of excessive amounts of sorbitol and fructose [119].

Taken together, the evidence indicates that there exists a delicate balance between the protective and damaging redox effects of glucose and chronic dietary carbohydrate overload may affect the regulatory mechanisms that developed during evolution to maintain redox homeostasis (Figure 3). Bell-shaped antioxidant activity of glucose explains well necessity to regulate it strictly within the narrow range, while both excessively high and/or low levels of glucose/carbohydrates lead to oxidative and metabolic stress that quickly becomes damaging [120]. The dysregulation of glucose metabolism that takes place in diabetes closes vicious cycle further exacerbating redox dysregulation that was actually meant to be fixed by induction of hyperglycemia.

Hypothesis on concentration-dependent influence of glucose on redox potential

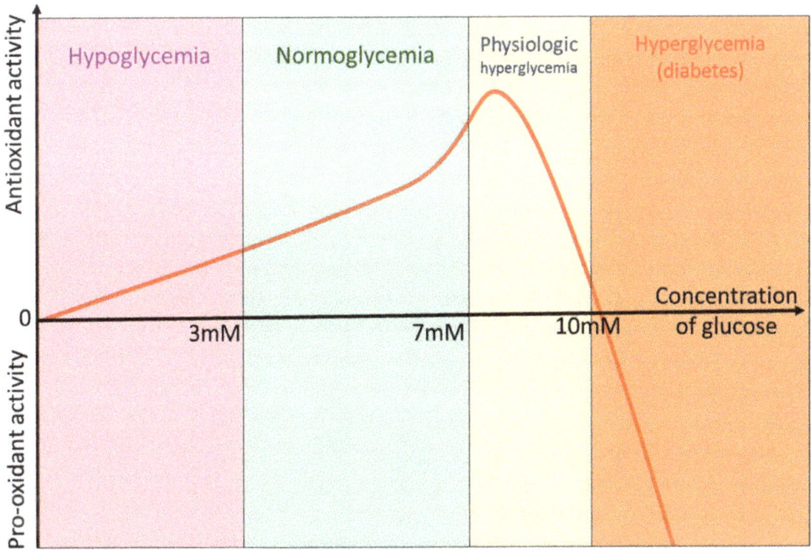

Figure 3. The dependence of redox effects of glucose on its concentration (hypothetic relationship suggested by the authors). Hypothetical simplified model describing the influence of the blood concentration of glucose on redox potential in human organisms. Multiple additional factors influencing redox potential such as concentration of oxygen, availability of amino and fatty acids, type of cells and effects of either hormones (insulin, glucagon) or cytokines are not considered. Concentrations of glucose as well as the shape of the curve are roughly estimated and not confirmed by actual experiments and may significantly vary depending on conditions and tissue type.

11. Important Implications

Glucose is a central metabolite and depending on its availability and metabolic need can play different roles in the organism. Glucose flows within the human body are strictly regulated and promptly adjustable in order to maintain metabolic flexibility and provide robust resistance to different types of stressing factors (Figure 4). The role of glucose is not limited to merely generating enough ATP (which is the case in conditions of glucose abundance and low stress), but more importantly, it is responsible for the emergency mechanisms of the maintenance of redox potential that is essential for survival in extreme situations (fight or flight reactions, infections, and poisoning).

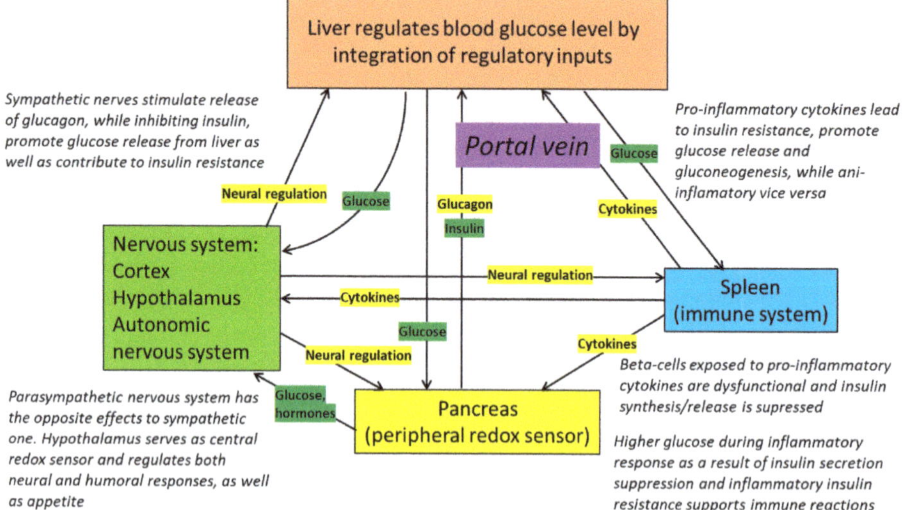

Figure 4. Role of redox sensors (pancreatic α and β cells), immune system and central nervous system in the regulation of blood glucose concentrations by liver. Glucose release or absorption by the liver integrates signals from nervous and immune systems, and peripheral redox sensors. The system is highly flexible and tunable, providing redox modulation that is dependent on actual needs. The other way around, glucose flows and redox state regulate the function of immune system [121]. Failure of feedback loops and distorted signaling—either from CNS (stress), the immune system (inflammation) or the malfunction of peripheral sensors—lead to excessive, uncontrolled (poorly controlled) glucose release and/or the activation of gluconeogenesis, leading to diabetes.

Since TCA-cycle reactions and oxidative phosphorylation can be effectively maintained in almost absence of glucose by oxidation of fatty and amino acids, its residual amounts are—in catabolic conditions—redirected into the PPP to maintain redox balance in the cells. At the same time, the anabolic state requires redirecting excess glucose into energy production and biosynthesis (growth, proliferation, hypertrophy), when the primary life-supporting and life preserving needs ensured by glucose have already been met ("survival—first, growth—follows"). The proposed principle is helpful for understanding the regulation of glucose metabolism aimed primarily to maintain redox balance, especially in acute extreme conditions requiring prompt and massive antioxidant responses. Oxidative stress, infections/inflammation, starvation, exercise, aging and many other pathological conditions or processes decrease GSH/GSSG and NADPH/NADP$^+$ ratios. Sensors sense these changes and drive glucose flows to compensate for these metabolic disturbances at the organismic level. In this regard insulin resistance develops as a protective "antioxidant" adaptive mechanism that stimulates glucose production and prevents its waste in order to cope with increased needs. When dysregulated and chronically over-activated, though, this leads to detrimental consequences caused by hyperglycemia [110].

The other important issue, which is often not taken into consideration by researchers, is that the presence of glucose stores in the form of glycogen or extracellular glucose availability provide enhanced resistance to oxidative stress. It is also possible that glucose—by strengthening the reductive power of NADPH-dependent antioxidant enzymes—prevents the activation of redox-sensitive regulatory factors such as Nrf2. That means that glucose nutritional overload affects physiological redox signaling, and the chronic over-activation of insulin signaling causes metabolic diseases, as described in detail [24]. On the other hand, severe glucose overload itself can serve as a source of oxidative stress in the cells through the participation of glucose in polyol pathway, over-the activation of NADPH oxidases and increased production of ROS by mitochondria. Exercise, caloric restriction, intermittent fasting, a ketogenic diet and some drugs have something in common—in that they deplete organisms' glycogen stores [44], reduce glucose availability for the cells, restore the physiological redox signaling suppressed by chronic excessive glucose consumption and lead to a dramatic improvement of life- and health-spans in model organisms and humans (Figure 2). Metabolic changes caused by carbohydrate overload in the general population often take place far before clinically significant changes occur [43]. That is why relevant and sensitive instruments for the early detection of these metabolic shifts are needed.

12. Limitations of the Analysis and Directions of Further Research

In this review we presented mainly general principles of glucose metabolism in the cells and its importance for maintenance of redox balance. However, different cells in different tissues have their specific metabolic patterns, specific functions, and own physiological peculiarities. We tried to focus on the most important findings and inconsistencies in the literature from our point of view rather than focusing on biochemical details, different pathways, tissue differences and to show how the clinical and experimental evidence may be interpreted when the glucose will be considered as redox mediator rather than simply fuel "burned" to generate ATP. Glucose metabolism may differ substantially depending on the specifics of tissues and cells, proliferation activity, redox balance required to maintain the functions, expression and activities of involved enzymes. It is not possible to accurately describe all the evidence that is published in the literature, and we are aware that many important details may be missing from our analysis. Nevertheless, we hypothesize that convincing literature evidence indicates that glucose flows in the organism are primarily targeted to maintain redox homeostasis and counteract possible oxidative challenges.

It may be argued that the actual flow through PPP is relatively low in the brain (as has been shown by Gaitonde M. et al. [122]) and it only increases substantially to approximately 20% of total glucose utilization by neurons in cases of severe oxidative stress induced by hydrogen peroxide [123] or during experimental brain injury [124]. However, a closer look into the methods used in the studies reveals that non-physiologically high concentrations of glucose were used, namely 22.3 mM in the medium and 50.0 mM for perfusion in [123] and 23.9–26.9 mM plasma glucose after infusion in [124]). The flow of glucose into the PPP is inhibited in conditions of high NADPH/NADP$^+$ ratio, therefore as soon as the levels of NADPH are restored, glucose is redirected into glycolysis or glycogen storage.

Our reconsiderations may provide better understanding of the physiology of glucose regulation in health and diseases and lead to the shift of the general paradigm of glucose-induced oxidative stress—which is, however true for hyperglycemia and diabetes—towards understanding the redox effects of glucose in a concentration-dependent manner. In other words, glucose maintenance at physiological levels is a fundamental mechanism of counteracting excessive oxidation due to its involvement in PPP and NADPH production. Endogenous antioxidant systems using glucose provide sufficient antioxidant defense in physiological conditions. Moreover, redox modulating agents that have some health benefits when supplemented often turn out to be rather prooxidant than antioxidant, and lead to the stimulation of endogenous antioxidant defenses and the improvement of glucose metabolism. Therapeutic approaches to applying antioxidant substances in order to reduce oxidative damage and the administration of compounds with the pure purpose to provide reducing equivalents appears weak [125] in contrast to often-high in vitro activities [126], and in comparison to existing endogenous antioxidant mechanisms based on glucose as a source of reducing power for the generation of NADPH and recycling oxidized glutathione and thiols, and can in some cases be deleterious since they may interfere with redox sensors—for example protein thiol groups, and cause dysregulations.

The other important issue is the lack of convenient, informative, sensitive and specific ways of glucose and/or glycogen determination in tissues for research and clinical use. As we suggested earlier, glycogen determination, preferably in a simple, affordable and noninvasive way could potentially be a good biomarker for redox biomedicine [44]. Unfortunately, there are too many technical issues preventing the development of such equipment for clinical use so far, but some efforts have been made to enable label-free glycogen estimation in *C. elegans*, an important animal model for metabolic and aging research [127].

13. Conclusions

Our critical analysis of the current literature unveiled significant controversies and an insufficient general understanding of metabolic role of glucose, particularly regarding redox homeostasis. The important function of glucose metabolism in the PPP for the maintenance of redox homeostasis is very often underestimated. In light of recent epidemiological evidence and advancements in the field of redox biology, it is hypothesized that a specific physiological function of glucose is its metabolism in the PPP, to provide stress resistance to unfavorable factors by the reduction in $NADP^+$ to NADPH in order to maintain redox homeostasis and the correct functioning of immune cells. Meanwhile, glycolysis takes place in favorable redox conditions when glucose is in abundance. This approach and interpretation explains—in a simple way—the adverse metabolic effects and detrimental health consequences of nutritional carbohydrate overload, and provides new details in explaining the positive metabolic effects of intermittent fasting, caloric restriction, exercise, and a ketogenic diet. A better understanding of the evolutionary adaptations and biological role of glucose may serve as an important theoretical background for future experimental and clinical studies related to glucose metabolism, aging, and diabetes, as well as other adjacent fields.

Author Contributions: Original idea and conceptualization A.C., original draft preparation A.C., S.H., T.M., R.G. and N.Z., review and editing A.C., S.H., T.M., R.G. and N.Z., preparation of the figures A.C. and S.H. All authors have read and agreed to the published version of the manuscript.

Funding: A.C. received Georg Forster (HERMES) Scholarship from Alexander von Humboldt Foundation (Bonn, Germany). This article/publication is based upon work from COST Action NutRedOx-CA16112 "Personalized Nutrition in aging society: redox control of major age-related diseases" supported by COST (European Cooperation in Science and Technology). AC is a participant in Hypo-RESOLVE (Hypoglycaemia—REdefining SOLutions for better liVEs), a project in the framework of Innovative Medicines Initiative (IMI) funded by European Commission and European Federation of Pharmaceutical Industries and Associations.

Acknowledgments: This paper is dedicated to the memory of our colleague and friend Peter Eckl from the University of Salzburg, who was our collaborator, research partner and scientific mentor for many years and passed away recently. He read the draft of this manuscript and gave valuable comments that were considered during the preparation of the submission, and we highly appreciate his contribution. We are very much grateful for his exceptional academic career and his particular role in our joint projects. The authors are grateful to Holger Steinbrenner (Department of Nutrigenomics, Institute of Nutrition, Friedrich Schiller University Jena, Germany) and Percy Herbert (Redlands, California, USA) for their valuable comments regarding the manuscript.

Conflicts of Interest: The authors declare no conflict of interest.

References

1. Wasserman, D.H. Four grams of glucose. *J. Physiol. Endocrinol. Metab.* **2009**, *296*, E11–E21. [CrossRef]
2. Aronoff, S.L.; Berkowitz, K.; Shreiner, B.; Want, L. Glucose metabolism and regulation: Beyond insulin and glucagon. *Diabetes Spectr.* **2004**, *17*, 183–190. [CrossRef]
3. Camandola, S.; Mattson, M.P. Brain metabolism in health, aging, and neurodegeneration. *EMBO J.* **2017**, *36*, 1474–1492. [CrossRef] [PubMed]
4. Lenzen, S. A fresh view of glycolysis and glucokinase regulation: History and current status. *J. Biol. Chem.* **2014**, *289*, 12189–12194. [CrossRef] [PubMed]
5. Cahill, G.F., Jr. Fuel metabolism in starvation. *Annu. Rev. Nutr.* **2006**, *26*, 1–22. [CrossRef] [PubMed]
6. Duran, J.; Guinovart, J.J. Brain glycogen in health and disease. *Mol. Asp. Med.* **2015**, *46*, 70–77. [CrossRef] [PubMed]
7. Soty, M.; Gautier-Stein, A.; Rajas, F.; Mithieux, G. Gut-Brain Glucose Signaling in Energy Homeostasis. *Cell Metab.* **2017**, *25*, 1231–1242. [CrossRef]
8. Herrero-Mendez, A.; Almeida, A.; Fernández, E.; Maestre, C.; Moncada, S.; Bolaños, J.P. The bioenergetic and antioxidant status of neurons is controlled by continuous degradation of a key glycolytic enzyme by APC/C-Cdh1. *Nat. Cell Biol.* **2009**, *11*, 747–752. [CrossRef]

9. Stincone, A.; Prigione, A.; Cramer, T.; Wamelink, M.M.; Campbell, K.; Cheung, E.; Olin-Sandoval, V.; Grüning, N.M.; Krüger, A.; Tauqeer Alam, M.; et al. The return of metabolism: Biochemistry and physiology of the pentose phosphate pathway. *Biol. Rev.* **2015**, *90*, 927–963. [CrossRef]

10. Fan, J.; Ye, J.; Kamphorst, J.J.; Shlomi, T.; Thompson, C.B.; Rabinowitz, J.D. Quantitative flux analysis reveals folate-dependent NADPH production. *Nature* **2014**, *510*, 298–302. [CrossRef]

11. Xiao, W.; Wang, R.S.; Handy, D.E.; Loscalzo, J. NAD(H) and NADP(H) Redox Couples and Cellular Energy Metabolism. *Antioxid. Redox Signal.* **2017**. [CrossRef] [PubMed]

12. Ralser, M.; Wamelink, M.M.; Kowald, A.; Gerisch, B.; Heeren, G.; Struys, E.A.; Klipp, E.; Jakobs, C.; Breitenbach, M.; Lehrach, H.; et al. Dynamic rerouting of the carbohydrate flux is key to counteracting oxidative stress. *J. Biol.* **2007**, *6*, 10. [CrossRef] [PubMed]

13. Kuehne, A.; Emmert, H.; Soehle, J.; Winnefeld, M.; Fischer, F.; Wenck, H.; Gallinat, S.; Terstegen, L.; Lucius, R.; Hildebrand, J.; et al. Acute Activation of Oxidative Pentose Phosphate Pathway as First-Line Response to Oxidative Stress in Human Skin Cells. *Mol. Cell* **2015**, *59*, 359–371. [CrossRef]

14. Dick, T.P.; Ralser, M. Metabolic Remodeling in Times of Stress: Who Shoots Faster than His Shadow? *Mol. Cell* **2015**, *59*, 519–521. [CrossRef]

15. Sthijns, M.M.; Weseler, A.R.; Bast, A.; Haenen, G.R. Time in Redox Adaptation Processes: From Evolution to Hormesis. *Int. J. Mol. Sci.* **2016**, *17*. [CrossRef]

16. Gebril, H.M.; Avula, B.; Wang, Y.H.; Khan, I.A.; Jekabsons, M.B. (13)C metabolic flux analysis in neurons utilizing a model that accounts for hexose phosphate recycling within the pentose phosphate pathway. *Neurochem. Int.* **2016**, *93*, 26–39. [CrossRef] [PubMed]

17. Yao, P.; Sun, H.; Xu, C.; Chen, T.; Zou, B.; Jiang, P.; Du, W. Evidence for a direct cross-talk between malic enzyme and the pentose phosphate pathway via structural interactions. *J. Biol. Chem.* **2017**, *292*, 17113–17120. [CrossRef] [PubMed]

18. Luzzatto, L.; Nannelli, C.; Notaro, R. Glucose-6-Phosphate Dehydrogenase Deficiency. *Hematol. Oncol. Clin.* **2016**, *30*, 373–393. [CrossRef]

19. Lai, Y.K.; Lai, N.M.; Lee, S.W. Glucose-6-phosphate dehydrogenase deficiency and risk of diabetes: A systematic review and meta-analysis. *Ann. Hematol.* **2017**, *96*, 839–845. [CrossRef] [PubMed]

20. Vincent, E.E.; Sergushichev, A.; Griss, T.; Gingras, M.C.; Samborska, B.; Ntimbane, T.; Coelho, P.P.; Blagih, J.; Raissi, T.C.; Choinière, L.; et al. Mitochondrial Phosphoenolpyruvate Carboxykinase Regulates Metabolic Adaptation and Enables Glucose-Independent Tumor Growth. *Mol. Cell* **2015**, *60*, 195–207. [CrossRef]

21. Hayes, J.D.; Dinkova-Kostova, A.T. The Nrf2 regulatory network provides an interface between redox and intermediary metabolism. *Trends Biochem. Sci.* **2014**, *39*, 199–218. [CrossRef] [PubMed]

22. Nielsen, J. Systems Biology of Metabolism. *Annu. Rev. Biochem.* **2017**, *86*, 245–275. [CrossRef] [PubMed]

23. Pomatto, L.C.D.; Dill, T.; Carboneau, B.; Levan, S.; Kato, J.; Mercken, E.M.; Pearson, K.J.; Bernier, M; de Cabo, R. Deletion of Nrf2 shortens lifespan in C57BL6/J male mice but does not alter the health and survival benefits of caloric restriction. *Free Radic. Biol. Med.* **2020**. [CrossRef] [PubMed]

24. Williams, K.J.; Wu, X. Imbalanced insulin action in chronic over nutrition: Clinical harm, molecular mechanisms, and a way forward. *Atherosclerosis* **2016**, *247*, 225–282. [CrossRef]

25. Dehghan, M.; Mente, A.; Zhang, X.; Swaminathan, S.; Li, W.; Mohan, V.; Iqbal, R.; Kumar, R.; Wentzel-Viljoen, E.; Rosengren, A.; et al. Associations of fats and carbohydrate intake with cardiovascular disease and mortality in 18 countries from five continents (PURE): A prospective cohort study. *Lancet* **2017**, *390*, 2050–2062. [CrossRef]

26. Shan, Z.; Guo, Y.; Hu, F.B.; Liu, L.; Qi, Q. Association of Low-Carbohydrate and Low-Fat Diets With Mortality among US Adults. *JAMA Int. Med.* **2020**. [CrossRef]

27. Ravichandran, M.; Grandl, G.; Ristow, M. Dietary Carbohydrates Impair Healthspan and Promote Mortality. *Cell Metab.* **2017**, *26*, 585–587. [CrossRef]

28. Muth, N.D.; Dietz, W.H.; Magge, S.N.; Johnson, R.K. Public Policies to Reduce Sugary Drink Consumption in Children and Adolescents. *Pediatrics* **2019**, *143*. [CrossRef]

29. Hardy, K.; Brand-Miller, J.; Brown, K.D.; Thomas, M.G.; Copeland, L. The Importance of Dietary Carbohydrate in Human Evolution. *Q. Rev. Biol.* **2015**, *90*, 251–268. [CrossRef]

30. Mosca, F.; Gianni, M.L. Human milk: Composition and health benefits. *Pediatr. Med. Chir.* **2017**, *39*, 155. [CrossRef]

31. Andreas, N.J.; Kampmann, B.; Mehring Le-Doare, K. Human breast milk: A review on its composition and bioactivity. *Early Hum. Dev.* **2015**, *91*, 629–635. [CrossRef]
32. Gromnatska, N.; Cherkas, A.; Lemishko, B.; Kulya, O. The Pattern of Metabolic Syndrome in Children with Abdominal Obesity. *Georgian. Med. News* **2019**, *289*, 68–72.
33. Beauchamp, G.K. Why do we like sweet taste: A bitter tale? *Physiol. Behav.* **2016**, *164*, 432–437. [CrossRef] [PubMed]
34. Damiani, C.; Colombo, R.; Gaglio, D.; Mastroianni, F.; Pescini, D.; Westerhoff, H.V.; Mauri, G.; Vanoni, M.; Alberghina, L. A metabolic core model elucidates how enhanced utilization of glucose and glutamine, with enhanced glutamine-dependent lactate production, promotes cancer cell growth: The WarburQ effect. *PLoS Comput. Biol.* **2017**, *13*, e1005758. [CrossRef]
35. Wu, D.; Hu, D.; Chen, H.; Shi, G.; Fetahu, I.S.; Wu, F.; Rabidou, K.; Fang, R.; Tan, L.; Xu, S.; et al. Glucose-regulated phosphorylation of TET2 by AMPK reveals a pathway linking diabetes to cancer. *Nature* **2018**, *559*, 637–641. [CrossRef]
36. Strongman, H.; Gadd, S.; Matthews, A.; Mansfield, K.E.; Stanway, S.; Lyon, A.R.; Dos-Santos-Silva, I.; Smeeth, L.; Bhaskaran, K. Medium and long-term risks of specific cardiovascular diseases in survivors of 20 adult cancers: A population-based cohort study using multiple linked UK electronic health records databases. *Lancet* **2019**, *394*, 1041–1054. [CrossRef]
37. Lau, E.; Paniagua, S.M.; Liu, E.; Jovani, M.; Li, S.; Takvorian, K.; Ramachandran, V.S.; Splansky, G.L.; Kreger, B.; Larson, M.; et al. *American Heart Association Scientific Sessions*; Circulation: Philadelphia, PA, USA, 2019; Volume 140, p. A12269.
38. Koene, R.J.; Prizment, A.E.; Blaes, A.; Konety, S.H. Shared Risk Factors in Cardiovascular Disease and Cancer. *Circulation* **2016**, *133*, 1104–1114. [CrossRef] [PubMed]
39. De Rosa, M.J.; Veuthey, T.; Florman, J.; Grant, J.; Blanco, M.G.; Andersen, N.; Donnelly, J.; Rayes, D.; Alkema, M.J. The flight response impairs cytoprotective mechanisms by activating the insulin pathway. *Nature* **2019**, *573*, 135–138. [CrossRef]
40. Schulz, T.J.; Zarse, K.; Voigt, A.; Urban, N.; Birringer, M.; Ristow, M. Glucose restriction extends *Caenorhabditis elegans* life span by inducing mitochondrial respiration and increasing oxidative stress. *Cell Metab.* **2007**, *6*, 280–293. [CrossRef]
41. Urban, N.; Tsitsipatis, D.; Hausig, F.; Kreuzer, K.; Erler, K.; Stein, V.; Ristow, M.; Steinbrenner, H.; Klotz, L.-O. Non-linear impact of glutathione depletion on *C. elegans* life span and stress resistance. *Redox Biol.* **2017**, *11*, 502–515. [CrossRef]
42. Laranjeiro, R.; Harinath, G.; Burke, D.; Braeckman, B.P.; Driscoll, M. Single swim sessions in *C. elegans* induce key features of mammalian exercise. *BMC Biol.* **2017**, *15*, 30. [CrossRef] [PubMed]
43. Cherkas, A.; Abrahamovych, O.; Golota, S.; Nersesyan, A.; Pichler, C.; Serhiyenko, V.; Knasmüller, S.; Zarkovic, N.; Eckl, P. The correlations of glycated hemoglobin and carbohydrate metabolism parameters with heart rate variability in apparently healthy sedentary young male subjects. *Redox Biol.* **2015**, *5*, 301–307. [CrossRef] [PubMed]
44. Cherkas, A.; Golota, S. An intermittent exhaustion of the pool of glycogen in the human organism as a simple universal health promoting mechanism. *Med. Hypotheses* **2014**, *82*, 387–389. [CrossRef] [PubMed]
45. Pereira, M.J.; Eriksson, J.W. Emerging Role of SGLT-2 Inhibitors for the Treatment of Obesity. *Drugs* **2019**, *79*, 219–230. [CrossRef]
46. Verma, S.; McMurray, J.J.V. SGLT2 inhibitors and mechanisms of cardiovascular benefit: A state-of-the-art review. *Diabetologia* **2018**, *61*, 2108–2117. [CrossRef] [PubMed]
47. Thomas, M.C.; Cherney, D.Z.I. The actions of SGLT2 inhibitors on metabolism, renal function and blood pressure. *Diabetologia* **2018**, *61*, 2098–2107. [CrossRef]
48. Yale, J.F.; Paty, B.; Senior, P.A. Hypoglycemia. *Can. J. Diabetes* **2018**, *42*, S104–S108. [CrossRef]
49. Ciarlone, G.E.; Hinojo, C.M.; Stavitzski, N.M.; Dean, J.B. CNS function and dysfunction during exposure to hyperbaric oxygen in operational and clinical settings. *Redox Biol.* **2019**, *27*, 101159. [CrossRef]
50. Gusarov, I.; Pani, B.; Gautier, L.; Smolentseva, O.; Eremina, S.; Shamovsky, I.; Katkova-Zhukotskaya, O.; Mironov, A.; Nudler, E. Glycogen controls *Caenorhabditis elegans* lifespan and resistance to oxidative stress. *Nat. Commun.* **2017**, *8*, 15868. [CrossRef]

51. Oliveira, J.M.; Losano, N.F.; Condessa, S.S.; de Freitas, R.; Cardoso, S.A.; Freitas, M.B.; de Oliveira, L.L. Exposure to deltamethrin induces oxidative stress and decreases of energy reserve in tissues of the Neotropical fruit-eating bat *Artibeus lituratus*. *Ecotoxicol. Environ. Saf.* **2017**, *148*, 684–692. [CrossRef] [PubMed]

52. Levin, E.; Lopez-Martinez, G.; Fane, B.; Davidowitz, G. Hawkmoths use nectar sugar to reduce oxidative damage from flight. *Science* **2017**, *355*, 733–735. [CrossRef] [PubMed]

53. Kaczor, J.J.; Robertshaw, H.A.; Tarnopolsky, M.A. Higher oxidative stress in skeletal muscle of McArdle disease patients. *Mol. Genet. Metab. Rep.* **2017**, *12*, 69–75. [CrossRef]

54. Philp, A.; Hargreaves, M.; Baar, K. More than a store: Regulatory roles for glycogen in skeletal muscle adaptation to exercise. *Am. J. Physiol. Endocrinol. Metab.* **2012**, *302*, E1343–E1351. [CrossRef]

55. Possik, E.; Pause, A. Glycogen: A must have storage to survive stressful emergencies. *Worm* **2016**, *5*, e1156831. [CrossRef]

56. LaMacchia, J.C.; Roth, M.B. Aquaporins-2 and -4 regulate glycogen metabolism and survival during hyposmotic-anoxic stress in *Caenorhabditis elegans*. *Am. J. Physiol. Cell Physiol.* **2015**, *309*, C92–C96. [CrossRef] [PubMed]

57. Anton, S.D.; Moehl, K.; Donahoo, W.T.; Marosi, K.; Lee, S.A.; Mainous, A.G., 3rd; Leeuwenburgh, C.; Mattson, M.P. Flipping the Metabolic Switch: Understanding and Applying the Health Benefits of Fasting. *Obesity* **2017**. [CrossRef]

58. Perry, R.J.; Wang, Y.; Cline, G.W.; Rabin-Court, A.; Song, J.D.; Dufour, S.; Zhang, X.M.; Petersen, K.F.; Shulman, G.I. Leptin Mediates a Glucose-Fatty Acid Cycle to Maintain Glucose Homeostasis in Starvation. *Cell* **2018**, *172*, 234–248.e17. [CrossRef]

59. Duerrschmid, C.; He, Y.; Wang, C.; Li, C.; Bournat, J.C.; Romere, C.; Saha, P.K.; Lee, M.E.; Phillips, K.J.; Jain, M.; et al. Asprosin is a centrally acting orexigenic hormone. *Nat. Med.* **2017**, *23*, 1444–1453. [CrossRef]

60. Romere, C.; Duerrschmid, C.; Bournat, J.; Constable, P.; Jain, M.; Xia, F.; Saha, P.K.; Del Solar, M.; Zhu, B.; York, B.; et al. Asprosin, a Fasting-Induced Glucogenic Protein Hormone. *Cell* **2016**, *165*, 566–579. [CrossRef]

61. Bordoni, L.; Gabbianelli, R. Primers on nutrigenetics and nutri(epi)genomics: Origins and development of precision nutrition. *Biochimie* **2019**, *160*, 156–171. [CrossRef] [PubMed]

62. Gabbianelli, R. Modulation of the Epigenome by Nutrition and Xenobiotics during Early Life and across the Life Span: The Key Role of Lifestyle. *Lifestyle Genom.* **2018**, *11*, 9–12. [CrossRef] [PubMed]

63. Gabbianelli, R.; Damiani, E. Epigenetics and neurodegeneration: Role of early-life nutrition. *J. Nutr. Biochem.* **2018**, *57*, 1–13. [CrossRef] [PubMed]

64. Perez, M.F.; Lehner, B. Intergenerational and transgenerational epigenetic inheritance in animals. *Nat. Cell Biol.* **2019**, *21*, 143–151. [CrossRef]

65. Afanas'ev, I. New nucleophilic mechanisms of ros-dependent epigenetic modifications: Comparison of aging and cancer. *Aging Dis.* **2014**, *5*, 52–62. [CrossRef] [PubMed]

66. Wu, Q.; Ni, X. ROS-mediated DNA methylation pattern alterations in carcinogenesis. *Curr. Drug Targets* **2015**, *16*, 13–19. [CrossRef]

67. Strzyz, P. A sugar rush of DNA methylation. *Nat. Rev. Mol. Cell Biol.* **2018**, *19*, 617. [CrossRef]

68. Bordoni, L.; Nasuti, C.; Di Stefano, A.; Marinelli, L.; Gabbianelli, R. Epigenetic Memory of Early-Life Parental Perturbation: Dopamine Decrease and DNA Methylation Changes in Offspring. *Oxidative Med. Cell. Longev.* **2019**, *2019*, 1472623. [CrossRef]

69. Liu, T.F.; McCall, C.E. Deacetylation by SIRT1 Reprograms Inflammation and Cancer. *Genes Cancer* **2013**, *4*, 135–147. [CrossRef]

70. Guillaumet-Adkins, A.; Yañez, Y.; Peris-Diaz, M.D.; Calabria, I.; Palanca-Ballester, C.; Sandoval, J. Epigenetics and Oxidative Stress in Aging. *Oxidative Med. Cell. Longev.* **2017**, *2017*, 9175806. [CrossRef]

71. Klotz, L.O.; Sánchez-Ramos, C.; Prieto-Arroyo, I.; Urbánek, P.; Steinbrenner, H.; Monsalve, M. Redox regulation of FoxO transcription factors. *Redox Biol.* **2015**, *6*, 51–72. [CrossRef] [PubMed]

72. Milkovic, L.; Cipak Gasparovic, A.; Cindric, M.; Mouthuy, P.A.; Zarkovic, N. Short Overview of ROS as Cell Function Regulators and Their Implications in Therapy Concepts. *Cells* **2019**, *8*. [CrossRef]

73. Zarkovic, N. 4-hydroxynonenal as a bioactive marker of pathophysiological processes. *Mol. Asp. Med.* **2003**, *24*, 281–291. [CrossRef]

74. Klotz, L.O.; Steinbrenner, H. Cellular adaptation to xenobiotics: Interplay between xenosensors, reactive oxygen species and FOXO transcription factors. *Redox Biol.* **2017**, *13*, 646–654. [CrossRef]

75. Monsalve, M.; Prieto, I.; de Bem, A.F.; Olmos, Y. Methodological Approach for the Evaluation of FOXO as a Positive Regulator of Antioxidant Genes. *Methods Mol. Biol.* **2019**, *1890*, 61–76. [CrossRef] [PubMed]
76. Milkovic, L.; Zarkovic, N.; Saso, L. Controversy about pharmacological modulation of Nrf2 for cancer therapy. *Redox Biol.* **2017**, *12*, 727–732. [CrossRef]
77. Cherkas, A.; Zarkovic, N. 4-Hydroxynonenal in Redox Homeostasis of Gastrointestinal Mucosa: Implications for the Stomach in Health and Diseases. *Antioxidants* **2018**, *7*. [CrossRef]
78. Jaganjac, M.; Milkovic, L.; Gegotek, A.; Cindric, M.; Zarkovic, K.; Skrzydlewska, E.; Zarkovic, N. The relevance of pathophysiological alterations in redox signaling of 4-hydroxynonenal for pharmacological therapies of major stress-associated diseases. *Free Radic. Biol. Med.* **2019**. [CrossRef] [PubMed]
79. Siler, U.; Romao, S.; Tejera, E.; Pastukhov, O.; Kuzmenko, E.; Valencia, R.G.; Meda Spaccamela, V.; Belohradsky, B.H.; Speer, O.; Schmugge, M.; et al. Severe glucose-6-phosphate dehydrogenase deficiency leads to susceptibility to infection and absent NETosis. *J. Allergy Clin. Immunol.* **2017**, *139*, 212–219. [CrossRef]
80. Ninfali, P.; Bresolin, N.; Baronciani, L.; Fortunato, F.; Comi, G.; Magnani, M.; Scarlato, G. Glucose-6-phosphate dehydrogenase Lodi844C: A study on its expression in blood cells and muscle. *Enzyme* **1991**, *45*, 180–187.
81. Demir, A.Y.; van Solinge, W.W.; van Oirschot, B.; van Wesel, A.; Vergouwen, P.; Thimister, E.; Maase, K.; Rijksen, G.; Schutgens, R.; van Wijk, R. Glucose 6-phosphate dehydrogenase deficiency in an elite long-distance runner. *Blood* **2009**, *113*, 2118–2119. [CrossRef] [PubMed]
82. Pandolfi, P.P.; Sonati, F.; Rivi, R.; Mason, P.; Grosveld, F.; Luzzatto, L. Targeted disruption of the housekeeping gene encoding glucose 6-phosphate dehydrogenase (G6PD): G6PD is dispensable for pentose synthesis but essential for defense against oxidative stress. *EMBO J.* **1995**, *14*, 5209. [CrossRef] [PubMed]
83. Stanton, R.C. Glucose-6-phosphate dehydrogenase, NADPH, and cell survival. *IUBMB Life* **2012**, *64*, 362–369. [CrossRef] [PubMed]
84. Legan, S.K.; Rebrin, I.; Mockett, R.J.; Radyuk, S.N.; Klichko, V.I.; Sohal, R.S.; Orr, W.C. Overexpression of glucose-6-phosphate dehydrogenase extends the life span of *Drosophila melanogaster*. *J. Biol. Chem.* **2008**, *283*, 32492–32499. [CrossRef] [PubMed]
85. Nobrega-Pereira, S.; Fernandez-Marcos, P.J.; Brioche, T.; Gomez-Cabrera, M.C.; Salvador-Pascual, A.; Flores, J.M.; Viña, J.; Serrano, M. G6PD protects from oxidative damage and improves healthspan in mice. *Nat. Commun.* **2016**, *7*, 10894. [CrossRef]
86. Monte Alegre, S.; Saad, S.T.; Delatre, E.; Saad, M.J. Insulin secretion in patients deficient in glucose-6-phosphate dehydrogenase. *Horm. Metab. Res.* **1991**, *23*, 171–173. [CrossRef]
87. Spegel, P.; Sharoyko, V.V.; Goehring, I.; Danielsson, A.P.; Malmgren, S.; Nagorny, C.L.; Andersson, L.E.; Koeck, T.; Sharp, G.W.; Straub, S.G.; et al. Time-resolved metabolomics analysis of beta-cells implicates the pentose phosphate pathway in the control of insulin release. *Biochem. J.* **2013**, *450*, 595–605. [CrossRef]
88. Kalwat, M.A.; Cobb, M.H. Mechanisms of the amplifying pathway of insulin secretion in the beta cell. *Pharmacol. Ther.* **2017**, *179*, 17–30. [CrossRef]
89. Gerber, P.A.; Rutter, G.A. The Role of Oxidative Stress and Hypoxia in Pancreatic Beta-Cell Dysfunction in Diabetes Mellitus. *Antioxid. Redox Signal.* **2017**, *26*, 501–518. [CrossRef]
90. Gylfe, E. Glucose control of glucagon secretion-'There's a brand-new gimmick every year. *Upsala J. Med. Sci.* **2016**, *121*, 120–132. [CrossRef]
91. Ashcroft, F.M.; Rohm, M.; Clark, A.; Brereton, M.F. Is Type 2 Diabetes a Glycogen Storage Disease of Pancreatic beta Cells? *Cell Metab.* **2017**, *26*, 17–23. [CrossRef] [PubMed]
92. Vergari, E.; Knudsen, J.G.; Ramracheya, R.; Salehi, A.; Zhang, Q.; Adam, J.; Asterholm, I.W.; Benrick, A.; Briant, L.J.B.; Chibalina, M.V.; et al. Insulin inhibits glucagon release by SGLT2-induced stimulation of somatostatin secretion. *Nat. Commun.* **2019**, *10*, 139. [CrossRef] [PubMed]
93. Rorsman, P.; Huising, M.O. The somatostatin-secreting pancreatic delta-cell in health and disease. *Nat. Rev. Endocrinol.* **2018**, *14*, 404–414. [CrossRef]
94. D'Alessio, D. Is GLP-1 a hormone: Whether and When? *J. Diabetes Investig.* **2016**, *7*, 50–55. [CrossRef]
95. Nauck, M.A.; Meier, J.J. Incretin hormones: Their role in health and disease. *Diabetes Obes. Metab.* **2018**, *20*, 5–21. [CrossRef] [PubMed]
96. Funcke, J.B.; Scherer, P.E. Beyond adiponectin and leptin: Adipose tissue-derived mediators of inter-organ communication. *J. Lipid Res.* **2019**, *60*, 1648–1684. [CrossRef] [PubMed]
97. Kalsbeek, A.; Bruinstroop, E.; Yi, C.X.; Klieverik, L.; Liu, J.; Fliers, E. Hormonal control of metabolism by the hypothalamus-autonomic nervous system-liver axis. *Front. Horm. Res.* **2014**, *42*, 1–28. [CrossRef] [PubMed]

98. Thorens, B. Neural regulation of pancreatic islet cell mass and function. *Diabetes Obes. Metab.* **2014**, *16*, 87–95. [CrossRef]

99. Yang, H.C.; Wu, Y.H.; Liu, H.Y.; Stern, A.; Chiu, D.T. What has passed is prolog: New cellular and physiological roles of G6PD. *Free Radic. Res.* **2016**, *50*, 1047–1064. [CrossRef]

100. Peiro, C.; Romacho, T.; Azcutia, V.; Villalobos, L.; Fernández, E.; Bolaños, J.P.; Moncada, S.; Sánchez-Ferrer, C.F. Inflammation, glucose, and vascular cell damage: The role of the pentose phosphate pathway. *Cardiovasc. Diabetol.* **2016**, *15*, 82. [CrossRef]

101. Okin, D.; Medzhitov, R. The Effect of Sustained Inflammation on Hepatic Mevalonate Pathway Results in Hyperglycemia. *Cell* **2016**, *165*, 343–356. [CrossRef] [PubMed]

102. Ying, W.; Riopel, M.; Bandyopadhyay, G.; Dong, Y.; Birmingham, A.; Seo, J.B.; Ofrecio, J.M.; Wollam, J.; Hernandez-Carretero, A.; Fu, W.; et al. Adipose Tissue Macrophage-Derived Exosomal miRNAs Can Modulate In Vivo and In Vitro Insulin Sensitivity. *Cell* **2017**, *171*, 372–384. [CrossRef] [PubMed]

103. Cherkas, A.; Golota, S.; Guéraud, F.; Abrahamovych, O.; Pichler, C.; Nersesyan, A.; Krupak, V.; Bugiichyk, V.; Yatskevych, O.; Pliatsko, M.; et al. A *Helicobacter pylori*-associated insulin resistance in asymptomatic sedentary young men does not correlate with inflammatory markers and urine levels of 8-iso-PGF2-alpha or 1,4-dihydroxynonane mercapturic acid. *Arch. Physiol. Biochem.* **2017**, 1–11. [CrossRef]

104. Ip, W.K.E.; Hoshi, N.; Shouval, D.S.; Snapper, S.; Medzhitov, R. Anti-inflammatory effect of IL-10 mediated by metabolic reprogramming of macrophages. *Science* **2017**, *356*, 513–519. [CrossRef] [PubMed]

105. Gotoh, K.; Fujiwara, K.; Anai, M.; Okamoto, M.; Masaki, T.; Kakuma, T.; Shibata, H. Role of spleen-derived IL-10 in prevention of systemic low-grade inflammation by obesity. *Endocr. J.* **2017**, *64*, 375–378. [CrossRef] [PubMed]

106. Cherkas, A.; Eckl, P.; Gueraud, F.; Abrahamovych, O.; Serhiyenko, V.; Yatskevych, O.; Pliatsko, M.; Golota, S. *Helicobacter pylori* in sedentary men is linked to higher heart rate, sympathetic activity, and insulin resistance but not inflammation or oxidative stress. *Croat. Med. J.* **2016**, *57*, 141–149. [CrossRef]

107. Gregory, J.M.; Rivera, N.; Kraft, G.; Winnick, J.J.; Farmer, B.; Allen, E.J.; Donahue, E.P.; Smith, M.S.; Edgerton, D.S.; Williams, P.E.; et al. Glucose autoregulation is the dominant component of the hormone-independent counterregulatory response to hypoglycemia in the conscious dog. *Am. J. Physiol. Endocrinol. Metab.* **2017**, *313*, E273–E283. [CrossRef]

108. Cherkas, A.; Zarkovic, K.; Cipak Gasparovic, A.; Jaganjac, M.; Milkovic, L.; Abrahamovych, O.; Yatskevych, O.; Waeg, G.; Yelisyeyeva, O.; Zarkovic, N. Amaranth oil reduces accumulation of 4-hydroxynonenal-histidine adducts in gastric mucosa and improves heart rate variability in duodenal peptic ulcer patients undergoing Helicobacter pylori eradication. *Free Radic. Res.* **2017**, 1–231. [CrossRef]

109. Lee, Y.S.; Wollam, J.; Olefsky, J.M. An Integrated View of Immunometabolism. *Cell* **2018**, *172*, 22–40. [CrossRef]

110. Tsatsoulis, A.; Mantzaris, M.D.; Bellou, S.; Andrikoula, M. Insulin resistance: An adaptive mechanism becomes maladaptive in the current environment—An evolutionary perspective. *Metabolism* **2013**, *62*, 622–633. [CrossRef]

111. Hohn, A.; Weber, D.; Jung, T.; Ott, C.; Hugo, M.; Kochlik, B.; Kehm, R.; König, J.; Grune, T.; Castro, J.P. Happily (n)ever after: Aging in the context of oxidative stress, proteostasis loss and cellular senescence. *Redox Biol.* **2017**, *11*, 482–501. [CrossRef] [PubMed]

112. Seo, Y.; Kingsley, S.; Walker, G.; Mondoux, M.A.; Tissenbaum, H.A. Metabolic shift from glycogen to trehalose promotes lifespan and healthspan in *Caenorhabditis elegans*. *Proc. Natl. Acad. Sci. USA* **2018**. [CrossRef] [PubMed]

113. Hibshman, J.D.; Doan, A.E.; Moore, B.T.; Kaplan, R.E.; Hung, A.; Webster, A.K.; Bhatt, D.P.; Chitrakar, R.; Hirschey, M.D.; Baugh, L.R. daf-16/FoxO promotes gluconeogenesis and trehalose synthesis during starvation to support survival. *eLife* **2017**, *6*. [CrossRef] [PubMed]

114. Song, S.B.; Hwang, E.S. A Rise in ATP, ROS, and Mitochondrial Content upon Glucose Withdrawal Correlates with a Dysregulated Mitochondria Turnover Mediated by the Activation of the Protein Deacetylase SIRT1. *Cells* **2018**, *8*. [CrossRef] [PubMed]

115. Stelmakh, A.; Abrahamovych, O.; Cherkas, A. Highly purified calf hemodialysate (Actovegin®) may improve endothelial function by activation of proteasomes: A hypothesis explaining the possible mechanisms of action. *Med. Hypotheses* **2016**, *95*, 77–81. [CrossRef] [PubMed]

116. da-Silva, W.S.; Gómez-Puyou, A.; de Gómez-Puyou, M.T.; Moreno-Sanchez, R.; De Felice, F.G.; de Meis, L.; Oliveira, M.F.; Galina, A. Mitochondrial bound hexokinase activity as a preventive antioxidant defense: Steady-state ADP formation as a regulatory mechanism of membrane potential and reactive oxygen species generation in mitochondria. *J. Biol. Chem.* **2004**, *279*, 39846–39855. [CrossRef]

117. Steinhorn, B.; Sartoretto, J.L.; Sorrentino, A.; Romero, N.; Kalwa, H.; Abel, E.D.; Michel, T. Insulin-dependent metabolic and inotropic responses in the heart are modulated by hydrogen peroxide from NADPH-oxidase isoforms NOX2 and NOX4. *Free Radic. Biol. Med.* **2017**, *113*, 16–25. [CrossRef]

118. Zhang, Y.; Murugesan, P.; Huang, K.; Cai, H. NADPH oxidases and oxidase crosstalk in cardiovascular diseases: Novel therapeutic targets. *Nat. Rev. Cardiol.* **2019**. [CrossRef]

119. Yan, L.J. Redox imbalance stress in diabetes mellitus: Role of the polyol pathway. *Anim. Models Exp. Med.* **2018**, *1*, 7–13. [CrossRef]

120. Rovenko, B.M.; Kubrak, O.I.; Gospodaryov, D.V.; Perkhulyn, N.V.; Yurkevych, I.S.; Sanz, A.; Lushchak, O.V.; Lushchak, V.I. High sucrose consumption promotes obesity whereas its low consumption induces oxidative stress in *Drosophila melanogaster*. *J. Insect Physiol.* **2015**, *79*, 42–54. [CrossRef] [PubMed]

121. Mullen, L.; Mengozzi, M.; Hanschmann, E.M.; Alberts, B.; Ghezzi, P. How the redox state regulates immunity. *Free Radic. Biol. Med.* **2019**. [CrossRef] [PubMed]

122. Gaitonde, M.K.; Evison, E.; Evans, G.M. The rate of utilization of glucose via hexosemonophosphate shunt in brain. *J. Neurochem.* **1983**, *41*, 1253–1260. [CrossRef] [PubMed]

123. Ben-Yoseph, O.; Boxer, P.A.; Ross, B.D. Noninvasive assessment of the relative roles of cerebral antioxidant enzymes by quantitation of pentose phosphate pathway activity. *Neurochem. Res.* **1996**, *21*, 1005–1012. [CrossRef] [PubMed]

124. Bartnik, B.L.; Sutton, R.L.; Fukushima, M.; Harris, N.G.; Hovda, D.A.; Lee, S.M. Upregulation of pentose phosphate pathway and preservation of tricarboxylic acid cycle flux after experimental brain injury. *J. Neurotrauma* **2005**, *22*, 1052–1065. [CrossRef] [PubMed]

125. Pastor, R.; Tur, J.A. Antioxidant Supplementation and Adaptive Response to Training: A Systematic Review. *Curr. Pharm. Des.* **2019**, *25*, 1889–1912. [CrossRef]

126. Kim, H.G.; Bae, J.H.; Jastrzebski, Z.; Cherkas, A.; Heo, B.G.; Gorinstein, S.; Ku, Y.G. Binding, Antioxidant and Anti-proliferative Properties of Bioactive Compounds of Sweet Paprika (*Capsicum annuum* L.). *Plant Foods Hum. Nutr.* **2016**, *71*, 129–136. [CrossRef]

127. Cherkas, A.; Mondol, A.S.; Rüger, J.; Urban, N.; Popp, J.; Klotz, L.O.; Schie, I.W. Label-free molecular mapping and assessment of glycogen in *C. elegans*. *Analyst* **2019**, *144*, 2367–2374. [CrossRef] [PubMed]

 *antioxidants* 

Article

Effect of Dipeptidyl-Peptidase 4 Inhibitors on Circulating Oxidative Stress Biomarkers in Patients with Type 2 Diabetes Mellitus

Elisabetta Bigagli [1,*], Cristina Luceri [1,*], Ilaria Dicembrini [2], Lorenzo Tatti [1], Francesca Scavone [1], Lisa Giovannelli [1], Edoardo Mannucci [2] and Maura Lodovici [1]

[1]   Department of Neurosciences, Psychology, Drug Research and Child Health (NEUROFARBA), Section of Pharmacology and Toxicology, University of Florence, 50139 Florence, Italy; lorenzo.tatti.95@gmail.com (L.T.); francesca.scavone@unifi.it (F.S.); lisa.giovannelli@unifi.it (L.G.); maura.lodovici@unifi.it (M.L.)

[2]   Department of Clinical and Experimental Biomedical Sciences, University of Florence and Diabetology Unit, Careggi Hospital, 50139 Florence, Italy; ilaria.dicembrini@unifi.it (I.D.); edoardo.mannucci@unifi.it (E.M.)

*   Correspondence: elisabetta.bigagli@unifi.it (E.B.); cristina.luceri@unifi.it (C.L.)

Received: 12 February 2020; Accepted: 9 March 2020; Published: 11 March 2020

Abstract: Pre-clinical studies suggested potential cardiovascular benefits of dipeptidyl peptidase-4 inhibitors (DPP4i), however, clinical trials showed neither beneficial nor detrimental effects in patients with type 2 diabetes mellitus (T2DM). We examined the effects of DPP4i on several circulating oxidative stress markers in a cohort of 32 T2DM patients (21 males and 11 post-menopausal females), who were already on routine antidiabetic treatment. Propensity score matching was used to adjust demographic and clinical characteristics between patients who received and who did not receive DPP4i. Whole-blood reactive oxygen species (ROS), plasma advanced glycation end products (AGEs), advanced oxidation protein products (AOPP), carbonyl residues, as well as ferric reducing ability of plasma (FRAP) and leukocyte DNA oxidative damage (Fpg sites), were evaluated. With the exception of Fpg sites, that showed a borderline increase in DPP4i users compared to non-users ($p = 0.0507$), none of the biomarkers measured was affected by DPP4i treatment. An inverse correlation between estimated glomerular filtration rate and AGEs ($p < 0.0001$) and Fpg sites ($p < 0.05$) was also observed. This study does not show any major effect of DPP4i on oxidative stress, assessed by several circulating biomarkers of oxidative damage, in propensity score-matched cohorts of T2DM patients.

Keywords: type 2 diabetes; dipeptidyl peptidase-4 inhibitors; biomarkers; oxidative stress

---

1. Introduction

Dipeptidyl peptidase-4 inhibitors (DPP4i) are oral agents used for the pharmacological treatment of adults with type 2 diabetes mellitus (T2DM). By preventing glucagon-like peptide 1 (GLP-1) breakdown, DPP4i enhance endogenous insulin secretion and suppress that of glucagon, resulting in the reduction of blood glucose levels [1]. Several clinical trials demonstrated that these agents are effective in reducing glycated hemoglobin (HbA1c) with a low risk of hypoglycemia and neutral effects on weight, compared to sulphonylureas [2–4].

A meta-analysis of short and medium-term trials with metabolic endpoints, showed that the treatment with DPP4i was associated with reduced incidence of major adverse cardiovascular events (MACE) and all-cause mortality [5]. Later on, trials with cardiovascular events as major endpoints, showed a neutral effect of DPP4i with respect to the incidence of MACE in T2DM patients with cardiovascular disease [6–8]. Moreover, a post hoc analysis from the SAVOR-TIMI 53, found a moderate increase in the risk of hospitalization due to hearth failure with saxagliptin compared to placebo [6], not observed in the EXAMINE and TECOS trials with alogliptin and sitagliptin [8,9]. More recently, in the CARMELINA trial, no increased risk of heart failure with linagliptin compared to placebo was reported in patients at high cardio renal risk [10]. The results of the CAROLINA trial also demonstrated that linagliptin was non-inferior to glimepiride on the prevention of MACE in patients with elevated cardiovascular risk [11].

Oxidative stress is a significant risk factor in the development and progression of T2DM and associated vascular complications [12–14]. Increased levels of circulating oxidative stress biomarkers have been consistently found in T2DM patients compared to healthy controls as well in T2DM patients with micro and macrovascular complications compared to those without [15].

Experimental in vitro and in vivo studies showed that DPP4i protect against oxidative stress, suggesting potential cardiovascular benefits that have not been observed in human studies. In endothelial and tubular cells, linagliptin inhibited reactive oxygen species (ROS) production and inflammation induced by advanced glycation end products (AGEs) [16,17]. Vildagliptin attenuated vascular injury by suppressing the AGEs-receptor for AGEs (RAGE)-oxidative stress axis in diabetic rats [18]. Saxagliptin prevented vascular remodeling and oxidative stress in T2DM mice by abolishing nicotinamide adenine dinucleotide phosphate NAD(P)H oxidase-driven endothelial nitric oxide synthase (eNOS) uncoupling [19] and prevented increased coronary artery stiffness and collagen deposition in a mini swine model of heart failure, partly by decreasing AGEs [20]. However, prolonged DPP4 inhibition let to the development of cardiac fibrosis in old, diabetic, high fat-fed mice [21].

Some studies were also conducted in T2DM patients, yielding conflicting results: isoprostanes, markers of lipid peroxidation, were not affected by the addition of sitagliptin to metformin [22]; similarly, no effect of vildagliptin on urinary isoprostanes levels [23] and of alogliptin on circulating AGEs, was observed [24]. Nomoto et al. demonstrated that sitagliptin improved antioxidant capacity but did not modify reactive oxygen metabolites-derived compounds (d-ROMs) levels [25]. However, some positive results were also reported: serum malondialdehyde-modified low-density lipoproteins and the soluble form of RAGE were reduced after treatment with linagliptin [26] and alogliptin [24], and both sitagliptin and vildagliptin reduced plasma nitrotyrosine levels [27].

These controversial results led us to further explore the effects of DPP4i on oxidative stress in patients with T2DM who were already on other antidiabetic treatment, evaluating a set of circulating biomarkers of oxidative damage to lipids, proteins and DNA.

2. Methods

2.1. Patients

Thirty-two T2DM patients (21 males and 11 post-menopausal females) from the Diabetology Unit, Careggi Teaching Hospital, Florence, Italy, were included in this study. Patients were given detailed explanations of the study protocol and all patients provided written informed consent prior to enrollment. The study protocol was approved by the Ethical Committee of Careggi Teaching Hospital, Florence, Italy on October, 25, 2018, registry number 13469_bio.

The inclusion criteria were as follows: (1) diagnosis of T2DM, (2) age $\geq$ 40 years, (3) treatment with diet and/or metformin and/or basal insulin analogues and/or DPP-4i and (4) no modifications of the antidiabetic therapy within the last 6 months.

The exclusion criteria were as follows: (1) use of antioxidant supplements (2) treatment with SGLT-2 inhibitors (3) treatment with GLP-1 receptor agonists.

Demographic and clinical characteristics obtained from the medical record included age, sex, height, weight, body mass index (BMI) and lipid profiles. The patients' history of coronary heart disease, peripheral vascular disease, diabetic retinopathy and nephropathy were also collected from the hospital records. Peripheral blood samples were collected by venipuncture into ethylenediaminetetraacetic acid (EDTA)-treated tubes and all plasma samples were stored at -20 °C until analysis, which was undertaken within 30 days.

2.2. Reactive Oxygen Species (ROS) Determination

ROS were determined using the Free Oxygen Radical Testing (Callegari 1930, Parma, Italy) according to [28]. Whole blood (20 µL) was added to acidic buffer and to phenylenediamine derivative [2CrNH2]. Samples were then centrifuged (3500 rpm) for 1 min and incubated for 6 min at room temperature. Absorbance was determined at 505 nm.

2.3. Ferric Reducing Ability of Plasma (FRAP)

The ferric reducing ability of plasma (FRAP) assay was performed to measure the total antioxidant capacity of plasma, according to the method by Benzie and Strain [29]. Plasma samples (30 µL) were added to 90 µL of distilled water and freshly prepared FRAP solution (300 mM acetate buffer (pH 3.6), 10 mM 2,4,6-tripyridyl-*S*-triazine (TPTZ) in 40 mM HCl and 20 mM FeCl3·6H2O (10:1:1)). The absorbance was measured at 595 nm. FRAP was estimated from a standard curve of $FeSO_4 \cdot 7H_2O$ and expressed in µM. All the reagents were purchased by Sigma Aldrich, Milan, Italy.

2.4. Advanced Oxidation Protein Product (AOPP)

Advanced oxidation protein product (AOPP) levels, markers of protein oxidation and inflammation, were determined by using 20 µL of plasma added to 980 µL of potassium phosphate buffer (PBS), 50 µL of KI 1.16 M and 100 µL of acetic acid according to the method by Witko-Sarsat et al. [30]. The absorbance was read at 340 nm. AOPP were expressed as µmol/mg of proteins. Chloramine-T (Sigma-Aldrich, Milan, Italy) was used for the calibration curve.

2.5. Carbonyl Residues

Carbonyl residues, markers of protein oxidation, were determined using the method of Correa-Salde and Albesa [31]. Plasma samples (100 µL) were derivatized with 900 µL of 0.1% dinitrophenylhydrazine in 2 M HCl (Sigma Aldrich, Milan, Italy city, country), for 1 h at room temperature (RT) followed by the addition of 400 µL of 10% trichloroacetic acid (TCA) and centrifugation at 10,000× *g* for 20 min at 4 °C. The pellets were washed three times with ethanol/ethyl acetate (1:1) and centrifuged at 10,000× *g* for 3 min at 4 °C. The pellets were then dissolved in 1.5 mL guanidine HCl (6M) in 20 mM phosphate-buffered saline (PBS) pH 7.5 and incubated at 37 °C for 30 min. Insoluble debris were removed by centrifugation. Carbonyl residues were calculated from absorbance readings at 370 nm using a molar absorption coefficient of 22,000 M^{-1} cm^{-1} and expressed as nmol/g of proteins. Protein content was measured with the Bio-Rad DC protein assay kit (Bio-Rad, Milan, Italy).

2.6. Thiobarbituric Acid Reactive Substances (TBARS)

Thiobarbituric acid reactive substances (TBARS), markers of lipid peroxidation, were evaluated using 100 µL of plasma, according to the method by [32]. Briefly, after the addition of 100 µL TCA, the resulting supernatant (160 µL) was added to 32 µL thiobarbituric acid, 0.12 M (Sigma-Aldrich, Milan, Italy city, country) and heated at 100 °C, for 15 min. The samples were then placed for 10 min in ice and centrifuged at 1600× *g* at 4 °C, for 10 min. The absorbance of the supernatants was measured at 532 nm by using a Wallac 1420 Victor3 Multilabel Counter (Perkin Elmer, Waltham, MA, USAmanufacturer, city, country). The amount of TBARS, expressed as µM, was calculated using a molar absorption coefficient of 1.56×10^{-5} M^{-1} cm^{-1}.

2.7. Advanced Glycated End-Products (AGEs)

Plasma samples (100 µL) were diluted in H_2O (1:5) and fluorescence intensity was read at 460 nm, after excitation at 355 nm. AGEs levels were expressed in arbitrary units (AU) [33].

2.8. FPG Sites

For the analysis of DNA damage in peripheral blood cells, whole blood aliquots (100 µL) were frozen at −80 °C as described in [34] and stored for 3 months before analysis. Briefly, the comet assay was performed as follows: 50 µL of cell suspension containing 10 µL of frozen whole blood and 40 µL Roswell Park Memorial Institute (RPMI) were mixed with 150 µL 1% low melting-point agarose in PBS in order to achieve 0.75% LMP agarose final concentration, and used to prepare duplicate gels for each patient (20,000 cells/gel) on Gelbond Films (Lonza, Basel, Switzerland). As a reference control for inter-experimental variability of electrophoresis conditions, we used HeLa cells aliquots prepared from a single culture and stored at −80 °C. One aliquot was used in each experiment to prepare 2 gels (20,000 cells/gel) and run with the experimental samples. We then followed the procedure described in [34]. For oxidatively generated damage detection, the enzyme formamidopyrimidine DNA glycosylase (Fpg) enzyme was used (crude E. Coli extract kindly provided by Prof. A.R. Collins, University of Oslo, Norway) at 1:1000 dilution, i.e., the optimum concentration established by titration.

2.9. Statistical Analyses

The Kolmogorov–Smirnov test was used to verify the normal distribution of the results. Continuous variables and those that resulted to be normally distributed were expressed as means ± standard deviation (SD). When data were not normally distributed, they were reported as median and interquartile range.

For the analysis, we used propensity score matching to adjust patient characteristics and disease severity between the two groups.

Among demographic and clinical characteristics at baseline of enrolled patients, putative determinants of referral to the DPP-4i treatment were identified in the whole sample by means of comparisons between treated and untreated subjects, using chi square tests for categorical variables and either unpaired Student's *t* test or Mann–Whitney test depending on the (normal or non-normal) distribution.

Variables significantly associated to the DPP-4i treatment were then inputted as covariates in a logistic regression model, with DPP-4i treatment (0/1) as dependent variable. The model was used to build an equation for the calculation of a propensity score. We estimated the propensity score using a logistic regression model, in which individuals at the time of hospital admission was regressed on measured baseline characteristics. Covariates for inclusion in the propensity score model were chosen based on their potential to be associated with the outcomes. Propensity score-matched analyses, using the 1:1 nearest neighbor technique with a small caliper of 0.05, were carried out to ensure better balance.

The propensity score-matched two cohorts were therefore compared for DPP-4i treatment in order to evaluate differences in the levels of a set of circulating biomarkers of oxidative damage to lipids, proteins and DNA.

Comparison of continuous variables between matched pairs were performed using the paired *t*-test (when data were normally distributed) or Wilcoxon test (when data were non-normally distributed). Differences between proportions were assessed using the Fisher exact test. p values < 0.05 were considered significant.

3. Results

3.1. Clinical Characteristics of the Patients

The clinical characteristics of the study groups after propensity score matching are shown in Table 1. After propensity score matching, there were no significant differences between the two groups in the levels of HbA1c and lipid profiles, body mass index (BMI) and in the male-to-female ratio. The distribution of diabetes-related micro and macro vascular complications was also similar. However, duration of diabetes was significantly higher in the DPP4i group compared to those who did not receive the treatment ($p < 0.01$). eGFR was also significantly reduced in T2DM patients treated with DPP4i compared to those not treated ($p < 0.01$).

Table 1. Demographic and clinical characteristics of the study groups after propensity score matching.

	−DDP4i	+DDP4i	*p* Value
Number of Patients	16	16	
Gender F/M	5/11	6/10	ns
Age (yrs)	69.8 ± 2.2	72.38 ± 2.98	ns
Body Mass Index (BMI)	28.12 ± 1.33	25.6 ± 1.26	ns
Duration of Diabetes (yrs)	7.87 ± 1.81	20.94 ± 3.18	0.0015
HbA1c (mmol/mol)	52 (43.75–63.25)	49 (45.25–54)	ns
Total Cholesterol (mg/dL)	166.2 ± 10.14	152.2 ± 8.07	ns
High Density Lipoprotein (HDL) (mg/dL)	47.5 (37.5–56)	49.5 (41.75–56)	ns
Triglycerides (mg/dL)	144.9 ± 13.48	122.1 ± 11.14	ns
Estimated glomerular filtration rate (eGRF) (mL/min)	79 (68.5–90)	54.5 (27.25–89.75)	0.0027
Micro Complications (yes/no)	4/12	10/6	ns
Macro Complications (yes/no)	4/12	5/11	ns

Data are expressed as means ± standard deviation (SD). Non-normally distributed variables are expressed as median and interquartile range.

3.2. Plasma Oxidative Stress Markers, Antioxidant Status and Oxidative DNA Damage

After propensity score matching, the plasma levels of AGEs, ROS, TBARS, AOPP and carbonyl residues were not significantly different in T2DM patients taking DPP4i compared with non-users. The plasma antioxidant capacity (FRAP) between patients taking DPP4i was also similar to those not treated (Figure 1, panels A,B,D,E,F,G). On the contrary, Fpg sites were borderline significantly higher in DPP4i users in comparison with matched non-users ($p = 0.0507$) (Figure 1, panel C). Several positive correlations among oxidative stress biomarkers were found; in particular, a strong correlation between TBARS and AOPP ($p < 0.001$) and between carbonyl residues and AOPP ($p < 0.0001$) were observed. Interestingly, Fpg sites and AGEs, were inversely associated with eGFR ($p < 0.05$ and $p < 0.0001$, respectively). Fpg sites were also correlated with the duration of diabetes and AGEs with the age of the patients (Table 2).

Table 2. Correlations between changes in oxidative stress-related biomarkers, duration of diabetes, eGRF and the age of the patients.

	Duration	eGRF	Age	AOPP	AGEs	TBARS	FRAP	ROS	Carbonyl Residues
Duration									
eGRF	−0.6934 ****								
Age	0.4243 *	−0.5970 ***							
AOPP	−0.2149	0.0421	0.2980						
AGEs	0.6587 ****	−0.6825 ****	0.5500 **	−0.1214					
TBARS	−0.3783 *	0.0814	0.1575	0.6358 ***	−0.2406				
FRAP	−0.0172	0.2041	0.0182	0.0411	0.0405	0.0592			
ROS	−0.1504	0.1554	−0.0942	−0.2483	−0.1995	−0.0253	−0.1191		
Carbonyl Residues	−0.2661	0.1154	0.2785	0.6948 ****	0.0514	0.5342 **	0.3504 *	−0.2148	
Fpg Sites	0.5659 ***	−0.4055 *	0.3473	−0.1654	0.3887 *	−0.2485	0.2160	0.0257	0.0206

**** $p < 0.0001$; *** $p < 0.001$; ** $p < 0.01$; * $p < 0.05$. eGRF: estimated glomerular filtration rate; AOPP: Advanced oxidation protein products; AGEs: advanced glycation end products; TBARS: thiobarbituric acid reactive substances; FRAP: ferric reducing ability of plasma; Fpg sites: formamidopyrimidine DNA glycosylase sites.

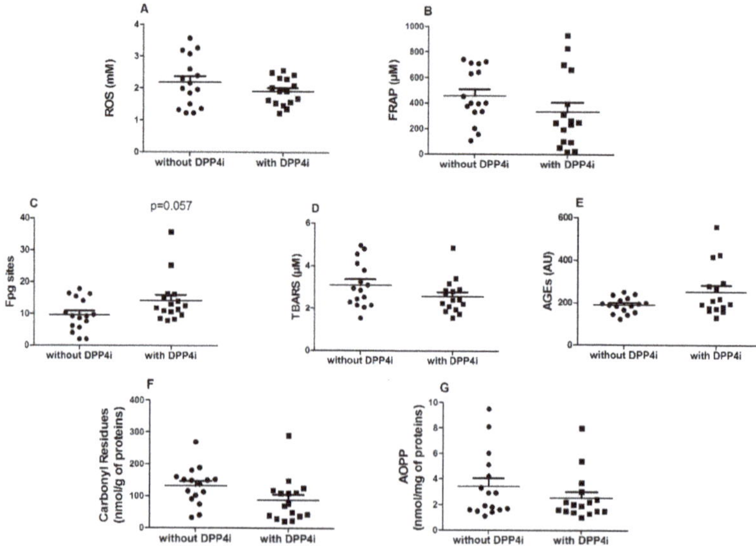

Figure 1. Scatter dot plot of reactive oxygen species (ROS) (panel **A**), ferric reducing ability of plasma (FRAP) (panel **B**) Fpg sites (panel **C**) thiobarbituric acid reactive substances (TBARS, panel **D**), advanced glycation end products (AGEs) (panel **E**), carbonyl residues (panel **F**), and advanced oxidation protein products (AOPP) levels (panel **G**) in type 2 diabetes mellitus (T2DM) patients with and without treatment with dipeptidyl peptidase-4 inhibitors (DPP4i). Comparison of continuous variables between matched pairs were performed using the paired *t*-test (when data were normally distributed) or Wilcoxon test (when data were non-normally distributed).

4. Discussion

The present data do not suggest any major effect of DPP4i on oxidative stress, as explored through the measurement of several circulating biomarkers. This result is at variance with those of several experimental studies, in vitro and in vivo on animal models of diabetes, showing a protective effect of DPP4i [16–20]. Differences from animal studies could be attributed either to diversities across species or to differences in circulating drug concentrations.

Some previous studies explored the effects of DPP4i on oxidative stress markers. In randomized, active-comparator-controlled trials, sitagliptin and vildagliptin failed to produce any effect on lipid peroxidation markers, in line with our results on TBARS [22,23]. Consistent with our findings, circulating levels of AGEs were not modified by alogliptin [24] and sitagliptin did not reduce ROS levels [25]. Conversely, a single-arm trial with linagliptin, revealed a beneficial effect of the treatment, when compared to baseline, on malondialdehyde-modified LDL; however, there was a concomitant reduction of mean glucose and HbA1c, which could explain the improvement of oxidative status [26]. Similarly, in the study by Rizzo and collaborators, the beneficial effects of sitagliptin and vildagliptin on nitrotyrosine plasma levels were associated to a significant decline in HbA1c and glucose levels and to a better control of daily glucose excursions [27]. Sakata et al. [24] demonstrated that alogliptin reduced AGE levels only in patients with higher values of AGEs at baseline and this decrease was positively correlated with that of HbA1c; however, when all patients were analyzed, no significant effect was observed.

In our study, the difference in glycemic control between cases and controls was very small, since both groups were extracted from a cohort receiving accurate treatment for T2DM. Therefore, the present study could underestimate the global effect of treatment with DPP4i, since the possible benefits of blood glucose reduction are blunted because of study design. The enrolled cohorts showed, on average, a satisfactory glycemic control, with a low dispersion of HbA1c values. This prevents any analysis on possible effects of hyperglycemia on oxidative stress. It is also important to highlight that the levels of oxidative stress biomarkers in the entire cohort of T2DM patients did not significantly differ from those measured in the serum of healthy controls [35]. We demonstrated previously that significant differences in FRAP levels were seen between heathy controls and poorly controlled T2DM patients but not between heathy controls and T2DM patients with good glycemic control [36]; similarly, Morsi et al. [37] did not find significant differences in malondialdehyde (MDA) levels between controls and well controlled T2DM patients. However, in heathy volunteers, Fpg sites, marker of oxidative damage, were reduced compared to well controlled T2DM patients [36]. Indeed, Fpg sites levels were the sole biomarker showing a borderline difference between patients taking DPP4i and those not treated.

Another limitation of this study, is represented by the fact that propensity score matching did not prevent a significant difference in estimated glomerular filtration rate between cases and controls. Renal failure is associated with oxidative stress [38]; this is consistent with the inverse correlation of estimated glomerular filtration rate with AGEs and Fpg sites observed in the present study. As a consequence, there could be a bias in favor of controls with respect to biomarkers of oxidative stress. It is also worth to highlight that although intriguing, the borderline increase in the levels of Fpg sites observed in patients taking DPP4i, is strongly linked to the presence of a single patient with particularly high levels of Fpg sites.

On the other hand, some strengths of this study should also be recognized. Propensity score matching warrants a greater reliability of results than traditional observational data. In addition, the selection of patients from a larger clinical cohort accounts for a greater representativeness of samples in comparison with randomized trials. Study procedures on patients were kept to a minimum (only one venipuncture with a blood sample), in order to remain close to routine clinical practice. A further strength is represented by the fact that several biomarkers of oxidative damage were assessed, with concordant results across different markers. Conversely, studies focused on the evaluation of a single oxidative damage marker may give incomplete information on the overall redox status. Notably, we measured various biomarkers reflecting oxidative damage to the main macromolecules such as lipids (TBARS), proteins (carbonyl residues and AOPPs) and DNA (Fpg sites) and found positive correlations among almost all of them, in line with our previous results [35].

All these biomarkers share some advantages including relatively easy detection and possibility to be used in large scale; carbonyl residues are highly stable products and AOPP are also considered inflammatory markers. However, some limitations should be also accounted: TBARS has been employed for MDA determination but thiobarbituric acid TBA may also react with other aldehydes; the FRAP assay is reliable for detecting uric acid, α-tocopherol, ascorbic acid and bilirubin, but may underestimate albumin and glutathione GSH content [39]. Another important issue also highlighted in our previous work [15], is the lack of standardization among methods employed for determining oxidative stress biomarkers, the different biological fluids used, the absence of reference range intervals and validation in prospective studies.

However, several data suggest that oxidative damage is associated to cardiovascular diseases: increased carbonyl residues, AGEs and AOPPs were found in T2DM patients with cardiovascular complications compared to those without [40,41]. Increased AOPP plasma levels were considered a risk factor for endothelial dysfunction in T2DM patients without albuminuria [42]. AOPP levels were also higher in hypertensive patients and in particular, in those with renal complications [43]. In a prospective study, oxidative DNA damage, measured as 8-hydroxy-2-deoxyguanosine (8-OHdG), predicted micro and macrovascular complications in T2DM patients [44]. Experimental studies also highlighted the pathological role of some of these markers: by binding to their common receptor RAGE, AOPP and AGEs activate inflammatory and apoptotic pathways leading to oxidative stress, inflammation and cell death both on cardiomyocytes and renal cells [45–47].

The results of the present study are consistent with those of available cardiovascular outcome trials, which failed to show either beneficial or detrimental effects of DPP4i on major cardiovascular events [6–8]. Considering that oxidative stress is a well-established risk factor for atherogenesis and cardiovascular diseases, pharmacological strategies able to limit oxidative stress in humans should theoretically produce a reduction of cardiovascular risk. Results from the Cardiovascular and Renal Microvascular Outcome Study with Linagliptin in Patients with Type 2 Diabetes Mellitus (CARMELINA) study [10], performed in patients at high renal risk, suggested a protective effect of linagliptin on microalbuminuria; this result was not confirmed by other large-scale trials enrolling patients at lower risk of renal disease [48]. Oxidative stress is thought to play a role in the pathogenesis of microvascular complications of diabetes [12,49]; however, our results suggest that the hypothetical protective effect of DPP4i on microalbuminuria does not appear to be mediated by a reduction of substrate oxidation.

In conclusion, this observational study assessing several biomarkers of oxidative damage in propensity score-matched cohorts does not suggest any major effect of DPP4i on oxidative stress in humans.

Author Contributions: Conceptualization, M.L., I.D. and E.M.; formal analysis, E.B., C.L. and I.D.; investigation, E.B., L.G., F.S. and L.T.; methods, M.L. and E.B.; original drafting of the manuscript, E.B.; review of the manuscript, all authors; patient qualification for the study and clinical material collection I.D. and E.M.; project administration, M.L.; funding acquisition, M.L. All authors have read and agreed to the published version of the manuscript.

Funding: Grants from Ente Cassa di Risparmio di Firenze (2017.0841) and from the University of Florence (Fondi di Ateneo) supported this research. The authors are grateful to the COST Action CA15132 'hCOMET' for supporting knowledge exchange among comet assay users.

Acknowledgments: We are grateful to all the patients for their participation in this trial.

Conflicts of Interest: I.D. has received speaking fees from Astra Zeneca, Bristol Myers Squibb, BoehringerIngelheim, Eli-Lilly, Merck, Novo Nordisk, Sanofi, and Novartis. EM has received speaking fees from from Astra Zeneca, Bristol Myers Squibb, Boehringer-Ingelheim, Eli-Lilly, Merck, Novo Nordisk, Sanofi, and Novartis and research grants from Merck, Novartis, and Takeda. The other authors declare that they have no conflicts of interest.

References

1. Ahrén, B.; Foley, J.E. Improved glucose regulation in type 2 diabetic patients with DPP-4 inhibitors: Focus on alpha and beta cell function and lipid metabolism. *Diabetologia* **2016**, *59*, 907–917. [CrossRef] [PubMed]

2. Scheen, A.J.; Paquot, N. Gliptin versus a sulphonylurea as add-on to metformin. *Lancet* **2012**, *380*, 450–452. [CrossRef]

3. Mishriky, B.M.; Cummings, D.M.; Tanenberg, R.J. The efficacy and safety of DPP4i compared to sulfonylureas as add-on therapy to metformin in patients with type 2 diabetes: A systematic review and meta-analysis. *Diabetes Res. Clin. Pract.* **2015**, *109*, 378–388. [CrossRef] [PubMed]

4. Chen, K.; Kang, D.; Yu, M.; Zhang, R.; Zhang, Y.; Chen, G.; Mu, Y. Direct head-to-head comparison of glycaemic durability of dipeptidyl peptidase-4 inhibitors and sulphonylureas in patients with type 2 diabetes mellitus: A meta-analysis of long-term randomized controlled trials. *Diabetes Obes. Metab.* **2018**, *20*, 1029–1033.

5. Monami, M.; Ahrén, B.; Dicembrini, I.; Mannucci, E. Dipeptidyl peptidase-4 inhibitors and cardiovascular risk: A meta-analysis of randomized clinical trials. *Diabetes Obes. Metab.* **2013**, *15*, 112–120. [CrossRef]

6. Scirica, B.M.; Bhatt, D.L.; Braunwald, E.; Steg, P.G.; Davidson, J.; Hirshberg, B.; Ohman, P.; Frederich, R.; Wiviott, S.D.; Hoffman, E.B.; et al. SAVOR-TIMI 53 Steering Committee and Investigators. Saxagliptin and cardiovascular outcomes in patients with type 2 diabetes mellitus. *N. Engl. J. Med.* **2013**, *369*, 1317–1326. [CrossRef]

7. White, W.B.; Cannon, C.P.; Heller, S.R.; Nissen, S.E.; Bergenstal, R.M.; Bakris, G.L.; Perez, A.T.; Fleck, P.R.; Mehta, C.R.; Kupfer, S.; et al. Alogliptin after acute coronary syndrome in patients with type 2 diabetes. *N. Engl. J. Med.* **2013**, *369*, 1327–1335. [CrossRef]

8. Green, J.B.; Bethel, M.A.; Armstrong, P.W.; Buse, J.B.; Engel, S.S.; Garg, J.; Josse, R.; Kaufman, K.D.; Koglin, J.; Korn, S.; et al. Effect of sitagliptin on cardiovascular outcomes in type 2 diabetes. *N. Engl. J. Med.* **2015**, *373*, 232–342. [CrossRef]

9.   Zannad, F.; Cannon, C.P.; Cushman, W.C.; Bakris, G.L.; Menon, V.; Perez, A.T.; Fleck, P.R.; Mehta, C.R.; Kupfer, S.; Wilson, C.; et al. Heart failure and mortality outcomes in patients with type 2 diabetes taking alogliptin versus placebo in EXAMINE: A multicentre, randomised, double-blind trial. *Lancet* **2015**, *385*, 2067–2076. [CrossRef]

10.   Rosenstock, J.; Kahn, S.E.; Johansen, O.E.; Zinman, B.; Espeland, M.A.; Woerle, H.J.; Pfarr, E.; Keller, A.; Mattheus, M.; Baanstra, D.; et al. Effect of linagliptin vs glimepiride on major adverse cardiovascular outcomes in patients with type 2 diabetes: The CAROLINA randomized clinical trial. *JAMA* **2019**. [CrossRef]

11.   Rosenstock, J.; Perkovic, V.; Johansen, O.E.; Cooper, M.E.; Kahn, S.E.; Marx, N.; Alexander, J.H.; Pencina, M.; Toto, R.D.; Wanner, C.; et al. Effect of linagliptin vs placebo on major cardiovascular events in adults with type 2 diabetes and high cardiovascular and renal risk: The CARMELINA randomized clinical trial. *JAMA* **2019**, *321*, 69–79. [CrossRef] [PubMed]

12.   Giacco, F.; Brownlee, M. Oxidative stress and diabetic complications. *Circ. Res.* **2010**, *107*, 1058–1070. [CrossRef]

13.   Lodovici, M.; Bigagli, E.; Luceri, C.; Mannucci, E.; Rotella, C.M.; Raimondi, L. Gender-related drug effect on several markers of oxidation stress in diabetes patients with and without complications. *Eur. J. Pharmacol.* **2015**, *766*, 86–90. [CrossRef] [PubMed]

14.   Schwartz, S.S.; Epstein, S.; Corkey, B.E.; Grant, S.F.; Gavin, J.R., III; Aguilar, R.B.; Herman, M.E. A unified pathophysiological construct of diabetes and its complications. *Trends Endocrinol. Metab.* **2017**, *28*, 645–655. [CrossRef] [PubMed]

15.   Bigagli, E.; Lodovici, M. Circulating oxidative stress biomarkers in clinical studies on type 2 diabetes and its complications. *Oxid. Med. Cell Longev.* **2019**, *2019*, 5953685. [CrossRef]

16.   Ishibashi, Y.; Matsui, T.; Maeda, S.; Higashimoto, Y.; Yamagishi, S. Advanced glycation end products evoke endothelial cell damage by stimulating soluble dipeptidyl peptidase-4 production and its interaction with mannose 6-phosphate/insulin-like growth factor II receptor. *Cardiovasc. Diabetol.* **2013**, *12*, 125. [CrossRef]

17.   Kaifu, K.; Ueda, S.; Nakamura, N.; Matsui, T.; Yamada-Obara, N.; Ando, R.; Kaida, Y.; Nakata, M.; Matsukuma-Toyonaga, M.; Higashimoto, Y.; et al. Advanced glycation end products evoke inflammatory reactions in proximal tubular cells via autocrine production of dipeptidyl peptidase-4. *Microvasc. Res.* **2018**, *120*, 90–93. [CrossRef]

18.   Matsui, T.; Nishino, Y.; Takeuchi, M.; Yamagishi, S. Vildagliptin blocks vascular injury in thoracic aorta of diabetic rats by suppressing advanced glycation end product-receptor axis. *Pharmacol. Res.* **2011**, *63*, 383–388. [CrossRef]

19.   Solini, A.; Rossi, C.; Duranti, E.; Taddei, S.; Natali, A.; Virdis, A. Saxagliptin prevents vascular remodeling and oxidative stress in db/db mice. Role of endothelial nitric oxide synthase uncoupling and cyclooxygenase. *Vascul. Pharmacol.* **2016**, *76*, 62–71. [CrossRef]

20.   Fleenor, B.S.; Ouyang, A.; Olver, T.D.; Hiemstra, J.A.; Cobb, M.S.; Minervini, G.; Emter, C.A. Saxagliptin prevents increased coronary vascular stiffness in aortic-banded mini swine. *Hypertension* **2018**, *72*, 466–475. [CrossRef]

21.   Mulvihill, E.E.; Varin, E.M.; Ussher, J.R.; Campbell, J.E.; Bang, K.W.; Abdullah, T.; Baggio, L.L.; Drucker, D.J. Inhibition of Dipeptidyl Peptidase-4 Impairs Ventricular Function and Promotes Cardiac Fibrosis in High Fat-Fed Diabetic Mice. *Diabetes* **2016**, *65*, 742–754. [CrossRef]

22.   Koren, S.; Shemesh-Bar, L.; Tirosh, A.; Peleg, R.K.; Berman, S.; Hamad, R.A.; Vinker, S.; Golik, A.; Efrati, S. The effect of sitagliptin versus glibenclamide on arterial stiffness, blood pressure, lipids, and inflammation in type 2 diabetes mellitus patients. *Diabetes Technol. Ther.* **2012**, *14*, 561–567. [CrossRef]

23.   Kim, G.; Oh, S.; Jin, S.M.; Hur, K.Y.; Kim, J.H.; Lee, M.K. The efficacy and safety of adding either vildagliptin or glimepiride to ongoing metformin therapy in patients with type 2 diabetes mellitus. *Expert Opin. Pharmacother.* **2017**, *18*, 1179–1186. [CrossRef]

24.   Sakata, K.; Hayakawa, M.; Yano, Y.; Tamaki, N.; Yokota, N.; Eto, T.; Watanabe, R.; Hirayama, N.; Matsuo, T.; Kuroki, K.; et al. Efficacy of alogliptin, a dipeptidyl peptidase-4 inhibitor, on glucose parameters, the activity of the advanced glycation end product (AGE)-receptor for AGE (RAGE) axis and albuminuria in Japanese type 2 diabetes. *Diabetes Metab. Res. Rev.* **2013**, *29*, 624–630. [CrossRef]

25. Nomoto, H.; Miyoshi, H.; Furumoto, T.; Oba, K.; Tsutsui, H.; Inoue, A.; Atsumi, T.; Manda, N.; Kurihara, Y.; Aoki, S.; et al. A randomized controlled trial comparing the effects of sitagliptin and glimepiride on endothelial function and metabolic parameters: Sapporo Athero-Incretin Study 1 (SAIS1). *PLoS ONE* **2016**, *11*, e0164255. [CrossRef]

26. Makino, H.; Matsuo, M.; Hishida, A.; Koezuka, R.; Tochiya, M.; Ohata, Y.; Tamanaha, T.; Son, C.; Miyamoto, Y.; Hosoda, K. Effect of linagliptin on oxidative stress markers in patients with type 2 diabetes: A pilot study. *Diabetol. Int.* **2018**, *10*, 148–152. [CrossRef]

27. Rizzo, M.R.; Barbieri, M.; Marfella, R.; Paolisso, G. Reduction of oxidative stress and inflammation by blunting daily acute glucose fluctuations in patients with type 2 diabetes: Role of dipeptidyl peptidase-IV inhibition. *Diabetes Care* **2012**, *35*, 2076–2082. [CrossRef]

28. Pavlatou, M.G.; Papastamataki, M.; Apostolakou, F.; Papassotiriou, I.; Tentolouris, N. FORT and FORD: Two simple and rapid assays in the evaluation of oxidative stress in patients with type 2 diabetes mellitus. *Metabolism* **2009**, *58*, 1657–1662. [CrossRef]

29. Benzie, I.; Strain, J. The ferric reducing ability of plasma as a measure of antioxodant. *Anal. Biochem.* **1996**, *239*, 70–76. [CrossRef]

30. Witko-Sarsat, V.; Friedlander, M.; Capeillère-Blandin, C.; Nguyen-Khoa, T.; Nguyen, A.T.; Zingraff, J.; Jungers, P.; Descamps-Latscha, B. Advanced oxidation protein products as a novel marker of oxidative stress in uremia. *Kidney Int.* **1996**, *49*, 1304–1313. [CrossRef]

31. Correa-Salde, V.; Albesa, I. Reactive oxidant species and oxidation of protein and heamoglobin as biomarkers of susceptibility to stress caused by chloramphenicol. *Biomed. Pharmacother.* **2009**, *63*, 100–104. [CrossRef]

32. Dietrich-Muszalska, A.; Kolińska-Łukaszuk, J. Comparative effects of aripiprazole and selected antipsychotic drugs on lipid peroxidation in plasma. *Psychiatry Clin. Neurosci.* **2018**, *72*, 329–336. [CrossRef]

33. Cournot, M.; Burillo, E.; Saulnier, P.; Planesse, C.; Gand, E.; Rehman, M.; Rondeau, P.; Gonthier, M.; Feigerlova, E.; Meilhac, O.; et al. Circulating concentrations of redox biomarkers do not improve the prediction of adverse cardiovascular events in patients with type 2 diabetes mellitus. *J. Am. Heart Assoc.* **2018**, *7*, e00007397. [CrossRef]

34. Ladeira, C.; Koppen, G.; Scavone, F.; Giovannelli, L. The comet assay for human biomonitoring: Effect of cryopreservation on DNA damage in different blood cell preparations. *Mutat. Res.* **2019**, *843*, 11–17. [CrossRef]

35. Luceri, C.; Bigagli, E.; Agostiniani, S.; Giudici, F.; Zambonin, D.; Scaringi, S.; Ficari, F.; Lodovici, M.; Malentacchi, C. Analysis of oxidative stress-related markers in Crohn's disease patients at surgery and correlations with clinical findings. *Antioxidants (Basel)* **2019**, *8*, 378. [CrossRef]

36. Lodovici, M.; Giovannelli, L.; Pitozzi, V.; Bigagli, E.; Bardini, G.; Rotella, C.M. Oxidative DNA damage and plasma antioxidant capacity in type 2 diabetic patients with good and poor glycaemic control. *Mutat. Res.* **2008**, *638*, 98–102. [CrossRef]

37. Morsi, H.K.; Ismail, M.M.; Gaber, H.A.; Elbasmy, A.A. Macrophage migration inhibitory factor and malondialdehyde as potential predictors of vascular risk complications in type 2 diabetes mellitus: Cross-sectional case control study in Saudi Arabia. *Mediators Inflamm.* **2016**, *2016*, 5797930. [CrossRef]

38. Honda, T.; Hirakawa, Y.; Nangaku, M. The role of oxidative stress and hypoxia in renal disease. *Kidney Res. Clin. Pract.* **2019**, *38*, 414–426. [CrossRef]

39. Gaggini, M.; Sabatino, L.; Vassalle, C. Conventional and innovative methods to assess oxidative stress biomarkers in the clinical cardiovascular setting. *Biotechniques* **2020**. [CrossRef]

40. Bigagli, E.; Raimondi, L.; Mannucci, E.; Colombi, C.; Bardini, G.; Rotella, C.M.; Lodovici, M. Lipid and protein oxidation products, antioxidant status and vascular complications in poorly controlled type 2 diabetes. *Brit. J. Diabetes Vasc. Dis.* **2012**, *12*, 33–39. [CrossRef]

41. Jakuš, V.; Sándorová, E.; Kalninová, J.; Krahulec, B. Monitoring of glycation, oxidative stress and inflammation in relation to the occurrence of vascular complications in patients with type 2 diabetes mellitus. *Physiol. Res.* **2014**, *63*, 297–309.

42. Liang, M.; Wang, J.; Xie, C.; Yang, Y.; Tian, J.W.; Xue, Y.M.; Hou, F.F. Increased plasma advanced oxidation protein products is an early marker of endothelial dysfunction in type 2 diabetes patients without albuminuria. *J. Diabetes* **2014**, *6*, 417–426. [CrossRef]

43. Conti, G.; Caccamo, D.; Siligato, R.; Gembillo, G.; Satta, E.; Pazzano, D.; Carucci, N.; Carella, A.; Campo, G.D.; Salvo, A.; et al. Association of higher advanced oxidation protein products (AOPPs) levels in patients with diabetic and hypertensive nephropathy. *Medicina (Kaunas)* **2019**, *55*, 675. [CrossRef]

44. Thomas, M.C.; Woodward, M.; Li, Q.; Pickering, R.; Tikellis, C.; Poulter, N.; Cooper, M.E.; Marre, M.; Zoungas, S.; Chalmers, J.; et al. Relationship between plasma 8-OH-deoxyguanosine and cardiovascular disease and survival in type 2 diabetes mellitus: Results from the ADVANCE trial. *J. Am. Heart Assoc.* **2018**, *7*, e008226. [CrossRef]

45. Valente, A.J.; Yoshida, T.; Clark, R.A.; Delafontaine, P.; Siebenlist, U.; Chandrasekar, B. Advanced oxidation protein products induce cardiomyocyte death via Nox2/Rac1/superoxide-dependent TRAF3IP2/JNK signaling. *Free Radic. Biol. Med.* **2013**, *60*, 125–135. [CrossRef]

46. Yamamoto, Y.; Yamamoto, H. Interaction of receptor for advanced glycation end products with advanced oxidation protein products induces podocyte injury. *Kidney Int.* **2012**, *82*, 733–735. [CrossRef]

47. Hou, X.; Hu, Z.; Xu, H.; Xu, J.; Zhang, S.; Zhong, Y.; He, X.; Wang, N. Advanced glycation endproducts trigger autophagy in cadiomyocyte via RAGE/PI3K/AKT/mTOR pathway. *Cardiovasc. Diabetol.* **2014**, *13*, 78. [CrossRef]

48. Groop, P.H.; Cooper, M.E.; Perkovic, V.; Hocher, B.; Kanasaki, K.; Haneda, M.; Schernthaner, G.; Sharma, K.; Stanton, R.C.; Toto, R.; et al. Linagliptin and its effects on hyperglycaemia and albuminuria in patients with type 2 diabetes and renal dysfunction: The randomized MARLINA-T2D trial. *Diabetes Obes. Metab.* **2017**, *19*, 1610–1619. [CrossRef]

49. Lodovici, M.; Bigagli, E.; Tarantini, F.; Di Serio, C.; Raimondi, L. Losartan reduces oxidative damage to renal DNA and conserves plasma antioxidant capacity in diabetic rats. *Exp. Biol. Med. (Maywood)* **2015**, *240*, 1500–1504. [CrossRef]

Article

Role of Sirt3 in Differential Sex-Related Responses to a High-Fat Diet in Mice

Marija Pinterić [1,†], Iva I. Podgorski [1,†], Marijana Popović Hadžija [1], Ivana Tartaro Bujak [3], Ana Dekanić [1], Robert Bagarić [2], Vladimir Farkaš [2], Sandra Sobočanec [1,*] and Tihomir Balog [1]

[1] Division of Molecular Medicine, Ruđer Bošković Institute, 10000 Zagreb, Croatia; mpinter@irb.hr (M.P.); iskrinj@irb.hr (I.I.P.); mhadzija@irb.hr (M.P.H.); adekanic@irb.hr (A.D.); balog@irb.hr (T.B.)

[2] Division of Experimental Physics, Ruđer Bošković Institute, 10000 Zagreb, Croatia; robert.bagaric@irb.hr (R.B.); vladimir.farkas@irb.hr (V.F.)

[3] Division of Materials Chemistry, Ruđer Bošković Institute,10000 Zagreb, Croatia; itartaro@irb.hr

* Correspondence: ssoboc@irb.hr; Tel.: +385-1-4561-172

† These authors contributed equally to this work.

Received: 7 February 2020; Accepted: 18 February 2020; Published: 20 February 2020

Abstract: Metabolic homeostasis is differently regulated in males and females. Little is known about the mitochondrial Sirtuin 3 (Sirt3) protein in the context of sex-related differences in the development of metabolic dysregulation. To test our hypothesis that the role of Sirt3 in response to a high-fat diet (HFD) is sex-related, we measured metabolic, antioxidative, and mitochondrial parameters in the liver of Sirt3 wild-type (WT) and knockout (KO) mice of both sexes fed with a standard or HFD for ten weeks. We found that the combined effect of Sirt3 and an HFD was evident in more parameters in males (lipid content, glucose uptake, *pparγ*, *cyp2e1*, *cyp4a14*, Nrf2, MnSOD activity) than in females (protein damage and mitochondrial respiration), pointing towards a higher reliance of males on the effect of Sirt3 against HFD-induced metabolic dysregulation. The male-specific effects of an HFD also include reduced Sirt3 expression in WT and alleviated lipid accumulation and reduced glucose uptake in KO mice. In females, with a generally higher expression of genes involved in lipid homeostasis, either the HFD or Sirt3 depletion compromised mitochondrial respiration and increased protein oxidative damage. This work presents new insights into sex-related differences in the various physiological parameters with respect to nutritive excess and Sirt3.

Keywords: sirtuin 3; high fat diet; sex differences; mice; oxidative stress; metabolic stress

---

1. Introduction

The metabolic syndrome is metabolic derangement involving a cluster of risk factors primarily associated with obesity, type two diabetes, and a high risk of cardiovascular events, but also with the development of inflammation, atherosclerosis, renal, liver and respiratory disease, cancer, and premature aging [1,2]. The prevalence of metabolic syndrome is dramatically on the rise in low- and mid-income countries and is one of the leading risks for global deaths representing serious threat to public health. The increase in caloric food intake or consumption of diets high in both fat and carbohydrates (also known as a "western diet") along with physical inactivity leads to increased obesity, a key factor in the development of metabolic syndrome. Human metabolic syndrome can be effectively mimicked in rodent models using dietary intervention, such as high-fat diets (HFDs) [3]. The nutrient overload generated by an HFD in mice leads to a chronic increase in reactive oxygen species (ROS) production, which leads to oxidative stress (rev. in [1]). Oxidative stress is associated with the metabolic syndrome, but whether it is the cause or the consequence is a matter of debate. Nevertheless, an HFD is capable of functioning as a metabolic stressor causing mitochondrial dysfunction and other metabolic changes that contribute to pathological state and may mimic age-related pathology [4].

In most mammals, including humans, life expectancy differs between sexes. Females show lower incidence of some age-related pathologies linked with oxidative stress conditions and this sex-difference disappears after menopause, which led to the conclusion that this protection is attributed to the effect of sex hormones (rev in [5]). The important approach to study ageing and age-linked pathologies is to investigate sex dimorphism in defense to metabolic stressors. Historically, the impact of sex on metabolic status in mouse models has been neglected [6]. Indeed, most in vivo studies are focused on male mice. One of the reasons for a strong male sex bias is the belief that female mammals are intrinsically more variable than males due to the estrogen cycle. However, this has been disproven and without foundation [7]. While numerous papers examined the metabolic profiles of inbred mouse strain 129 [8–10], little attention has been paid to the impact of sex on the development of metabolic syndrome caused by HFD.

Sirtuin 3 (Sirt3) is the only member of sirtuin family that has been linked to longevity in humans [11]. In addition, Sirt3 integrates cellular energy metabolism and various mitochondrial processes including ROS generation. The excessive production of ROS leads to oxidative stress, a crucial event that contributes to mitochondrial dysfunction and age-related pathologies [12]. Sirt3 plays a role in preventing metabolic syndrome [10], and is found to be upregulated in response to caloric restriction and exercise [13], while being downregulated with age, upon an HFD, and in diabetes [14]. Sirt3 is dispensable at a young age under homeostatic conditions but is essential under stress conditions or at an old age, making it a potential regulator of aging process. Although Sirt3 mediates oxidative stress suppression during caloric restriction [15], it remains to be elucidated whether Sirt3 may alleviate chronic oxidative stress initiated by excessive caloric intake in the form of an HFD. Many mitochondrial proteins are targets for deacetylation and activation by Sirt3, including proteins of oxidative phosphorylation, fatty acid oxidation, and the citric acid (TCA) cycle [16]. These data suggest that fatty acid metabolism and the TCA cycle are among the pathways that are tightly regulated by Sirt3. Moreover, Sirt3 is found to be regulated by a redox-sensitive, Keap1/Nrf2/ARE signaling axis, which facilitates transcription of Sirt3 and other antioxidant-response genes under stress conditions [17].

Little is currently known about Sirt3 expression in the context of sex-related differences in the development of metabolic syndrome. The understanding of sex differences in physiology and disease is of fundamental importance with regard to preventing metabolic disease. Therefore, our aim was to determine the role of Sirt3 in sex-related responses to a high-fat diet in 129S mice. In this study, we found that an HFD reduces hepatic Sirt3 expression only in wild-type (WT) males, while HFD-induced lipid accumulation is alleviated in Sirt3 knockout (KO) males, with impaired glucose uptake and increased reliance on fatty acids. Moreover, females had more efficient lipid metabolism but also displayed higher oxidative stress following an HFD in both genotypes with compromised mitochondrial respiration and increased protein oxidative damage. These data present new insights into sex-related differences in the metabolic parameters with respect to nutritive excess and Sirt3.

2. Materials and Methods

2.1. Animal Model and Experimental Design

Sirt3 WT (hereafter WT) and Sirt3 KO (hereafter KO) mice on a 129/SV background (stock no. 027975) were purchased from the Jackson Laboratory (Bar Harbor, ME, USA). WT and KO mice of both sexes at of 8 weeks of age, were fed with either a standard fat diet (SFD, 11.4% fat, 62.8% carbohydrates, 25.8% proteins; Mucedola, Settimo Milanese, Italy) or a high fat diet (HFD, 58% fat, 24% carbohydrates, 18% proteins; Mucedola, Settimo Milanese, Italy) during 10 weeks. The mice were age-matched and housed in standard conditions (3 mice per cage, 22 °C, 50–70% humidity, and a cycle of 12 h light and 12 h darkness). The animals were divided into 8 groups (6 mice per group): SFD-fed WT males, HFD-fed WT males, SFD-fed KO males, HFD-fed KO males, SFD-fed WT females, HFD-fed WT females, SFD-fed KO females, and HFD-fed KO females. Body weight was measured once a week, as well as glucose level, which was measured by glucometer (StatStrip Xpress-I, Nova Biomedical,

GmbH, Mörfelden-Walldorf, Germany) in a blood drop taken from the tail vain. Before glucose measurements, mice were fasted for 6 h. After 10 weeks, mice were anesthetized by intraperitoneal injection of ketamine/xylazine (Ketamidor 10%, Richter pharma Ag, Wels, Austria; Xylazine 2%, Alfasan International, Woerden, Netherlands) and organs of interest were obtained and stored in liquid nitrogen until analysis. Animal experiments were done within the project funded by Croatian Science Foundation, project ID: IP-014-09-4533, approved on 01/09/2015. All procedures were approved by the Ministry of Agriculture of Croatia, (No: UP/I-322-01/15-01/25 525-10/0255-15-2 from 20 July 2015) and carried out in accordance with the associated guidelines EU Directive 2010/63/EU.

2.2. Histology and Oil Red O Staining

A histological analysis of the liver sections from all experimental groups was performed in order to determine lipid accumulation caused by HFD. At the end of the feeding period, the anesthetized animals were perfused via the left ventricle of the heart with ice-cold phosphate-buffered saline (PBS; 137 mM NaCl, 2.7 mM KCl, 8 mM Na_2HPO_4, and 2 mM K_2PO_4, pH 7.4) for 2–3 min (to remove blood via the incised abdominal vena cava). Liver tissue was fixed by immersion in 4% paraformaldehyde in 0.1 M PBS (pH 7.4), left overnight at 4 °C, and was then washed with 1× PBS and cryoprotected in 30% sucrose in PBS until the sectioning on cryomicrotome at 8 μm. Fat vacuoles in hepatocytes of frozen sections were visualized by Oil Red O dye (Sigma Aldrich, St. Louis, MO, USA) prepared in propylene glycol (0.5% Oil Red O solution) according to the following protocol. The slides were dried for 60 min at room temperature (RT) and then fixed in ice-cold 10% formalin for 10 min followed by another 60 min of drying. The slides were then incubated for 5 min in absolute propylene glycol followed by staining in pre-warmed (60 °C) Oil Red O solution for 10 min and incubation for 5 min in 85% propylene glycol. Before staining with hematoxylin for 30 s, the slides were rinsed 2× in distilled water. After hematoxylin, the stained slides were thoroughly washed in running tap water followed by distilled water and mounted in mounting medium. An analysis of the stained liver sections was done using an Olympus BX51 microscope (Tokyo, Japan) with associated software analysis. The quantification of the lipid accumulation signal was done using ImageJ software (U.S. National Institutes of Health, Bethesda, MD, USA).

2.3. Total Lipid Extraction

Total lipids were extracted from liver tissue according to a modified Folch procedure [18]. Briefly, 0.1 g of liver tissue was homogenized in MeOH, followed by the addition of 2 mL of $CHCl_3$ and vigorous shaking. The mixture was filtered, and the remaining mixture was resuspended in $CHCl_3$-MeOH (2:1). The mixture was filtered and washed again with fresh solvent, followed by washing the solution with 1.5 mL of 0.88% KCl and drawing off the aqueous layer using a Pasteur pipette. The washing step was repeated with 1.5 mL of KCl/MeOH (4:1, *v/v*), and the organic layer containing lipids was carefully transferred to a glass tube. A rotary evaporator (IKA Rotary evaporator RV 10 digital, Staufen, Germany) was used to remove the solvent, and the total lipids were determined by gravimetric analysis.

2.4. RNA Isolation and Quantitative Real-Time PCR Analysis

The TRIzol reagent (Invitrogen, Waltham, MA, USA) was used for total RNA extraction from a mouse liver (5% extract). The isolated RNA was treated with DNAse (TURBO DNA-free Kit, Thermo Fisher Scientific, Waltham, MA, USA) followed by reverse transcription using a High-Capacity cDNA Reverse Transcription Kit (Thermo Fisher Scientific) according to the manufacturer's recommendations. For real-time PCR analysis, an ABI 7300 sequence detection system was used. To quantify the relative mRNA expression of *cyp2a4*, *cyp2e1*, *cyp4a14*, *hnf4α*, *pparα*, and *pparγ* in the livers of mice, the comparative C_T ($\Delta\Delta C_T$) method according to the Taqman® Gene Expression Assays Protocol (Applied Biosystems, Foster City, CA, USA) was used. The assays' ID used for the analyses are shown in Supplementary Table S1. The data on the graphs are shown as the fold-change in gene expression, which is normalized to the endogenous reference gene (*β-actin*) and relative to the Sirt3 WT males fed with a SFD.

2.5. Protein Carbonylation

Protein carbonylation experiments were performed with an ELISA-based assay. Liver homogenates (5%) were prepared in PBS with protease inhibitors (Roche, Basel, Switzerland) and incubated for 1 h at RT with a lipid removal agent (13360-U, Sigma Aldrich, St. Louis, MO, USA), followed by centrifugation for 20 min at 16,000× g. A Pierce™ BCA Protein Assay Kit (Thermo Fischer Scientific) was used to determine the protein concentration, and 100 μL of 1 μg/μL lysate was incubated overnight at 4 °C using Maxisorb wells (Sigma Aldrich). 2,4-dinitrophenylhydrazine (DNPH; 12 μg/mL) was used for derivatization of adsorbed proteins, and rabbit anti-DNP primary (D9656, Sigma Aldrich) followed by goat anti-rabbit secondary antibody conjugated to HRP (NA934, GE Healthcare, Chicago, IL, USA) were used to detect the derivatized dinitrophenol (DNP)-carbonyl. Enzyme substrate 3,3′,5,5′-tetramethylbenzidine (Sigma Aldrich) was added into samples and incubated until color developed, followed by stopping the reaction using 0.3 M H_2SO_4. The absorbance was measured at 450 nm on a microplate reader (Bio-Tek Instruments, Winooski, VT, USA).

2.6. Analysis of Antioxidative Enzyme Activities

Antioxidative enzyme activities were analyzed in liver tissue lysates homogenized in PBS supplemented with protease inhibitors (Roche, Basel, Switzerland) using an ice-jacketed Potter-Elvehjem homogenizer (Thomas Scientific, Swedesboro, NJ, USA), at 1300 rpm. Superoxide dismutase (SOD) activities were determined measuring inhibition of the xanthine/xanthine oxidase-mediated reduction of 2-(4-iodophenyl)-3-(4-nitrophenyl)-5-phenyltetrazolium chloride (I.N.T.) using a Ransod kit (Randox Laboratories, Crumlin, UK) according to the manufacturer's recommendations. This reaction is inhibited by the conversion of the superoxide radical to hydrogen peroxide and oxygen as a consequence of SOD activity. A total of 4 mM of KCN for 30 min was used to selectively inhibit CuZnSOD activity and obtain the MnSOD activity. CuZnSOD activity was obtained by subtracting the MnSOD activity from the total SOD activity. The Catalase (Cat) activity was done, as previously described [19], by measuring the change in absorbance (at 240 nm) in the reaction mixture during the interval of 30 s following sample addition. The final concentrations of 10 mM H_2O_2 and 50 mM PBS (pH 7.0) were used. Glutathione peroxidase (Gpx) activity was measured at 340 nm, as previously described [20], using a RANSEL kit (Randox Laboratories). In kit, a decrease in absorbance at 340 nm is accompanied by NADPH oxidation to $NADP^+$ as an indirect measure of Gpx activity.

2.7. Protein Isolation and Western Blot Analysis

Liver samples were homogenized in an ice-cold lysis buffer (RIPA buffer supplemented with protease inhibitors (Roche)) using an ice-jacketed Potter-Elvehjem homogenizer (1300 rpm). Liver homogenates (5%) were centrifuged at 2000× g for 15 min at 4 °C, and the supernatant was sonicated (3 × 30 s) and centrifuged at 16,000× g for 20 min at 4 °C. The resulting lysate was transferred to a new tube and the protein concentration was estimated by the Pierce™ BCA Protein Assay Kit (Thermo Fischer Scientific). Proteins (40 μg/μL) were resolved by SDS-PAGE and were electrotransferred onto a PVDF membrane (Roche, Basel, Switzerland). Membranes were blocked in a I-Block™ Protein-Based Blocking Reagent (Invitrogen, Waltham, MA, USA) for 1 h at RT and were incubated with primary polyclonal or monoclonal antibodies overnight at 4 °C. For chemiluminiscence detection, an appropriate horseradish peroxidase (HRP)-conjugated secondary antibody was used. The list of primary and secondary antibodies is in Supplementary Table S2. AmidoBlack (Sigma Aldrich) was used for total protein normalization. The Alliance 4.7 Imaging System (UVITEC, Cambridge, UK) was used for the detection of immunoblots using an enhanced chemiluminscence kit Kit (Thermo Fischer Scientific).

2.8. Mitochondria Isolation and Oxygen Consumption

Mice liver mitochondria were isolated by differential centrifugation as described previously [21], with the following modification: liver was homogenized at a ratio of 100 mg tissue/mL of isolation buffer (10% liver homogenate). Isolated mitochondria were kept in the isolation buffer (250 mM sucrose, 2 mM EGTA, 0.5% fatty acid-free BSA, 20 mM Tris-HCl, pH 7.4) until the experiment on the Clark-type electrode (Oxygraph, Hansatech Instruments Ltd., Pentney, UK) in an airtight 1.5 mL chamber at 35°C. The protein concentration was determined with a Pierce™ BCA Protein Assay Kit. For the determination of oxygen consumption, mitochondria (800 μg protein) were resuspended in a 500 μL respiration buffer (200 mM sucrose, 20 mM TrisHCl, 50 mM KCl, 1 mM $MgCl_2 \cdot 6H_2O$, 5 mM KH_2PO_4, pH 7,0). Complex I assessment samples were incubated with 2.5 mM glutamate and 1.25 mM malate. Mitochondrial respiration was accelerated by the addition of 2 mM ADP for state 3 respiration measurements. Then, ATP synthesis was terminated by adding 5 μg/mL of oligomycin to achieve state 4 rate. To inhibit the mitochondrial respiration, 2 μM antimycin A was used. Oxygen uptake is calculated in nmol/min/mg protein.

2.9. PET Analysis

For [18]FDG-microPET imaging, animals have been anesthetized in induction chamber with 4% isoflurane (Forane, Abbott Laboratories, Chicago, IL, USA) and intraperitoneally injected with 100–200 μL of solution containing 25 MBq of radiotracer [18F] fluoro-2-deoxy-2-D-glucose ([18]FDG). To avoid the influence of warming on [18]FDG biodistribution in mice injected intravenously, in our experiments we used the model of intraperitoneal FDG administration described in [22]. [18]FDG-microPET imaging, along with [18]FDG liver uptake data analysis, was performed according to our previous model [23]. The co-registration of PET images was made in PMOD FUSION software mode (PMOD Technologies LLC, Zürich, Switzerland). The final result is given in standardized uptake value units (SUV).

2.10. Statistical Analysis

For the statistical analysis of data, SPSS for Windows (17.0, IBM, Armonk, NY, USA) was used. A Shapiro–Wilk test was used before all analyses to test the samples for normality of distribution. Since all data followed normal distribution, parametric tests for multiple comparisons were performed: a student's *t*-test for comparisons between males and females, and a two-way ANOVA for the interaction effect of Sirt3 × diet within each sex. If a significant interaction was observed, all pairwise comparisons were made between groups, using Tukey's post-hoc test with Bonferroni's correction. Significance was set at $p < 0.05$. On graphical displays, the indicator of the differences between males and females was marked as x; the indicator of differences between SFD and HFD (the effect of diet) was marked as a letter (a, b, etc.); the indicator of differences between WT and KO (the effect of Sirt3) was marked as *.

3. Results

3.1. HFD Reduces Hepatic Sirt3 Protein Expression in Males Only

To investigate if the hepatic expression of Sirt3 was altered in a sex- or diet-dependent manner, we first detected Sirt3 protein expression in all groups. Expectedly, in KO mice, Sirt3 protein level was undetectable. In WT mice, HFD partially (24%) reduced Sirt3 protein expression in males but did not affect Sirt3 in females. Therefore, HFD-fed males had lower Sirt3 expression than females (Figure 1A,B). These data suggest that ten weeks of HFD feeding reduces Sirt3 protein expression in a male-specific manner.

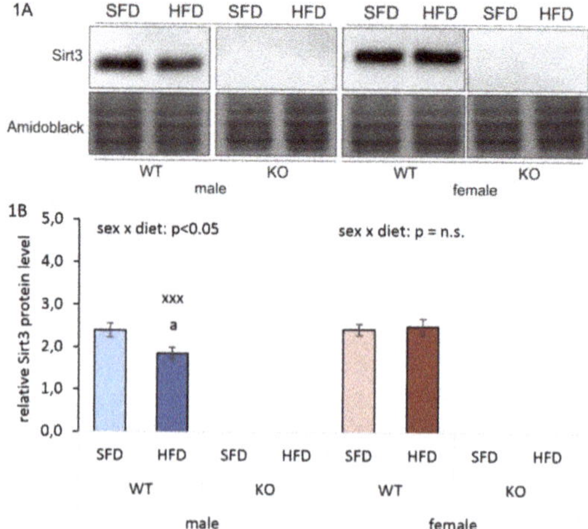

Figure 1. Hepatic Sirtuin 3 (Sirt3) protein expression in Sirt3 wild-type (WT) and knockout (KO) mice fed with a standard fat diet (SFD) or a high fat diet (HFD) for 10 weeks. (**A**) A representative immunoblot of the hepatic Sirt3 protein expression level. (**B**) A graphical display of the averaged densitometry values of immunoblots in (A). Males: HFD-fed vs. SFD-fed WT mice ([a] $p < 0.001$); Females: no changes. Males vs. females: HFD-fed WT mice ([xxx] $p < 0.001$). The data are shown as mean ± SD. $n = 6$. Amidoblack was used as a loading control.

3.2. HFD Has no Infuence on Body Weight and Glucose Level

It has been shown that the decreased level of Sirt3 contributes to impaired glucose metabolism [24], and since we noticed reduced Sirt3 expression upon HFD in WT males, we next tested whether differential Sirt3 expression and HFD would affect body weight and fasting glucose levels. After 10 weeks of feeding, treatment with an HFD did not affect whole body weight or glucose level in either sex, irrespective of Sirt3 (Figure 2A,B, Figure S1A,B). However, at the end of the 10th week, males had a higher body weight and glucose level compared to female mice (Figure 2A,B).

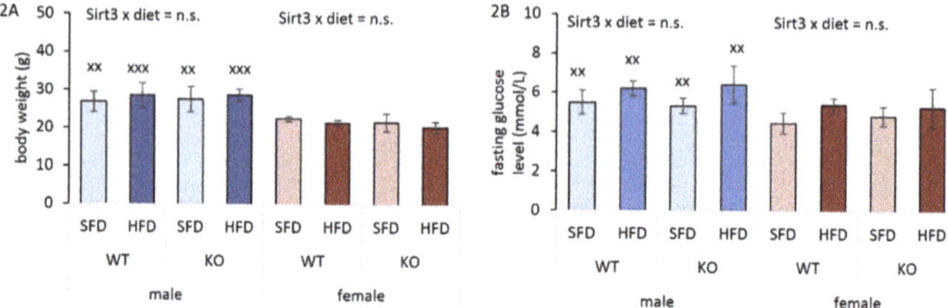

Figure 2. Body weight and fasting glucose level in Sirt3 wild-type (WT) and knockout (KO) mice fed with a standard fat diet (SFD) or a high fat diet (HFD) for 10 weeks. (**A**) A graphical display of body weight. Males: no changes. Females: no changes. Males vs. females: SFD-fed mice ([xx] $p < 0.01$); HFD-fed mice ([xxx] $p < 0.001$). (**B**) A graphical display of fasting glucose levels. Males: no changes. Females: no changes. Males vs. females: [xx] $p < 0.01$. The data are shown as mean ± SD. $n = 6$ per group.

3.3. Differential Hepatic Fat Accumulation in Males Upon HFD Depends on Sirt3

While glucose levels showed a male-specific pattern, the interesting fact was that the HFD failed to cause weight gain compared to SFD in both sexes. To investigate whether HFD-fed mice developed a fatty liver, we determined the hepatic accumulation of lipids. Expectedly, fat vacuoles were predominately present in the livers of HFD-fed mice of both genotypes, confirming that HFD treatment caused lipid accumulation in mice livers (Figure 3A–H, Supplementary Figure S2). Folch extraction was performed to measure lipid content in larger parts of liver tissue. In males, an HFD caused significant lipid accumulation in both WT and KO mice. In SFD conditions, WT males had less, and, in HFD conditions, WT males had more hepatic lipids than KO males. In females, an HFD also induced significant lipid accumulation regardless of Sirt3. While in WT mice there were no differences in lipid content between males and females, more lipids were observed in SFD-fed, and less in HFD-fed KO males compared to KO females (Figure 3I). These results suggest that, in males, the hepatic lipid accumulation was partially influenced by Sirt3.

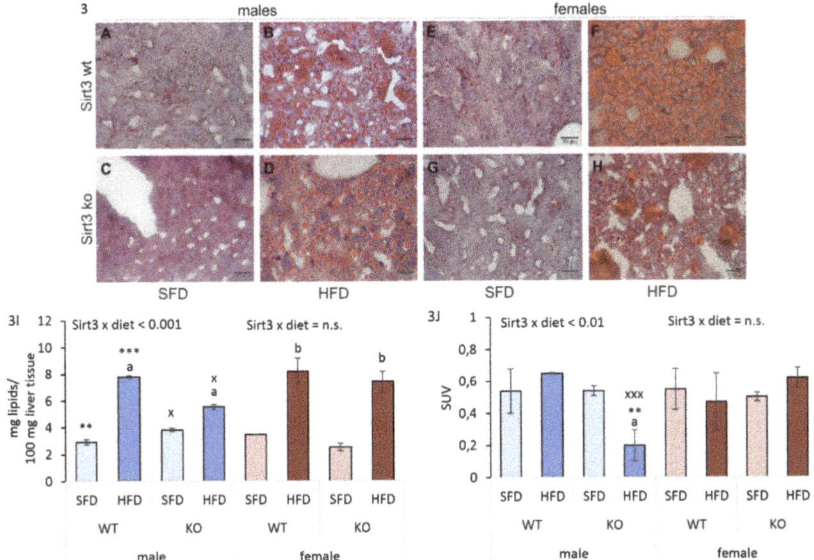

Figure 3. Sex-related differences in hepatic lipid accumulation and glucose ([18]FDG) uptake in Sirt3 wild-type (WT) and knockout (KO) mice of both sexes fed with a standard fat diet (SFD) or a high fat diet (HFD) for 10 weeks. (**A–H**) Liver cryo-sections stained with Oil Red O. Representative images show lower fat content in mice fed with a SFD (A, C, E, G), and promoted lipid accumulation in HFD-fed mice of both sexes and genotypes (B,D,F,H). (**I**) A graphical display of total lipid content in liver samples. Males: HFD-fed vs. SFD-fed mice ([a] $p < 0.001$); SFD-fed WT vs. KO mice (** $p < 0.01$); HFD-fed WT vs. KO mice (*** $p < 0.001$). Females: HFD-fed vs. SFD-fed mice ([b] $p < 0.001$). Males vs. females: SFD-fed KO mice ([x] $p < 0.05$); HFD-fed KO mice ([x] $p < 0.05$). (**J**) A graphical display of hepatic glucose uptake expressed as standardized uptake value (SUV). Males: SFD-fed vs. HFD-fed KO mice ([a] $p < 0.001$); HFD-fed WT vs. KO mice (** $p < 0.01$). Females: no changes. Males vs. females: HFD-fed KO mice ([xxx] $p < 0.001$). Data are shown as mean ± SD. $n = 3$–4 per group. scale bar = 20 µm.

3.4. HFD Reduces Hepatic Glucose Uptake in Sirt3 KO Males

Because the presence of non-alcoholic fatty liver disease (NAFLD) is closely associated with decreased glucose metabolism, we determined hepatic glucose uptake using ^{18}F-FDG PET. In KO males, an HFD significantly reduced glucose uptake. Also, HFD-fed KO males had lower glucose uptake than WT mice. On the contrary, in females, an HFD had no effect on glucose uptake across groups. Reduced hepatic glucose uptake in HFD-fed KO males resulted in significantly lower values compared to HFD-fed KO females. These data indicate the importance of Sirt3 in the maintenance of hepatic glucose uptake following HFD, but only in males (Figure 3J).

3.5. HFD Further Suppresses the Male-Specific Reduction of Hepatic hnf4α, pparα, pparγ, and cyp2a4 mRNA Level

Since it was observed that an HFD causes lipid accumulation in liver, we wanted to investigate several important genes involved in the lipid metabolism that are known to be expressed in sex-related manner and may be responsible for the observed phenotype, such as *hnf4α, pparα, pparγ, cyp2a4, cyp2e1,* and *cyp4a14.*

Hepatocyte nuclear factor 4α (hnf4α) is a master regulator of many genes involved in lipid, glucose, and drug metabolism. In our study, no change in *hnf4a* transcript level was found either within the male or female group of mice. The only differences were found between WT males and females, where males had less *hnf4α* transcript compared to females (Figure 4A,C).

The peroxisome proliferator-activated receptor α (pparα) is shown to be upregulated during HFD as it serves as a ligand for free fatty acids [25]. In males, we observed no change in the transcription level of *pparα*, while in females, *pparα* was reduced in the absence of Sirt3. Interestingly, HFD had no effect on the *pparα* in any sex. WT females had higher *pparα* than males. These data indicate sex-related differences in *pparα* gene expression that are lost in the absence of Sirt3 (Figure 4A,D).

Hepatic *pparγ* gene expression is upregulated in animal models of severe obesity and lipoatrophy [26]. Since in our study HFD-fed mice did not have increased weight gain compared to SFD, we checked whether this was associated with the altered expression of *pparγ* in liver. In WT males, an HFD strongly reduced *pparγ*, which was maintained in KO males as well. On the contrary, females exhibited no change in *pparγ*. Overall, HFD-fed WT males had reduced *pparγ* compared to females. These data indicate that both HFD and Sirt3 depletion reduce *pparγ* only in males (Figure 4A,E).

Cyp2a4 (steroid 15α-hydroxylase) is the enzyme that is constitutively expressed in the livers of female mice while in males its expression is hormonally regulated [27]. Considering its sex-related expression and regulation, we wanted to check whether lack of Sirt3 or change in diet affects the expression of this gene. In WT males, an HFD reduced *cyp2a4*, whereas in KO mice, an HFD rescued the reduced *cyp2a4* observed in SFD-fed mice. In SFD conditions, WT males had more *cyp2a4*, and in HFD conditions less than KO males. In females, HFD induced *cyp2a4*. Additionally, higher *cyp2a4* was observed in SFD-fed KO females compared to WT. Overall, males had lower *cyp2a4* than females across all groups (Figure 4A,F).

Cyp2e1 expression and activity in the liver are increased in humans and in animal models of NAFLD [7,10,11]. Similar to *cyp2a4*, in WT males, an HFD reduced *cyp2e1*, whereas in KO mice an HFD rescued reduced *cyp2e1* observed in SFD-fed mice. SFD-fed KO males displayed lower *cyp2e1* than WT mice. In females, HFD induced *cyp2e1*. The differences in *cyp2e1* between males and females were observed only in SFD-fed KO mice, with lower *cyp2e1* expression in males (Figure 4A,G).

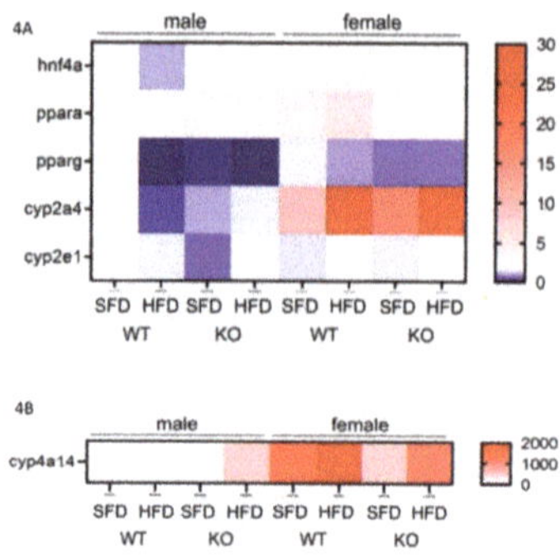

Figure 4. *Cont.*

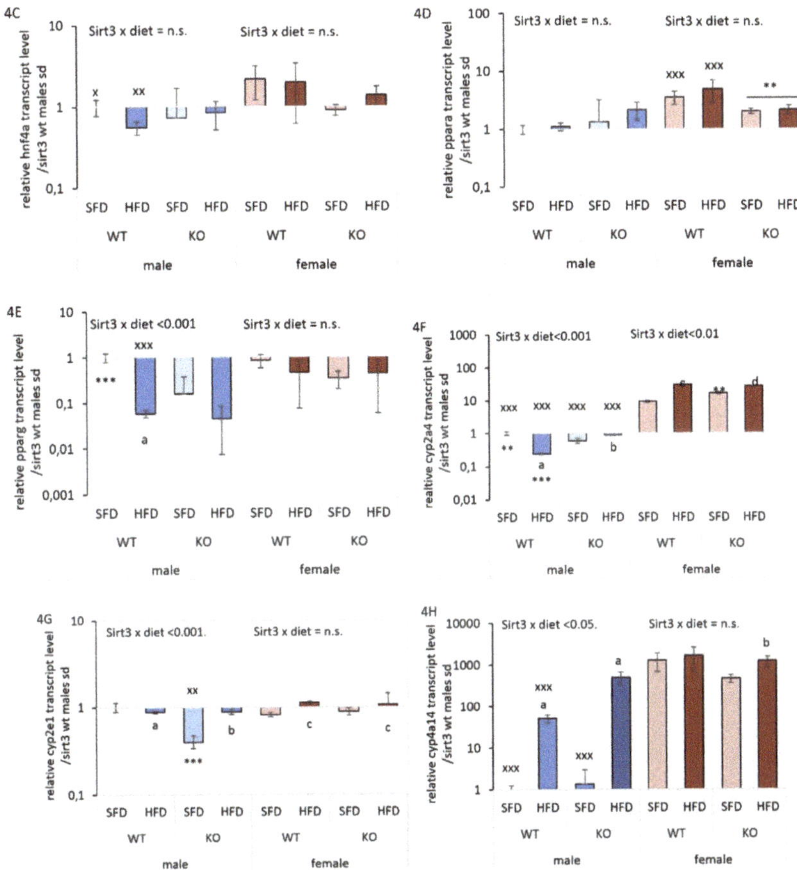

Figure 4. Gene expression of *hnf4α*, *pparα*, *pparγ*, *cyp2a4*, *cyp2e1*, and *cyp4a14* in the livers of Sirt3 wild-type (WT) and knockout (KO) mice fed with a standard fat diet (SFD) or a high fat diet (HFD) for 10 weeks. (**A**) A heatmap of the mRNA levels of *hnf4α*, *pparα*, *pparγ*, *cyp2a4*, *cyp2e1*, and (**B**) *cyp4a14* genes. The color of the squares on the heat map corresponds to the mean value of the log fold change from three biological and three technical replicates. (**C**) A graphical display of *hnf4a* mRNA level. Males: no changes. Females: no changes. Males vs. females: SFD-fed ([x] $p < 0.05$) and HFD-fed ([xx] $p < 0.01$). (**D**) A graphical display of *pparα* mRNA levels. Males: no changes. Females: KO vs. WT mice (** $p < 0.01$). Males vs. females: WT mice ([xxx] $p < 0.001$). (**E**) A graphical display of *pparγ* mRNA levels. Males: HFD-fed vs. SFD-fed WT mice ([a] $p < 0.001$); SFD-fed WT vs. KO mice (*** $p < 0.001$). Females: no changes. Males vs. females: HFD-fed WT mice ([xxx] $p < 0.001$). (**F**) A graphical display of *cyp2a4* mRNA levels. Males: HFD-fed vs. SFD-fed WT mice ([a] $p < 0.001$); HFD-fed vs. SFD-fed KO mice ([b] $p < 0.01$); SFD-fed WT vs. KO mice (** $p < 0.01$) and HFD-fed WT vs. KO mice (*** $p < 0.001$); Females: HFD-fed vs. SFD-fed WT ([c] $p < 0.001$) and KO mice ([d] $p < 0.01$); SFD-fed KO vs. WT mice (** $p < 0.01$); Males vs. females: [xxx] $p < 0.001$. (**G**) A graphical display of *cyp2e1* mRNA levels. Males: HFD-fed vs. SFD-fed WT mice ([a] $p < 0.05$); HFD-fed vs. SFD-fed KO mice ([b] $p < 0.001$); SFD-fed KO vs. WT mice (*** $p < 0.001$). Females: HFD-fed vs. SFD-fed mice ([c] $p < 0.01$). Males vs. females: SFD-fed KO mice ([xx] $p < 0.01$). (**H**) A graphical display of *cyp4a14* mRNA levels. Males: HFD-fed vs. SFD-fed mice ([a] $p < 0.001$). Females: HFD-fed vs. SFD-fed KO mice ([b] $p < 0.01$). Males vs. females: WT mice and SFD-fed KO mice ([xxx] $p < 0.001$). β-actin was used for normalization. The data are shown as mean ± SD. $n = 3$ per group in technical triplicates.

Mouse Cyp4a14 catalyzes the omega-hydroxylation of saturated and unsaturated fatty acids in mice and shows female-predominant expression in liver, where it is induced by pparα [28]. In WT males, HFD strongly increased otherwise very low *cyp4a14* by 51-fold, and in KO males by almost 400-fold. In females, HFD upregulated *cyp4a14* by 2.5-fold in KO mice. Overall, females displayed higher *cyp4a14* expression than males, except in HFD-fed KO mice (Figure 4B,H).

3.6. Sirt3 KO Mice Exhibit Increased Protein Oxidative Damage and Upregulated Keap1-Nrf2-Ho1 Axis in Liver

Following the observed differences between males and females in lipid accumulation and the genes involved in lipid homeostasis, along with the fact that sensitivity towards the oxidative stress is sex-related as well [29,30], we next examined sensitivity to oxidative stress with respect to Sirt3 or diet by measuring protein carbonylation (PC), a marker of protein oxidative damage. In males, higher PC levels were observed in the absence of Sirt3. Contrary to that, in WT females, an HFD increased PC, resulting in levels similar to KO females in both diets. Moreover, SFD-fed KO females had higher PC than WT mice. Overall, the sex-specific differences evident in HFD-fed WT mice were abrogated in the KO mice, where similar PC levels were observed in both sexes (Figure 5A). These data indicate that, besides protein oxidative damage caused by the absence of Sirt3 in both sexes, females are also susceptible to protein damage caused by an HFD only.

We next aimed to investigate whether major proteins involved in antioxidative response pathway, such as Keap1/Nrf2/Ho1, were altered. Nrf2 is a transcription factor that induces the gene expression of antioxidant enzymes and many other cytoprotective enzymes. Upon oxidative stress, the interaction between Keap1 and Nrf2 is disrupted, which induces Nrf2-dependent gene expression [31]. WT males had higher Keap1 compared to KO mice and HFD partially rescued Keap1 in KO mice. Likewise, in females, Keap1 was also reduced in KO mice. There were no differences in Keap1 between males and females (Figure 5C). In WT males, HFD reduced Nrf2. Moreover, Nrf2 was higher in KO mice compared to WT mice of both sexes, which is in accordance with reduced Keap1. Differences between males and females were evident only in SFD-fed WT mice, where males had higher Nrf2 (Figure 5D). In males, an HFD had the opposite effect on Ho1 expression, where it decreased the expression of the Ho1 protein in WT and increased it in KO males. In WT females, an HFD increased Ho-1 protein level, which remained higher in KO mice, irrespective of diet. Differences in Ho1 between males and females were only observable in WT mice, with higher Ho1 in SFD-fed and lower in HFD-fed males (Figure 5E). This finding indicates the presence of the adaptive stress response in conditions of nutritional excess only in the absence of Sirt3.

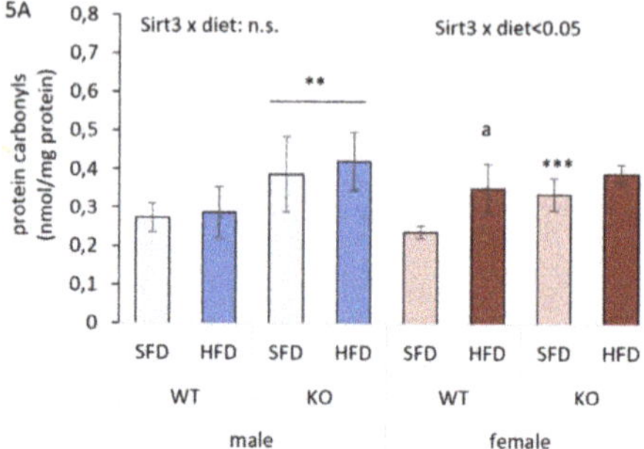

Figure 5. *Cont.*

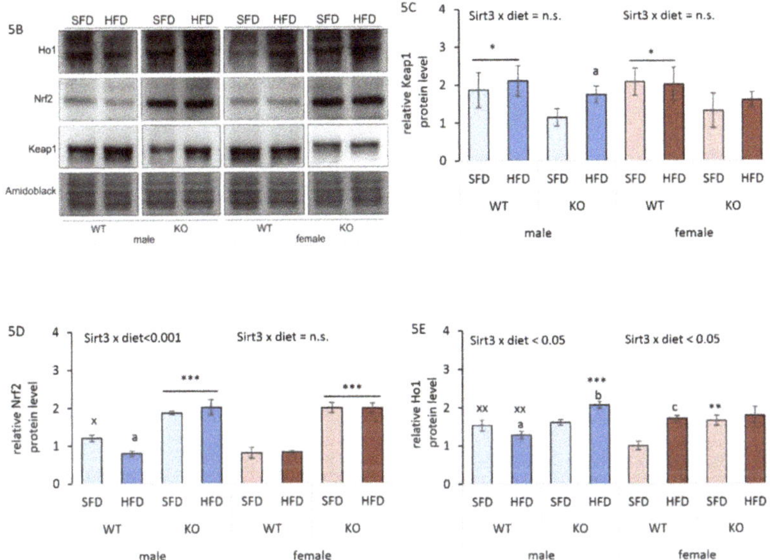

Figure 5. Protein oxidative damage and antioxidant response in the livers of Sirt3 wild-type (WT) and knockout (KO) mice fed with a standard fat diet (SFD) or a high fat diet (HFD) for 10 weeks. (**A**) The total amount of protein carbonyls (PC) measured with an ELISA-based assay at 450 nm. Males: KO vs. WT mice (** $p < 0.01$). Females: HFD-fed vs. SFD-fed WT mice ([a] $p < 0.01$); SFD-fed KO vs. WT mice (*** $p < 0.001$). Males vs. females: no changes. The data are shown as mean ± SD. $n = 6$ per group. (**B**) The representative immunoblots of hepatic Keap1, Nrf2, and Ho1 protein expression. Amidoblack was used as a loading control. (**C**) A graphical display of the averaged densitometry values for Keap-1 protein expression. Males: HFD-fed vs. SFD-fed KO mice ([a] $p < 0.05$); WT vs. KO mice (* $p < 0.05$). Females: WT vs. KO mice (* $p < 0.05$). Males vs. females: no changes. (**D**) A graphical display of the averaged densitometry values for Nrf-2 protein expression. Males: HFD-fed vs. SFD-fed WT mice ([a] $p < 0.001$); KO vs. WT mice (*** $p < 0.001$). Females: KO vs. WT mice (*** $p < 0.001$). Males vs. females: SFD-fed WT mice ([x] $p < 0.05$). (**E**) A graphical display of averaged densitometry values of Ho-1 protein expression. Males: HFD-fed vs. SFD-fed WT mice ([a] $p < 0.05$); HFD-fed vs. SFD-fed KO mice ([b] $p < 0.01$); HFD-fed KO vs. WT mice (*** $p < 0.001$). Females: HFD-fed vs. SFD-fed WT mice ([c] $p = 0.002$); SFD-fed KO vs. WT mice (** $p < 0.01$). Males vs. females: WT mice ([xx] $p < 0.05$). The data are shown as mean ± SD. $n = 6$.

3.7. HFD-Induced Reduction in Sirt3/MnSOD Axis is Male-Specific

Sirt3 mediates the reduction of oxidative and metabolic damage, while exposure to HFD leads to reduced Sirt3 expression, consequent hyper acetylation of manganese superoxide dismutase (AcSOD2), and the reduction of SOD2 activity (hereafter MnSOD) [32]. Since we found that an HFD reduced Sirt3 protein level only in males, we further examined whether this pattern affects AcSOD2 protein and MnSOD activity in the same way. HFD-fed WT males displayed increased AcSOD2 protein levels along with decreased MnSOD activity. SFD-fed WT males had reduced AcSOD2 protein and higher MnSOD activity than KO mice. HFD-fed WT females also displayed increased AcSOD2 protein, but without change in MnSOD activity. SFD-fed WT females had reduced AcSOD2 protein but higher MnSOD activity on both diets compared to KO mice (Figure 6A–D). This suggests that the acetylation status of MnSOD acetyl-K68 that regulates MnSOD activity in males under HFD conditions, does not contribute to alteration of MnSOD activity in females. Both KO males and females had similar AcSOD2 levels, with reduced MnSOD activity, thus confirming the importance of Sirt3 in regulation of MnSOD activity.

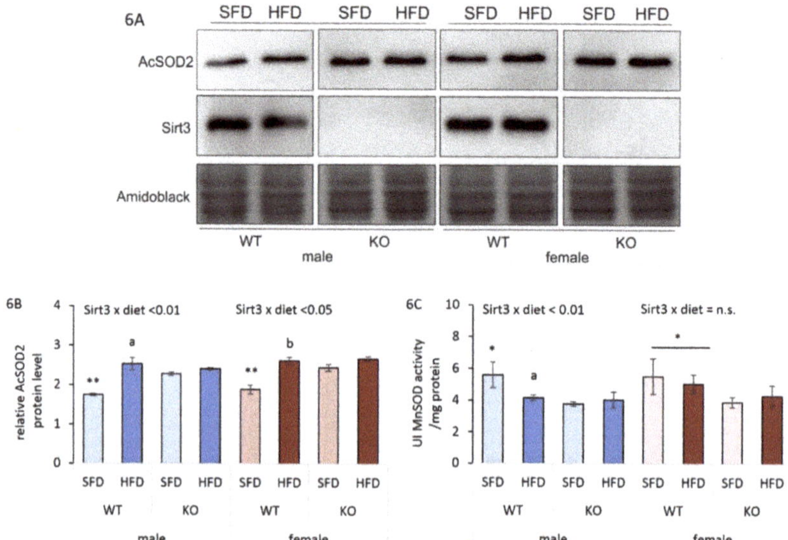

Figure 6. Sirt3/MnSOD axis in liver of Sirt3 wild-type (WT) and knockout (KO) mice fed with a standard fat diet (SFD) or a high fat diet (HFD) for 10 weeks. (**A**) A representative immunoblot of hepatic Sirt3 and AcSOD2 protein expression. Amidoblack was used as a loading control. (**B**) A graphical display of averaged densitometry values for AcSOD2 protein expression. Males: HFD-fed vs. SFD-fed WT mice ([a] $p < 0.01$); SFD-fed WT vs. KO mice (** $p < 0.01$). Females: HFD-fed vs. SFD-fed WT mice ([b] $p < 0.01$); SFD-fed WT vs. KO mice (** $p < 0.01$). Males vs. females: no change. (**C**) MnSOD activity. Males: HFD-fed vs. SFD-fed WT mice ([a] $p < 0.01$); SFD-fed WT vs. KO mice (* $p < 0.05$). Females: WT vs. KO mice (* $p < 0.05$). Males vs. females: no change. The data are shown as mean ± SD. $n = 6$ per group.

3.8. HFD Affects Antioxidant Enzyme Activities Differently in Males and Females

Beside major mitochondrial antioxidant enzyme MnSOD, we also determined other antioxidant enzymes, such as copper zinc superoxide dismutase (CuZnSOD, SOD1), catalase (Cat), and glutathione peroxidase (Gpx1) at protein level (Supplementary Figure S3) and their activities (Figure 7A–C). We observed a discrepancy between protein expression and antioxidant enzyme activity, indicating the complex regulatory mechanisms of enzyme activities. Therefore, the results of enzyme activities are more informative and are shown here. In males, an HFD reduced very mildly CuZnSOD activity in both WT (14%) and KO (10%) mice. Additionally, SFD-fed WT mice had marginally higher CuZnSOD activity compared to KO mice (14.2%). On the contrary, females displayed no changes in CuZnSOD activity (Figure 7A). Cat activity was increased following an HFD in both sexes and genotypes (Figure 7B). The inducing effect of HFD on Cat activity in both genotypes and sexes suggests that increased Cat activity is needed to effectively degrade excess of H_2O_2 produced by increased lipid metabolism, irrespective of sex or Sirt3. Gpx is a cytosolic enzyme that also functions in the detoxification of H_2O_2, specifically by catalyzing the reduction of hydrogen peroxide to water. In WT males, HFD increased Gpx activity to levels of KO mice. SFD-fed WT males also displayed lower Gpx activity compared to KO mice. In females, Gpx activity was lower in WT mice compared to KO mice. The differences in Gpx activity between males and females were evident only in HFD-fed WT mice, where males had higher Gpx activity (Figure 7C). These results point toward differential sex-related effect of HFD on antioxidant enzyme activities.

3.9. HFD and Sirt3 Depletion Affect Mitochondrial Respiration Differently in Males and Females

Sirt3 is an important regulator of mitochondrial respiration, and its targets include subunits of the respiratory chain complexes. Studies showed that KO mice display decreased oxygen consumption in liver [33] and reduced glucose tolerance when placed on an HFD [10]. To examine whether the loss of Sirt3 would impair respiration in a sex- and diet-dependent manner, we measured oxygen consumption in mitochondria from liver using a Clark type electrode.

WT males had higher malate/glutamate (MG) + ADP respiration than KO males, suggesting defective CI-driven respiration in the absence of Sirt3. Interestingly, an HFD partially restored low CI-driven respiration in KO males. In WT females, an HFD significantly decreased CI-driven respiration, which remained low in the absence of Sirt3. Finally, we observed that females had higher CI-driven respiration compared to males, except in HFD-fed WT mice, where this difference was reversed in favor of males (Figure 8A).

The respiratory control ratio (RCR) is defined as the ratio of the state 4 respiratory rate (termination of ATP synthesis by addition of oligomycin) to the state 3 (ADP-stimulated respiration) respiratory rate and indicates how well the electron transport system is coupled to ATP synthesis. WT males had higher RCR than KO males on a SFD, while an HFD partially restored low the RCR observed in the absence of Sirt3. In WT females, an HFD decreased RCR, which remained low in the absence of Sirt3. Females had higher RCR than males in SFD-fed conditions, while the opposite was found in HFD-fed WT mice. These results collectively indicate that the effective CI respiration in both sexes depends on Sirt3. Moreover, both CI-driven respiration and RCR display a sex-specific effect of HFD and Sirt3, with an inducing effect of HFD in KO males and a suppressive effect in WT females.

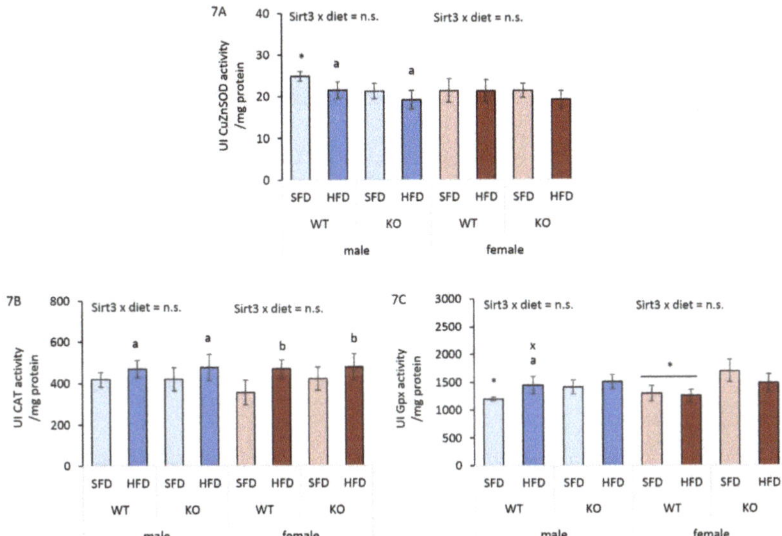

Figure 7. Analysis of CuZnSOD, Cat, and Gpx1 activities in the livers of Sirt3 wild-type (WT) and knockout (KO) mice of both sexes fed with a standard fat diet (SFD) or a high fat diet (HFD) for 10 weeks. (**A**) CuZnSOD activity. Males: HFD-fed vs. SFD-fed mice ([a] $p < 0.05$); SFD-fed WT vs. KO mice (* $p < 0.05$); Females: no changes. Males vs. females: no changes. (**B**) Cat activity. Males: HFD-fed vs. SFD-fed mice ([a] $p < 0.05$). Females: HFD-fed vs. SFD-fed mice ([b] $p < 0.01$). Males vs. females: no changes. (**C**) Gpx1 activity. Males: HFD-fed vs. SFD-fed WT mice ([a] $p < 0.05$); SFD-fed WT vs. KO mice (* $p < 0.05$). Females: WT vs. KO mice (* $p < 0.05$). Males vs. females: HFD-fed WT mice ([x] $p < 0.05$). The data are shown as mean ± SD. $n = 6$ per group.

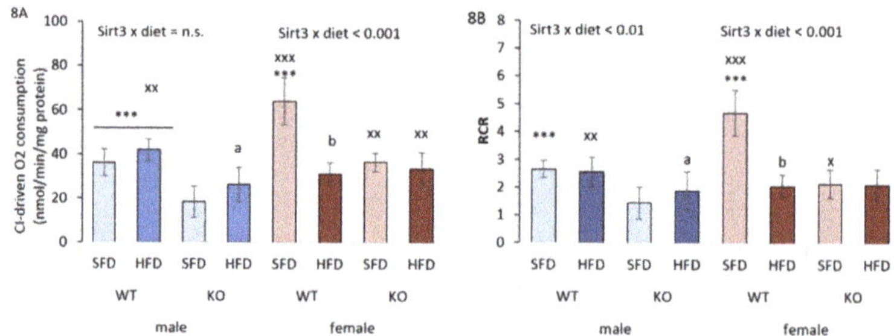

Figure 8. Complex I (CI)-driven respiration and the respiratory control ratio (RCR) of liver mitochondria of Sirt3 wild-type (WT) and knockout (KO) mice of both sexes fed with a standard fat diet (SFD) or a high fat diet (HFD) for 10 weeks. (**A**) CI-driven respiration. Males: HFD-fed vs. SFD-fed KO mice ([a] $p < 0.05$); WT vs. KO mice (*** $p < 0.001$). Females: HFD-fed vs. SFD-fed WT mice ([b] $p < 0.001$); SFD-fed WT vs. KO mice (*** $p < 0.001$). Males vs. females: SFD-fed WT mice ([xxx] $p < 0.001$); HFD-fed WT mice ([xx] $p < 0.01$); KO mice ([xx] $p < 0.01$). (**B**) The respiratory control ratio (RCR). Males: HFD-fed vs. SFD-fed KO mice ([a] $p < 0.001$); SFD-fed WT vs. KO mice (*** $p < 0.001$). Females: HFD-fed vs. SFD-fed WT mice ([b] $p < 0.001$); SFD-fed WT vs. KO mice (*** $p < 0.001$). Males vs. females: SFD-fed WT mice ([xxx] $p < 0.001$); HFD-fed WT mice ([xx] $p < 0.01$); SFD-fed KO mice ([x] $p < 0.05$). The data are shown as mean ± SD. $n = 4$ per group.

4. Discussion

Sexual dimorphism exists in various physiological processes, with males and females differing with respect to their regulation of energy homeostasis, metabolic rate, or body weight gain [34]. For example, HFD feeding leads to larger body weight gain in male rats/mice than in females [35]. In this study, we report the following novel observations in conditions of HFD feeding and Sirt3 depletion in liver of 129S mice: (a) HFD reduces hepatic Sirt3 expression solely in males, but only partially; (b) HFD-induced lipid accumulation is alleviated in the absence of Sirt3 only in males, which may be attributed to impaired glucose uptake and an increased reliance on fatty acids; (c) in females, either an HFD or the absence of Sirt3 compromised mitochondrial respiration and increased protein oxidative damage, which could be associated with more efficient lipid metabolism.

Sirt3 plays an important role in regulating lipid homeostasis by ameliorating HFD-induced inflammation, liver fibrosis, and steatosis. Reports show that HFD feeding in WT male mice results in a reduction of Sirt3 expression in the liver [36] and that deleterious effects of HFD feeding are exacerbated in male mice lacking Sirt3 [10]. Accordingly, we found that Sirt3 was reduced upon HFD feeding in male mice, however, without any change in females (Figure 1A,B). This result suggests that the regulation of Sirt3 in a sexually dimorphic manner following HFD in males may contribute to the observed sex-related differences in the development of metabolic dysregulation.

It has been shown that a decreased level of Sirt3 contributes to impaired glucose metabolism [24]. However, in our study, we did not observe the effect of Sirt3 on differences in glucose levels, only the effect of sex, where females had lower fasting glucose levels compared to males (Figure 2B), suggesting a possible role of female sex hormones in maintaining low glucose in females [37]. The interesting fact is that we did not observe gained weight in mice fed with an HFD (Figure 2A). The observed phenotype upon HFD feeding comes from the specific characteristics of this particular strain of mice (129S), which are not prone to the development of obesity when fed with an HFD [38]. However, despite the fact that mice did not gain weight following HFD feeding, both sexes displayed excessive hepatic lipid accumulation (Figure 3), indicating a fatty liver phenotype after ten weeks of an HFD.

Sirt3 deficiency reduces fatty acid oxidation and results in the accumulation of hepatic lipids in SFD-fed male mice [39], which is confirmed in our study. On the contrary, this effect was not evident in Sirt3-depleted females, which showed less accumulation of lipids than males in SFD conditions

(Figure 3I). Thus, disturbances in fatty acid oxidation in standard conditions and upon the depletion of Sirt3 may result in elevated lipids and impaired metabolism, thus supporting the critical role of Sirt3 in the maintenance of metabolic homeostasis in males. Studies have shown that an HFD acts as a metabolic stressor causing mitochondrial dysfunction and other metabolic changes that contribute to pathological conditions [4]. In HFD conditions, Sirt3-depleted males displayed the lowest total lipid content in liver along with the lowest hepatic glucose uptake (Figure 3I,J). These results indicate that in the absence of Sirt3 in males, HFD may cause impairments in hepatic glucose uptake, creating an increased reliance on fatty acids. This was previously also shown in the skeletal muscles of Sirt3 KO male mice [33]. Again, the observed differences in males were not evident in Sirt3-depleted females. The absence of these effects in WT mice indicates that sex-related differences in lipid accumulation and glucose uptake are present only in Sirt3-depleted mice.

In tissues that have a high fatty acid uptake, such as adipocytes and liver tissue, the capacity for the β oxidation of fatty acids, the major lipid catabolic pathways, is regulated at the level of gene expression in response to various physiologic stimuli [28]. Moreover, the maintenance of lipid homeostasis relies on cytochrome P450 (P450) enzymes, the majority of which are regulated in a sex-dependent manner [40]. To test if hepatic P450 enzymes are involved in sex-related differences in lipid accumulation in the absence of Sirt3, we measured the expression of several genes involved in fatty acid oxidation and lipid homeostasis (Figure 4). One important factor is *pparγ*, which is predominantly expressed in adipose tissue but also in the liver and muscle, and promotes the storage of lipids [41]. Since in our study HFD-fed mice did not have increased weight compared to SFD-fed mice, we tested whether this was due to the altered expression of *pparγ*, which is essential for adipogenesis [42]. Indeed, both sexes displayed downregulated levels of *pparγ* (Figure 4E), allowing us to hypothesize that the downregulation of hepatic *pparγ* may be related to the fact that mice did not gain weight after feeding with HFD. Furthermore, we observed sex-related differences in *cyp2e1*, *cyp2a4*, and *cyp4a14* transcripts of P450 genes involved in the maintenance of lipid homeostasis (Figure 4A,F–H). Females generally responded to an HFD with an upregulated expression of these involved genes, while males had a different response to an HFD with respect to Sirt3. Considering their role in lipid metabolism, the downregulation of *cyp2a4* and *cyp2e1* in male KO mice fed with SFD may be associated with their tendency towards accumulating more hepatic fat than females. In HFD conditions, these genes tend to be upregulated in KO males, therefore suggesting more effective lipid metabolism in association with lower accumulation of total lipids in liver.

Since we found sex-related differences in several parameters involved in lipid homeostasis upon different types of diet, we next investigated whether they reflected sex-related sensitivity to oxidative stress under the same conditions. Protein carbonylation (PC) is considered to be an important marker of oxidative stress resulting from HFD-induced oxidative damage to proteins [43,44]. Here, we show that an HFD induced a female-specific increase in the PC of WT mice (Figure 5A) as a possible consequence of the upregulated expression of genes involved in fatty acid oxidation, that is associated with the generation of hydrogen peroxide [45] involved in protein oxidative damage. Unlike WT males, HFD-fed WT females also displayed increased Ho1 protein expression (Figure 5B,E), which is an indicator of the presence of oxidative stress [46]. However, they failed to counteract oxidative damage by the activation of an antioxidative response, since no induction of the Keap1/Nrf2 axis (Figure 5C,D) and major antioxidative enzyme activities (MnSOD, CuZnSOD, Gpx1) was observed (Figures 6C and 7A,C), resulting in greater protein oxidative damage (Figure 5A). On the other hand, higher PC levels in KO mice of both sexes were associated with upregulated Keap1/Nrf2 axis, but with no response in antioxidant enzyme activities. This shows that in WT mice, an HFD induces oxidative stress only in females, pointing at their increased susceptibility towards the oxidative stress in conditions of nutritive stress. Furthermore, Sirt3 deficiency increases susceptibility to protein oxidative damage equally in both sexes, indicating that Sirt3 is important for protection against the oxidative damage of proteins.

Several studies demonstrated that lipid metabolism has a potential to generate ROS [47,48]. For example, both peroxisomal and mitochondrial β-oxidation produce H_2O_2 and O_2^- as byproducts of fatty acid degradation. During nutritional excess, the imbalance in lipid metabolism along with upregulated key genes involved in fatty acid β-oxidation, such as *ppara*, contributes to increased ROS and oxidative stress [45]. Catalase (Cat), a peroxisomal enzyme, has also been found in cardiac mitochondria with significantly increased activity during HFD feeding [48]. We observed the inducing effect of HFD on Cat activity in both genotypes and sexes (Figure 7B), suggesting that increased Cat activity is needed to effectively degrade excess of H_2O_2 produced by increased fat metabolism in HFD-fed mice.

Mitochondrial function is tightly associated with the activity of the respiratory chain complexes, and depends on the degree of coupling of oxidative phosphorylation [49]. Fatty acids are metabolic fuels and β-oxidation represents their main degradation pathway in mitochondria [50]. Sirt3 WT females had significantly higher CI-driven respiration and RCR than Sirt3 WT males in standard conditions, suggesting a more efficient function of Complex I in the electron transport chain of hepatic mitochondria in females. Contrary to males, in WT females, HFD feeding dramatically decreased MG-ADP state respiration, indicating female-specific impairment in CI-driven respiration and RCR following an HFD (Figure 8A,B). The observed defective CI-driven respiration in HFD-fed females could be due to their higher fatty acid oxidation, in synchronization with upregulated genes involved in the lipid oxidation process (*ppara, cyp2a4, cyp2e1, cyp4a14*), resulting in increased oxidative damage of mitochondrial proteins, thereby affecting mitochondrial respiration. Indeed, fatty acids can act as inhibitors of mitochondrial respiration, either by the partial inhibition of electron transport within CI or III, or by a decrease in proton motive force [51,52]. Based on this result, we hypothesize that in conditions of nutritive excess, more efficient lipid metabolism in WT females may cause higher oxidative stress in association with reduced mitochondrial function.

Sirt3 KO male mice display decreased CI-driven oxygen consumption [33], which is in agreement with our results. In addition, we show for the first time that, despite reduced respiration in Sirt3-depleted mice, females have higher respiration than males. However, since in KO males an HFD increased respiration, we hypothesize that they compensate impaired glucose uptake by increasing their reliance on fatty acids to provide substrates for the respiratory chain. This is in accordance to [53,54], where HFD-fed male rodents exhibited increased mitochondrial capacity and respiration [55]. Our results collectively indicate that the CI-driven respiration displays a sex-specific effect with respect to both HFD and Sirt3, with an inducing effect of HFD on respiration in KO males and a suppressive effect in WT females.

Sexual dimorphism exists in various physiological processes, which also includes sex-related prevalence towards metabolic dysregulation. In this study, we found significant sex differences at the level of metabolic, oxidant/antioxidant, and mitochondrial parameters. In addition, this study points towards a different role of Sirt3 in males and females under the conditions of nutritive stress. This is an important step that adds to previous knowledge that can be studied to prevent metabolic dysfunction, improve preclinical research, and allow for the development of sex-related therapeutic agents for obesity and diabetes.

Supplementary Materials: The following are available online at http://www.mdpi.com/2076-3921/9/2/174/s1, Figure S1: Body weight gain of Sirt3 WT and KO mice fed with standard fat diet (SFD) or high fat diet (HFD) for 10 weeks. Figure S2: Quantification of hepatic lipid accumulation signal (from Figure S2A–H) in Sirt3 WT and KO mice of both sexes fed with standard fat diet (SFD) or high fat diet (HFD) for 10 weeks. Figure S3: Western blot analysis of antioxidative enzymes in Sirt3 WT and KO mice of both sexes fed with standard fat diet (SFD) or high fat diet (HFD) for 10 weeks. Table S1: Assays (Taqman® Applied Biosystems, Foster City, CA, USA) used for real time quantitative PCR analyses, Table S2: Antibodies used in this study for the Western blot analyses.

Author Contributions: Conceptualization, M.P., I.I.P. and S.S.; Data curation, M.P., I.I.P., M.P.H. and S.S.; Formal analysis, I.I.P., I.T.B., A.D., R.B., V.F. and S.S.; Funding acquisition, T.B.; Investigation, S.S.; Methodology, M.P., I.I.P., M.P.H., I.T.B., A.D., R.B., V.F. and S.S.; Software, R.B. and V.F.; Supervision, T.B.; Validation, I.I.P.; Visualization, M.P.H.; Writing—original draft, M.P., I.I.P. and S.S.; Writing—review & editing, M.P., I.I.P., M.P.H., I.T.B., A.D., R.B., V.F., S.S. and T.B. All authors have read and agreed to the published version of the manuscript.

Funding: This work was supported by the Croatian Science Foundation (HRZZ), Grant no. [IP-2014-09-4533] "SuMERA".

Acknowledgments: The authors would like to thank Iva Pešun Međimorec and Marina Marš for their excellent technical contribution.

Conflicts of Interest: The authors declare no conflict of interest.

References

1. Bonomini, F.; Rodella, L.F.; Rezzani, R. Metabolic syndrome, aging and involvement of oxidative stress. *Aging Dis.* **2015**, *6*, 109–120. [CrossRef]
2. Nunn, A.V.; Guy, G.W.; Brodie, J.S.; Bell, J.D. Inflammatory modulation of exercise salience: Using hormesis to return to a healthy lifestyle. *Nutr. Metab. (Lond).* **2010**, *7*, 87. [CrossRef] [PubMed]
3. Kakimoto, P.A.; Kowaltowski, A.J. Effects of high fat diets on rodent liver bioenergetics and oxidative imbalance. *Redox Biol.* **2016**, *8*, 216–225. [CrossRef] [PubMed]
4. Guarner-Lans, V.; Rubio-Ruiz, M.E.; Pérez-Torres, I.; Baños de MacCarthy, G. Relation of aging and sex hormones to metabolic syndrome and cardiovascular disease. *Exp. Gerontol.* **2011**, *46*, 517–523. [CrossRef] [PubMed]
5. Lejri, I.; Grimm, A.; Eckert, A. Mitochondria, Estrogen and Female Brain Aging. *Front. Aging Neurosci.* **2018**, *10*, 124. [CrossRef]
6. Beery, A.K.; Zucker, I. Sex bias in neuroscience and biomedical research. *Neurosci. Biobehav. Rev.* **2011**, *35*, 565–572. [CrossRef] [PubMed]
7. Meziane, H.; Ouagazzal, A.-M.; Aubert, L.; Wietrzych, M.; Krezel, W. Estrous cycle effects on behavior of C57BL/6J and BALB/cByJ female mice: Implications for phenotyping strategies. *Genes Brain Behav.* **2007**, *6*, 192–200. [CrossRef]
8. Kahle, M.; Horsch, M.; Fridrich, B.; Seelig, A.; Schultheiß, J.; Leonhardt, J.; Irmler, M.; Beckers, J.; Rathkolb, B.; Wolf, E.; et al. Phenotypic comparison of common mouse strains developing high-fat diet-induced hepatosteatosis. *Mol. Metab.* **2013**, *2*, 435–446. [CrossRef]
9. Narvik, J.; Vanaveski, T.; Innos, J.; Philips, M.-A.; Ottas, A.; Haring, L.; Zilmer, M.; Vasar, E. Metabolic profile associated with distinct behavioral coping strategies of 129Sv and Bl6 mice in repeated motility test. *Sci. Rep.* **2018**, *8*, 3405. [CrossRef]
10. Hirschey, M.D.; Shimazu, T.; Jing, E.; Grueter, C.A.; Collins, A.M.; Aouizerat, B.; Stančáková, A.; Goetzman, E.; Lam, M.M.; Schwer, B.; et al. SIRT3 Deficiency and Mitochondrial Protein Hyperacetylation Accelerate the Development of the Metabolic Syndrome. *Mol. Cell* **2011**, *44*, 177–190. [CrossRef]
11. Bellizzi, D.; Rose, G.; Cavalcante, P.; Covello, G.; Dato, S.; De Rango, F.; Greco, V.; Maggiolini, M.; Feraco, E.; Mari, V.; et al. A novel VNTR enhancer within the SIRT3 gene, a human homologue of SIR2, is associated with survival at oldest ages. *Genomics* **2005**, *85*, 258–263. [CrossRef]
12. Barja, G. Updating the Mitochondrial Free Radical Theory of Aging: An Integrated View, Key Aspects, and Confounding Concepts. *Antioxid. Redox Signal.* **2013**, *19*, 1420–1445. [CrossRef]
13. Vargas-Ortiz, K.; Pérez-Vázquez, V.; Macías-Cervantes, H.M. Exercise and Sirtuins: A Way to Mitochondrial Health in Skeletal Muscle. *Int. J. Mol. Sci.* **2019**, *20*, 2717. [CrossRef]
14. Brown, K.; Xie, S.; Qiu, X.; Mohrin, M.; Shin, J.; Liu, Y.; Zhang, D.; Scadden, D.T.T.; Chen, D. SIRT3 Reverses Aging-Associated Degeneration. *Cell Rep.* **2013**, *3*, 319–327. [CrossRef]
15. Qiu, X.; Brown, K.; Hirschey, M.D.; Verdin, E.; Chen, D. Calorie Restriction Reduces Oxidative Stress by SIRT3-Mediated SOD2 Activation. *Cell Metab.* **2010**, *12*, 662–667. [CrossRef]
16. Hebert, A.S.; Dittenhafer-Reed, K.E.; Yu, W.; Bailey, D.J.; Selen, E.S.; Boersma, M.D.; Carson, J.J.; Tonelli, M.; Balloon, A.J.; Higbee, A.J.; et al. Calorie Restriction and SIRT3 Trigger Global Reprogramming of the Mitochondrial Protein Acetylome. *Mol. Cell* **2013**, *49*, 186–199. [CrossRef]
17. Calabrese, V.; Cornelius, C.; Cuzzocrea, S.; Iavicoli, I.; Rizzarelli, E.; Calabrese, E.J. Hormesis, cellular stress response and vitagenes as critical determinants in aging and longevity. *Mol. Aspects Med.* **2011**, *32*, 279–304. [CrossRef]
18. Ways, P.; Hanahan, D.J. Characterization and quantification of red cell lipids in normal man. *J. Lipid Res.* **1964**, *5*, 318–328.
19. Aebi, H.B.T.-M. Catalase in vitro. In *Oxygen Radicals in Biological Systems*; Academic Press: Cambridge, MA, USA, 1984; Volume 105, pp. 121–126. ISBN 0076-6879.

20. Paglia, D.E.; Valentine, W.N. Studies on the quantitative and qualitative characterization of erythrocyte glutathione peroxidase. *J. Lab. Clin. Med.* **1967**, *70*, 158–169.

21. Šarić, A.; Crnolatac, I.; Bouillaud, F.; Sobočanec, S.; Mikecin, A.M.; Mačak Šafranko, Ž.; Delgeorgiev, T.; Piantanida, I.; Balog, T.; Petit, P.X. Non-toxic fluorescent phosphonium probes to detect mitochondrial potential. *Methods Appl. Fluoresc.* **2017**, *5*, 15007. [CrossRef] [PubMed]

22. Fueger, B.J.; Czernin, J.; Hildebrandt, I.; Tran, C.; Halpern, B.S.; Stout, D.; Phelps, M.E.; Weber, W.A. Impact of Animal Handling on the Results of 18F-FDG PET Studies in Mice. *J. Nucl. Med.* **2006**, *47*, 999–1006.

23. Šaric, A.; Sobočanec, S.; Mačak Šafranko, Ž.; Popović Hadžija, M.; Bagarić, R.; Farkaš, V.; Švarc, A.; Marotti, T.; Balog, T. Diminished Resistance to Hyperoxia in Brains of Reproductively Senescent Female CBA/H Mice. *Med. Sci. Monit. Basic Res.* **2015**, *21*, 191–199. [CrossRef]

24. Jing, E.; Emanuelli, B.; Hirschey, M.D.; Boucher, J.; Lee, K.Y.; Lombard, D.; Verdin, E.M.; Kahn, C.R. Sirtuin-3 (Sirt3) regulates skeletal muscle metabolism and insulin signaling via altered mitochondrial oxidation and reactive oxygen species production. *Proc. Natl. Acad. Sci. USA* **2011**, *108*, 14608–14613. [CrossRef] [PubMed]

25. Patsouris, D.; Reddy, J.K.; Müller, M.; Kersten, S. Peroxisome Proliferator-Activated Receptor α Mediates the Effects of High-Fat Diet on Hepatic Gene Expression. *Endocrinology* **2006**, *147*, 1508–1516. [CrossRef] [PubMed]

26. Liss, K.H.H.; Finck, B.N. PPARs and nonalcoholic fatty liver disease. *Biochimie* **2017**, *136*, 65–74. [CrossRef]

27. Waxman, D.J.; Holloway, M.G. Sex differences in the expression of hepatic drug metabolizing enzymes. *Mol. Pharmacol.* **2009**, *76*, 215–228. [CrossRef] [PubMed]

28. Kersten, S.; Rakhshandehroo, M.; Knoch, B.; Müller, M. Peroxisome proliferator-activated receptor alpha target genes. *PPAR Res.* **2010**, *2010*, 612089.

29. Sobočanec, S.; Šarić, A.; Mačak Šafranko, Ž.; Popović Hadžija, M.; Abramić, M.; Balog, T. The role of 17β-estradiol in the regulation of antioxidant enzymes via the Nrf2-Keap1 pathway in the livers of CBA/H mice. *Life Sci.* **2015**, *130*, 57–65. [CrossRef] [PubMed]

30. Kander, M.C.; Cui, Y.; Liu, Z. Gender difference in oxidative stress: A new look at the mechanisms for cardiovascular diseases. *J. Cell. Mol. Med.* **2017**, *21*, 1024–1032. [CrossRef] [PubMed]

31. Tu, W.; Wang, H.; Li, S.; Liu, Q.; Sha, H. The Anti-Inflammatory and Anti-Oxidant Mechanisms of the Keap1/Nrf2/ARE Signaling Pathway in Chronic Diseases. *Aging Dis.* **2019**, *10*, 637–651. [CrossRef]

32. Tyagi, A.; Nguyen, C.U.; Chong, T.; Michel, C.R.; Fritz, K.S.; Reisdorph, N.; Knaub, L.; Reusch, J.E.B.; Pugazhenthi, S. SIRT3 deficiency-induced mitochondrial dysfunction and inflammasome formation in the brain. *Sci. Rep.* **2018**, *8*, 1–16. [CrossRef] [PubMed]

33. Lantier, L.; Williams, A.S.; Williams, I.M.; Yang, K.K.; Bracy, D.P.; Goelzer, M.; James, F.D.; Gius, D.; Wasserman, D.H. SIRT3 is crucial for maintaining skeletal muscle insulin action and protects against severe insulin resistance in high-fat-fed mice. *Diabetes* **2015**, *64*, 3081–3092. [CrossRef] [PubMed]

34. Shi, H.; Seeley, R.J.; Clegg, D.J. Sexual differences in the control of energy homeostasis. *Front. Neuroendocrinol.* **2009**, *30*, 396–404. [CrossRef] [PubMed]

35. Yang, Y.; Smith, D.L., Jr.; Keating, K.D.; Allison, D.B.; Nagy, T.R. Variations in body weight, food intake and body composition after long-term high-fat diet feeding in C57BL/6J mice. *Obesity* **2014**, *22*, 2147–2155. [CrossRef]

36. Bao, J.; Scott, I.; Lu, Z.; Pang, L.; Dimond, C.C.; Gius, D.; Sack, M.N. SIRT3 is regulated by nutrient excess and modulates hepatic susceptibility to lipotoxicity. *Free Radic. Biol. Med.* **2010**, *49*, 1230–1237. [CrossRef]

37. Mauvais-Jarvis, F. Gender differences in glucose homeostasis and diabetes. *Physiol. Behav.* **2018**, *187*, 20–23. [CrossRef]

38. Ussar, S.; Griffin, N.W.; Bezy, O.; Fujisaka, S.; Vienberg, S.; Softic, S.; Deng, L.; Bry, L.; Gordon, J.I.; Kahn, C.R. Interactions between Gut Microbiota, Host Genetics and Diet Modulate the Predisposition to Obesity and Metabolic Syndrome. *Cell Metab.* **2015**, *22*, 516–530. [CrossRef]

39. Chen, T.; Liu, J.; Li, N.; Wang, S.; Liu, H.; Li, J.; Zhang, Y.; Bu, P. Mouse SIRT3 attenuates hypertrophy-related lipid accumulation in the heart through the deacetylation of LCAD. *PLoS ONE* **2015**, *10*, e0118909. [CrossRef]

40. Yang, L.; Li, Y.; Hong, H.; Chang, C.-W.; Guo, L.-W.; Lyn-Cook, B.; Shi, L.; Ning, B. Sex Differences in the Expression of Drug-Metabolizing and Transporter Genes in Human Liver. *J. Drug Metab. Toxicol.* **2012**, *3*, 1000119. [CrossRef]

41. Sugii, S.; Olson, P.; Sears, D.D.; Saberi, M.; Atkins, A.R.; Barish, G.D.; Hong, S.H.; Castro, G.L.; Yin, Y.Q.; Nelson, M.C.; et al. PPARγ activation in adipocytes is sufficient for systemic insulin sensitization. *Proc. Natl. Acad. Sci. USA* **2009**, *106*, 22504–22509. [CrossRef]

42. Rosen, E.D.; Sarraf, P.; Troy, A.E.; Bradwin, G.; Moore, K.; Milstone, D.S.; Spiegelman, B.M.; Mortensen, R.M. PPARγ is required for the differentiation of adipose tissue in vivo and in vitro. *Mol. Cell* **1999**, *4*, 611–617. [CrossRef]

43. Krisko, A.; Radman, M. Phenotypic and genetic consequences of protein damage. *PLoS Genet.* **2013**, *9*, e1003810. [CrossRef] [PubMed]

44. Rincón-Cervera, M.A.; Valenzuela, R.; Hernandez-Rodas, M.C.; Marambio, M.; Espinosa, A.; Mayer, S.; Romero, N.; Barrera Cynthia, M.S.; Valenzuela, A.; Videla, L.A. Supplementation with antioxidant-rich extra virgin olive oil prevents hepatic oxidative stress and reduction of desaturation capacity in mice fed a high-fat diet: Effects on fatty acid composition in liver and extrahepatic tissues. *Nutrition* **2016**, *32*, 1254–1267. [CrossRef] [PubMed]

45. Chen, Q.; Tang, L.; Xin, G.; Li, S.; Ma, L.; Xu, Y.; Zhuang, M.; Xiong, Q.; Wei, Z.; Xing, Z.; et al. Oxidative stress mediated by lipid metabolism contributes to high glucose-induced senescence in retinal pigment epithelium. *Free Radic. Biol. Med.* **2019**, *130*, 48–58. [CrossRef] [PubMed]

46. Kamalvand, G.; Pinard, G.; Ali-Khan, Z. Heme-oxygenase-1 response, a marker of oxidative stress, in a mouse model of AA amyloidosis. *Amyloid* **2003**, *10*, 151–159. [CrossRef] [PubMed]

47. Flor, A.C.; Wolfgeher, D.; Wu, D.; Kron, S.J. A signature of enhanced lipid metabolism, lipid peroxidation and aldehyde stress in therapy-induced senescence. *Cell Death Discov.* **2017**, *3*, 17075. [CrossRef]

48. Tahara, E.B.; Navarete, F.D.T.; Kowaltowski, A.J. Tissue-, substrate-, and site-specific characteristics of mitochondrial reactive oxygen species generation. *Free Radic. Biol. Med.* **2009**, *46*, 1283–1297. [CrossRef]

49. Crescenzo, R.; Bianco, F.; Mazzoli, A.; Giacco, A.; Liverini, G.; Iossa, S. Alterations in proton leak, oxidative status and uncoupling protein 3 content in skeletal muscle subsarcolemmal and intermyofibrillar mitochondria in old rats. *BMC Geriatr.* **2014**, *14*, 79. [CrossRef]

50. Aon, M.A.; Bhatt, N.; Cortassa, S. Mitochondrial and cellular mechanisms for managing lipid excess. *Front. Physiol.* **2014**, *5*, 282. [CrossRef]

51. Schönfeld, P.; Reiser, G. Rotenone-like Action of the Branched-chain Phytanic Acid Induces Oxidative Stress in Mitochondria. *J. Biol. Chem.* **2006**, *281*, 7136–7142. [CrossRef]

52. Schönfeld, P.; Wojtczak, L. Fatty acids decrease mitochondrial generation of reactive oxygen species at the reverse electron transport but increase it at the forward transport. *Biochim. Biophys. Acta-Bioenerg.* **2007**, *1767*, 1032–1040. [CrossRef] [PubMed]

53. Hoeks, J.; Briedé, J.J.; de Vogel, J.; Schaart, G.; Nabben, M.; Moonen-Kornips, E.; Hesselink, M.K.C.; Schrauwen, P. Mitochondrial function, content and ROS production in rat skeletal muscle: Effect of high-fat feeding. *FEBS Lett.* **2008**, *582*, 510–516. [CrossRef] [PubMed]

54. Turner, N.; Bruce, C.R.; Beale, S.M.; Hoehn, K.L.; So, T.; Rolph, M.S.; Cooney, G.J. Excess Lipid Availability Increases Mitochondrial Fatty Acid Oxidative Capacity in Muscle. *Diabetes* **2007**, *56*, 2085–2092. [CrossRef] [PubMed]

55. Crescenzo, R.; Bianco, F.; Mazzoli, A.; Giacco, A.; Liverini, G.; Iossa, S. Mitochondrial efficiency and insulin resistance. *Front. Physiol.* **2015**, *5*, 512. [CrossRef]

MDPI
St. Alban-Anlage 66
4052 Basel
Switzerland
Tel. +41 61 683 77 34
Fax +41 61 302 89 18
www.mdpi.com

Antioxidants Editorial Office
E-mail: antioxidants@mdpi.com
www.mdpi.com/journal/antioxidants